Extraction, Utilization and Conversion of Woody Biomass

Extraction, Utilization and Conversion of Woody Biomass

Guest Editors

Ali Umut Şen
Catarina Pereira Nobre
Terencio Rebello de Aguiar Junior

MDPI
Basel • Beijing • Wuhan • Barcelona • Belgrade • Novi Sad • Cluj • Manchester

Guest Editors

Ali Umut Şen
Department of Energy Systems Engineering
Izmir Institute of Technology
Izmir
Turkey

Catarina Pereira Nobre
CoLAB BIOREF-Collaborative Laboratory for Biorefineries
S. Mamede Infesta
Portugal

Terencio Rebello de Aguiar Junior
Polytechnic School
Federal University of Bahia
Salvador
Brazil

Editorial Office

MDPI AG
Grosspeteranlage 5
4052 Basel, Switzerland

This is a reprint of the Special Issue, published open access by the journal *Processes* (ISSN 2227-9717), freely accessible at: https://www.mdpi.com/journal/processes/special_issues/utilization_woody_biomass.

For citation purposes, cite each article independently as indicated on the article page online and as indicated below:

Lastname, A.A.; Lastname, B.B. Article Title. *Journal Name* **Year**, *Volume Number*, Page Range.

ISBN 978-3-7258-6003-6 (Hbk)
ISBN 978-3-7258-6004-3 (PDF)
https://doi.org/10.3390/books978-3-7258-6004-3

Contents

About the Editors

Ali Umut Şen

Ali Umut Şen holds an integrated MSc in Chemical Engineering and a PhD in Forest Products Chemistry from Technical University of Lisbon (Portugal), specializing in wood chemistry and biomass valorization. He is currently an Assistant Professor at Izmir Institute of Technology (Turkey), where he teaches courses in Mass and Energy Balances, Waste-to-Energy systems, and Biomass Energy and Technologies. He also supervises both undergraduate and graduate theses. During his PhD and postdoctoral research at University of Lisbon, he focused on lignocellulosic biomass valorization, with extensive work on the pyrolysis kinetics, extractive chemistry, and advanced characterization of barks and other underutilized biomasses. His research interests include thermochemical conversion processes (pyrolysis, gasification, combustion), kinetic modeling and reactor design, biochar and activated carbon production, and the extraction and valorization of bioactive compounds from diverse biomass sources. He has published extensively in international journals and is the author of more than 60 journal articles and conference papers and actively participates in national and international projects, including EU collaboration networks. ORCID: 0000-0003-3734-466X

Catarina Pereira Nobre

Catarina Pereira Nobre holds a degree in Biochemistry, a MSc, and a PhD in Energy and Bioenergy from NOVA School of Science and Technology (NOVA-SST, Portugal), where she also carried out postdoctoral research on effluent bioremediation. During her PhD, she collaborated with companies in the bioenergy and waste management sectors, taught graduate courses on biomass energy and biofuels, and co-supervised several MSc theses. From 2020 to 2022, she was a researcher and team leader at CoLAB BIOREF, focusing on project development and industry-oriented services. She is currently an Auxiliary Researcher at VALORIZA – Research Center for Endogenous Resource Valorization, at Portalegre Polytechnic University (PPU, Portugal), where she participates in ten international projects, coordinating one of them, and also serves as Science Communication Coordinator of the WIRE COST Action (CA20127). At PPU, she lectures in postgraduate programs on hydrogen and renewable energies. Her research interests include thermochemical conversion processes such as gasification, pyrolysis, and hydrothermal treatments of lignocellulosic biomass and waste-derived fuels, as well as the production of renewable gases like hydrogen and biomethane. She is the author of more than 90 scientific outputs, including journal articles, book chapters, and conference papers. ORCID: 0000-0001-5733-902X

Terencio Rebello de Aguiar Junior

Terencio Rebello de Aguiar Junior is a researcher with a bachelor's degree in Biological Sciences and a master's in Natural Resources Management. He holds a doctorate in River Ecosystem Restoration and Management from the University of Lisbon, with a sandwich period in Environmental Sciences at the University of California/Polytechnic University of Madrid, in addition to a postdoctoral fellowship in Toxicology. His experience covers water resource contamination, watershed management, ecology, and toxicology. Currently, he is a collaborating professor at the Federal University of Bahia, a researcher at the Technical University of Lisbon, and a public servant in Porto Seguro. He is the author of several scientific publications and serves as a reviewer for international journals. ORCID: 0000-0002-7080-1618

processes MDPI

Review

Life Cycle Assessment (LCA) of Biochar Production from a Circular Economy Perspective

Joana Carvalho [1,2], Lucas Nascimento [1,2], Margarida Soares [1], Nádia Valério [1,2], André Ribeiro [1,2], Luciana Faria [1], André Silva [1], Nuno Pacheco [1,2], Jorge Araújo [1] and Cândida Vilarinho [2,*]

[1] CVR—Centre for Waste Valorisation, University of Minho, 4800042 Guimarães, Portugal
[2] MEtRICs—Mechanical Engineering and Resource Sustainability Center, Campus de Azurém, Universidade do Minho, 4800058 Guimarães, Portugal
* Correspondence: candida@dem.uminho.pt

Highlights:

Among several approaches to circular economy and zero-waste concepts, biochar production is a great example and might be a way to push the economy to a carbon-neutral balance. Overall, despite all the differences in assumptions and methodologies adopted, LCA proved that biochar is a very promising way of contributing to carbon-efficient resource circulation, mitigation of climate change, and economic sustainability.

What are the main findings?

- Biochar is considered a black porous and carbon-rich matter;
- Biochar is a promising source of alternative energy.

What is the implication of the main finding?

- It can be concluded that the costs are closely related to the technologies used in biochar production and also to the feedstock used.

Abstract: Climate change and environmental sustainability are among the most prominent issues of today. It is increasingly fundamental and urgent to develop a sustainable economy, capable of change the linear paradigm, actively promoting the efficient use of resources, highlighting product, component and material reuse. Among the many approaches to circular economy and zero-waste concepts, biochar is a great example and might be a way to push the economy to neutralize carbon balance. Biochar is a solid material produced during thermochemical decomposition of biomass in an oxygen-limited environment. Several authors have used life cycle assessment (LCA) method to evaluate the environmental impact of biochar production. Based on these studies, this work intends to critically analyze the LCA of biochar production from different sources using different technologies. Although these studies reveal differences in the contexts and characteristics of production, preventing direct comparison of results, a clear trend appears. It was proven, through combining life cycle assessment and circular economy modelling, that the application of biochar is a very promising way of contributing to carbon-efficient resource circulation, mitigation of climate change, and economic sustainability.

Keywords: life cycle assessment (LCA); biochar; biomass; circular economy

1. Introduction

As climate change is threatening the world and society grows exponentially, with more and more waste being generated, environmental sustainability is being questioned. It has been proven that linear models promote huge negative impacts on the environment. This occurs by extraction and landfilling at the end-of-life, and also in the economy, since not only when discarded is the economic value of products zero, but the value of finite resources also increases [1,2]. The transition to a circular economy is connected to great expectations of ecological and economic benefits, helping in the clear separation between economic growth and the use of resources, the building block of the linear economy and its respective impacts, and additionally promoting the perspective of sustainable growth and egalitarian society. The circular models can potentially improve the efficiency of using primary raw materials and allow the waste to return to production as high-quality secondary raw materials [1,3]. In addition, a circular economy can also provide a platform for pioneering methodologies, technologies and business models that create improved economic value from limited natural resources, helping industry to become more resilient to external impacts and improve its global competitiveness.

Among several approaches to circular economy and zero-waste concepts, biochar production is a great example and might be a way to push the economy to a carbon neutral balance [4]. The assembly and appliance of biochar has been widely developed all over the world. Biochar is a solid material formed during the thermochemical decomposition of biomass in an oxygen-limited environment. It is defined, by the International Biochar Initiative, as "a solid material obtained from the carbonization of biomass" [5,6]. Biochar can be produced through several techniques, such as pyrolysis, torrefaction, and gasification. It can also be obtained by various biomass feedstocks, such as wood, agro-residues, or wastewater sludge.

Several studies report the environmental benefits of using biochar in the most diverse industrial areas. However, the safest way to affirm this is through LCA studies that assess the most diverse environmental impacts such as climate change and ozone layer degradation, among others. LCA methodology is a technical tool that allows the systematic analysis and assessment of the environmental aspects and potential impacts associated with products or services throughout its life cycle. In this context, the aim of this work is to perform an overview on biochar concepts and applications and mostly on LCA of biochar production, analyzing several study cases, from the perspective of circular economy.

2. Biochar

Biochar is considered a black porous and carbon-rich matter. This material can be produced with little or unavailable air, through a thermochemical conversion of biomass. Chemical, biological and physical properties of biochar make it a great material with many purposes [7]. The following Figure 1 presents some benefits of biochar.

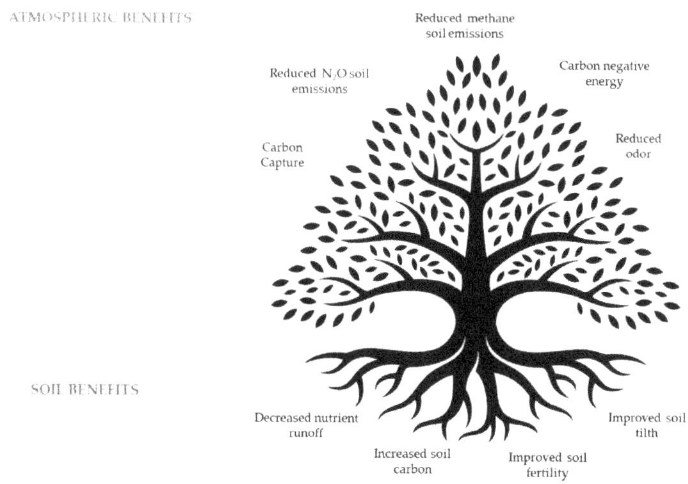

Figure 1. Biochar benefits (adapted from [6]).

2.1. Biochar Production

For biochar production, several thermochemical processes are used based on different reaction conditions (amount of oxygen available) including pyrolysis, gasification, and torrefaction. In this way, several types of reactors were developed for biomass production, with the objective of achieving the highest quality and yield of the product. These reactors differ in certain thermochemical parameters, such as oxygen availability, temperature, and the rate of heating itself. Thus, these parameters vary the physical and chemical properties of the biochar [8]. The following Figure 2 shows these several methods [8].

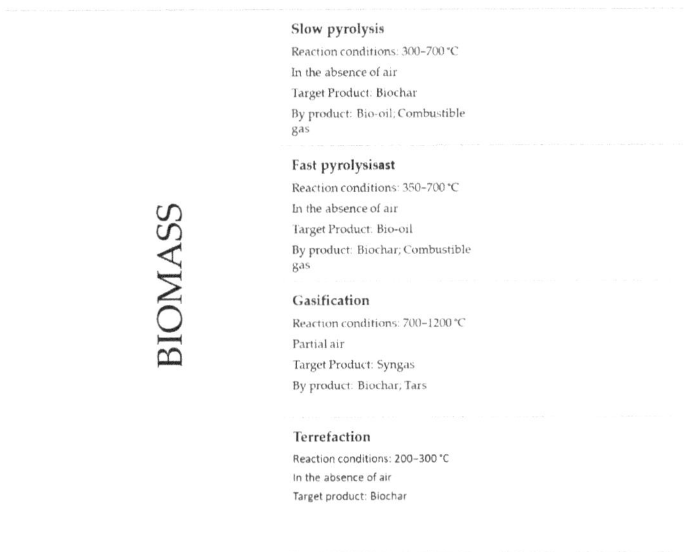

Figure 2. Biomass thermochemical conversion technologies for biochar production (adapted from [8]).

Biochar production quality heavily depends on the biomass' compositions and properties. These parameters define the application fields, production, quality and toxicity of biochar. Their stability is influenced by pyrolysis temperature, this being the main parameter [9]. Biochar production involves a complex biochemical reaction process where biomass undertakes decomposition, depolymerization and condensation in anoxic high temperature conditions [4,8,10]. Several factors define the quality of biochar, such as pH, specific surface area, porosity, nutrients, and carbon content, the latter being one of the main factors. High carbon content is directly linked to a high-quality biochar [11].

2.1.1. Slow Pyrolysis and Fast Pyrolysis

Different studies have demonstrated that slow pyrolysis allows the production of a quality biochar. This technique is characterized by using moderate temperatures (350–500 °C) and low heating rates [8]. In addition, other operational parameters strongly influence this type of pyrolysis, such as particle size, the atmosphere itself, the use of a catalyst, etc. Veses et al. (2015), stated that improving the residence time of the raw material and the pyrolysis vapor in contact, as well as the particle size and the ratio of catalyst in the biomass can favor the quality and production yield of biochar. Furthermore, it was mentioned that higher pyrolysis temperatures are very important for improving the quality of biochar (expanding carbon content), and reduced heating rate can increase its production in this type of pyrolysis [12].

Unlike the slow process, fast pyrolysis is characterized by the application of high heating rates, temperatures around 500 °C, and residence times shorter than 2 s, with rapid decomposition of the biomass [13]. As mentioned above, higher pyrolysis temperatures increase the carbon content of the product and its specific surface area [14]. However, Chen et al. (2016), mentioned that increases in pyrolysis temperature, such as an increase in the heating rate, reduces the production yield due to the release of volatile gases. Since there is a shorter residence time, the amount of carbon deposition is reduced too [8].

2.1.2. Gasification

Regarding gasification, this process involves an incomplete combustion of biomass using several gasifying agents (air, pure oxygen, steam) that occur at 700–1000 °C [8]. Briefly, the process involves three main reactions: the devolatilization of the biomass, the combustion, and the gasification itself. Char from gasification is different from that obtained through pyrolysis, essentially due to the oxidizing environment of the gasifier. This environment affects the physical, chemical and morphological properties of biochar [15]. As mentioned above, the quality of biochar is directly related to the carbon content present, which is essentially affected by gasification conditions. The main parameters are the equivalence ratio, the properties of the raw material, pressure and the gasifying agent [16]. Several studies have shown that production yield decreases with increasing equivalence ratio, since the gasification temperature increases and carbon content also decreases [17,18].

2.1.3. Torrefaction

In turn, the torrefaction process is a modern method of obtaining biochar from biomass. This process is considered a mild pyrolysis, carried out at temperatures between 200–300 °C, under anaerobic conditions [18]. This method uses low heating rates and long residence times [8]. As for fast pyrolysis, increasing the processing temperature leads to a reduction in production yield. However, it is possible to achieve an energetically denser material, correlated with larger destruction of the structural elements. In turn, increasing the process time increases the calorific value of the biomass and has positive effects on carbon content [19,20]. The structure and biomass composition have influence on the process. For example, the size particles and the presence of heavy metals affect the torrefaction mechanism [20].

In short, the higher the carbon content, the more superior the quality of the biochar. In turn, the carbon content is higher when all volatile compounds are released from the biomass. This occurs when slow pyrolysis is carried out, being considered a deeper pyrolysis, using moderate temperatures over a long period of time. In this way, slow pyrolysis yields a high-quality biochar [8]. In torrefaction, the biochar undergoes only a lighter pyrolysis, with a low content of volatile compounds released, and there are not so many chemical reactions [21].

2.2. Biochar Applications

The interest in conversion of biomass waste from agriculture and forestry to energy production and carbon sequestration has been growing [22–24]. Carbon and energy content in these wastes make them potential candidates for thermochemical process to produce bioenergy and bioproducts [24]. In general, the many approaches to using biochar facilitate zero waste and the interest of circular economy values [25]. Pyrolysis of biomass waste produces biochar with great capacity in agricultural, industrial and construction applications. In addition, a bio-oil that can be burned to generate hot water, steam and/or electricity, is also produced in the process, thus reducing amount of waste discarded [4]. Recently, the manufacture and use of biochar has been widely developed worldwide. However, there are great uncertainties surrounding the operations, costs and emissions associated with wood processing [26].

As already mentioned, raw material quality, production technology and method conditions mainly define the efficiency, quality and biochar toxicity, and thus affect the subsequent biochar application strategy [4,27,28]. The physicochemical characteristic of biochar reveal its distinct application perspectives, involving soil conditioning, composting additive, building material, activated carbon, promoting anaerobic digestion and a list of potential uses that continues to expand [5,29–32]. As the number of biochar purposes increases, so does the number of producers. However, it is neither financially nor energetically viable to generate and apply biochar without standards or regulations for its production and application [4]. To solve this problem, many countries are developing their own standards. So far, the International Biochar Initiative (IBI) in the United States of America and the European Biochar Certificate (EBC) are the most-usually used standards all over the world. These two standards were built helping to decrease the health and environmental risks correlated with the production and use of biochar, particularly in agriculture. However, the two standards are voluntary industry guidelines. As biochar has excellent potential for sustainability in several industries, these two guidelines appear to be unsatisfactory to regulate the quality of biochar manufactured as a whole. With the growing importance in biochar, many countries have their individual biochar policies in line with IBI and EBC standards. Other nations, which presently do not have a biochar standard, are controlling biochar usage with fertilizer or compost specifications [4].

Figure 3 shows some examples of promising biochar applications.

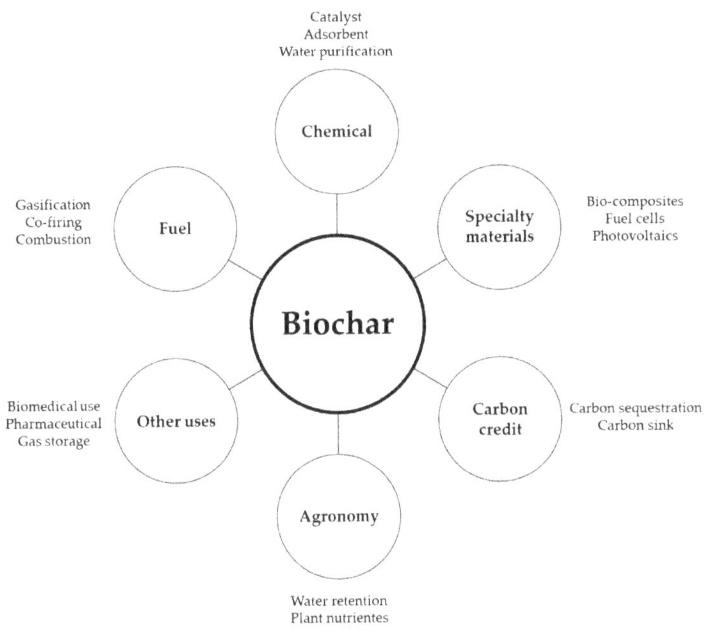

Figure 3. Promising applications of biochar (adapted from [33]).

2.2.1. Agriculture

Biochar application in agriculture has been extensively tested in the laboratory and field, and it has been usually employed as a chemical component [34–36], or as a soil corrective to enhance crop production, increasing nutrient availability [35,36], water retention capacity [36] and soil microbial activity [37]. Biochar can increase the pH of highly acidic soils [38–40] and mitigate the release of heavy metals into the soil [40–42]. Several studies have shown the advantages of biochar application as a soil conditioner and have demonstrated its potential in that respect. Some of the main benefits are the decreasing of greenhouse gas emissions and the mitigation soil infertility and desertification, improving soil quality and crop yield [43]. This is justified by the inorganic content present in biochar that acts as a nutrient and aids in fertilization [44]. As previously mentioned, biochar is produced by the thermal decomposition of biomass in an oxygen-limited ecosystem while composting is the natural biodegradation of organic material by the microbial community in aerobic conditions. Carbon-based compounds in compost degrades rapidly and mineralizes so its valuable effects are quite short-lived, which is distinct from biochar, which can endure in soil for a long time [45]. Biochar has also been applied as an additive to enhance the composting rate e.g., a combination of 35% over mushroom compost with 20% biochar has been demonstrated to decrease the composting time to 24 days in comparison to the more common 90–270 days [46]. The elevated porosity of biochar can also enhance microbial development in the compost stockpile and consequently speed up nutrient recovery [47]. The incorporation of biochar in poultry manure for composting condensed the thermophilic stage and increased the highest temperature reached [48]. To manage the quality of biochar, it is extremely essential to determine its organic and inorganic pollutions [49]. General organic and inorganic contaminants that affect biochar quality are polycyclic aromatic hydrocarbons (PAHs), polychlorinated biphenols (PCBs), dioxins, furans and heavy metals [50]. In addition to biochar contaminations, it is also necessary to guarantee that biochar brings more advantages to its applications, such as cation exchange capacity, moisture retention, nutrient retention, plant growth promotion, heavy metals stabilization from soil or construction material, etc. The effectiveness of biochar stabilization of heavy metals can be assigned to the high surface area of the biochar, ability to raise the pH value, small-scale particle, etc. Nevertheless, if biochar is used precisely for its toxicity assessment without soil, a toxic influence is commonly found in living organisms [51]. This negative impact is basically caused by its elevated pH value and potential contaminants, but once biochar is blended with soil in a correct percentage (for example 1% or less), it generally has no noticeable toxic effect and it can improve plant biomass and amount of microorganisms in the soil, which are positive impacts. Studies show that the positive impacts are attributable to the biochar raw material, pyrolysis temperature, biochar percentage and type of soil. The cleaner the biochar raw material is, the less toxic the final biochar product is.

Another biochar application studied is soil remediation and amendment. Biochar has been considered an appropriate matter for this purpose, allowing the absorption and modification of pollutants, through their properties previously mentioned, associated with physical adsorption on the surface and in micropore structures [52].

Biochar has great potential to sequester carbon from materials based on plants, binding CO_2 through long-term storage of carbon in soil. This ability is associated with the yield of biochar, the content of stable carbon in biochar and the biochar stability in several circumstances and timeframes of soils [53].

2.2.2. Wastewater Treatments

Regarding water purification, biochar can be used in wastewater treatments. Biochar is efficient in the removal of several pollutants, as well as organic compounds (dyes, phenolics, pesticides, polynuclear aromatics and antibiotic), nutrients and heavy metals. Heavy metals are toxic, carcinogenic and non-biodegradable components, so their removal is fundamental [54]. This capacity is justified by their absorbent ability in aqueous solutions due their properties, such as porosity, pH, surface area, surface charge and mineral composition [55,56]. Concerning nutrients (nitrogen, phosphorus, sodium, potassium, etc.), if their concentrations exceed the limits, these components become dangerous to the aquatic ecosystem. Physical and chemical properties associated to biochar are of great significance to the removal of these components, through various removal methods, involving electrostatic interaction, ligand exchange, precipitations with Mg^{2+} or Ca^{2+}, complexation with functional hydroxyl, surface sorption, etc. [56–58]. The same is observed in the removal of other organic pollutants, since the efficiency is strongly affected by physical properties, such as specific surface areas and pore-size distribution, surface functional groups and hydrophobic nature [59].

2.2.3. Building Materials

Another application of biochar is as a partial component of cementitious building materials, specifically as a sustainable alterative to concrete cement, sand or another energy-intensive additives used for concrete production [60–63]. Recent studies [61–65] have established that biochar, made from several raw biomass materials under ideal pyrolysis conditions, can lead to an enhancement in strength and biochar-mortar by 15–20% when evaluated to control mixture. The filling of biochar particles can also promote a reduction in capillary water absorption and water infiltration in biochar-mortar mix by 30–40%, thus indicating reduction in water penetration and improved durability of the composite [64]. Heavy metal leaching is a common problem in waste heat treatment products, including ashes and biochar, with a high level of heavy metal content being determined in sources [66]. Heavy metal contained in the biochar can run like leachate when structure builds are exposed to rain and other weathering agents [67].

2.2.4. Activated Carbon

Biochar, with a large surface area and pore formation, can potentially be employed as a low-cost activated carbon [68]. Nevertheless, beyond typical pyrolysis treatment of waste biomass, the resulting biochar frequently shows poorer pore behavior than commercial activated carbon. Therefore, several physical and chemical treatment processes were proposed to convert crude biomass or low-grade coal into activated carbon [69]. Normally, physical activation involves elevated temperature in the presence of steam, air, CO_2, N_2 and inert gases, and chemical activation was carried out utilizing nitric acid, phosphoric acid, sulfuric acid and potassium hydroxide, among others [70–72].

2.2.5. Anaerobic Digestion (AD)

AD is an important expertise for organic waste remediation and bioenergy recovery, thus performing a vital role in development the global circular economy [72]. In fact, biochar has been supported as capable of increasing the AD process through quite a few mechanisms. More precisely, it was found that biochar performs a crucial role in increasing AD through several mechanisms, including promoting direct interspecies electron transfer among different microbial species due to biochar's good conductivity [73]; increasing microbial growth through biochar's immobilization effect [74]; adsorbing inhibitory compounds such as heavy metals, ammonia and volatile fatty acids in anaerobic bioreactor [75] and increasing the buffering capacity of the bioreactor due to the relatively high alkalinity of the biochar [75–77].

3. Life Cycle Assessment (LCA) Definitions

The concern for the environmental impacts generated by the supply of products and services to society has led to the growth of new tools and methods that aim to understand, control and reduce these impacts. Life cycle assessment (LCA) can be characterized as a compilation and assessment of inputs, outputs and potential environmental impacts of a product or system throughout its life cycle. It is the most widely used assessment type with wide international approval to measure environmental impacts [78,79]. LCA includes entire supply chains, representing all the impacts that occur at different steps and locations throughout the life cycle, regardless of particular processes' physical locations [79].

For an LCA study to be validate, it is necessary to follow the steps stipulated by the ISO 14040 standard, which consists of (i) defining an objective and scope; (ii) carrying out an inventory with as much information as possible about inputs and outputs related to the product or service; (iii) calculating environmental impacts; (iv) systematic and interactive review to validate all information.

Several studies report the environmental benefits of using biochar in the most diverse industrial areas. However, the safest way to affirm this is through LCA studies that assess various environmental impacts, such as climate change and ozone layer degradation among others.

According to Matuski et al. (2020), many authors have used the LCA method to assess biochar projects environmental impact, and these articles are systematically revised to discover a general trend or some pattern. The differences in these studies' contexts and characteristics do not allow a direct comparison of results, limiting the spectrum of LCA studies. However, biochar application brings substantial benefits, either by neutralizing the emission of greenhouse gases from agricultural production or as a carbon sequestration approach. There is also a great capacity for energy production using synthesis gas and bio-oil byproducts. The advantages of carbon sequestration in biochar and energy production generally overbalance the greenhouse gas emissions produced through the production and handling of the raw material. On the other hand, the effect on another types of environmental impact needs to be assessed and normalized with the intuition of observing some negative effect to guarantee the project's economic sustainability [80].

3.1. Scope and Objective

The LCA concept's definition includes the full definition and description of the product, process or activity, the choice of the functional unit to be used and establishes the context in which the valuation must be made, identifying the limits, environmental effects and the methodologies considered for the evaluation.

According to Zhu et al. (2022), for the particular case of LCA of biochar, the objectives and scope are, in general, divided into two fields of research, the first one related to the assessment of the impacts of manufacture and use with association of carbon footprint, midpoint impacts, air pollutants, acidification and other parameters. On the other hand, the second field of research is focused on energy consumption efficiency and economic aspects [81].

Table 1 presents some articles that assess the environmental impacts of biochar production and use. They are divided by a functional unit, reference system, software and impact methodologies used and other focuses.

Table 1. LCA case studies for biochar (adapted from [82]).

Region Country	Functional Unit	Reference System	Allocation + System Boundaries	Ref.
Denmark	1 ton of dry seed	Typical Danish rapeseed to production.	Crop cultivation included.	[83]
Belgium, Spain, Italy	1 kg product 1 ha/yr	Compost, compost blend, mineral fertilizer.	System expansion, cut-off (feedstock), construction and pesticide omitted.	[84]
Zambia	Preparation and utilization of 1 kg biochar	x	System expansion—where applicable avoided electricity production (diesel fuel generator wood burning).	[85]
Canada	Production of 1 ton (Mg) biochar	Compare two different temperature scenarios.	System expansion, switchgrass cultivation included, energy production offset.	[86]
Belgium	1 ton of biochar	x	System expansion, energy and fertilizer offset, cut-off (manure) þ all (willow cultivation in marginal soils), pyrolysis plant construction included.	[87]
Vietnam	1 ton of rice straw	Open burning of straw, two seasons (spring and summer) modelled, comparison with enriched biochar.	System expansion, open burning of straw eschewed.	[88]
Finland	1 ton oat flakes	Oat flows used as feed or for energy.	System expansion.	[89]
China	1 ton of odt straw	Three straw utilization scenarios: briquetting gasification, pyrolysis two baseline (reference) scenarios: reincorporation, burning	System expansion, offset from avoided fuel consumption.	[90]

Some authors used LCA study to measure the potential environmental impacts of recovering nutrients from the soil using biochar [84]. Erison et al. (2022), performed an LCA study envisaging the production of biodiesel using biochar as a catalyst [91]. Lefebvre et al. (2021), evaluated the differences between sugarcane biomass and biochar to assess which energy source emitted less CO_2-eq in energy supply [92]. The objectives related to the use of LCA in biochar are diversified, and often adapted to particular needs of the research.

The functional unit used varies within each study, e.g., in the study performed by Field et al. (2013), the functional unit was the conversion of 1 ton of dry biomass into biochar. In Hamedani et al.'s (2019), LCA study of biochar, the production of 1 ton of biochar as functional unit was used. Xu et al. (2019), defined the functional unit of their study as a "hectare of agricultural area used for one year" [87,93,94].

When considering the biochar systems' main objective is the use or management of biomass waste streams, upstream functional units are generally used, such as a dry or wet raw material [90]. Another widely applied approach is related to downstream flows, where functional units are primarily specified as the mass of biochar produced or the mass of a crop produced in the treated field. Additionally, it is also common the combination of the two methodologies, for example, a quantity of raw material that produces a certain amount of biochar [95], or even multiple functional units [96]. In the study of Zhu et al. (2022), the author infers that a commonly used functional unit can be the production of 1 kWh of produced energy, stating that, in terms of comparison, one of the most recommended functional units to be used is 1 ton of feed and 1 ton of biochar. [81]

When referring to the boundaries, in the scope of LCA, they can be called "cradle to gate", "gate to gate", "cradle to grave" and "cradle to cradle". Each type of boundary system is described in the literature in accordance with the detail level of study. In brief, the boundary system "cradle to gate" analyses environmental impacts from the extraction of the raw material to the "entrance" at the factory, while the "gate to gate" system measures the environmental impacts of one or more manufacturing processes. The "cradle to grave" expression is used in the assessment of the impacts that cover the extraction of raw materials, the use phase and the final disposal of the product. On its turn, the "cradle-to-cradle" boundary methodology involves recycling and reuse processes [97], as shown in Figure 4.

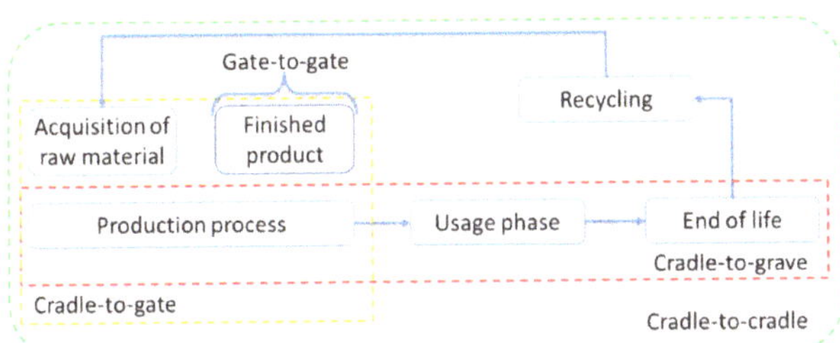

Figure 4. LCA boundary system approach.

In a process of biochar production from agro-wastes obtained through pyrolysis, Tiegam et al. (2021), stated that this system boundary can be classified as cradle-to-grave or cradle-to-gate [98]. On the other hand, in the study carried out by Hamedani et al. (2019), involving pyrolysis of pig manure residues for biochar production, its boundary was characterized as cradle-to-grave [87].

In the study performed by Hammond et al. (2011), the frontier used was gate-to-gate with the aim of calculating the impacts of producing 1 kWh of biochar. In another study conducted by Dutta and Raghavam, (2014), the authors infer in their work that the most-indicated boundary system for LCA studies focused on evaluating greenhouse gas was the cradle-to-grave one [99].

3.2. Life Cycle Inventory

Life cycle inventory (LCI) is the collection and quantification of all natural resources consumed and all substances that are emitted into the environment by the life cycle within life cycle phases expressed in the goal and scope definition of this study. The procedure of collecting inventory data begins on the basis of the data requisite for each life cycle stage, in particular the type of data needed for the stage: foreground or background. Foreground data mean that the data are attributable to the target products precisely, while background data can be obtained from special or temporal averages. LCIs are usually based on average data of material and energy inputs (resources and energy used) and outputs (product, waste and emissions produced) collected at a real site or estimated from the literature or from developing studies performed prior to the LCA study [100]. The LCI has a necessary step where elementary flows are included over time and space and is a demanding activity that requires time and attention [101]. It is essential to understand that LCA and LCI are both supported by concepts and assessment of energy consumption and mass balance, based on the first law of thermodynamics related to the principle of energy conservation, by which energy cannot be created or destroyed, but converted into another form. The second law of thermodynamics is also a fundamental concept for understanding LCA, being related to energy degradation; that is, some losses can be due to heat, radiation and others. In this context, LCA makes an energy balance of materials between inputs and outputs and calculates the environmental impacts through the difference between the loss and energy efficiency [102,103].

Some authors declare that the inventory related to biochar production usually focuses on feedstock, transport, pyrolysis process, use of biochar and end-of-life [104]. Dutta and Raghavam (2014), in their work consider inventory, obtaining feedstock analysis, all transports associated with biochar, pyrolysis process to obtain biochar and use of biochar for energy production [99]. In the study carried out by Llorach-Massana et al. (2017), the inventory was focused on the feedstock, where the feasibility of using tomato plant residues from an urban garden was studied [105]. In other studies, the benefits of producing biochar from rice field residues were analyzed [88].

In the study performed by Matuštík et al. (2020), LCI for biochar must contain information on the raw material, land use, transport, energy used in pyrolysis, pyrolysis gas emissions and, depending on the functional unit and borders, waste must be also accounted for [82].

3.3. Life Cycle Impact Assessment (LCIA)

To understand the potential environmental impact of the use of natural resources and the attempts created by the system, the LCIA stage is essential. In this methodology, flows resulting from a product's life cycle are aggregated into impact categories [82].

The same studies presented in Table 1 were compiled in Table 2 with LCIA methodologies, impact assessment and GWP results.

According to de Vries and de Boer, (2010), the environmental impact types considered during the life cycle of a product are related to the use of resources such as earth or fossil fuels and the emission of the pollutant's ammonia or methane. Ferrão, (2009), and Santero et al. (2011), considered categories of impacts on climate change, eutrophication, acidification, depletion of resources and ecotoxicity, among others, where each type of emission is related to one or more impacts [102,106,107]. The impact assessment uses the inventory results to evaluate the potential of environmental impacts while providing information for interpretation. In this phase, statistical techniques, such as weighting, normalization and aggregation, are essential to comparing impact types and measuring their relevance [100].

Table 2. Study cases with LCIA methodologies, impact categories analyzed and GWP results (adapted from [80]).

Study	LCIA Methodology	Impact Assessment	Results	Ref.
Thers et al. (2019)	IPCC (2013)	GWP 100 yr, 20 yr	171 kg CO_2-eq/Mg dry seed (400 °C) and 111 kg CO_2-eq/Mg dry seed (800 °C) compared to 638 kg CO_2-eq/Mg dry seed for oilseed rape cultivation without biochar amendment = significant GHG reduction.	[83]
Oldfield et al. (2018)	CML	GWP, AP, EP	Net negative—single value not presented.	[84]
Smebye et al. (2017)	ReCiPe 2016 endpoint	GWP, PMF, land occupation, FDP, þ all ReCiPe 1 endpoint categories	Only endpoint GWP value calculated.	[85]
Brassard et al. (2018)	IPCC 2007	GWP	Scenario A (459 °C) −2110 kg CO_2-eq/Mg biochar, Scenario B (591 °C) −2561 kg CO_2-eq/Mg biochar.	[86]
Hamedani et al. (2019)	IMPACT 2002+, CML	All midpoint (IMPACT) and endpoint (CML)	IMPACT 2002+: −2063 kg CO_2-eq/t biochar (willow) and −472 kg CO_2-eq/t biochar (pig manure); CML: −2089.65 kg CO_2-eq/t biochar (willow) and −466 kg CO_2-eq/t biochar (pig manure).	[87]
Uusitalo and Leino, (2019)	IPCC (2013)	GWP	Biochar production from oat side-flows leads to GHG emission reduction of 350 k gCO_2-eq/t oat flake, buffer zone biomass biochar potential of 390 kg CO_2-eq/t oat flake.	[89]
Clare et al. (2015)	IPCC 2007	GWP	−1.35 Mg CO_2 e/odt straw	[90]

There are several methodologies for calculating environmental impacts such as Recipe, CML, Usertox, Carbon Footprint IPCC for parameters and others. According to Zhu et al. (2022), the choice of impact category must converge with the goals. Yang et al. (2018), used the CML methodology; Thers et al. (2019), used the IPCC methodology; Oldefild et al. (2018), applied CML methodology and Smebye et al. (2018), assessed the impact through ReCiPe methodology [82–85,103].

Several authors studying LCA on biochar have used methods such as Eco Indicator 99 and the ReCiPe midpoint approach to evaluate the entire biochar chain [108]. In the study carried out by Hamedani et al. (2019), the use of two methodologies, IMPACT 2002+ and CML, was applied to calculate environmental impacts [87].

When looking at Table 2 and considering all the differences in the limits of the system such as functional units, among other parameters, the LCA results are not comparable. The GWP results presented make a general balance of GHG that varies in value according to the methodology used. According to a study by Brassard et al. (2018), in their work they used biochar for soil correction and observed a GWP impact of up to 2561 mg CO_2-eq/t of biochar [86]; Mohammadi et al. (2016), revealed the carbon footprint of rice production with a change in the biochar of 3.85 kg CO_2-eq/kg. Nevertheless, the results showed similar tendencies. However, when looking closely, the results show a certain tendency. It is remarkable that the biochar–soil correction systems show a clear advantage from the point of view of climate change [88]. Biochar production appraised the neutralization's impact of agricultural production when GHG emissions were positive, such as in the case presented by Thers et al. (2019), and other studies. The processing of raw materials and especially the pyrolysis system are the most important causes of GHG emissions, not including agricultural production's impact. In relation to the impact, it is normally considered residual. The benefit of carbon capture and energy production from biochar, in addition to the advantages associated with the co-products generated, compensate the GHG emissions caused by the biochar production itself [83].

Biochar application can contribute to carbon sequestration by increasing production, reducing the use of fertilizers and reducing CH_4 or N_2O emissions, among other factors [84].

Other studies have been carried out with the aim of evaluating the influence of biochar production on the impact categories. Concerning acidification and eutrophication potential, agricultural processes, such as the application of fertilizers, are the most important sources of negative impacts [84], as well as with regard to the impact of ecotoxicity categories [87]. Electricity from the network for pyrolysis and agricultural operation was reported as another source of negative impact on acidification and eutrophication potential [109].

4. Life Cycle Cost Assessment (LCCA)

For a process to be viable, in addition to the environmental aspects, it is mandatory to assess their associated costs. Life cycle cost assessment (LCCA) is a method of assessing the total cost of a process. It takes into account all costs of acquiring, owning, operating and disposing. LCCA is especially useful when project alternatives that fulfill the same performance requirements differ with respect to initial costs and operating costs and should be compared in order to select the one that maximizes net savings best.

In this context, conducting a complete LCCA of biochar-based products is an essential step in the breakeven and viability of this technology during the project or industrial application period.

Durairaj et al. (2002), reviewed models for LCCA [110], namely: (i) LCCA model of Fabrycky and Blanchard; (ii) LCCA model of Woodward [111]; (iii) LCCA model of Dahlen and Bolmsjo [112]; (iv) the activity-based costing (ABC) model [113]; (v) the economic input–output LCA model [114]; (vi) the design-to-cost model (DTC) [115]; (vii) the product life cycle costing (PLCC) to manufacturing system [116]; (viii) the total cost assessment (TCA) model.

Fabrycky and Blanchard presented an elaborated LCCA model to approach detailed cost analysis of all the costs related to the entire life cycle of either product.

The major advantage of this model is found in its detailed cost breakdown structure (CBS). First, they split the total cost of a product or a system into four categories, respectively: (1) research and development costs; (2) production and construction costs; (3) operation and maintenance costs; (4) retirement and disposal cost. This model comprises the essential features of a holistic methodology to assess the life cycle and determine the total cost of a product.

Woodward's LCCA model was made to focus on planning and monitoring assets over their entire life cycle—from the development stage through to disposal. The minimum life cycle cost of the asset is given by the optimization of the trade-off among the cost factors. This process requires the estimation of whole-life costs prior to making a purchase choice for an asset from the accessible alternatives. This approach emboldens a long-term perspective on the investment decision-making procedure [111].

Dahlen and Bolmsjö's LCCA model aims to extend the application's field and perform an investment analysis when raising the production factor—labor. It considers all the costs associated with an employee over their employment time. The costs of labor can be graphed like the costs over the life cycle for production equipment [112]. In turn, the ABC Model has the most likely-looking cost-effective evaluation in lifecycle design. Respecting environmental matters, uncertainty must be considered due to the predominant lack of hard data. When there is a lack of information and the existence of unexpected activities, uncertainty conditions must be also used [113]. The economic input–output LCA model is a new tool that can complement the conventional LCA and overcome its limitations. The objective is to augment conventional economic input–output tables with appropriate sectored environmental impacts indices, which are then used to analyze economywide, direct and indirect environmental impacts of changes in the output of selected industries.

This model represents all the supplier relationships in the supply chain for industrial production. Its application was shown to be feasible, rapid and inexpensive [114].

In relation to the DTC model for manufacture systems, it provides a common methodology to merge cost modelling and quality function deployment (QFD) to assess the possible trade-offs among the costs and performance of product alternatives in the initiation of the production system's design. The design-to-cost approach has a plan for choosing a system design [115].

The PLCC model determines the life cycle expenses of capital goods such as machines and manufacturing systems. In this methodology, single processes connected to the product's life cycle are expressed. With an intent to the redesign of present product structures, it is feasible to derive approaches out the cost structures of the life cycle. The early stages of the product life cycle are production, use and disposal or de-production. The design of the product should be directed upon the needs of the use phase to reduce costs in the several phases of the life cycle. A similar dependency comes up in the disposal phase [116].

The TCA model seems to be the most helpful and practical tool for small manufacturers. It supplies a streamlined approach to identifying and quantifying the costs of pollution prevention investments. It expands the scope of capital budgeting to comprise indirect benefits, thereby increasing the magnitude of savings originated from pollution prevention investments. The information requirements of TCA can be readily implemented in the small business scenario.

It is not possible to create a single LCCA model that considers all the requirements, yet it is possible to develop models to answer specific needs, such as the elaboration of an ecological, sustainable and economical product.

Most of the related studies in the literature focus on the economic assessment of biochar systems, such as the studies carried out by Roberts et al. (2010), and Galinato et al. (2011). In these studies, is typical to find that the potential profitability of biochar production systems is highly dependent on the feedstock used. In their study, Yoder et al. (2011), proceed to model the trade-off between product yield and product quality as the conversion temperature increases, exploring the implications of different production techniques and resulting variations in the properties of biochar for overall system performance [117–119].

In another study, Homagain et al. (2016), found that, within the limit of life cycle analysis, the economic viability of the biochar-based bioenergy production system is directly dependent on pyrolysis costs and raw material processing (drying, grinding and pelletizing), in on-site collection, and also in total carbon offset amount provided by the system. Through a sensitivity analysis of the transport distance and the displacement values, it was shown that the system is profitable in the case of high biomass availability within 200 km and when the cost of carbon sequestration exceeds the 60 Canadian dollars-per-ton carbon equivalent (CO_2-eq) [120].

In the study performed by Clare et al. (2015), it was found that straw briquetting for thermal energy is the most economical carbon reduction technology, requiring a subsidized CAD 7 mgCO_2-eq. However, China's current bioelectricity subsidy scheme makes gasification (net present value (NPV) CAD 12.6 million) more financially attractive to investors when compared to briquetting (NPV CAD 7.34 million) and pyrolysis (NPV CAD 1.84 million). The potential for direct carbon reduction from pyrolysis (1.06 mgCO_2-eq per odt straw) is also less than briquetting (1.35 mgCO_2-eq per odt straw) and gasification (1.16 mgCO_2-eq per odt straw). The authors conclude that the indirect carbon reduction processes that results from biochar utilization can significantly improve the pyrolysis scenario and carbon reduction potential, bearing in mind that improving the agronomic benefit of biochar is essential for the pyrolysis scenario to compete as an economically viable and cost-effective mitigation technology [90].

According to the study performed by Cleary, (2018), the added value of pyrolysis for biochar production is more profitable than selling highest quality wood chips for cellulose. The modelled biochar price ranged from CAD 3 to CAD 4/kg, quoted for 10 to 20 kg of biochar packages. The pyrolysis cost was estimated at about CAD 150,000.00, the operating cost around CAD 78,840.00, including labor and electricity. Thus, it can be concluded that the costs are closely related to the technologies used in biochar production and also to the feedstock used [121].

5. Final Remarks

Two of the most important domains facing specific challenges within the circular economy are biomass and bio-based products. Materials based on biological resources can be used for a wide range of products and energy uses. The bioeconomy offers alternatives to products and energies based on fossil fuels and can contribute to the circular economy. Bio-based materials can also have advantages linked to their renewability, biodegradability, or the possibility of composting. On the other hand, the use of biological resources requires attention to be paid to the environmental impacts of its life cycle and to its sustainable supply. From a circular economy perspective, the cascading use of renewable resources should be promoted, where appropriate, with various cycles of reuse and recycling. Bio-based materials can be used in multiple ways, with the possibility of reusing and recycling them several times, which is consistent with the application of the waste hierarchy and, more generally, with options that lead to the best overall result for the environment.

Biochar is a solid, carbon-rich material generally obtained from thermochemical conversion of biomass and respective carbonization in oxygen-limited environments and has been proposed as a potential solution to climate change, energy security, degradation of natural resources, food security and catastrophic forest fires worldwide. Biochar production implies a complex chemical reaction process where biomass undergoes decomposition, depolymerization and condensation in anoxic high temperature conditions. As was extensively discussed, the many strategies for using biochar facilitate zero waste and the development of the circular economy. In addition, LCA has proven an excellent tool for quantifying the potential of biochar utilization, as well as for fostering and managing its production. Nevertheless, as was shown, multi-purpose applications of biochar make the functional units and system boundaries of different cases variable. Additionally, although these variables can be contextualized, they can barely be eliminated. Therefore, extensive system boundaries and more inclusive inventory considerations must be integrated and comprehensively analyzed in the LCA of biochar production. Although solid international guidelines and frameworks are available to promote consistency, and most of the studies follow the ISO standard for LCA, they still may consider different criteria and assumptions, resulting in different outcomes, so the methodology should be unified allowing to compare the results, at least to some extent. In addition, economic analysis through LCCA should be encouraged to optimize the flows of sustainable biochar production.

Overall, despite all the differences in assumptions and methodologies adopted, LCA proves that biochar is a very promising way of contributing to carbon-efficient resource circulation, mitigation of climate change and economic sustainability.

Author Contributions: Conceptualization J.C. and L.N.; writing—original draft preparation, J.C., L.N., M.S. and N.V.; writing—review and editing, A.R., L.F., A.S. and N.P.; project administration, J.C. and J.A.; funding acquisition, C.V. All authors have read and agreed to the published version of the manuscript.

Funding: This work was co-financed by Compete 2020, Portugal 2020 and the European Union through the European Regional Development Fund—FEDER within the scope of the project WAST'AWARENESS—Technology Transfer in Waste Valorization and Sustainability (POCI-01-0246-FEDER-181304).

Conflicts of Interest: The authors declare no conflict of interest.

Abbreviation

AD	aerobic digestion
EBC	European Biochar Certificate
GHG	greenhouse gas
GWP	global warming potential
IBI	International Biochar Certificate
LCA	life cycle assessment
LCI	life cycle inventory
LCIA	life cycle impact assessment
PAHs	polycycle aromatic hydrocarbons
PCBs	polychlorinated biphenyls

References

1. Korhonen, J.; Honkasalo, A.; Seppälä, J. Circular Economy: The Concept and its Limitations. *Ecol. Econ.* **2018**, *143*, 37–46. [CrossRef]
2. Beaulieu, L.; Durne, G.; Arpin, M. Circular Economy: A Critical Literature Review of Concepts. *CIRAIG-Int. Ref. Cent. Life Cycle Prod. Process. Serv.* **2015**, *91*. [CrossRef]
3. Rashid, A.; Asif, F.M.A.; Krajnik, P.; Nicolescu, C.M. Resource Conservative Manufacturing: An essential change in business and technology paradigm for sustainable manufacturing. *J. Clean. Prod.* **2013**, *57*, 166–177. [CrossRef]
4. Hu, Q.; Jung, J.; Chen, D.; Leong, K.; Song, S.; Li, F.; Mohan, B.C.; Yao, Z.; Prabhakar, A.K.; Lin, X.H.; et al. Biochar industry to circular economy. *Sci. Total Environ.* **2021**, *757*, 143820. [CrossRef]
5. Cha, J.S.; Park, S.H.; Jung, S.C.; Ryu, C.; Jeon, J.K.; Shin, M.C.; Park, Y.K. Production and utilization of biochar: A review. *J. Ind. Eng. Chem.* **2016**, *40*, 1–15. [CrossRef]
6. Biochar Is a Valuable Soil Amendment. WOKA Foundation. Available online: https://biochar-international.org/biochar/ (accessed on 7 September 2022).
7. Xie, Y.; Wang, L.; Li, H.; Westholm, L.J.; Carvalho, L.; Thorin, E.; Yu, Z.; Yu, X.; Skreiberg, Ø. A critical review on production, modification and utilization of biochar. *J. Anal. Appl. Pyrolysis* **2022**, *161*, 105405. [CrossRef]
8. Wang, D.; Jiang, P.; Zhang, H.; Yuan, W. Biochar production and applications in agro and forestry systems: A review. *Sci. Total Environ.* **2020**, *723*, 137775. [CrossRef]
9. Leng, L.; Huang, H. An overview of the effect of pyrolysis process parameters on biochar stability. *Bioresour. Technol.* **2018**, *270*, 627–642. [CrossRef]
10. Zhao, L.; Cao, X.; Mašek, O.; Zimmerman, A. Heterogeneity of biochar properties as a function of feedstock sources and production temperatures. *J. Hazard. Mater.* **2013**, *256–257*, 1–9. [CrossRef]
11. Yao, Z.; You, S.; Ge, T.; Wang, C.-H. Biomass gasification for syngas and biochar co-production: Energy application and economic evaluation. *Appl. Energy* **2018**, *209*, 43–55. [CrossRef]
12. Veses, A.; Aznar, M.; López, J.M.; Callén, M.S.; Murillo, R.; García, T. Production of upgraded bio-oils by biomass catalytic pyrolysis in an auger reactor using low cost materials. *Fuel* **2015**, *141*, 17–22. [CrossRef]
13. Choi, J.H.; Kim, S.-S.; Ly, H.V.; Kim, J.; Woo, H.C. Effects of water-washing Saccharina japonica on fast pyrolysis in a bubbling fluidized-bed reactor. *Biomass Bioenergy* **2017**, *98*, 112–123. [CrossRef]
14. Zhao, B.; O'Connor, D.; Zhang, J.; Peng, T.; Shen, Z.; Tsang, D.C.W.; Hou, D. Effect of pyrolysis temperature, heating rate, and residence time on rapeseed stem derived biochar. *J. Clean. Prod.* **2018**, *174*, 977–987. [CrossRef]
15. Hernández, J.J.; Saffe, A.; Collado, R.; Monedero, E. Recirculation of char from biomass gasification: Effects on gasifier performance and end-char properties. *Renew. Energy* **2020**, *147*, 806–813. [CrossRef]
16. Benedetti, V.; Patuzzi, F.; Baratieri, M. Characterization of char from biomass gasification and its similarities with activated carbon in adsorption applications. *Appl. Energy* **2018**, *227*, 92–99. [CrossRef]
17. Muvhiiwa, R.; Kuvarega, A.; Llana, E.M.; Muleja, A. Study of biochar from pyrolysis and gasification of wood pellets in a nitrogen plasma reactor for design of biomass processes. *J. Environ. Chem. Eng.* **2019**, *7*, 103391. [CrossRef]
18. Krysanova, K.; Krylova, A.; Kulikova, M.; Kulikov, A.; Rusakova, O. Biochar characteristics produced via hydrothermal carbonization and torrefaction of peat and sawdust. *Fuel* **2022**, *328*, 125220. [CrossRef]
19. Krysanova, K.; Krylova, A.; Zaichenko, V. Properties of biochar obtained by hydrothermal carbonization and torrefaction of peat. *Fuel* **2019**, *256*, 115929. [CrossRef]
20. Wang, Z.; Lim, C.J.; Grace, J.R.; Li, H.; Parise, M.R. Effects of temperature and particle size on biomass torrefaction in a slot-rectangular spouted bed reactor. *Bioresour. Technol.* **2017**, *244*, 281–288. [CrossRef] [PubMed]
21. Barskov, S.; Zappi, M.; Buchireddy, P.; Dufreche, S.; Guillory, J.; Gang, D.; Hernandez, R.; Bajpai, R.; Baudier, J.; Cooper, R.; et al. Torrefaction of biomass: A review of production methods for biocoal from cultured and waste lignocellulosic feedstocks. *Renew. Energy* **2019**, *142*, 624–642. [CrossRef]
22. Thengane, S.K.; Bandyopadhyay, S. Biochar mines: Panacea to climate change and energy crisis? *Clean Technol. Environ. Policy* **2020**, *22*, 5–10. [CrossRef]

23. Malmsheimer, R.W.; Bowyer, J.L.; Fried, J.S.; Gee, E.; Izlar, R.L.; Miner, R.A.; Munn, I.A.; Oneil, E.; Stewart, W.C. Managing Forests because Carbon Matters: Integrating Energy, Products, and Land Management Policy. *J. For.* **2011**, *109*, S7–S50.
24. Sanchez, D.L.; Nelson, J.H.; Johnston, J.; Mileva, A.; Kammen, D.M. Biomass enables the transition to a carbon-negative power system across western North America. *Nat. Clim. Chang.* **2015**, *5*, 230–234. [CrossRef]
25. Matrapazi, V.K.; Zabaniotou, A. Experimental and feasibility study of spent coffee grounds upscaling via pyrolysis towards proposing an eco-social innovation circular economy solution. *Sci. Total Environ.* **2020**, *718*, 137316. [CrossRef] [PubMed]
26. Thengane, S.K.; Kung, K.; York, R.; Sokhansanj, S.; Lim, C.J.; Sanchez, D.L. Technoeconomic and emissions evaluation of mobile in-woods biochar production. *Energy Convers. Manag.* **2020**, *223*, 113305. [CrossRef]
27. Yang, X.; Ng, W.; Wong, B.S.E.; Baeg, G.H.; Wang, C.-H.; Ok, Y.S. Characterization and ecotoxicological investigation of biochar produced via slow pyrolysis: Effect of feedstock composition and pyrolysis conditions. *J. Hazard. Mater.* **2019**, *365*, 178–185. [CrossRef]
28. Pecchi, M.; Baratieri, M. Coupling anaerobic digestion with gasification, pyrolysis or hydrothermal carbonization: A review. *Renew. Sustain. Energy Rev.* **2019**, *105*, 462–475. [CrossRef]
29. Woolf, D.; Amonette, J.E.; Street-Perrott, F.A.; Lehmann, J.; Joseph, S. Sustainable biochar to mitigate global climate change. *Nat. Commun.* **2010**, *1*, 56. [CrossRef]
30. Wang, L.; Chen, L.; Tsang, D.C.W.; Kua, H.W.; Yang, J.; Ok, Y.S.; Ding, S.; Hou, D.; Poon, C.S. The roles of biochar as green admixture for sediment-based construction products. *Cem. Concr. Compos.* **2019**, *104*, 103348. [CrossRef]
31. Pan, J.; Ma, J.; Zhai, L.; Luo, T.; Mei, Z.; Liu, H. Achievements of biochar application for enhanced anaerobic digestion: A review. *Bioresour. Technol.* **2019**, *292*, 122058. [CrossRef]
32. Kah, M.; Sigmund, G.; Xiao, F.; Hofmann, T. Sorption of ionizable and ionic organic compounds to biochar, activated carbon and other carbonaceous materials. *Water Res.* **2017**, *124*, 673–692. [CrossRef]
33. Kang, K.; Nanda, S.; Hu, Y. Current trends in biochar application for catalytic conversion of biomass to biofuels. *Catal. Today* **2022**, *404*, 3–18. [CrossRef]
34. Glaser, B.; Lehmann, J.; Zech, W. Ameliorating physical and chemical properties of highly weathered soils in the tropics with charcoal—A review. *Biol. Fertil. Soils* **2002**, *35*, 219–230. [CrossRef]
35. Pandit, N.R.; Mulder, J.; Hale, S.E.; Martinsen, V.; Schmidt, H.P.; Cornelissen, G. Biochar improves maize growth by alleviation of nutrient stress in a moderately acidic low-input Nepalese soil. *Sci. Total Environ.* **2018**, *625*, 1380–1389. [CrossRef]
36. Mohamed, B.A.; Ellis, N.; Kim, C.S.; Bi, X.; Emam, A.E. Engineered biochar from microwave-assisted catalytic pyrolysis of switchgrass for increasing water-holding capacity and fertility of sandy soil. *Sci. Total Environ.* **2016**, *566–567*, 387–397. [CrossRef]
37. Yang, X.; Tsibart, A.; Nam, H.; Hur, J.; El-Naggar, A.; Tack, F.M.G.; Wang, C.-H.; Lee, Y.H.; Tsang, D.C.W.; Ok, Y.S. Effect of gasification biochar application on soil quality: Trace metal behavior, microbial community, and soil dissolved organic matter. *J. Hazard. Mater.* **2019**, *365*, 684–694. [CrossRef]
38. Ruzickova, J.; Koval, S.; Raclavska, H.; Kucbel, M.; Svedova, B.; Raclavsky, K.; Juchelkova, D.; Scala, F. A comprehensive assessment of potential hazard caused by organic compounds in biochar for agricultural use. *J. Hazard. Mater.* **2021**, *403*, 123644. [CrossRef]
39. Purakayastha, T.J.; Bera, T.; Bhaduri, D.; Sarkar, B.; Mandal, S.; Wade, P.; Kumari, S.; Biswas, S.; Menon, M.; Pathak, H.; et al. A review on biochar modulated soil condition improvements and nutrient dynamics concerning crop yields: Pathways to climate change mitigation and global food security. *Chemosphere* **2019**, *227*, 345–365. [CrossRef]
40. Rizwan, M.; Ali, S.; Zia ur Rehman, M.; Rinklebe, J.; Tsang, D.C.W.; Bashir, A.; Maqbool, A.; Tack, F.M.G.; Ok, Y.S. Cadmium phytoremediation potential of Brassica crop species: A review. *Sci. Total Environ.* **2018**, *631–632*, 1175–1191. [CrossRef]
41. Rizwan, M.; Ali, S.; Qayyum, M.F.; Ibrahim, M.; Zia-ur-Rehman, M.; Abbas, T.; Ok, Y.S. Mechanisms of biochar-mediated alleviation of toxicity of trace elements in plants: A critical review. *Environ. Sci. Pollut. Res.* **2016**, *23*, 2230–2248. [CrossRef]
42. Qin, J.; Niu, A.; Liu, Y.; Lin, C. Arsenic in leafy vegetable plants grown on mine water-contaminated soils: Uptake, human health risk and remedial effects of biochar. *J. Hazard. Mater.* **2021**, *402*, 123488. [CrossRef]
43. Kapoor, A.; Sharma, R.; Kumar, A.; Sepehya, S. Biochar as a means to improve soil fertility and crop productivity: A review. *J. Plant Nutr.* **2022**, *45*, 2380–2388. [CrossRef]
44. Ding, Y.; Liu, Y.; Liu, S.; Li, Z.; Tan, X.; Huang, X.; Zeng, G.; Zhou, L.; Zheng, B. Biochar to improve soil fertility. A review. *Agron. Sustain. Dev.* **2016**, *36*, 36. [CrossRef]
45. Tratsch, M.V.M.; Ceretta, C.A.; da Silva, L.S.; Ferreira, P.A.A.; Brunetto, G. Composition and mineralization of organic compost derived from composting of fruit and vegetable waste. *Rev. Ceres* **2019**, *66*, 307–315. [CrossRef]
46. Zhang, L.; Sun, X. Changes in physical, chemical, and microbiological properties during the two-stage co-composting of green waste with spent mushroom compost and biochar. *Bioresour. Technol.* **2014**, *171*, 274–284. [CrossRef]
47. Sanchez-Monedero, M.A.; Cayuela, M.L.; Roig, A.; Jindo, K.; Mondini, C.; Bolan, N. Role of biochar as an additive in organic waste composting. *Bioresour. Technol.* **2018**, *247*, 1155–1164. [CrossRef]
48. Czekała, W.; Malińska, K.; Cáceres, R.; Janczak, D.; Dach, J.; Lewicki, A. Co-composting of poultry manure mixtures amended with biochar—The effect of biochar on temperature and C-CO_2 emission. *Bioresour. Technol.* **2016**, *200*, 921–927. [CrossRef]
49. Sørmo, E.; Silvani, L.; Thune, G.; Gerber, H.; Schmidt, H.P.; Smebye, A.B.; Cornelissen, G. Waste timber pyrolysis in a medium-scale unit: Emission budgets and biochar quality. *Sci. Total Environ.* **2020**, *718*, 137335. [CrossRef]

50. Weng, Z.; Van Zwieten, L.; Singh, B.P.; Tavakkoli, E.; Joseph, S.; Macdonald, L.M.; Rose, T.J.; Rose, M.T.; Kimber, S.W.L.; Morris, S.; et al. Biochar built soil carbon over a decade by stabilizing rhizodeposits. *Nat. Clim. Chang.* **2017**, *7*, 371–376. [CrossRef]
51. Godlewska, P.; Ok, Y.S.; Oleszczuk, P. THE DARK SIDE OF BLACK GOLD: Ecotoxicological aspects of biochar and biochar-amended soils. *J. Hazard. Mater.* **2021**, *403*, 123833. [CrossRef]
52. He, M.; Xiong, X.; Wang, L.; Hou, D.; Bolan, N.S.; Ok, Y.S.; Rinklebe, J.; Tsang, D.C.W. A critical review on performance indicators for evaluating soil biota and soil health of biochar-amended soils. *J. Hazard. Mater.* **2021**, *414*, 125378. [CrossRef]
53. Wang, J.; Wang, S. Preparation, modification and environmental application of biochar: A review. *J. Clean. Prod.* **2019**, *227*, 1002–1022. [CrossRef]
54. Kimbell, L.K.; Tong, Y.; Mayer, B.K.; McNamara, P.J. Biosolids-Derived Biochar for Triclosan Removal from Wastewater. *Environ. Eng. Sci.* **2018**, *35*, 513–524. [CrossRef]
55. Li, H.; Dong, X.; da Silva, E.B.; de Oliveira, L.M.; Chen, Y.; Ma, L.Q. Mechanisms of metal sorption by biochars: Biochar characteristics and modifications. *Chemosphere* **2017**, *178*, 466–478. [CrossRef]
56. Beckinghausen, A.; Reynders, J.; Merckel, R.; Wu, Y.W.; Marais, H.; Schwede, S. Post-pyrolysis treatments of biochars from sewage sludge and A. mearnsii for ammonia (NH4-n) recovery. *Appl. Energy* **2020**, *271*, 115212. [CrossRef]
57. Zhou, J.; Liu, S.; Zhou, N.; Fan, L.; Zhang, Y.; Peng, P.; Anderson, E.; Ding, K.; Wang, Y.; Liu, Y.; et al. Development and application of a continuous fast microwave pyrolysis system for sewage sludge utilization. *Bioresour. Technol.* **2018**, *256*, 295–301. [CrossRef]
58. Nobaharan, K.; Bagheri Novair, S.; Asgari Lajayer, B.; van Hullebusch, E. Phosphorus Removal from Wastewater: The Potential Use of Biochar and the Key Controlling Factors. *Water* **2021**, *13*, 517. [CrossRef]
59. Ambaye, T.G.; Vaccari, M.; van Hullebusch, E.D.; Amrane, A.; Rtimi, S. Mechanisms and adsorption capacities of biochar for the removal of organic and inorganic pollutants from industrial wastewater. *Int. J. Environ. Sci. Technol.* **2021**, *18*, 3273–3294. [CrossRef]
60. Dixit, A.; Gupta, S.; Pang, S.D.; Kua, H.W. Waste Valorisation using biochar for cement replacement and internal curing in ultra-high performance concrete. *J. Clean. Prod.* **2019**, *238*, 117876. [CrossRef]
61. Akhtar, A.; Sarmah, A.K. Novel biochar-concrete composites: Manufacturing, characterization and evaluation of the mechanical properties. *Sci. Total Environ.* **2018**, *616–617*, 408–416. [CrossRef]
62. Gupta, S.; Kua, H.W. Factors Determining the Potential of Biochar As a Carbon Capturing and Sequestering Construction Material: Critical Review. *J. Mater. Civ. Eng.* **2017**, *29*, 04017086. [CrossRef]
63. Gupta, S.; Kua, H.W. Carbonaceous micro-filler for cement: Effect of particle size and dosage of biochar on fresh and hardened properties of cement mortar. *Sci. Total Environ.* **2019**, *662*, 952–962. [CrossRef] [PubMed]
64. Gupta, S.; Kua, H.W.; Koh, H.J. Application of biochar from food and wood waste as green admixture for cement mortar. *Sci. Total Environ.* **2018**, *619–620*, 419–435. [CrossRef] [PubMed]
65. Gupta, S.; Kua, H.W. Effect of water entrainment by pre-soaked biochar particles on strength and permeability of cement mortar. *Constr. Build. Mater.* **2018**, *159*, 107–125. [CrossRef]
66. Kim, J.-Y.; Oh, S.; Park, Y.-K. Overview of biochar production from preservative-treated wood with detailed analysis of biochar characteristics, heavy metals behaviors, and their ecotoxicity. *J. Hazard. Mater.* **2020**, *384*, 121356. [CrossRef]
67. Lee, J. Characterization of urban stormwater runoff. *Water Res.* **2000**, *34*, 1773–1780. [CrossRef]
68. Inyang, M.I.; Gao, B.; Yao, Y.; Xue, Y.; Zimmerman, A.; Mosa, A.; Pullammanappallil, P.; Ok, Y.S.; Cao, X. A review of biochar as a low-cost adsorbent for aqueous heavy metal removal. *Crit. Rev. Environ. Sci. Technol.* **2016**, *46*, 406–433. [CrossRef]
69. Jin, H.; Wang, X.; Gu, Z.; Polin, J. Carbon materials from high ash biochar for supercapacitor and improvement of capacitance with HNO3 surface oxidation. *J. Power Sources* **2013**, *236*, 285–292. [CrossRef]
70. Coetzee, G.H.; Sakurovs, R.; Neomagus, H.W.J.P.; Everson, R.C.; Mathews, J.P.; Bunt, J.R. Particle size influence on the pore development of nanopores in coal gasification chars: From micron to millimeter particles. *Carbon N. Y.* **2017**, *112*, 37–46. [CrossRef]
71. Ge, X.; Wu, Z.; Wu, Z.; Yan, Y.; Cravotto, G.; Ye, B.-C. Microwave-assisted modification of activated carbon with ammonia for efficient pyrene adsorption. *J. Ind. Eng. Chem.* **2016**, *39*, 27–36. [CrossRef]
72. Fagerström, A.; Al Seadi, T.; Rasi, S.; Briseid, T. The role of anaerobic digestion and biogas in the circular economy. *IEA Bioenergy* **2018**, *8*, 1–24.
73. Liu, F.; Rotaru, A.-E.; Shrestha, P.M.; Malvankar, N.S.; Nevin, K.P.; Lovley, D.R. Promoting direct interspecies electron transfer with activated carbon. *Energy Environ. Sci.* **2012**, *5*, 8982. [CrossRef]
74. Fidel, R.B.; Laird, D.A.; Thompson, M.L.; Lawrinenko, M. Characterization and quantification of biochar alkalinity. *Chemosphere* **2017**, *167*, 367–373. [CrossRef]
75. Mumme, J.; Srocke, F.; Heeg, K.; Werner, M. Use of biochars in anaerobic digestion. *Bioresour. Technol.* **2014**, *164*, 189–197. [CrossRef]
76. Sunyoto, N.M.S.; Zhu, M.; Zhang, Z.; Zhang, D. Effect of biochar addition on hydrogen and methane production in two-phase anaerobic digestion of aqueous carbohydrates food waste. *Bioresour. Technol.* **2016**, *219*, 29–36. [CrossRef]
77. Ismaeel, W.S.E. Assessing and Developing the Application of LEED Green Building Rating System as a Sustainable Project Management and Market Tool in the Italian Context. *J. Eng. Proj. Prod. Manag.* **2016**, *6*, 136–152. [CrossRef]
78. Klöpffer, W. *Background and Future Prospects in Life Cycle Assessment*; Springer Science & Business Media: Berlin, Germany, 2014.
79. Souza, D.M.; Teixeira, R.F.M.; Ostermann, O.P. Assessing biodiversity loss due to land use with Life Cycle Assessment: Are we there yet? *Glob. Chang. Biol.* **2015**, *21*, 32–47. [CrossRef]

80. Matuštík, J.; Hnátková, T.; Kočí, V. Life cycle assessment of biochar-to-soil systems: A review. *J. Clean. Prod.* **2020**, *259*, 120998. [CrossRef]
81. Zhu, X.; Labianca, C.; He, M.; Luo, Z.; Wu, C.; You, S.; Tsang, D.C.W. Life-cycle assessment of pyrolysis processes for sustainable production of biochar from agro-residues. *Bioresour. Technol.* **2022**, *360*, 127601. [CrossRef]
82. Miller-Robbie, L.; Ulrich, B.A.; Ramey, D.F.; Spencer, K.S.; Herzog, S.P.; Cath, T.Y.; Higgins, C.P. Life cycle energy and greenhouse gas assessment of the co-production of biosolids and biochar for land application. *J. Clean. Prod.* **2015**, *91*, 118–127. [CrossRef]
83. Thers, H.; Djomo, S.N.; Elsgaard, L.; Knudsen, M.T. Biochar potentially mitigates greenhouse gas emissions from cultivation of oilseed rape for biodiesel. *Sci. Total Environ.* **2019**, *671*, 180–188. [CrossRef] [PubMed]
84. Oldfield, T.L.; Sikirica, N.; Mondini, C.; López, G.; Kuikman, P.J.; Holden, N.M. Biochar, compost and biochar-compost blend as options to recover nutrients and sequester carbon. *J. Environ. Manag.* **2018**, *218*, 465–476. [CrossRef] [PubMed]
85. Smebye, A.B.; Sparrevik, M.; Schmidt, H.P.; Cornelissen, G. Life-cycle assessment of biochar production systems in tropical rural areas: Comparing flame curtain kilns to other production methods. *Biomass Bioenergy* **2017**, *101*, 35–43. [CrossRef]
86. Brassard, P.; Godbout, S.; Pelletier, F.; Raghavan, V.; Palacios, J.H. Pyrolysis of switchgrass in an auger reactor for biochar production: A greenhouse gas and energy impacts assessment. *Biomass Bioenergy* **2018**, *116*, 99–105. [CrossRef]
87. Hamedani, R.; Kuppens, T.; Malina, R. Life cycle assessment and environmental valuation of biochar production: Two case studies in Belgium. *Energies* **2019**, *12*, 2166. [CrossRef]
88. Mohammadi, A.; Cowie, A.; Mai, T.L.A.; de la Rosa, R.A.; Brandao, M.; Kristiansen, P.; Joseph, S. Quantifying the greenhouse gas reduction benefits of utilising straw biochar and enriched biochar. *Energy Procedia* **2016**, *97*, 254–261. [CrossRef]
89. Uusitalo, V.; Leino, M. Neutralizing global warming impacts of crop production using biochar from side flows and buffer zones: A case study of oat production in the boreal climate zone. *J. Clean. Prod.* **2019**, *227*, 48–57. [CrossRef]
90. Clare, A.; Shackley, S.; Joseph, S.; Hammond, J.; Pan, G.; Bloom, A. Competing uses for China's straw: The economic and carbon abatement potential of biochar. *GCB Bioenergy* **2015**, *7*, 1272–1282. [CrossRef]
91. anak Erison, A.E.; Tan, Y.H.; Mubarak, N.M.; Kansedo, J.; Khalid, M.; Abdullah, M.O.; Ghasemi, M. Life cycle assessment of biodiesel production by using impregnated magnetic biochar derived from waste palm kernel shell. *Environ. Res.* **2022**, *214*, 114149. [CrossRef]
92. Lefebvre, D.; Williams, A.; Kirk, G.J.D.; Meersmans, J.; Sohi, S.; Goglio, P.; Smith, P. An anticipatory life cycle assessment of the use of biochar from sugarcane residues as a greenhouse gas removal technology. *J. Clean. Prod.* **2021**, *312*, 127764. [CrossRef]
93. Field, J.L.; Keske, C.M.H.; Birch, G.L.; DeFoort, M.W.; Cotrufo, M.F. Distributed biochar and bioenergy coproduction: A regionally specific case study of environmental benefits and economic impacts. *GCB Bioenergy* **2013**, *5*, 177–191. [CrossRef]
94. Xu, X.; Cheng, K.; Wu, H.; Sun, J.; Yue, Q.; Pan, G. Greenhouse gas mitigation potential in crop production with biochar soil amendment—A carbon footprint assessment for cross-site field experiments from China. *GCB Bioenergy* **2019**, *11*, 592–605. [CrossRef]
95. Barry, D.; Barbiero, C.; Briens, C.; Berruti, F. Pyrolysis as an economical and ecological treatment option for municipal sewage sludge. *Biomass Bioenergy* **2019**, *122*, 472–480. [CrossRef]
96. Hammond, J.; Shackley, S.; Sohi, S.; Brownsort, P. Prospective life cycle carbon abatement for pyrolysis biochar systems in the UK. *Energy Policy* **2011**, *39*, 2646–2655. [CrossRef]
97. Baumann, H.; Tillman, A.-M. The Hitch Hiker´s Guide to LCA—An orientation in LCA methodology and application. *Int. J. Life Cycle Assess.* **2006**, *11*, 142. [CrossRef]
98. Tiegam, R.F.T.; Tchuifon, D.R.T.; Santagata, R.; Nanssou, P.A.K.; Anagho, S.G.; Ionel, I.; Ulgiati, S. Production of activated carbon from cocoa pods: Investigating benefits and environmental impacts through analytical chemistry techniques and life cycle assessment. *J. Clean. Prod.* **2021**, *288*, 125464. [CrossRef]
99. Dutta, B.; Raghavan, V. A life cycle assessment of environmental and economic balance of biochar systems in Quebec. *Int. J. Energy Environ. Eng.* **2014**, *5*, 106. [CrossRef]
100. Ferreira, J.V.R. *Análise de Ciclo de Vida de Produtos*; Instituto Politécnico de Viseu: Viseu Dão Lafões, Portugal, 2004; 80p. [CrossRef]
101. Ismaeel, W.S.E. Midpoint and endpoint impact categories in Green building rating systems. *J. Clean. Prod.* **2018**, *182*, 783–793. [CrossRef]
102. Ferrão, P.C. Ecologia industrial: Princípios e ferramentas. *Eng. Sanit. Ambient.* **2009**, *17*. [CrossRef]
103. Yang, Q.; Zhou, H.; Zhang, X.; Nielsen, C.P.; Li, J.; Lu, X.; Yanga, H.; Chen, H. Hybrid life-cycle assessment for energy consumption and greenhouse gas emissions of a typical biomass gasification power plant in China. *J. Clean. Prod.* **2018**, *205*, 661–671. [CrossRef]
104. Zhang, C.; Liu, L.; Zhao, M.; Rong, H.; Xu, Y. The environmental characteristics and applications of biochar. *Environ. Sci. Pollut. Res.* **2018**, *25*, 21525–21534. [CrossRef] [PubMed]
105. Llorach-Massana, P.; Lopez-Capel, E.; Peña, J.; Rieradevall, J.; Montero, J.I.; Puy, N. Technical feasibility and carbon footprint of biochar co-production with tomato plant residue. *Waste Manag.* **2017**, *67*, 121–130. [CrossRef] [PubMed]
106. Santero, N.J.; Masanet, E.; Horvath, A. Life-cycle assessment of pavements Part II: Filling the research gaps. *Resour. Conserv. Recycl.* **2011**, *55*, 810–818. [CrossRef]
107. De Vries, M.; de Boer, I.J.M. Comparing environmental impacts for livestock products: A review of life cycle assessments. *Livest. Sci.* **2010**, *128*, 1–11. [CrossRef]
108. Muñoz, E.; Curaqueo, G.; Cea, M.; Vera, L.; Navia, R. Environmental hotspots in the life cycle of a biochar-soil system. *J. Clean. Prod.* **2017**, *158*, 1–7. [CrossRef]

109. Ilankoon, I.; Ghorbani, Y.; Chong, M.N.; Herath, G.; Moyo, T.; Petersen, J. E-waste in the international context–A review of trade flows, regulations, hazards, waste management strategies and technologies for value recovery. *Waste Manag.* **2018**, *82*, 258–275. [CrossRef]
110. Durairaj, S. Evaluation of Life Cycle Cost Analysis Methodologies. *Corp. Environ. Strateg.* **2002**, *9*, 30–39. [CrossRef]
111. Woodward, D.G. Life cycle costing—Theory, information acquisition and application. *Int. J. Proj. Manag.* **1997**, *15*, 335–344. [CrossRef]
112. Dahlén, P.; Bolmsjö, G.S. Life-cycle cost analysis of the labor factor. *Int. J. Prod. Econ.* **1996**, *46–47*, 459–467. [CrossRef]
113. Bras, B.; Emblemsvåg, J. Designing for the Life-Cycle: Activity-Based Costing and Uncertainty. In *Design for X*; Springer: Dordrecht, The Netherlands, 1996; pp. 398–423.
114. Cobas-Flores, E.; Hendrickson, C.T.; Lave, L.B.; McMichael, F.C. Life cycle analysis of batteries using economic input-output analysis. In Proceedings of the 1996 IEEE International Symposium on Electronics and the Environment, Dallas, TX, USA, 6–8 May 1996; pp. 130–134.
115. Eversheim, W.; Neuhausen, J.; Sesterhenn, M. Design-to-Cost for Production Systems. *CIRP Ann.* **1998**, *47*, 357–360. [CrossRef]
116. Westkämper, E.; Osten-Sacken, D.v.d. Product Life Cycle Costing Applied to Manufacturing Systems. *CIRP Ann.* **1998**, *47*, 353–356. [CrossRef]
117. Roberts, K.G.; Gloy, B.A.; Joseph, S.; Scott, N.R.; Lehmann, J. Life cycle assessment of biochar systems: Estimating the energetic, economic, and climate change potential. *Environ. Sci. Technol.* **2010**, *44*, 827–833. [CrossRef]
118. Galinato, S.P.; Yoder, J.K.; Granatstein, D. The economic value of biochar in crop production and carbon sequestration. *Energy Policy* **2011**, *39*, 6344–6350. [CrossRef]
119. Yoder, J.; Galinato, S.; Granatstein, D.; Garcia-Pérez, M. Economic tradeoff between biochar and bio-oil production via pyrolysis. *Biomass Bioenergy* **2011**, *35*, 1851–1862. [CrossRef]
120. Homagain, K.; Shahi, C.; Luckai, N.; Sharma, M. Life cycle cost and economic assessment of biochar-based bioenergy production and biochar land application in Northwestern Ontario, Canada. *For. Ecosyst.* **2016**, *3*, 21. [CrossRef]
121. Cleary, J. Life Cycle Impacts and Costs of Biochar Production Using Sawmill Residues. Available online: http//www.biofuelnet.ca/nce/wp-content/uploads/2015/08/Pathways_Clear.pdf (accessed on 7 October 2022).

processes

MDPI

Article

Techno-Economic Assessment of Solar-Driven Steam Gasification of Biomass for Large-Scale Hydrogen Production

Houssame Boujjat [1], Sylvain Rodat [2] and Stéphane Abanades [2,*]

[1] CEA-LITEN Solar and Thermodynamic Systems Laboratory (L2ST), 38054 Grenoble, France; houssame.boujjat@cea.fr

[2] Processes, Materials and Solar Energy Laboratory, PROMES-CNRS, 7 Rue du Four Solaire, 66120 Font-Romeu, France; sylvain.rodat@promes.cnrs.fr

* Correspondence: stephane.abanades@promes.cnrs.fr; Tel.: +33-(0)4-68-30-77-30

Abstract: Solar biomass gasification is an attractive pathway to promote biomass valorization while chemically storing intermittent solar energy into solar fuels. The economic feasibility of a solar gasification process at a large scale for centralized H_2 production was assessed, based on the discounted cash-flow rate of return method to calculate the minimum H_2 production cost. H_2 production costs from solar-only, hybrid and conventional autothermal biomass gasification were evaluated under various economic scenarios. Considering a biomass reference cost of 0.1 €/kg, and a land cost of 12.9 €/m^2, H_2 minimum price was estimated at 2.99 €/kg$_{H2}$ and 2.48 €/kg$_{H2}$ for the allothermal and hybrid processes, respectively, against 2.25 €/kg$_{H2}$ in the conventional process. A sensitivity study showed that a 50% reduction in the heliostats and solar tower costs, combined with a lower land cost of below 0.5 €/m^2, allowed reaching an area of competitiveness where the three processes meet. Furthermore, an increase in the biomass feedstock cost by a factor of 2 to 3 significantly undermined the profitability of the autothermal process, in favor of solar hybrid and solar-only gasification. A comparative study involving other solar and non-solar processes led to conclude on the profitability of fossil-based processes. However, reduced CO_2 emissions from the solar process and the application of carbon credits are definitely in favor of solar gasification economics, which could become more competitive. The massive deployment of concentrated solar energy across the world in the coming years can significantly reduce the cost of the solar materials and components (heliostats), and thus further alleviate the financial cost of solar gasification.

Keywords: solar gasification; hybridization; hydrogen; biomass/waste conversion; syngas; economics

1. Introduction

Today, around 96% of hydrogen is generated from fossil fuels (78% from natural gas and liquid hydrocarbons and 18% from coal) and only a low proportion of 4% is generated from water electrolysis [1]. In industry, hydrogen is mostly generated using carbon-based CO_2-emitting methods, such as steam reforming of light hydrocarbons, partial oxidation (POX) and autothermal reforming (AR) (which is a combination of the two previous processes) followed by coal gasification. Although extremely dependent on the price of natural gas, steam reforming remains the most preferred pathway for H_2 production given that it reached a high state of maturity outlined by lower production costs, usually below 2 \$/kg of H_2, including CO_2 capture and sequestration [2,3].

Currently, the largest volumes of hydrogen, produced or commercially available, are consumed in the chemical industry with a share of 63% for the production of ammonia, methanol, polymer and resin industries. Refineries are the second largest hydrogen consumers with a share of more than 30%, mainly for hydrocracking and crude oil hydrotreatment. Metallurgical industry consumes around 6% of the share. It is followed by general industries such as semiconductor, glass production, hydrogenation of vegetable oils and fats, etc., with a minor share of 1% [4].

Hydrogen, used as an intermediate chemical species for the above-mentioned processes, is also seen as a promising zero-carbon footprint energy vector for massive storage of intermittent renewable energies. Clean, i.e., CO_2-neutral hydrogen, can be produced using biomass or/and water as primary feedstocks. The most mature methods for decarbonized H_2 production are water electrolysis and steam biomass gasification [5]. Possibly powered by a renewable energy source such as solar or wind, electrolysis uses an electrical current to split water electrochemically into separate streams of H_2 and O_2. Being commercially available for over a century, current commercial electrolysers reach single-stack/module capacities of several megawatts with conversion efficiencies up to 85% [6]. Biomass gasification has advantages from the extensive accumulated experience of fossil fuels thermochemical gasification, which represents the state-of-the-art for industrial-scale H_2 generation. Biomass steam gasification produces a synthesis gas composed of both H_2 and CO at a high temperature (>900 °C). The syngas, therefore, needs to be upgraded (shifted to hydrogen) and purified in downstream equipment. According to the IEA Bioenergy's report in 2018 [5], hydrogen production from biomass, as a complementary route to increase the share of renewables, cannot be accomplished without the full-process chain validation at a large scale, involving an optimal biomass gasification technology capable of treating and converting a wide range of feedstocks.

Considering the growing demand of biomass in the future, the optimization of the conversion systems to make the best use of biomass is an absolute necessity. A promising way to save the biomass resource while maximizing the yield, quality and purity of the synthesis gas consists of using concentrated sunlight as an external source of energy to drive the endothermal thermochemical reactions instead of continually burning a part of the feedstock. The process goal is to replace feedstock combustion totally (solar-only systems) or partially (hybrid solar/autothermal systems), thanks to external heating using high-temperature concentrating solar technologies.

The solar process viability for coal and biomass gasification has been thoroughly studied at both laboratory [7–13] and pilot scales [14–16]. Accordingly, the extrapolation of these solar technologies to larger scales for semi-central or centralized green solar hydrogen production is auspicious in the future, in view of the increasing decarbonized hydrogen demand. Although more environmentally friendly than the conventional autothermal biomass gasification process (because it avoids the use of part of the biomass source for process heat), the question of the solar process's economic feasibility and competitiveness arises. On the one hand, the solar process allows for the production of a high-quality synthesis gas with a higher gas output per unit of feedstock, and on the other hand, the solar process is highly dependent on an intermittent heat source, which requires an initial substantial investment. The question is, therefore, not simple, and requires detailed investigation to highlight both technical aspects related to the management of the heat source variability [17–20] and economic and financial aspects for accurate cost evaluation [21–23]. In a previous work [19], a dynamic mathematical model of an up-scaled MW steam solar gasifier was developed. The model was used to assess the transient behavior of the reactor for three successive days with and without cloud cover. Different reactants' feeding management strategies were proposed and compared, with the aim of achieving enhanced syngas productivity and optimized use of solar energy. The OPTI mode controlled the supplies (biomass and steam) in order to stabilize the reactor temperature around a set point value (assumed to be 1200 °C) for as long as possible. The HYB production mode used the OPTI mode when the solar irradiation was sufficiently high, to gasify a minimum biomass flow rate (e.g., 1 t/h). Otherwise, the solar heating was assisted by in situ injection of O_2 to counteract the solar power decline and to maintain the reactor temperature constant all day long. Annual data were generated thanks to the dynamic model to estimate the feedstock consumption and syngas productivity under real solar irradiation conditions.

In the present work, a techno-economic study was carried out using the dynamic model predictions regarding the yearly gas production with the two recalled control strategies (OPTI and HYB). H_2 cost at plant gate for the autothermal (non-solar), hybrid (solar/optimized-combustion) and allothermal (solar-only) processes operating at different design capacities is evaluated using the DOE's H2A tool for hydrogen cost analysis [24]. Furthermore, a sensitivity study is performed to establish the impact of different factors on the cost of hydrogen. Finally, the cost of hydrogen is compared with other solar and non-solar processes for H_2 generation.

2. Solar Hydrogen Cost Model

2.1. General Principle

The DOE's H2A tool used in this study is based on the Discounted Cash Flow (DCF) rate of return method. DCF analysis finds the present value of expected future cash flows using a discount rate. The Internal Rate of Return (IRR) is the discount rate that cancels the Net Present Value (NPV). The *NPV* calculation (Equation (1)) converts all the expected future cash flows of a project into their "present value", i.e., their value at the initial time, at the very beginning of the project. Then, all the present values are added together to characterize the overall value of the company's project, in other words, the profitability of the project. The *NPV* is the cash flow generated at the end of the project

$$NPV = -I + \sum_{p=1}^{p=N} \frac{F_p}{(1+i)^p} \tag{1}$$

with I, the investment, F_p, the cash flow for year p, N, the total duration of the project (years), 'i', the discount rate (it reflects the cost of capital, so it may take the value of the market interest rate for a comparable duration, even though this value is often discussed). In the DOE's tool, 'i' is fixed and the model calculates the minimum hydrogen price so that the *NPV* cancels.

The starting point is a reference conventional (non-solar) biomass gasification process, previously developed by Mann and Steward [25]. The minimum cost of hydrogen was calculated with an indirectly heated steam woody biomass gasifier (based on a dual fluidized bed technology). The process model included biomass treatment and injection units, the reactor, gas compressors and scrubbing units, followed by a steam methane reformer (SMR), water gas shift reactors (WGS), and a Pressure Swing Absorption (PSA) unit to reach a hydrogen purity above 99.9%. Hydrogen was thereafter compressed to 7 MPa prior to shipment through a pipeline. In order to minimize the plant water consumption, the water contained in the syngas was recovered at different points of the cycle. Moreover, part of the electricity needed by the chemical plant was generated by recovering heat from the high-temperature syngas. A heat-recovery system using a steam turbine and a generator was, therefore, coupled to the chemical units. More details about the energy/material inflows and outflows can be found in [25]. The solarization of such a chemical process impacts a number of factors, including the capital investment, the O&M costs, and the plant biomass, water and electricity consumptions. These factors were estimated and integrated with the previously developed cost model using the dynamic simulation results (for yearly productivity estimation), as well as the previously reported Concentrated Solar Tower (CST) plants' operational costs.

2.2. Cost Model Assumptions

2.2.1. Basic Flow Diagram

The solar powered chemical process was modeled using the measured solar data (averaged over a 19-year period: 1991–2010) in Odeillo, in France. This region is characterized by a high duration and quality of sunshine (more than 2.2 MWh/m^2.year) with a great clarity of atmosphere.

The basic process flow diagram is described in Figure 1. It consisted of a solar plant composed of a heliostat field and reflecting towers (beam-down technology), and of a chemical plant for biomass gasification and gas processing/purification. The gasifier was fed by both steam and air. Air injection was considered only when operating in full autothermal or in hybrid (solar-combustion) modes. In the model, gas cleaning and upgrading chemical units (such as WGS and PSA) were assumed to be able to withstand rapid changes in gas flows and composition. WGS aims to react CO with H_2O to produce H_2 and CO_2, while PSA is used to obtain a pure H_2 stream as a process output. The PSA off-gas thus has no calorific value because CO is first shifted to H_2 in the WGS unit, which means the calorific value of CO contained in the produced syngas is recovered in the form of H_2 after WGS. As the aim was here to yield pure H_2 (for purpose of comparison with other H_2 production methods), the remaining PSA off-gas was not further considered in the study (the remaining PSA off-stream would consist of mainly CO_2 and N_2, but further post-processing was not included).

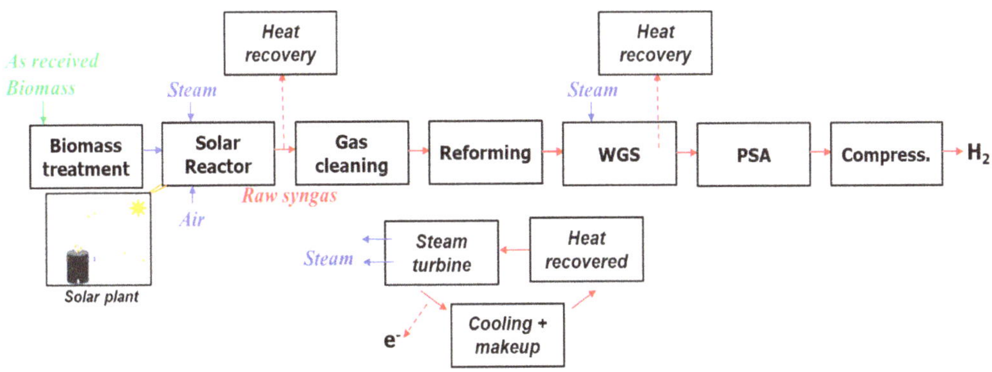

Figure 1. Biomass solar gasification flow diagram.

2.2.2. Capital costs

(i) Direct capital costs:

The major chemical component costs (installed) at the chemical baseline (defined by a design capacity *DC* of 155,236 kg of H_2 per day) are presented in Table 1.

Table 1. Direct capital costs at chemical baseline (installed, to be scaled).

	M€
Feed Handling and Drying	24.57
Gasification, Tar Reforming, and Quench	21.84
Compression and Sulfur Removal	20.28
Steam Methane Reforming, Shift, and PSA	39.39
Steam System and Power Generation	18.72
Cooling Water and Other Utilities	4.42
Buildings and Structures	19.26

These costs were scaled to different design points using Equation (2).

$$Component'i'cost_{at\ DC} = Component_{\ at\ chemical\_baseline}\ cost\cdot\left(\frac{DC}{Chemical\_baseline\ DC}\right)^{n_{chemical}} \quad (2)$$

with $n_{chemical}$ = 0.78 [25] and *DC* in kg of H_2 per day.

The *heliostat field cost* at *DC* was calculated by Equation (3).

$$Heliost\ cost_{at\ DC} = Mirror\ reference\ cost.field\ surface_{\ at\ DC} \quad (3)$$

$$field\ surface_{at\ DC} = field\ surface_{at\ solar\_baseline} \cdot \frac{DC}{Solar\_baseline_{OPTI,HYB}\ DC} \tag{4}$$

The *mirror reference cost* in €/m^2 was assumed to be equal to 120 €/m^2 based on reference [26]. The solar baseline was defined by a thermal power input on the solar field of 10 MW$_{th}$ (at a Direct Normal Irradiance *DNI* of 1000 W.m^{-2}). Thus, *Solar_baseline* $_{OPTI,\ HYB}$ DC was directly deduced from the previously developed dynamic model [19]. It was estimated at 1402.0 kg H_2/day for the OPTI mode and 4196.1 kg H_2/day for the HYB mode. The *field surface* $_{at\ solar\_baseline}$ in m^2 was calculated by Equation (5).

$$field\ surface_{at\ solar\_baseline} = \frac{Q_{sun\rightarrow field,max}}{DNI_{max}} \tag{5}$$

with $Q_{sun\rightarrow field,max}$ = 10 MW and DNI_{max} = 1000 W.m^{-2}.

Similarly, the *tower cost* $_{at\ DC}$ was calculated by Equation (6).

$$tower\ cost_{at\ DC} = tower\ cost_{at\ solar\_baseline} \cdot \frac{DC}{Solar\_baseline_{OPTI,HYB}\ DC} \tag{6}$$

As tower costs in the literature are often expressed in € per MWe, *tower cost* $_{at\ solar\_baseline}$ was deduced from Equation (7), assuming a solar-to-electric efficiency $\eta_{solar\text{-}to\text{-}electric}$ of 30% [27] including the receiver thermal efficiency and a field efficiency $\eta_{opt,field}$ of 70% [28].

$$tower\ cost_{at\ solar\_baseline} = tower\ cost\ in\ /10\ MW_e \cdot (\eta_{opt,field} \cdot \eta_{solar-to-electric}) \tag{7}$$

Tower cost per 10 MW_e was considered equal to 2 M€ based on the data provided by Becker et al. [29] in the Ecostar roadmap;

(ii) Indirect depreciable capital costs:

The site preparation, engineering and design, project contingency and up-front permitting costs were calculated by applying a percentage to the sum of the direct capital costs of the overall plant (solar and chemical). These percentages were, respectively, 2%, 10%, 15%, 15%;

(iii) Non-depreciable capital costs:

The cost of land, which can greatly vary depending on the plant's location, was varied from 0.5 to 50 €/m^2. A typical serviced land in Odeillo costs up to 150 €/m^2 while bare land in rural regions costs only few cents to few euros per m^2. Equation (8) was used to estimate the plant *land cost*.

$$Land\ cost = Chemical\ land\ cost + Solar\ land\ cost \tag{8}$$

The *chemical land cost* was calculated by Equation (9).

$$Chemical\ land\ cost = land\ cost\ per\ m^2 . \left(\frac{DC}{Chemical\_baseline\ DC}\right)^{n_{chemical}} .Chemical.\ land\ required_{at\ Chemical\_\ baseline} \tag{9}$$

where the chemical plant land required at baseline design capacity was assumed equal to 20.2 hectares. The *solar land cost* was calculated by Equation (10).

$$Solar\ land\ cost = land\ cost\ per\ m^2 .solar\ field\ size_{at\ DC} \tag{10}$$

The *solar field size* $_{at\ DC}$ was considered seven times the *field surface* $_{at\ DC}$ (Equation (4)) according to the PS10 plant data in Spain [30].

2.2.3. Fixed Costs

The total *plant staff* was calculated by Equation (11).

$$Plant\ staff = Chemical\ plant\ staff + Solar\ plant\ staff \quad (11)$$

$$Chemical\ plant\ staff = chemical\ plant\ staff_{at\ Chemical\_baseline} \cdot \left(\frac{DC\ (kg\ H_2/day)}{Chemical\_baseline\ DC\ (kg\ H_2/day)}\right)^{0.25} \quad (12)$$

The *chemical plant staff* $_{at\ chemical\_baseline}$ was considered equal to 54. The CST plant associated one operator for each 6.25 hectares of mirrors, following the equation provided by Sargent and Lundy [31]. The total plant staffing cost was thus deduced, assuming a burdened labor cost of 54 €/person/h. The general and administrative expenses were estimated as 20% of the total plant staffing cost. The property taxes and insurance were assumed to be equal to 2% of the total capital costs, and the material maintenance costs and repairs were assumed to be equal to 0.5% of the project direct capital costs.

2.2.4. Utilities, Feedstock and Variable Costs

(i) Water:

Water was used for different purposes, beyond its main role as a biomass oxidizer. Water was used to clean (by removing impurities such as particulates and tars residuals) and to shift the syngas into hydrogen. An important amount of water was also used for the cooling of syngas at the exit of the gasification unit and after the last stages of compression. It was also substantially used for heat rejection in the condenser and as a makeup for the steam cycle. Design calculations allowed for estimation of the process water consumption at about 3.8 L/kg of H_2. The cooling water consumption was considerably higher, at around 300.0 L/kg of H_2 [25]. Additional washing water was required for the solar-powered chemical plant due to the periodic cleaning of the mirrors. Considering a washing water consumption V_{water} of the heliostat field of 18 L/$MWh_{th,on\ field}$ [32], the amount of required water per kg of H_2 was deduced from Equation (13). The total cost of water per kg of H_2 was hence calculated by Equation (14)

$$Washing\ water = V_{water} \cdot \frac{Q_{sun \rightarrow field,max}}{Solar\_baseline_{OPTI,HYB}\ DC} \quad (13)$$

$$Cost\ of\ water = (Washing\ water + Process\ water).\ C2 + Cooling\ water.\ C3 \quad (14)$$

with $C2 = 6.1 \times 10^{-4}$ €/L and $C3 = 3.0 \times 10^{-5}$ €/L;.

(ii) Electricity:

The different chemical plant sections consumed electricity to different extents. The compression of syngas was the most energy-demanding step in the process, representing up to 60% of the total electricity requirement. The heat recovery system generated most of the power. The deficit in electricity was, therefore, directly supplied by the grid. In conventional CST plants, the electrical requirement comprises the Heat Transfer Fluid (HTF) pumping, along with the electricity used for tracking the solar rays, which remains very low. As there is no HTF in the proposed solar gasification concept, and as the energy of the tracking is of minor significance [33], the electricity requirement of the solar plant was neglected. The overall process electricity requirement (supplied by the grid) was estimated in the previous cost model [25] at 0.98 kWh_e/kg of H_2, with a cost of electricity of 0.1 €/kWh;

(iii) Biomass:

Biomass consumption varies depending on how the gasifier is heated. In solar gasification, the available solar energy is collected, then concentrated by a field of mirrors and towers to ensure the complete and total conversion of the biomass load. In a purely autothermal mode, the reactor is heated solely by burning part of the feedstock. In the hybrid mode, the biomass is partially burned, but to a lesser extent than in pure autothermal mode, as it is complemented by solar power when available. The prediction of the biomass consumption (expressed in $kg_{biomass,dry}/kg$ of H_2) for the three modes, OPTI, HYB and autothermal, was done based on the annual simulations. The biomass consumption of the three modes (OPTI, HYB, autothermal) used in the economic analysis are, respectively, 5.8, 8.7, and 9.7 $kg_{biomass,dry}/ kg_{H2}$;

(iv) Other costs:

Other costs include catalysts and bed materials, environmental surcharges, waste treatment and solid waste disposal. These costs are recalled in Table 2 at the chemical baseline design capacity.

Table 2. Other variable operating costs.

	M€
Other materials	7.00
Waste treatment	1.20
Solid waste disposal	0.73
Environmental surcharges	0.13

The scaling to different design capacities followed Equation (15) [25].

$$Scaled\ variable\ cost = 1.426.Baseline\ cost \cdot \frac{DC\ (kg\ H_2/day)}{Chemical\_baseline\ DC\ (kg\ H_2/day)} \tag{15}$$

3. Results and Discussion

3.1. Design Parameters

Table 3 shows the calculated design parameters in the three plant configurations.

Table 3. Comparison between the three studied processes at DC = 150,000 kg_{H2}/day.

	Autothermal	Hybrid	Allothermal
Plant size (hectares)	19.7	250.2	748.9
Solar power on field (MW)	-	357.5	1069.9
Biomass consumption (t_{dry}/day)	1.45×10^3	1.30×10^3	0.87×10^3
Water consumption (m^3/day)	4.58×10^4	4.60×10^4	4.63×10^4
Annual Cold Gas efficiency [19]	0.80	0.93	1.34
CO_2 emissions (t/day) [19]	1.04×10^3	5.78×10^2	29.5

The plant land surface area dramatically increases by 12 times in the hybrid mode and by up to 37 times in the allothermal mode. The solar power is, therefore, 67% lower in the hybrid mode compared to the allothermal mode, at the expense of a greater biomass requirement. In fact, around 0.43×10^3 t/day more biomass is needed to power the reactor during hybrid and full-autothermal phases, which represents about 30% of the total feedstock consumed by the hybrid process. The interest in the allothermal process lies in its high CGE, which by far exceeds those of the hybrid and autothermal processes. The process water requirement, which includes the heat-recovery system for local power generation, is hardly impacted by solarization, as water consumption due to mirror cleaning represents only a small proportion of the total plant water requirement. CO_2 direct emissions due to gasification process (reactor heating and/or gasification reaction) are, on the other hand, 35 times lower in allothermal solar gasification because no combustion is used for process heat supply in this case. The overall carbon balance includes additional greenhouse gas emissions, which are released during the different phases of the solar plant's life-cycle, i.e., during raw material extraction, manufacturing and assembly, transport, construction, site improvement, maintenance, replacements, dismantling/disposal and/or recycling. The application of credits for CO_2 mitigation and pollution avoidance will further enable the solar thermochemical technologies to compete favorably with fossil-fuel-based processes or autothermal technologies.

3.2. Cost Assessment

3.2.1. Key Parameters

The project was assumed to start in 2030 with a construction period of three years. It was financed through equity contributions and debt. All the financial inputs used in the economic study are recapped in Table 4. Hydrogen cost evaluation was carried out with a fixed operating capacity factor (CF) of 80% (accounting for possible maintenance and outage times). Due to the novelty of the project that may discourage some of the investors, the IRR was varied from 8% to 16%. The impact of equity financing (%) and the IRR on the cost of hydrogen for the three presented configurations is shown in Figure 2 (at a DC of 150,000 kg H_2/day, a biomass reference price of 0.10 €/kg and a land cost of 12.9 €/m^2). Due to the possible variability of the reference values for investment and operational costs estimations (depending on seasonal variations, biomass type, plant location, exploitation period, etc.), several sensitivity studies are proposed in the following discussion.

Table 4. Financial inputs.

Start-up time (years)	1
Analysis period and plant life (years)	30
Length of construction period (years)	3
% of Capital spent in 1st, 2nd and 3rd year of construction	8%, 60%, 32%
Depreciation Schedule Length (years)	20
Depreciation Type	MACRS
% of Fixed Operating Costs During Start-up (%)	100%
% of Revenues During Start-up (%)	50%
% of Variable Operating Costs During Start-up	75%
Decommissioning costs (% of depreciable capital investment)	10%
Salvage value (% of total capital investment)	10%
Inflation rate (%)	1.9%
Interest rate on debt	3.7%
Total Tax Rate (%)	25.7%
Working Capital (%)	15%

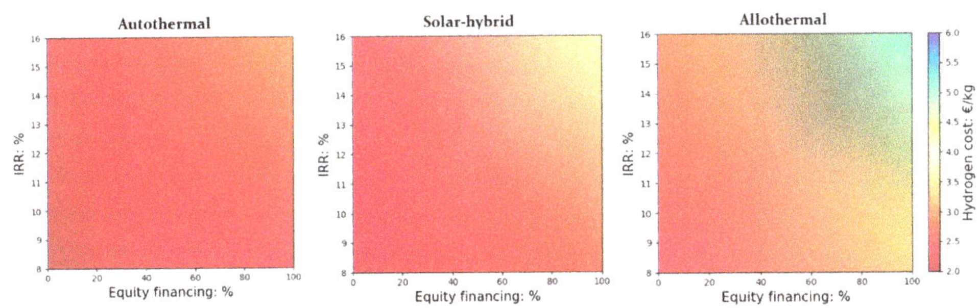

Figure 2. Impact of equity financing (%) and IRR on the cost of H_2 (DC = 150,000 kg H_2/day and a biomass reference price of 0.1 €/kg and land cost of 12.9 €/m^2).

It can be seen that these two purely financial parameters have a considerable impact on the price of hydrogen, which varies from 2.41 to 5.15 €/kg_{H2} for the solar-only heated process, from 2.15 to 3.72 €/kg_{H2} for the hybrid process, and from 2.03 to 3.02 €/kg_{H2} for the autothermal (non-solar) process. The increase in equity financing (at the expense of less incurred debt) drives up the cost of hydrogen markedly; for instance, for an IRR of 8%, an equity increases from 0% to 100% raises the cost of hydrogen by 37%, 25% and 18% (solar-only, hybrid, and autothermal, respectively). The influence of equity percentage on hydrogen price is, therefore, much greater for the allothermal process that required the largest initial capital investment. In such cases, the capital investment of the major pieces of equipment (installed) at *DC* = 150,000 kg H_2/day is 335.7 M€, while it is around 201.0 M€ for the hybrid process and only about 137.01 M€ for the non-solar autothermal process. The new shares issued by the increase in equity contributions hence provide more room to launch the investments, but, on the other hand, imply greater production costs.

In the following, a percentage of equity of 40% with an IRR of 10% is considered. The overall study reference assumptions are recapped in Table 5. The breakdown of the direct capital costs of the studied plants is shown in Figure 3. In the solar-driven processes (solar-only and hybrid solar/autothermal), the heliostat fields hold the largest share of the investment. It contributes to approximately 44% (solar-only) and 24% (hybrid) of the overall direct costs, in agreement with previously reported conventional CST plant values [34,35]. The smallest solar plant size required for the hybrid process leads to a reduced hydrogen price from 2.99 (allothermal) to 2.48 €/kg_{H2} (hybrid). The autothermal process is the cheapest, with a hydrogen price of 2.25 €/kg_{H2}.

Table 5. Reference assumptions.

Biomass cost (€/kg)	0.10
Land cost (€/m^2)	12.9
Mirror cost (€/m^2)	120
Tower cost (M€/MW_{th})	0.42
DC (kg H_2/day)	150,000
Electricity cost (€/kWh)	0.10
Water cost (€/m^3)	0.61 €/m^3 (process), 0.03 €/m^3 (cooling)

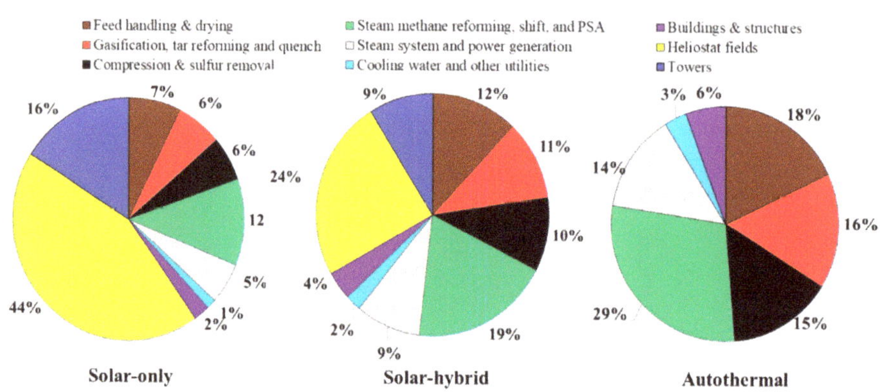

Figure 3. Direct capital investment breakdown (at reference assumptions, Table 5).

Figure 4 shows, in detail, the specific contribution of each item in the plant on the total cost of hydrogen. Solarization respectively increases the capital and the O&M costs by more than three times and up to 46% (in the allothermal configuration). Moreover, the feedstock cost for the hybrid and the autothermal processes is the most predominant, and contributes to nearly 37% and 39% of the total hydrogen production cost (at plant gate). The allothermal process consumes less biomass, and, therefore, the feedstock cost is lower, representing barely 20%. This is approximately 1.6 times less than for the autothermal process. The impact on the plant variable costs and utilities remains very limited, showing a relative variation of only 2%.

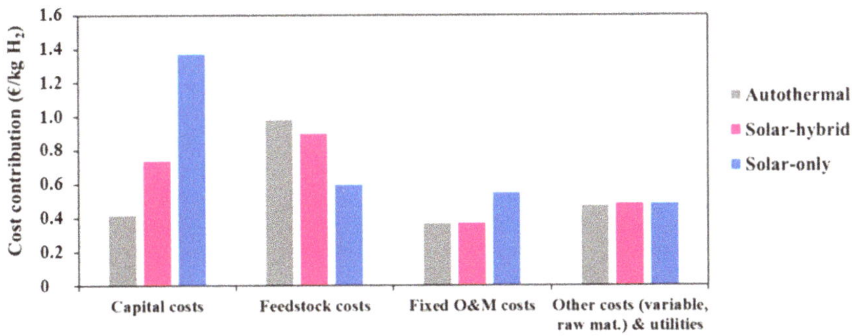

Figure 4. Specific contribution of each project component.

In the following sections, the influence of solar investment and biomass feedstock costs, as major economic factors affecting the minimum hydrogen price, is studied. A sensitivity analysis is carried out on these parameters to examine their impact on the profitability, the competitiveness and the relevance of the projects.

3.2.2. Impact of Solar Investment

The solar investment represents a high proportion of the overall project expenditure, which may be high enough to undermine the economic attractivity and viability of solar processes. In coming years, and in view of the increasing deployment of solar energy worldwide, the solar investment effort is expected to drop appreciably [26]. In fact, innovative designs and new technological solutions are studied intensely at present in many research laboratories with the objective of increasing the concentration efficiency and the durability of the materials. In conjunction with the economy of scale, this should reduce the solar costs to a certain extent, for the benefit of solar and solar hybrid gasification. In this respect, the influence of a possible cost reduction of the main solar compounds (i.e., heliostat field and towers) on the minimum hydrogen price was studied (Figure 5).

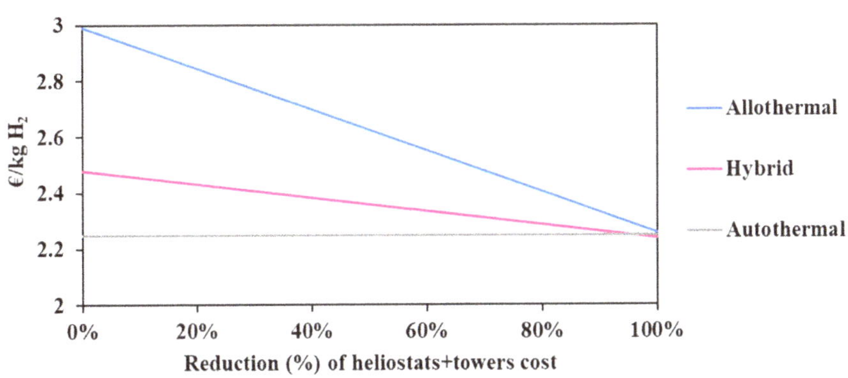

Figure 5. Impact of solar technologies cost reduction on the hydrogen minimum price.

It can be observed in Figure 5 that the solar allothermal process is the most costly. Moreover, the decline in the heliostat and tower costs results in reducing the hydrogen minimum price in a more pronounced fashion in the allothermal configuration. In fact, a 50% cost reduction declines the price of hydrogen by 0.37 €/kg_{H2} for the allothermal process, and by 0.12 €/kg_{H2} for the hybrid process. The intersection between the two curves is achieved only when the cost reduction is beyond 95%, which is practically unattainable. In any case, and whatever the cost reduction, solar and solar hybrid hydrogen generation remains more expensive as compared to the conventional autothermal process, which is due to major additional costs related to the heliostat field and tower, plant land, maintenance and staffing. Design calculations allowed for estimating the total area occupied by the solar plants (allothermal and hybrid): the heliostat field surface at DC = 150,000 kg H_2/day is estimated at 11.5 km^2 for the allothermal process and at 3.9 km^2 for the hybrid process. This represents nearly 57 and 20 times the required chemical plant land surface. Figure 6 shows the impact of the land cost (varied between 0.5 and 50 €/m^2) on the hydrogen minimum price. The graphic shows the importance of the choice of land, which, apart from being highly irradiated and allowing for continuous biomass supply, must be economically profitable. In fact, a quite significant decrease in the cost of hydrogen from 2.99 €/kg_{H2} at reference land cost (12.9 €/m^2) to 2.63 €/kg_{H2} at 0.5 €/m^2 is observed for the allothermal process. As the hybrid plant occupies a smaller area, the hydrogen minimum price decreases less markedly, by 0.16 €/kg_{H2} against 0.36 €/kg_{H2}, in the allothermal process. On the other hand, the autothermal configuration is almost insensitive to land cost, showing a relative H_2 price variation of less than 0.5%. Additionally, a 50% reduction in the heliostats and solar towers costs, combined with a lower land cost below 0.5 €/m^2, allows for reaching an area of competitiveness where the three processes meet. This could also correspond to a more favorable plant site in which the solar resource is greater, e.g., Chilean desert, although the cost of water may be somewhat higher in desert locations.

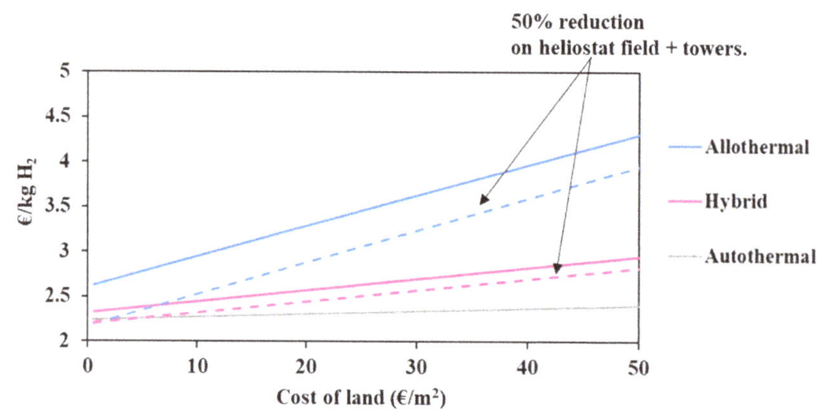

Figure 6. Impact of land cost on the hydrogen minimum price (solid lines: reference assumptions, Table 5; Chain-dotted lines: 50% cost reduction on heliostats and towers).

Another important parameter affecting the level of solar investment is the plant design capacity. This parameter was varied in Figure 7 from 20,000 to 150,000 kg H_2/day to analyze its impact on the solar/chemical direct costs and on the hydrogen minimum price.

It can be observed that the solar direct costs of allothermal process grow at least two times faster than the hybrid process solar direct costs and the chemical facilities costs (reaching a maximum value of 202.3 M€). Conversely, the increase in plant design capacity reduces the hydrogen minimum price by up to 25%, 34%, and 39% for the allothermal, hybrid, and autothermal processes, respectively. Additionally, due to the sharp rise in the solar facility costs, the relative difference in hydrogen minimum price between the solar processes and the autothermal process goes up from 18% (allothermal) and 6% (hybrid) at 20,000 kg H_2/day to 32% (allothermal) and 10% (hybrid) at 150,000 kg H_2/day. This suggests that, upon scale-up, the competitive gap between the conventional and the solar processes increases. However, this may be due solely to the linearity assumption that was adopted in (Equations (3)–(6)) between the solar costs and the production capacity. As a matter of fact, solar costs prediction is not straightforward and relies on uncertain data. Generally speaking, heliostats field scale-up depends on many factors, such as the design of individual mirrors, their number, their arrangement, their sub-composition and their reflective properties. Larger solar fields impair the quality of concentration and suffer from amplified atmospheric attenuation (due to a greater heliostat-to-receiver slant path) [36]. At the same time, larger solar fields involve higher solar power inputs that allow the use of larger cavity receivers (i.e., gasification reactors), which reduce the energy losses (due to a better absorption of radiation), and thus positively impact the solar costs. A power law with a global scaling exponent factor of 0.7 accounting for these trends was previously used by Kromer et al. [37] for assessing hydrogen cost of several solar thermochemical processes. The impact of this value on the solar costs and on hydrogen minimum price is shown in Figure 7 (scaling exponent factor = 1 in the base case and 0.7 for the dashed lines). The results show that the trends strongly vary with the scaling exponent. The solar costs decline by approximately 72% for the allothermal process and by 60% for the hybrid process at 150,000 kg H_2/day. In the same way, the hydrogen price for the allothermal process sharply drops to 2.09 €/kg_{H2} at 150,000 kg/day, against 2.22 €/kg_{H2} for the hybrid process and 2.25 €/kg_{H2} for the autothermal process. This highlights the necessity of a proper field layout optimization during the scale-up to maximize the energy/materials savings and further reduce the hydrogen cost.

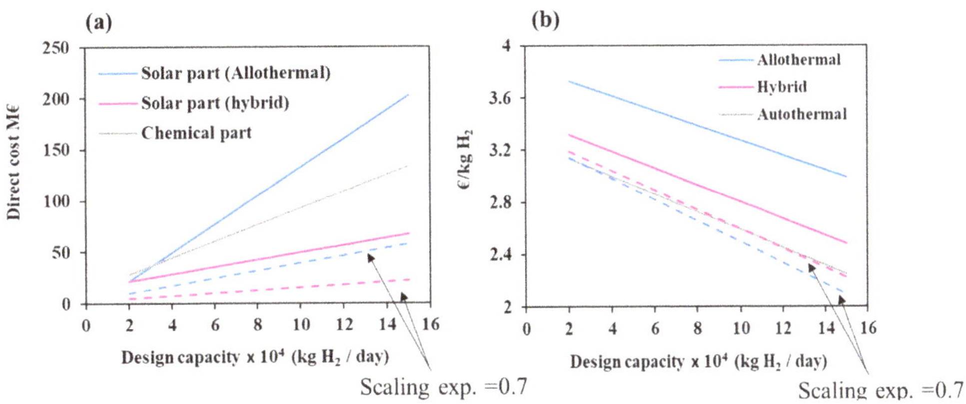

Figure 7. (**a**) Direct costs at different design capacities separated in two parts: solar and chemical and impact of scaling exponent factor; (**b**) Hydrogen minimum price for different design capacities and impact of scaling exponent factor. For the base case (solid lines), the scaling exponent factor is taken equal to 1 for the solar part.

The feedstock cost is another crucial parameter to be studied (Figure 4). Its impact on hydrogen cost is presented in the following section.

3.2.3. Impact of Feedstock Cost

The cost of the feedstock is a key dynamic parameter that evolves with different factors such as local supply chains, resource availability, sustainability criteria, political choices or competing uses of biomass. In this part of the study, the biomass price was varied in the 0–1 €/kg interval to cover a large range of woody and non-woody biomasses (such as food waste, agricultural and crop waste, and Solid Recovered Fuels) and also a potential increase in the resource price (due, for instance, to the increasing pressure on the resource in the incoming years or to extra-cost associated with additional required feedstock pre-treatment). Figure 8 shows the hydrogen cost as a function of the biomass price for the three studied processes (at reference conditions represented by solid lines). Two additional scenarios are considered: the first one assumes a 50% cost reduction of the heliostat field and towers (at reference land cost) and the second one assumes (in addition) a land cost of 0.5 €/m^2. Two zones on these graphs can be observed for each of the considered solar scenarios: one zone when the biomass price is below a critical value and one zone when the resource price is above. In the first zone, the autothermal mode prevails and imposes lower production costs. In the second zone, a significant reversal trend occurs in favor of the solar processes. Table 6 shows the critical biomass prices at the intersection between the autothermal and solar processes curves. It can be seen that the trend turnaround occurs faster in the allothermal process than in the hybrid one at reference assumptions. In fact, it takes place at a biomass critical price of 0.29 €/kg (for the allothermal mode, which represents three times the reference biomass price), against 0.37 €/kg for the hybrid configuration. By reducing the solar equipment cost by 50%, the turnaround biomass price decreases by about 31% for the allothermal mode and by 38% for the hybrid mode (at reference land cost). It decreases even more, by a total of 69% (allothermal) and 84% (hybrid), when the land cost is set to 0.5 €/m^2. In the latter scenario, the turnaround occurs earlier in the hybrid process.

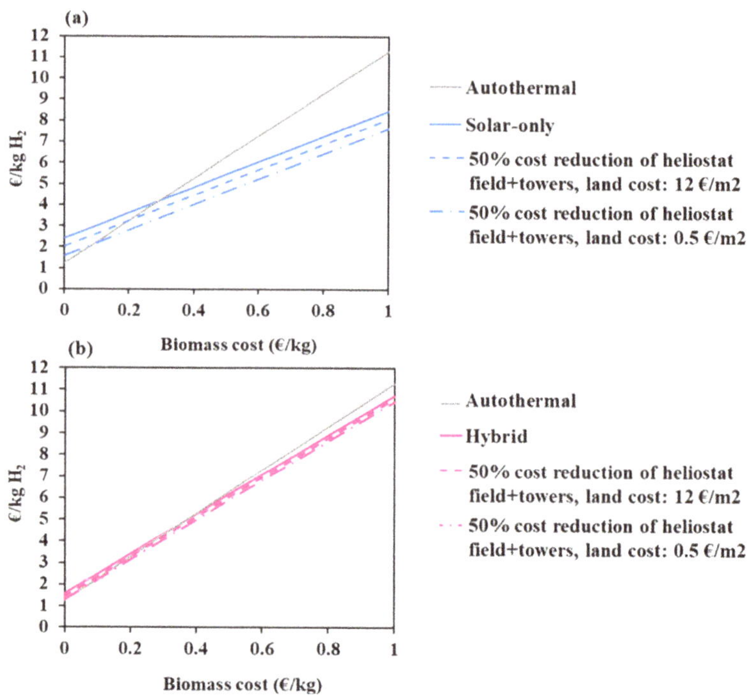

Figure 8. Impact of biomass cost on hydrogen minimum price; (**a**)-solar-only, (**b**)-hybrid.

Table 6. Turnaround biomass price (€/kg): autothermal/solar, DC = 150,000 kg H_2/day.

	Autothermal/Allothermal	Autothermal/Hybrid
Reference assumptions	0.29	0.35
+50% reduction on (heliostats + towers), land cost = 12.9 €/m^2	0.20	0.23
+50% reduction on (heliostats + towers), land cost = 0.50 €/m^2	0.09	0.06

In summary, the analysis shows that a slight increase in the price of the feedstock undermines the autothermal process. The better use of biomass provided by the solar processes clearly limits the increase in hydrogen cost, especially when combined with lower solar plant and land costs. Zero or even negative-priced feedstocks remain more profitable using the conventional autothermal process. However, in the case of waste gasification, CO_2 emissions and environmental impact should be considered if carbon is not renewable (e.g., plastic waste). Direct CO_2 emissions released by the solar processes are negligible or significantly lower than those from the current autothermal processes. Solar gasification processes thus have favorable long-term prospects because they avoid or reduce costs for CO_2 mitigation and pollution abatement. Moreover, due to the more heterogeneous nature of waste, its conversion implies additional costs, to deal with syngas impurities, especially the H_2S content that is a major corrosive constituent. More costly reactor materials and gas cleaning units are thus required for waste feedstock.

3.2.4. Impact of Environmental Subsidies

As shown in Table 3, solar gasification avoids, respectively, 3.08 and 6.73 kg of CO_2 (due to reaction) per kilogram of H_2 in hybrid and allothermal processes, which is significant. Although a detailed analysis has not yet been published comparing the three processes based on environmental criteria, important and achievable CO_2 emission mitigation is greatly expected thanks to solar heating. In fact, a conventional CSP tower plant generates barely 38 g of CO_2/kWh_e [38], which is far (more than 10 times) lower than the 750–900 g of CO_2/kWh_e generated by conventional Integrated Gasification Combined Cycles (IGCC) power plants when no CO_2 capture is considered. The capture/sequestration of 80% of CO_2 during operation decreases the net emissions to about 200 g of CO_2/kWh_e, resulting in a total saving of more than 550 g of CO_2/kWh_e [39]. In this sense, solar-driven processes can drastically reduce the Greenhouse gas (GHG) emissions, which allows them to take advantage of carbon pricing and environmental subsidies to improve their economic balance and their competitiveness. Indeed, the application of credits for CO_2 mitigation and pollution avoidance will further enable the solar thermochemical technologies to compete favorably with conventional processes. Carbon price varies from country to country [40], and, in France, it is estimated to be 100 €/t CO_2 in 2030 according to Quinet report [41]. In Europe, carbon prices are expected to double by 2021, and even quadruple to reach up to 55 €/t CO_2, as stipulated in the Paris climate agreement [42]. In this section, the impact of possible capital subsidies due to CO_2 emission reduction is studied. The total subventions were calculated on the basis of the amount of CO_2 that would have been emitted by the conventional process. For allothermal gasification, the subvention was estimated at 14.74 M€ (6.73 × 150,000 × 80% × 365 × 50/1000) and at about 6.74 M€ for the hybrid process (3.08 × 150,000 × 80% × 365 × 50/1000) for DC = 150 000 kg H_2/day, a capacity factor of 80% and a capital subsidy of 50 €/tCO_2. Figure 9 shows the impact of CO_2 subsidies on hydrogen cost.

At reference conditions, the intersection between the curves (autothermal and solar processes) takes place at CO_2 subventions of 82 €/tCO_2 (allothermal) and 55 €/tCO_2 (hybrid). Considering a cost reduction for heliostats and towers of 50% and a fixed environmental subsidy of 30 €/tCO_2, the biomass turnaround price goes down to 0.14 €/kg and 0.09 €/kg (lower than reference cost ~0.1 €/kg) for the allothermal and hybrid processes, respectively. This confirms that subsidies can play a key role in the reduction in solar hydrogen costs.

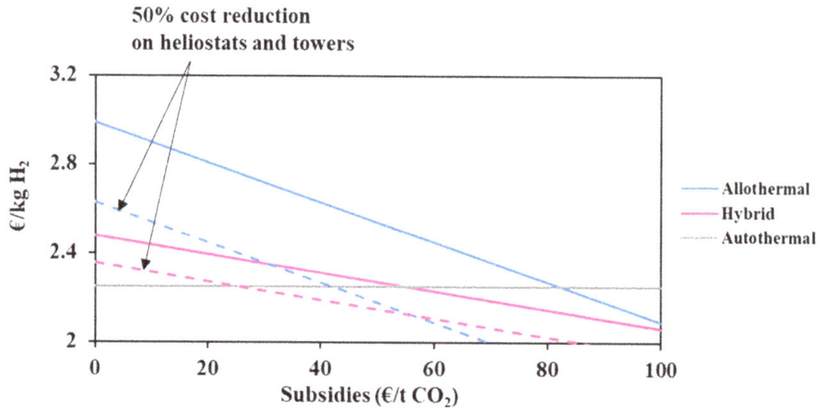

Figure 9. Impact of CO_2 subsidies on hydrogen cost

3.3. Comparison with Other Hydrogen Production Methods

This part of the study gives insights into hydrogen production costs with different technologies, such as biomass gasification, coal gasification, natural gas reforming and water electrolysis (based on a solid oxide technology). Previously developed NREL cost models were used for this purpose [43]. The financial inputs of all the technologies are the same as those presented in Table 4, and the operating capacity factor is fixed to 80%. To focus on the comparison with low-carbon technologies, the considered coal gasification and natural gas-reforming models integrate carbon capture and sequestration units that remove CO_2 from syngas before storing it in underground reservoirs. The reference primary resources costs used in this section are: biomass cost = 0.1 €/kg, coal cost = 0.04 €/kg, NG cost = 0.01 €/kWh, electricity cost for electrolysis = 0.10 €/kWh. Figure 10 shows the hydrogen minimum price of the different technologies. The grey bars show the sensitivity to the primary resource price (electricity price for electrolysis; process water was fixed to 0.61 €/m^3) when it increases from zero to twice the reference cost.

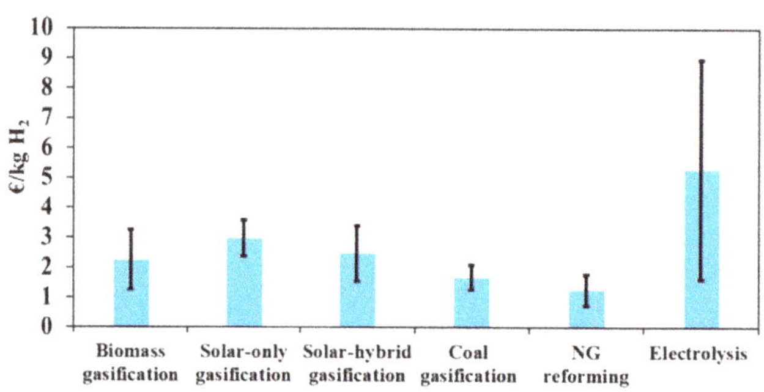

Figure 10. Hydrogen production cost and sensitivity on the primary resource cost (electricity for electrolysis) when ranging from zero to twice the reference cost (Table 4).

It can be seen that natural gas reforming is the most profitable process so far, with a hydrogen production cost of only 1.28 €/kg. This process is followed by coal (1.69 €/kg) and biomass gasification. The difference between these three processes is mainly due to two factors: the cost of the primary resource and the capacity of the plant. Clearly, due to the low fossil fuel cost, fossil-based processes are the most competitive on the market.

Moreover, very large plants can be envisaged, which lowers the price of hydrogen even further thanks to the economy of scale. The cost of hydrogen produced by electrolysis is much higher than that of the other processes (5.48 €/kg), and shows a greater sensitivity to primary resource cost. It can be seen that, for a zero-resource cost, the price of hydrogen produced by electrolysis decreases drastically to 1.68 €/kg, making this technology potentially more competitive when recovering and valorizing unusable electricity (due, for example, to lack of demand and storage). Although coal gasification plant capacity is seven times larger than that of biomass gasification (and, therefore, benefits from the economy of scale), it appears that, for low feedstock costs, hydrogen price is almost the same in both configurations. This is due to the extra costs entailed by CO_2 capture and sequestration operations (which are not considered in biomass gasification). Finally, NG reforming with CO_2 capture and sequestration is the cheapest process with a hydrogen cost below than that of coal and biomass gasification (solar and non-solar), making this technology the most economically attractive option at present for decarbonized hydrogen generation.

It is essential to recall that other solar thermochemical processes are prospected for the generation of solar fuels. In view of the increasingly stringent environmental regulations, and noting the urgency of climate change, the sustainable paths have undergone extensive research and development to increase efficiency and cost-effectiveness. The number of publications in this field considerably increased, by more than five times since 2000 [44]. This brought significant insights regarding the technical feasibility and the possibilities of scale up. A number of economic studies were carried out to estimate the cost of hydrogen using different solar thermochemical technologies. Although initial assumptions differ from one study to another regarding plant site, solar material cost, operating hours, installation costs and optical/thermal efficiencies, the calculated values estimate the prices and their sensitivity to the input parameters. Möller et al. [45] analyzed the solar steam reforming of natural gas for the production of 103.8 Million Nm^3/year (i.e., ~25,594 kg/day) of hydrogen. The study showed that the solar process allows 40% of the fuel to be saved compared to the conventional process, with a hydrogen cost of less than 0.05 €/kWh LHV of H_2 (~1.67 €/kg of H_2). Similarly, Rodat et al. [46] studied solar thermal decomposition of natural gas at a plant design capacity of 436 kg of H_2/day. Hydrogen cost was about 1.42 $/kg and showed great sensitivity to carbon blacks' (which are the process byproducts) selling price. The study determined carbon blacks' minimum cost, making solar thermal decomposition of natural gas competitive with solar and conventional reforming processes. Baykara and Bilgen [47] compared commercial, hybrid and solar coal gasification processes for the production of hydrogen. The plants were designed to produce 10^7 GJ of H_2 per year (~228,310 kg/day). The study showed that the commercial process (based on partial feedstock combustion) is the most cost-effective, with a hydrogen price of 0.94 $/kg, which is 5.2% and 6.3% lower than the hybrid and solar (only) process and is, in some respects, consistent with the present study outcomes. This means that solarization is still economically challenging and requires incentive environmental policies. Other researchers took a keen interest in the thermochemical splitting of water using high-temperature solar heat. Given that single-step direct thermolysis, at temperatures beyond 2500 °C, was hardly feasible, thermochemical cycles technologies were rather considered. This process involves several reactional intermediates, which are regenerated during the cycles to lower the water dissociation temperature. Over 280 cycles were developed and screened to select the most suitable ones for coupling with concentrated solar thermal energy [48]. Among the most promising cycles studied from an economic perspective, zinc, ferrite, and sulphur cycles were proposed. Charvin et al. [49,50] studied two-step (ZnO/Zn and Fe_3O_4/FeO) and three-step (Fe_2O_3/Fe_3O_4) thermochemical cycles driven by concentrated solar energy. The economic study was performed with a design capacity ranging from 50 to 250 kg/h of H_2. The analysis gave a hydrogen production cost between 7.98 and 14.75 $/kg of H_2 depending on process intensification and on the targeted hydrogen productivity. In a similar study, Steinfeld [51] analyzed hydrogen production cost via two-step water-splitting ZnO/Zn redox system, and the hydrogen price was around 4–5 €/kg at a design capacity of 61 Million-kWh/year (~5014 kg/day), which is somewhat larger than Charvin et al. [49] design capacity and therefore lowers hydrogen prices. Solar hydrogen production costs from the hybrid-sulfur cycle and a metal oxide-based cycle were studied and compared to that of commercial electrolysis (powered by a CSP plant with a thermal storage capacity of 4.5 h) by Graf et al. [52]. The metal-oxide-based cycle hydrogen cost ranged between 3.5 and 12.8 €/kg, and thus covered the ranges calculated Charvin et al. [49] and Steinfeld [51]. It showed the greatest cost variability due to the high demand of the metal oxide and its cost dependence. Hydrogen costs for hybrid sulfur cycle were the lowest and ranged between 3.9 and 5.6 €/kg. Water electrolysis was highly influenced by the cost of electricity, with a hydrogen price between 2.1 and 6.8 €/kg.

Overall, it appears from these results that, to date, solar thermochemical processes are far from being competitive with conventional processes based on fossil fuels (coal and natural gas). Major challenges remain to improve the efficiency of the processes. They concern the cost of the solar concentrators, which represents a significant part of the investment, and the cost of the receiver, which, in many cases, must withstand high temperatures in the presence of highly corrosive chemical species. Another challenge concerns the solar reactor design, which should minimize the heat losses and maximize the chemical conversion for a better use of the solar resource. The recycling of chemicals in thermochemical cycles that impose a high degree of purity and a precise control of phases and constituent separation is another issue that needs to be properly managed and solved. Carbon-based solar thermochemical technologies generally show lower hydrogen production costs. These processes, which, by definition, are less harmful to the environment, offer the possibility to extend the lifespan of fossil resources on earth and can play a role in the transition towards a zero-carbon economy.

4. Conclusions

A techno-economic study of solar and solar hybrid gasification was carried out. The study was based on the discounted cash-flow rate of return method to calculate the minimum hydrogen production cost. At first, the most important solar parameters were identified, and they were then integrated to a previously developed autothermal gasification cost model. The new solarized cost model was thereafter used to examine the profitability and the cost-effectiveness of each of the studied heating configurations (i.e., allothermal, hybrid and autothermal). A sensitivity analysis of the main cost-influencing factors was carried out. The analysis showed that at the current biomass reference cost (considered equal to 0.1 €/kg), the most competitive scenario (in which solar hydrogen cost is lower than conventional hydrogen), assumes a cost reduction of 50% of the heliostats and towers costs, with a land cost of 0.5 €/m^2, which is clearly challenging at present and requires an important economic effort. However, the analysis also showed that an increase in the biomass cost by a factor of 2 to 3 significantly undermines the profitability of the autothermal process, in favor of solar gasification, which becomes more competitive without any substantial economic and financial efforts. A comparative analysis with other solar and non-solar clean technologies was carried out. This confirmed that the two most economically favorable processes for hydrogen generation are those based on fossil fuels with CO_2 capture and sequestration. These processes will, therefore, make the greatest contribution to the hydrogen market in the near future. Nonetheless, fossil fuels are neither universally available nor inexhaustible, and depend on a large number of strategic and geopolitical parameters that remain uncertain, sensitive and hardly predictable. Moreover, carbon sequestration is not without risk for the environment and human health. Leakage during transport and storage is possible, and the long-term process performance is uncertain, especially in cases of large-scale development. These represent major constraints to circumvent in order to ensure security and sustainability. Renewable technologies are hardly competitive with fossil-based technologies at present, and require more effort to gain in efficiency, durability and cost-effectiveness. Government policy incentives have also a major role to play through the use of mechanisms like carbon credits, renewable energy credits, capital subsidies, and reverse auctions. This way, the financial viability of the sustainable path can be improved.

Author Contributions: Conceptualization, H.B., S.R. and S.A.; methodology, H.B., S.R. and S.A.; validation, H.B., S.R. and S.A.; formal analysis, H.B., S.R. and S.A.; investigation, H.B., S.R. and S.A.; writing—original draft preparation, H.B., S.R. and S.A.; writing—review and editing, S.R. and S.A.; visualization, H.B.; supervision, S.R. and S.A.; project administration, S.R. and S.A. All authors have read and agreed to the published version of the manuscript.

Funding: This research received no external funding.

Institutional Review Board Statement: Not applicable.

Informed Consent Statement: Not applicable.

Data Availability Statement: The data presented in this study are available on request from the corresponding author.

Acknowledgments: The support of ADEME (French Environment and Energy Management Agency) is gratefully acknowledged.

Conflicts of Interest: The authors declare no conflict of interest.

References

1. Hydrogen—Chemical Economics Handbook (CEH) I IHS Markit. Available online: https://ihsmarkit.com/products/hydrogen-chemical-economics-handbook.html (accessed on 14 July 2020).
2. Safari, F.; Dincer, I. A review and comparative evaluation of thermochemical water splitting cycles for hydrogen production. *Energy Convers. Manag.* **2020**, *205*, 112182. [CrossRef]
3. Ursua, A.; Gandia, L.M.; Sanchis, P. Hydrogen Production from Water Electrolysis: Current Status and Future Trends. *Proc. IEEE* **2011**, *100*, 410–426. [CrossRef]
4. Fraile, D.; Lanoix, J.C.; Maio, P.; Rangel, A.; Torres, A. Overview of the Market Segmentation for Hydrogen across Potential Customer Groups, Based on Key Application Areas. 2015. Available online: https://www.certifhy.eu/images/D1_2_Overview_of_the_market_segmentation_Final_22_June_low-res.pdf (accessed on 1 January 2021).
5. Binder, M.; Kraussler, M.; Kuba, M.; Luisser, M. *IEA Bioenergy Hydrogen from Biomass Gasification*; Reinhard Rauch, Ed.; IEA Bioenergy, December 2018; ISBN 978-1-910154-59-5. Available online: https://www.ieabioenergy.com/blog/publications/hydrogen-from-biomass-gasification/ (accessed on 26 February 2020).
6. Schmidt, O.; Gambhir, A.; Staffell, I.; Hawkes, A.; Nelson, J.; Few, S. Future cost and performance of water electrolysis: An expert elicitation study. *Int. J. Hydrogen Energy* **2017**, *42*, 30470–30492. [CrossRef]
7. Boujjat, H.; Rodat, S.; Chuayboon, S.; Abanades, S. Numerical simulation of reactive gas-particle flow in a solar jet spouted bed reactor for continuous biomass gasification. *Int. J. Heat Mass Transf.* **2019**, *144*, 118572. [CrossRef]
8. Chuayboon, S.; Abanades, S.; Rodat, S. Experimental analysis of continuous steam gasification of wood biomass for syngas production in a high-temperature particle-fed solar reactor. *Chem. Eng. Process. Process. Intensif.* **2018**, *125*, 253–265. [CrossRef]
9. Gokon, N.; Ono, R.; Hatamachi, T.; Liuyun, L.; Kim, H.-J.; Kodama, T. CO_2 gasification of coal cokes using internally circulating fluidized bed reactor by concentrated Xe-light irradiation for solar gasification. *Int. J. Hydrogen Energy* **2012**, *37*, 12128–12137. [CrossRef]
10. Kodama, T.; Kondoh, Y.; Tamagawa, T.; Funatoh, A.; Shimizu, K.-I.; Kitayama, Y. Fluidized Bed Coal Gasification with CO_2 under Direct Irradiation with Concentrated Visible Light. *Energy Fuels* **2002**, *16*, 1264–1270. [CrossRef]
11. Kruesi, M.; Jovanovic, Z.R.; Dos Santos, E.C.; Yoon, H.C.; Steinfeld, A. Solar-driven steam-based gasification of sugarcane bagasse in a combined drop-tube and fixed-bed reactor—Thermodynamic, kinetic, and experimental analyses. *Biomass Bioenergy* **2013**, *52*, 173–183. [CrossRef]
12. Muroyama, A.P.; Guscetti, I.; Schieber, G.L.; Haussener, S.; Loutzenhiser, P.G. Design and demonstration of a prototype 1.5 kWth hybrid solar/autothermal steam gasifier. *Fuel* **2018**, *211*, 331–340. [CrossRef]
13. Taylor, R.; Berjoan, R.; Coutures, J. Solar gasification of carbonaceous materials. *Sol. Energy* **1983**, *30*, 513–525. [CrossRef]
14. Gregg, D.; Taylor, R.; Campbell, J.; Taylor, J.; Cotton, A. Solar gasification of coal, activated carbon, coke and coal and biomass mixtures. *Sol. Energy* **1980**, *25*, 353–364. [CrossRef]
15. Piatkowski, N.; Wieckert, C.; Steinfeld, A. Experimental investigation of a packed-bed solar reactor for the steam-gasification of carbonaceous feedstocks. *Fuel Process. Technol.* **2009**, *90*, 360–366. [CrossRef]
16. Vidal, A.; Denk, T.; Steinfeld, L.; Zacarías, L.; Almería, T. Upscaling of a 500 kW Solar Gasification Plant. *Proc. WHEC* **2010**, *78*, 177–181.

17. Rodat, S.; Abanades, S.; Boujjat, H.; Chuayboon, S. On the path toward day and night continuous solar high temperature thermochemical processes: A review. *Renew. Sustain. Energy Rev.* **2020**, *132*, 110061. [CrossRef]
18. Boujjat, H.; Rodat, S.; Chuayboon, S.; Abanades, S. Experimental and numerical study of a directly irradiated hybrid solar/combustion spouted bed reactor for continuous steam gasification of biomass. *Energy* **2019**, *189*, 116118. [CrossRef]
19. Boujjat, H.; Junior, G.M.Y.; Rodat, S.; Abanades, S. Dynamic simulation and control of solar biomass gasification for hydrogen-rich syngas production during allothermal and hybrid solar/autothermal operation. *Int. J. Hydrogen Energy* **2020**, *45*, 25827–25837. [CrossRef]
20. Saw, W.L.; Guo, P.; Van Eyk, P.J.; Nathan, G.J. Approaches to accommodate resource variability in the modelling of solar driven gasification processes for liquid fuels synthesis. *Sol. Energy* **2017**, *156*, 101–112. [CrossRef]
21. Nickerson, T.A.; Hathaway, B.J.; Smith, T.M.; Davidson, J.H. Economic assessment of solar and conventional biomass gasification technologies: Financial and policy implications under feedstock and product gas price uncertainty. *Biomass Bioenergy* **2015**, *74*, 47–57. [CrossRef]
22. Rahbari, A.; Shirazi, A.; Venkataraman, M.B.; Pye, J. A solar fuel plant via supercritical water gasification integrated with Fischer–Tropsch synthesis: Steady-state modelling and techno-economic assessment. *Energy Convers. Manag.* **2019**, *184*, 636–648. [CrossRef]
23. Saw, W.; Kaniyal, A.; Van Eyk, P.; Nathan, G.; Ashman, P. Solar Hybridized Coal-to-liquids via Gasification in Australia: Techno-economic Assessment. *Energy Procedia* **2015**, *69*, 1819–1827. [CrossRef]
24. NREL H2A. H2A: Hydrogen Analysis Production Models | Hydrogen and Fuel Cells | NREL. Available online: https://www.nrel.gov/hydrogen/h2a-production-models.html (accessed on 7 July 2020).
25. Spath, P.; Aden, A.; Eggeman, T.; Ringer, M.; Wallace, B.; Jechura, J. *Biomass to Hydrogen Production Detailed Design and Economics Utilizing the Battelle Columbus Laboratory Indirectly Heated Gasifier*; NREL Technical Report; NREL: Golden, CO, USA, 2005. Available online: https://www.nrel.gov/docs/fy05osti/37408.pdf (accessed on 1 January 2021).
26. Kolb, G.J.; Jones, S.A.; Donnelly, M.W.; Gorman, D.; Thomas, R.; Davenport, R.; Lumia, R. *Heliostat Cost Reduction Study*; Sandia Report; Sandia National Laboratories: Albuquerque, NM, USA, June 2007. Available online: https://prod.sandia.gov/techlib-noauth/access-control.cgi/2007/073293.pdf (accessed on 1 January 2021).
27. Mendelsohn, M.; Lowder, T.; Canavan, B. *Utility-Scale Concentrating Solar Power and Photovoltaics Projects: A Technology and Market Overview*; NREL Technical Report; NREL: Golden, CO, USA, April 2012. Available online: https://www.nrel.gov/docs/fy12osti/51137.pdf (accessed on 1 January 2021).
28. Sudiro, M.; Bertucco, A. Synthetic Fuels by a Limited CO_2 Emission Process Which Uses Both Fossil and Solar Energy. *Energy Fuels* **2007**, *21*, 3668–3675. [CrossRef]
29. Becker, M.; Klimas, P.; Chavez, J.; Kolb, G.; Meinecke, W. *Second Generation Central Receiver Technologies*; Germany; p. 1993, ISBN 3-7880-7482-5. Available online: https://www.osti.gov/etdeweb/biblio/5295357 (accessed on 26 February 2021).
30. SOLARPACES PS10. Available online: https://solarpaces.nrel.gov/planta-solar-10 (accessed on 2 August 2020).
31. Sargent & Lundy LLC Consulting Group Chicago, Illinois. *Executive Summary: Assessment of Parabolic Trough and Power Tower Solar Technology Cost and Performance Forecasts*; NREL: Golden, CO, USA, October 2003. Available online: https://www.nrel.gov/docs/fy04osti/35060.pdf (accessed on 1 January 2021).
32. Bracken, N.; Macknick, J.; Tovar-Hastings, A.; Komor, P.; Gerritsen, M.; Mehta, S. Concentrating Solar Power and Water Issues in the U.S. Southwest. *Conc. Sol. Power Water Issues U.S. Southwest* **2015**. [CrossRef]
33. Hinkley, J.T.; Hayward, J.A.; Curtin, B.; Wonhas, A.; Boyd, R.; Grima, C.; Tadros, A.; Hall, R.; Naicker, K. An analysis of the costs and opportunities for concentrating solar power in Australia. *Renew. Energy* **2013**, *57*, 653–661. [CrossRef]
34. Pfahl, A. Survey of Heliostat Concepts for Cost Reduction. *J. Sol. Energy Eng.* **2013**, *136*, 014501. [CrossRef]
35. Pidaparthi, A.; Hoffmann, J. Effect of heliostat size on the levelized cost of electricity for power towers. *AIP Conf. Proc.* **2017**, *1850*, 030038.
36. Lovegrove, K.; Stein, W. *Concentrating Solar Power Technology*; Woodhead Publishing, October 2012; Available online: https://www.sciencedirect.com/book/9781845697693/concentrating-solar-power-technology (accessed on 14 July 2020).
37. Kromer, M.; Roth, K.; Takata, R.; Chin, P. *Support for Cost Analyses on Solar-Driven High Temperature Thermo-Chemical Water-Splitting Cycles*; TIAX LLC: Lexington, MA, USA, 2011.
38. Burkhardt, J.J.; Heath, G.; Cohen, E. Life Cycle Greenhouse Gas Emissions of Trough and Tower Concentrating Solar Power Electricity Generation. *J. Ind. Ecol.* **2012**, *16*, 93. [CrossRef]
39. Ordorica-Garcia, G.; Douglas, P.; Croiset, E.; Zheng, L. Technoeconomic evaluation of IGCC power plants for CO_2 avoidance. *Energy Convers. Manag.* **2006**, *47*, 2250–2259. [CrossRef]
40. Carbon Pricing Dash Board. Available online: https://carbonpricingdashboard.worldbank.org/map_data (accessed on 2 August 2020).
41. Quinet, A. Report of the Commission Chaired by A. Quinet, State-imposed Value of Carbon, Center for Strategic Analysis. 2009. Available online: https://www.vie-publique.fr/sites/default/files/rapport/pdf/094000195.pdf (accessed on 1 January 2021).
42. Carbon Tracker REPORT Carbon Countdown: Prices and Politics in the EU-ETS. Available online: https://carbontracker.org/reports/carbon-countdown/ (accessed on 20 July 2020).

43. Penev, M.; Saur, G.; Hunter, C.; Zuboy, J. NREL, Hydrogen & Fuel Cells, H2A Hydrogen Production Model: Version 3.2018 User Guide. Available online: https://www.nrel.gov/hydrogen/assets/pdfs/h2a-production-model-version-3-2018-user-guide-draft.pdf (accessed on 1 January 2021).
44. Yadav, D.; Banerjee, R. A review of solar thermochemical processes. *Renew. Sustain. Energy Rev.* **2016**, *54*, 497–532. [CrossRef]
45. Möller, S.; Kaucic, D.; Sattler, C. Hydrogen Production by Solar Reforming of Natural Gas: A Comparison Study of Two Possible Process Configurations. *J. Sol. Energy Eng.* **2006**, *128*, 16–23. [CrossRef]
46. Rodat, S.; Abanades, S.; Flamant, G. Methane Decarbonization in Indirect Heating Solar Reactors of 20 and 50 kW for a CO_2-Free Production of Hydrogen and Carbon Black. *J. Sol. Energy Eng.* **2011**, *133*, 031001. [CrossRef]
47. Baykara, S.; Bilgen, E. Synthesis gas and H_2 production from solar gasification of albertan coal. *Energy Convers. Manag.* **1985**, *25*, 391–398. [CrossRef]
48. Abanades, S.; Charvin, P.; Flamant, G.; Neveu, P. Screening of water-splitting thermochemical cycles potentially attractive for hydrogen production by concentrated solar energy. *Energy* **2006**, *31*, 2805–2822. [CrossRef]
49. Charvin, P.; Abanades, S.; Lemort, F.; Flamant, G. Analysis of solar chemical processes for hydrogen production from water splitting thermochemical cycles. *Energy Convers. Manag.* **2008**, *49*, 1547–1556. [CrossRef]
50. Charvin, P.; Abanades, S.; Lemort, F.; Flamant, G. Hydrogen Production by Three-Step Solar Thermochemical Cycles Using Hydroxides and Metal Oxide Systems. *Energy Fuels* **2007**, *21*, 2919–2928. [CrossRef]
51. Steinfeld, A. Solar hydrogen production via a two-step water-splitting thermochemical cycle based on Zn/ZnO redox reactions. *Int. J. Hydrogen Energy* **2002**, *27*, 611–619. [CrossRef]
52. Graf, D.; Monnerie, N.; Roeb, M.; Schmitz, M.; Sattler, C. Economic comparison of solar hydrogen generation by means of thermochemical cycles and electrolysis. *Int. J. Hydrogen Energy* **2008**, *33*, 4511–4519. [CrossRef]

processes

Article

Extraction of Added-Value Triterpenoids from *Acacia dealbata* Leaves Using Supercritical Fluid Extraction

Vítor H. Rodrigues, Marcelo M. R. de Melo, Inês Portugal and Carlos M. Silva *

CICECO—Aveiro Institute of Materials, Department of Chemistry, University of Aveiro, Campus Universitário de Santiago, 3810-193 Aveiro, Portugal; vitorhrodrigues@ua.pt (V.H.R.); marcelo.melo@ua.pt (M.M.R.d.M.); inesport@ua.pt (I.P.)

* Correspondence: carlos.manuel@ua.pt

Abstract: Forestry biomass is a by-product which commonly ends up being burnt for energy generation, despite comprising valuable bioactive compounds with valorisation potential. Leaves of *Acacia dealbata* were extracted for the first time by supercritical fluid extraction (SFE) using different conditions of pressure, temperature and cosolvents. Total extraction yield, individual triterpenoids extraction yields and concentrations were assessed and contrasted with Soxhlet extractions using solvents of distinct polarity. The extracts were characterized by gas chromatography coupled to mass spectrometry (GC-MS) and target triterpenoids were quantified. The total extraction yields ranged from 1.76 to 11.58 wt.% and the major compounds identified were fatty acids, polyols, and, from the triterpenoids family, lupenone, α-amyrin and β-amyrin. SFE was selective to lupenone, with higher individual yields (2139–3512 mg kg_{leaves}^{-1}) and concentrations (10.1–12.4 wt.%) in comparison to Soxhlet extractions, which in turn obtained higher yields and concentrations of the remaining triterpenoids.

Keywords: *Acacia dealbata*; GC-MS; leaves; lupenone; supercritical fluid extraction; Soxhlet extraction; triterpenoids

1. Introduction

The genus *Acacia* is widespread through the Portuguese landscape, consisting of three main species: *Acacia dealbata*, *Acacia longifolia* and *Acacia melanoxylon* [1]. *A. dealbata* was introduced for dune erosion protection as well as ornamental and wood supply purposes during the 19th and 20th century [2]. Currently, it is considered a plague due to its fast growth and dominance over the natural flora [1,3]. From 2005 to 2015, the occupied area of *Acacia* species increased 4000 ha in Portugal, corresponding to an estimated total arboreal biomass growth of 2 Mt [4]. The removal of these trees generates forest biomass that, under the Renewable Energy Directive II of the European Union Commission [5], can be utilized for the production of liquid and gaseous biofuels. However, it is a common practice to leave these residues in the forest for soil remediation.

The research towards *A. dealbata* biomass extraction has focused on several morphological parts, namely wood [6–10], bark [6–8,10–14], flowers [11,15–20] and leaves [6,8,11,17,21,22]. The explored extraction methods so far consist of solid-liquid extraction with organic solvents, such as dichloromethane, ethanol, methanol, hexane, acetone and some hydroalcoholic mixtures. Extraction of essential oils by steam distillation has been applied only to flowers [16]. Besides these conventional methods, there are few works on greener and more innovative extraction procedures, such as the work of Borges et al. [22], who applied microwave and ultrasound-assisted extraction to the leaves, and Lopez-Hortas et al. [16], using microwave hydrodiffusion to obtain the flower essential oil. One alternative technique for the extraction of vegetable biomass is supercritical fluid extraction (SFE) [23]. It is mainly employed with carbon dioxide (CO_2) as solvent due to its low cost, safety, availability and low critical point conditions, which allows extraction at near room temperatures [24]. The manipulation of temperature and pressure allows the tuning of CO_2 properties (density, viscosity and diffusivity) that maximize the desired responses, such as total extraction yield or the selective uptake of target chemical families or compounds. For example, triterpenic acids in the case of *Eucalyptus globulus* leaves [25,26] and bark [27–29], triterpenes from *Vitis vinifera* leaves [30], friedelin from *Quercus cerris* cork [31] and sterols from *Eichhornia crassipes* [32].

The phytochemistry of *A. dealbata* biomass (bark, leaves, wood, flowers and seeds) includes several families of compounds, such as alkaloids [8], amines [33], phenolics [8,10,13, 15], polysaccharides [9], chalcone glycosides [18], steryl glucosides [7], tannins [10,11,13], caffeic acid esters [12], sterols [6] and triterpenes [6,17]. SFE is a proven technology for the selective removal of triterpenes and sterols from many vegetable matrices [23,30–32,34,35]. For instance, compounds such as lupenone, lupeol, lupenyl palmitate, lupenyl cinnamate, squalene, β-amyrone, α-amyrin, β-amyrin and 22,23-dihydrospinasterol have already been identified and quantified [6,17]. These have been reported for several potential bioactive properties, namely anti-inflammatory, anti-virus, anti-diabetes, anti-cancer and antiproliferative, among others [36–45] which may explain the association of *Acacia* species with traditional medicine practices [46,47]. The wide range of biological activities potentiates the interest for multiple applications of the extracts, for example, to obtain active pharmaceutical ingredients or for incorporation in nutraceuticals, food, animal feed and cosmetic products.

This work focuses on the SFE of triterpenoids from *Acacia dealbata* leaves under different experimental conditions of pressure, temperature and cosolvents content, and its comparison with conventional Soxhlet extraction using organic solvents of distinct polarity. The extracts were characterized by gas chromatography coupled to mass spectrometry (GC-MS) and triterpenoids contents were determined. To the best of our knowledge, this is the first time SFE is applied to *Acacia dealbata* leaves aiming for the extraction of potential bioactive compounds.

2. Materials and Methods

2.1. Chemicals

Carbon dioxide (CO_2, purity 99%) was supplied by Air Liquide (Algés, Portugal). Dichloromethane (purity 99.98%), *n*-hexane (purity 99%) and ethanol (purity 99.5%) were supplied by Fisher Scientific (Leicestershire, UK). Ethyl acetate (purity 99%) was supplied by VWR International (Fontenay-sous-Bois, France). Pyridine (purity 99.5%), tetracosane (purity 99%) N,O-Bis(trimethylsilyl)trifluoroacetamide (BSTFA, purity 98%) and chlorotrimethylsilane (TMSCl, purity 99%) were supplied by Sigma Aldrich (Madrid, Spain). Betulinic, oleanolic and ursolic acids (purity 98%) were supplied by AK Scientific (Union City, CA, USA).

2.2. Acacia Dealbata Biomass

The *A. dealbata* leaves were supplied by RAIZ—Forest and Paper Research Institute (Eixo, Portugal). The leaves were collected from 8-year-old *Acacia dealbata* trees located in Porto/Valongo (Portugal) region during the winter season. The leaves (see Figure 1) were manually separated from the branches and dried at 35 °C for 72 h in a forced convection oven, reducing their moisture content from 65.6 to 4.5 wt.%.

Figure 1. Oven-dried *Acacia dealbata* leaves.

2.3. Soxhlet Extraction

The leaves of *A. dealbata* were extracted with *n*-hexane, dichloromethane, ethyl acetate and ethanol (Table 1). In each assay, extraction of the leaves (ca. 3 g) was performed with 180 mL of each solvent for 6 h. The produced extracts were evaporated to dryness in a rotary evaporator, weighed for the determination of total extraction yield (η_{Total}, wt.%) and analyzed by GC-MS to evaluate triterpenoids individual yields (η_i, mg kg^{-1}_{leaves}) and concentrations (C_i, wt.%), as follows:

$$\eta_{Total} = \frac{m_{extract}}{m_{dry\ leaves}} \times 100 \quad (1)$$

$$\eta_i = \frac{m_i}{m_{dry\ leaves}} \times 10^6 \quad (2)$$

$$C_i = \frac{m_i}{m_{extract}} \times 100 \quad (3)$$

where $m_{dry\ leaves}$ is the mass of dry leaves, $m_{extract}$ corresponds to extract mass free of solvent and m_i is the mass of triterpenoids measured by GC-MS.

Table 1. List of Soxhlet and SFE assays with the respective operating conditions.

Run	Method	Solvent	T (°C)	P (bar)	ρ_f (kg m^{-3})
SX1	Soxhlet	*n*-Hexane	68.5 *	1	-
SX2		Dichloromethane	39.6 *	1	-
SX3		Ethyl acetate	77.1 *	1	-
SX4		Ethanol	78.4 *	1	-
SFE1	SFE	CO_2	40	200	840.6 [48]
SFE2		CO_2	80	200	594.9 [48]
SFE3		CO_2	60	300	830.4 [48]
SFE4		CO_2:Ethanol (95:5 wt.%)	80	300	764.6 [49]
SFE5		CO_2:Ethyl acetate (95:5 wt.%)	80	300	761.4 [50]

* boiling point of the pure solvents.

2.4. Supercritical Fluid Extraction

The SFE assays were performed in a lab scale Spe-ed SFE unit, a model of Helix SFE System-Applied Separations, Inc., (Allentown, PA, USA) schematically presented in Figure 2. In each run, ca. 25 g of leaves were loaded into the extractor while the supercritical fluid flowed upwards at constant flow rate (Q_{CO_2}) of 12 g min^{-1} for 6 h. The experimental conditions of pressure, temperature and cosolvent content are presented in Table 1 (runs SFE1 to SFE3). The detailed procedure is described elsewhere [25]. The total extraction yield, the individual compound concentrations and respective yields were determined according to Equations (1)–(3), respectively.

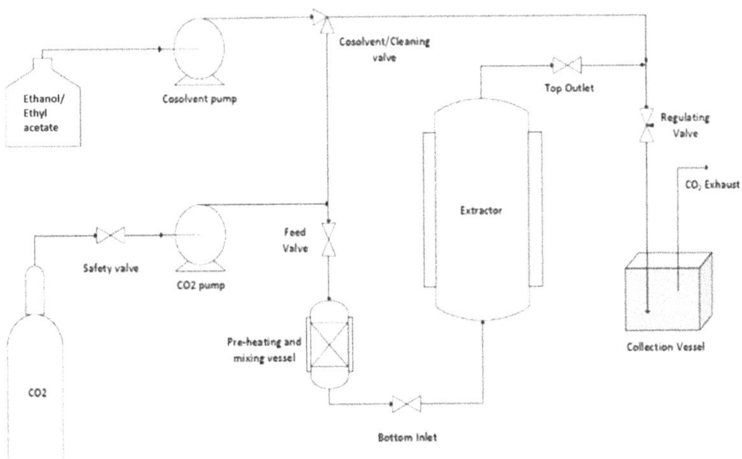

Figure 2. Simplified scheme of the SFE installation. Reprinted with permission from [25]. Copyright 2021, Elsevier.

For the runs, SFE4 and SFE5 ethanol and ethyl acetate were added as cosolvent to modify the supercritical fluid polarity and the solubility of solutes. In these runs, the cosolvent was fed to the pre-heating vessel using a HPLC pump, as presented in Figure 2.

The densities of the supercritical fluids (ρ_f), both pure and modified supercritical carbon dioxide (SC-CO_2), at each experimental condition are presented in Table 1. They were obtained using the equation of state of Pitzer and Schreiber for pure SC-CO_2 [48], Falco and Kiran for SC-CO_2 modified with ethyl acetate [50] and Pöhler and Kiran for SC-CO_2 modified with ethanol [49].

2.5. Gas Chromatography Coupled to Mass Spectrometry

The extracts were analyzed by GC–MS using a Trace Gas Chromatograph Ultra equipped with a DB-1 J&W capillary column (30 m × 0.32 mm i.d., 0.25 μm film thickness) and coupled with a Thermo DSQ mass spectrometer. The extracts were prepared and analyzed following a procedure previously published [28,29]. For the quantification of individual triterpenoids in the extracts, tetracosane and pure betulinic, oleanolic and ursolic acids were selected as internal and external standards, respectively. The identification of the compounds was performed with the aid of reverse match factors (RSI) from Wiley 9 library, which defines thresholds of mass spectral match: 900 and above is considered excellent; 800–900 is considered good, 700–800 is considered fair, and below 700 is considered a poor match [51].

3. Results and Discussion

3.1. Total Extraction Yield

The total extraction yield (η_{Total}, Equation (1)) measures the total extract amount produced independently of its composition. The results obtained with Soxhlet and SFE can be visualized in Figure 3. Concerning Soxhlet extraction, the η_{Total} values increased with the polarity of the solvent and ranged from 3.60 to 11.58 wt.% for dichloromethane and ethanol, respectively. The second highest value (7.97 wt.%) was obtained by ethyl acetate, and *n*-hexane achieved an equivalent value to dichloromethane (3.64 wt.%). Literature results on Soxhlet extractions of *A. dealbata* leaves present unequal yield scores. For instance, Oliveira et al. [6] obtained a η_{Total} of 6.2 wt.% with dichloromethane, almost double of the value obtained in this work, which may be due to the particle size reduction performed. On the other hand, Luís et al. [8] obtained a η_{Total} of 6.75 wt.% with ethanol, which is considerably lower than the value obtained in this work, and may be due to a different time of extraction, as it was stopped as soon as the extraction solvent became colorless. Borges et al. [22] obtained a η_{Total} of 13 wt.% using water and 16 h of Soxhlet extraction, which can compare with the value obtained with ethanol (11.58 wt.%), the closest solvent in terms of polarity. However, it is noteworthy that extraction with ethanol (6 h at 78.4 °C) would be more energy efficient than with water (16 h at 100 °C).

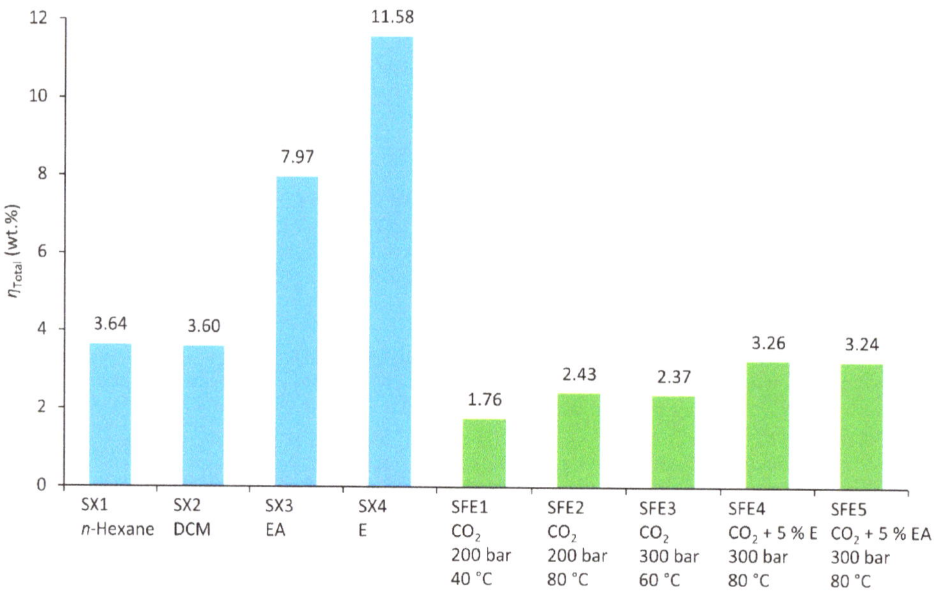

Figure 3. Total extraction yields (η_{Total}) obtained by Soxhlet using *n*-hexane, dichloromethane (DCM), ethyl acetate (EA), ethanol (E) and SFE at different conditions.

Regarding SFE assays, the η_{Total} results varied from 1.76 to 3.26 wt.% for SFE1 (200 bar, 40 °C, no cosolvent) and SFE5 (300 bar, 80 °C, 5 wt.% ethanol), respectively. Three experimental parameters were tested in these runs: pressure, temperature and cosolvent addition. Setting SFE1 (200 bar, 40 °C, no cosolvent) as the reference run, the effect of temperature on η_{Total} can be assessed by comparison with run SFE2 (200 bar, 80 °C, no cosolvent), where an increase of 38% was observed. Even though higher temperatures lower the solvent power due to SC-CO_2 density decrease (840.6 kg m^{-3} for SFE1 and 594.6 kg m^{-3} for SFE2), they increase the vapour pressure of solutes (i.e., their solubility in the supercritical solvent). These opposing effects can lead to different results from system to system, and in this case, the solubility enhancement effect was prevalent. Moving to Run SFE3 (300 bar, 60 °C, no cosolvent), this assay was performed at more 100 bar and less 20 °C of SFE2, but yielded an identical η_{Total} (2.37 wt.%). This occurs possibly because the higher SC-CO_2 density in Run 3 (830.4 kg m^{-3}, i.e., 40% more than in SFE2) compensated the thermal penalization on the vapor pressure side of solutes. Furthermore, the addition of cosolvents was tested at 300 bar and 80 °C in runs SFE4 (5 wt.% of ethanol) and SFE5 (5 wt.% of ethyl acetate). The results were identical for both runs (3.24–3.26 wt.%), almost doubling the η_{Total} value of SFE1. The different cosolvents did not significantly affect the fluid densities between each other (764.6 kg m^{-3} for SFE4 and 761.4 kg m^{-3} for SFE5) and imposed an increase of only 2.4% in relation to pure SC-CO_2 under the same $P-T$ conditions (746.2 kg m^{-3}) [48]. Hence, the yield gain in runs SFE4 and SFE5 was attributed to a greater affinity of the solute to solvent mixtures of higher polarity.

Overall, it is clear that Soxhlet extraction produces higher η_{Total}, especially when employing high polarity solvents (i.e., ethanol or ethyl acetate). When employing low polarity solvents (i.e., dichloromethane or *n*-hexane), the η_{Total} values are analogous to those of modified SC-CO_2 and higher than those of SFE without entrainers.

3.2. Volatile Extractives

The extracts were analyzed by GC-MS, and the chromatograms of runs SX2 (A), SX4 (B) and SFE1 (C) are presented in Figure 4A–C. These are representative of the Soxhlet extractions with low and higher polarity solvents, and SFE runs, respectively. It is possible to observe that the chromatograms of runs SX2 and SFE1 (Figure 4A,C) are considerably similar, as the same peaks appear in both, which in turn confirms the affinity of dichloromethane and SC-CO_2 to similar solutes. On the contrary, SX4 (Figure 4B) stands out due to the proliferation of peaks in the region on the left of the internal standard—i.e., at retention time (Rt) lower than 38.76 min—and also due to the very sharp peak at Rt = 27.55 min, identified as *myo*-inositol (whose structure is disclosed in Figure 5). The latter forced a rescaling of the relative absorbance axis to a comparable range (from 100% in SX2 and SFE1 to 6% in SX4). All but one of the identified compounds in these peaks encompassed RSI scores comprehended between 700 and 942, which correspond to a matching quality from fair to excellent. In fact, the only exception to this was α-amyrin, whose RSI score was 691.

The full list of identified compounds for all Soxhlet extractions and SFE runs is reported in Table 2, altogether with the respective retention times and maximum RSI scores within the analyzed extracts. As previously observed in Figure 4B, run SX4 shows more peaks in the left half of the chromatogram, and that observation can be confirmed in Table 2, also for runs SX3 (ethyl acetate) and SX4 (ethanol), specifically amid retention times of 10.77 to 44.37 min. The said peaks consist mainly of polyols (P) and monosaccharides (M). After these come the fatty acids (FA), long-chain aliphatic alcohols (LCAA) and triterpenoids (TT), although these were also found in every extract. Furthermore, SX1 (*n*-hexane) and SX2 (dichloromethane) present a very similar pattern of detected compounds, as well as the runs SFE1-SFE5. This corroborates what was observed in the analysis of Figure 4A,C. Such similarities show that the polarity of the solvent strongly influences the compounds extracted, and that, in the case of SFE, the modification of SC-CO_2 with 5 wt.% of ethanol or ethyl acetate was not enough to lead to the extraction of volatile compounds with higher affinity to polar organic solvents.

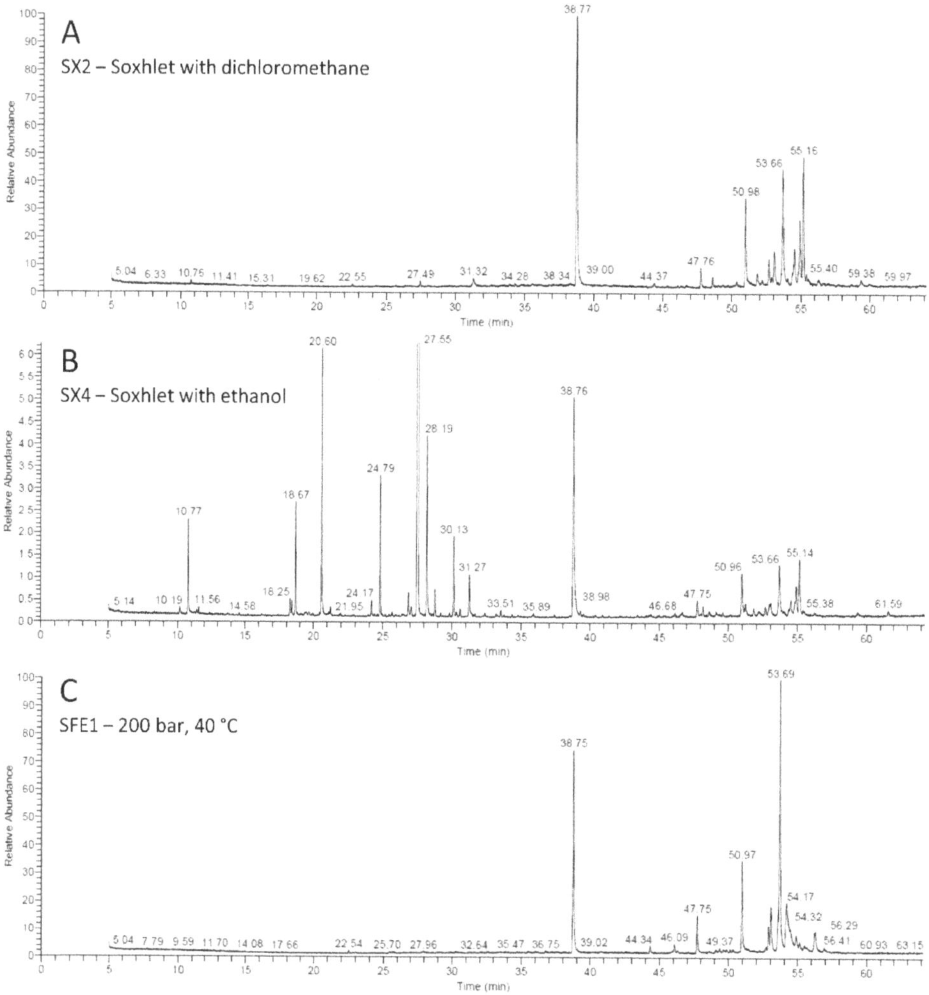

Figure 4. Chromatograms of the extracts from runs (**A**) SX2 (dichloromethane), (**B**) SX4 (ethanol) and (**C**) SFE1 (200, bar 40 °C). Internal standard (tetracosane) appears at 38.7 min.

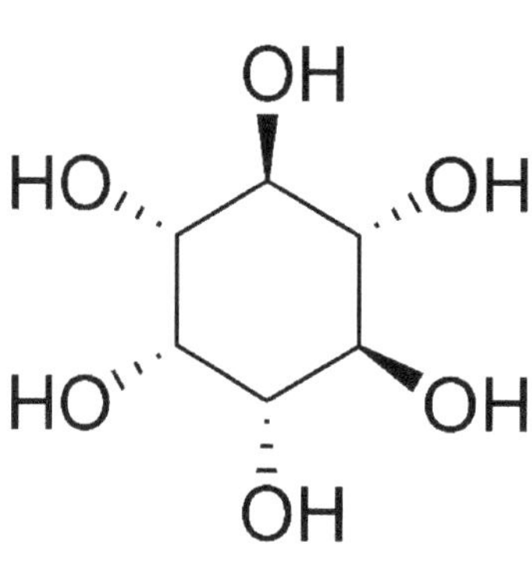

Figure 5. Structural formula of *myo*-inositol.

Table 2. List of identified compounds and respective retention times and reverse match factors (RSI) for all Soxhlet and SFE runs.

Rt (min)	Compound	Family	RSI	SX1	SX2	SX3	SX4	SFE1	SFE2	SFE3	SFE4	SFE5
10.77	Glycerol	P	909	-	+	+	+	-	-	-	-	-
18.67	Erythritol	P	942	-	-	+	+	-	-	-	-	-
20.6	5-Hydroxypipecolic acid	CA	770	-	-	-	+	-	-	-	-	-
24.17	Xylitol	P	800	-	-	+	+	-	-	-	-	-
24.79	Ribitol	P	887	-	-	+	+	-	-	-	-	-
27.55	*myo*-Inositol	P	803	-	-	+	+	-	-	-	-	-
28.19	Tyramine	A	890	-	-	-	+	-	-	-	-	-
28.79	D-Mannose	M	803	-	-	+	+	-	-	-	-	-
30.13	Mannitol	P	851	-	-	+	+	-	-	-	-	-
31.27	Glucose	M	827	-	-	+	+	-	-	-	-	-
44.37	Docosanoic acid	FA	729	+	+	-	-	+	+	+	+	+
46.09	Squalene	TT	833	-	-	-	-	+	+	+	+	+
47.76	Pentacosanoic acid	FA	829	+	+	+	+	+	+	+	+	+
48.61	Hexadecanoic acid	FA	705	+	+	+	-	-	-	-	-	-
50.98	Octacosanoic acid	FA	849	+	+	+	+	+	+	+	+	+
51.82	Octacosan-1-ol	LCAA	811	+	-	+	-	-	-	-	-	-
52.66	4′-OH,5-OH,7-Di-O-Glucoside	F	688	+	+	+	+	+	+	+	+	+
52.86	α-Amyrone	TT	786	+	+	-	-	+	+	+	+	+
53.06	β-Amyrone	TT	738	+	+	+	+	+	+	+	+	+
53.66	Lupenone	TT	843	+	+	+	+	+	+	+	+	+
54.92	β-Amyrin	TT	755	+	+	+	+	+	+	+	+	+
55.16	α-Amyrin	TT	691	+	+	+	+	+	+	+	+	+
56.28	Lupenyl acetate	TT	742	+	+	-	-	+	+	+	+	+

P—polyol; CA—carboxylic acid; A—amine; M—monosaccharide; FA—fatty acid; TT—triterpenoid; LCAA—long-chain aliphatic alcohol; F—flavonoid.

As observed in Figure 4B, *myo*-inositol (see Figure 5) peak has an area of different magnitude from the targeted triterpenoids, thirteen times higher (run SX4) than the internal standard. Even though it is a known constituent of plant and animal cells, the results obtained demonstrate the potential of ethanol and ethyl acetate, and potentially other polar organic solvents, for the production of *myo*-inositol-rich extracts. According to the literature, this compound and its derivatives have been identified and quantified in several plant species [52–56], including *Acacia* trees, such as in *Acacia pennata* and *Acacia farnesiana* leaves [57], and *Acacia mangium* and *Acacia maidenii* seeds [58]. Moreover, *myo*-inositol plays an important role in several cell functions, such as growth, development and reproduction, among others [59]. As a dietary supplement, it can be beneficial for human disorders associated with insulin resistance, such as polycystic ovary syndrome, gestational diabetes mellitus or metabolic syndrome, and the prevention or treatment of some diabetic complications, namely, neuropathy, nephropathy and cataract [59]. Even though this represents a promising result for the valorisation of *Acacia dealbata* biomass, the main focus of this work is the triterpenoid fraction attainable by SFE, which leaves *myo*-inositol out of the work scope. Nevertheless, it might open the way to sequential extraction strategies.

The main triterpenoids (TT) identified in the produced extracts were squalene, α-amyrone, β-amyrone, lupenone, β-amyrin, α-amyrin and lupenyl acetate. These contain 30 carbon atoms, except for lupenyl acetate which has 32, and all of these compounds are interrelated by known biosynthesis pathways, as summarized in Figure 6. Here, it can be seen that squalene is the prime precursor, having been originated by successive condensation reactions of the isomers of isopentenyl diphosphate and dimethylallyl diphosphate [60–62]. Eventually, squalene can be oxidized to 2,3-oxidosqualene, which in turn is the direct precursor of tricyclic, tetracyclic or pentacyclic triterpenoids. When under the chair-chair-chair conformation, 2,3-oxidosqualene can undergo cyclization reactions forming the tetracyclic dammarenyl cation, which, after ring expansions, originates diverse skeletons of pentacyclic triterpenoids, such as the lupane, oleanane and ursane types (see Figure 6). The latter three types are the precursors of lupeol, β-amyrin and α-amyrin, respectively [60–63]. Upon undergoing further rearrangements, and/or oxidation, substitution or glycosylation reactions, other triterpenoids are generated, namely lupenone, β-amyrone and α-amyrone (as the ketone versions of lupeol, β-amyrin and α-amyrin, respectively), and lupenyl acetate (as the acetylated version of lupeol). To conclude, the majority of these compounds were also identified in previous extraction works of *A. dealbata* biomass [6,17], especially in the leaves, bark and other external parts, since they are thought to provide protection against insects and microbes [63].

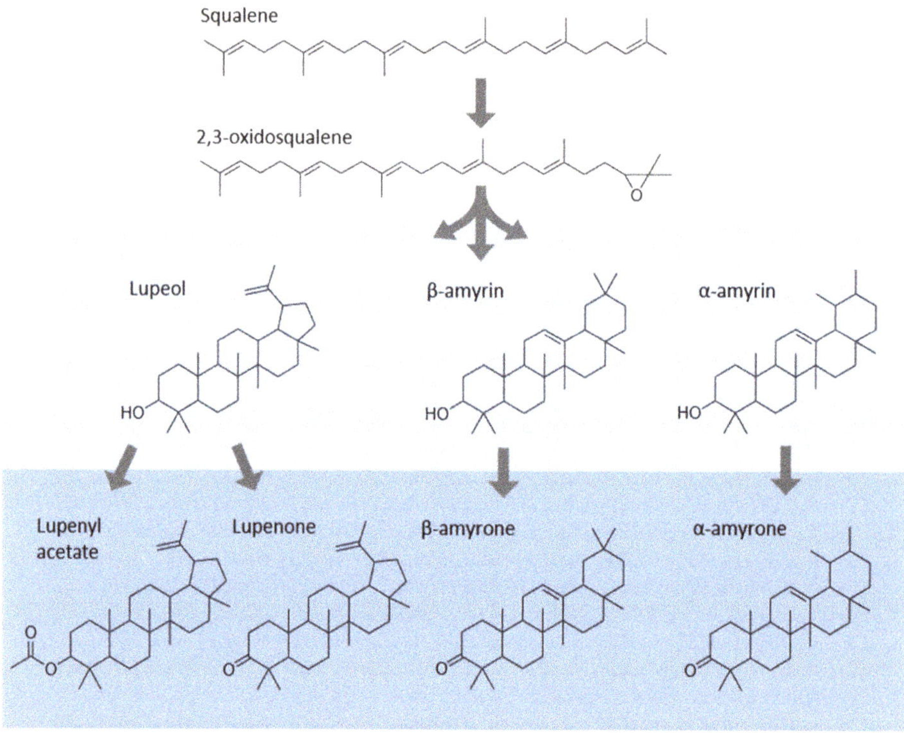

Figure 6. Scheme representative of synthesis pathways of triterpenoids from squalene, namely the lupane, oleanane and ursane series.

3.3. Triterpenoid Extraction Yields

The individual (η_i) and total triterpenoid ($\eta_{Total\ TT}$) extraction yields of Soxhlet and SFE assays are presented in Figure 7A,B, respectively. Regarding Soxhlet extractions (Figure 7A), lupenone was the most extracted compound of the four organic solvents tested, with $\eta_{lupenone}$ values ranging from 2114 to 2994 mg kg_{leaves}^{-1} for *n*-hexane and ethyl acetate extraction, respectively. It was followed by α-amyrin, with yields from 1249 to 2851 mg kg_{leaves}^{-1} for dichloromethane and ethyl acetate, respectively. Although $\eta_{\beta-amyrin}$ was generally lower than $\eta_{\alpha-amyrin}$, this difference is especially evident in the dichloromethane Soxhlet extract, where the latter yielded circa 2.7 times more. In turn, ethyl acetate attenuated the two amyrin yields, with their ratio falling to ca. 1.6. For extracts produced with more polar solvents, the two amyrin yields approached the values of $\eta_{lupenone}$. This was translated to the magnitude of $\eta_{Total\ TT}$, which incremented from the minimum of 4908 mg kg_{leaves}^{-1} for *n*-hexane to 8201 mg kg_{leaves}^{-1} for ethyl acetate and to 6259 mg kg^{-1}_{leaves} for ethanol. The remaining triterpenoids did not differ significantly between organic solvents, and their individual yields were markedly low.

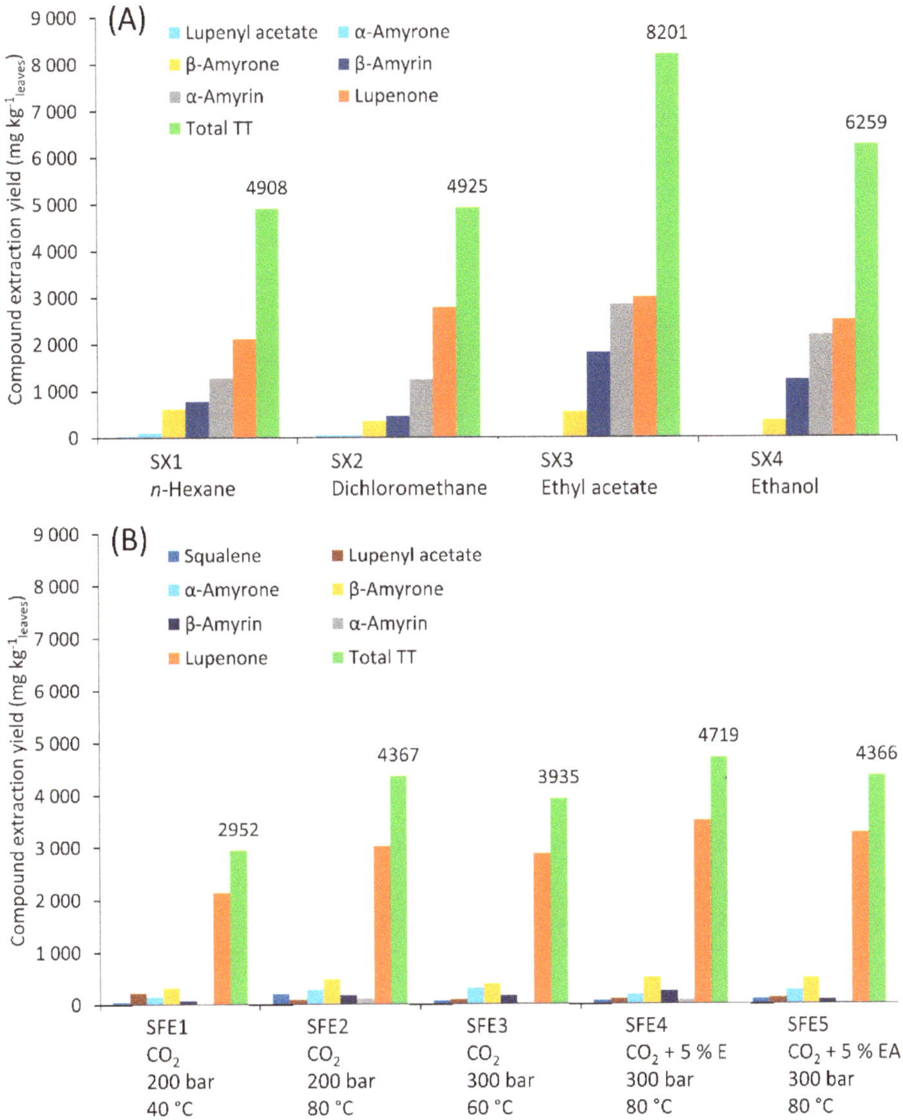

Figure 7. Individual and total triterpenoid (Total TT) extraction yields in: (**A**) Soxhlet extracts using different organic solvents, and (**B**) SFE at different conditions of temperature, pressure, and ethanol (E) or ethyl acetate (EA) content as modifiers.

Concerning uptake of triterpenoids by SFE (see Figure 7B), lupenone confirmed its leading individual yield also in this separation method, which varied from 2139 to 3512 mg kg^{-1}_{leaves} for SFE1 (200 bar, 40 °C, no cosolvent) and SFE4 (300 bar, 80 °C, 5 wt.% of ethanol), respectively. None of the remaining triterpenoids surpassed 500 mg kg^{-1}_{leaves} threshold. The increase of temperature from 40 °C to 80 °C of runs SFE1 and SFE2, respectively, improved the $\eta_{lupenone}$ by 41% and, if contrasted with the 38% increase verified in the η_{Total} (recall Figure 3), represents a proportional increase. Furthermore, run SFE3 (300 bar, 60 °C, no cosolvent) shows that even though it produced a similar η_{Total} of SFE2, the joint $P - T$ change decreased the lupenone uptake to 2879 mg kg_{leaves}^{-1}. This confirms that the favorable effect of temperature on solubility prevailed over the loss of SC-CO_2 density. With the employment of ethanol and ethyl acetate at 300 bar and 80 °C (runs SFE4 and SFE5), higher lupenone yields were obtained, namely 3512 mg kg_{leaves}^{-1} and 3273 mg kg_{leaves}^{-1}, respectively. This suggests that the joint optimization of P, T and cosolvent content can be determinant to potentiate the removal of this compound by SFE.

The low yields for the other triterpenoids explain the lower $\eta_{Total\ TT}$ obtained by SFE. These were only slightly inferior to the Soxhlet extractions with *n*-hexane and dichloromethane, but substantially lower in relation to ethyl acetate and ethanol, whose $\eta_{Total\ TT}$ values scored, respectively, 74% and 33% higher than of SFE4 (the richest in $\eta_{Total\ TT}$ with 4719 mg kg_{leaves}^{-1}).

3.4. Triterpenoid Concentration in Extracts

To assess the effect of distinct extraction methods and operating conditions on the selectivity of the identified triterpenoids, their individual concentrations were determined and are presented in Figure 8A,B for Soxhlet and SFE assays, respectively.

Regarding the Soxhlet extracts (Figure 8A), the concentration trend does not follow those of total or individual triterpenoid yields. The highest extract concentration of lupenone amounted 7.7 wt.% and occurred for dichloromethane assay (SX2), followed by *n*-hexane (SX1) with 5.8 wt.%, ethyl acetate (SX3) with 3.8 wt.%, and finally, ethanol (SX4) with 2.2 wt.%. Even though $\eta_{Total\ TT}$ and $\eta_{lupenone}$ of dichloromethane and *n*-hexane Soxhlet extraction were lower than those of ethyl acetate and ethanol, they resulted in higher $C_{lupenone}$. This is due to the fact that, as discussed previously (see Table 2), the more polar solvents are able to coextract other families of compounds, thus diluting the content of triterpenoids and fading the selectivity towards them, namely to lupenone. The same can be observed for the remaining triterpenoids. Another interesting result is the levelled scores of $C_{\alpha-amyrin}$ among the four organic solvents studied, ranging from 1.89 to 3.57 wt.%, which were far from expected given the significantly higher $\eta_{\alpha-amyrin}$ obtained with ethyl acetate and ethanol (see Figure 7A). The observed yield/concentration nuances are ultimately reflected in the $C_{Total\ TT}$ response, which was as low as 10.3% and 5.4 wt.% for ethyl acetate and ethanol, respectively, against 13.4 and 14.4 wt.% for dichloromethane and *n*-hexane, respectively. As a result, the use of the two non-polar solvents are better choices than ethanol or ethyl acetate when seeking a higher selectivity to lupenone, despite having lower yields of this compound as counterpart.

The SFE results (see Figure 8B) show higher $C_{lupenone}$ than any of the Soxhlet extracts, ranging from 10.1 to 12.4 wt.%, for runs SFE5 (300 bar, 80 °C, 5 wt.% of ethyl acetate) and SFE2 (200 bar, 80 °C, no cosolvent), respectively. Once again, one can observe that the discussed trends on $\eta_{Total\ TT}$ and $\eta_{lupenone}$ are not verified for the concentration values. Accordingly, even though run SFE2 showed the highest selectivity towards lupenone, it only attained the third highest $\eta_{lupenone}$ (see Figure 7B), and the same applies to run SFE1 (200 bar, 40 °C, no cosolvent). In turn, the inclusion of polar modifiers in runs SFE4 (300 bar, 80 °C, 5 wt.% of ethanol) and SFE5 (300 bar, 80 °C, 5 wt.% of ethyl acetate) created the same penalization observed in Soxhlet: lower $C_{lupenone}$ was attained despite the higher $\eta_{Total\ TT}$ and $\eta_{lupenone}$ scores (see Figures 3 and 7B). As a result, the SFE results show that pure SC-CO_2 was the most selective to lupenone, but the method was not able to selectively coextract other triterpenoids. In terms of $C_{Total\ TT}$, similarly to what was observed for the triterpenoid yield (see Figure 7B), the values follow the lupenone trend since the remaining triterpenoids were extracted in significantly lower amounts. Accordingly, the maximum was attained by run SFE2 (200 bar, 80 °C, no cosolvent), where $C_{Total\ TT}$ is worth 18.0 wt.%. The individual contribution of other triterpenoids in the extract for this score did not surpass 2 wt.%.

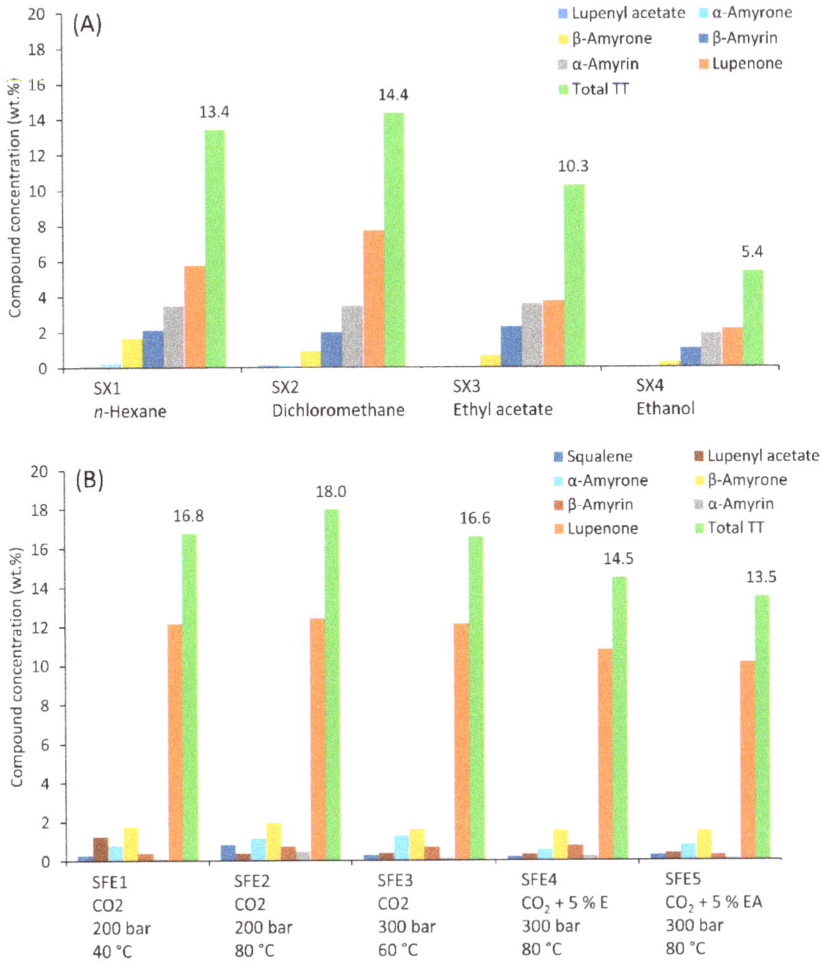

Figure 8. Plots of the concentrations of individual triterpenoids identified in (**A**) Soxhlet extracts and respective solvents, and (**B**) SFE at different conditions of temperature, pressure, and addition of ethanol (E) and ethyl acetate (EA) as modifiers.

Overall, the attained SFE results show that the method can selectively extract lupenone from the leaves of *A. dealbata*, even though similar triterpenoids are only coextracted on a negligible basis. Furthermore, lupenone therapeutic potential for inflammation, virus, infection, diabetes, cancer, and treatment of Chagas disease [36] justifies its valorisation in an industrial process for pharmaceutical and food applications.

4. Conclusions

Soxhlet extraction of *Acacia dealbata* leaves provided higher total extraction yields than SFE, and the yield increased with the polarity of the organic solvent. In turn, SFE yields were favored by increasing temperature and pressure, and by the addition of polar cosolvents. In both cases the main triterpenoids extracted were squalene, β-amyrone, α-amyrone, β-amyrin, α-amyrin, lupenyl acetate and lupenone, with the latter exhibiting the highest content. Overall, Soxhlet extracts exhibited higher amounts of total triterpenoids in comparison to SFE extracts. Interestingly SFE selectively extracted lupenone, reaching higher contents than those obtained with Soxhlet.

Even though the increase of pressure and temperature combined with the addition of cosolvents favored the SFE total yield, the same effect was not so evident for lupenone yield. In fact, pure CO_2 attained a lupenone yield comparable to the maximum value achieved at high pressure with SC-CO_2 modified with ethanol. Furthermore, the highest lupenone concentration was also obtained for pure CO_2 extracts.

Considering the anti-inflammatory, anti-virus, anti-diabetes, and anti-cancer properties of lupenone, as well as its potential for the treatment of Chagas disease, one may state that SFE contributes to the valorisation of *A. dealbata* leaves in the production of lupenone-enriched extracts for the pharmaceutical, nutraceutical or food industries.

Globally, this work envisions SFE technology as a tool to address the current challenges associated to the management of *A. dealbata* spread, namely by pointing to biorefinery opportunities for its leaves. Even though the present study indicates that high selectivity to lupenone may be achieved by SFE, optimization of the extraction conditions and a preliminary economic evaluation of the process are highly recommended.

Author Contributions: Conceptualization, C.M.S.; writing—original draft preparation, V.H.R. and M.M.R.d.M.; methodology, V.H.R. and M.M.R.d.M.; investigation, V.H.R.; formal analysis, M.M.R.d.M., I.P. and C.M.S.; writing—review and editing, I.P. and C.M.S.; supervision, I.P. and C.M.S.; funding acquisition, C.M.S.; resources, C.M.S. All authors have read and agreed to the published version of the manuscript.

Funding: This work was developed within the scope of the project CICECO-Aveiro Institute of Materials, UIDB/50011/2020 & UIDP/50011/2020, financed by national funds through the FCT/MEC and, when appropriate, co-financed by FEDER under the PT2020 Partnership Agreement. Authors want to thank Project inpactus—innovative products and technologies from eucalyptus, Project N° 21874 funded by Portugal 2020 through European Regional Development Fund (ERDF) in the frame of COMPETE 2020 n°246/AXIS II/2017. Authors want to thank the funding from Project AgroForWealth (CENTRO-01-0145-FEDER-000001), funded by Centro2020, through FEDER and PT2020.

Conflicts of Interest: The authors declare no conflict of interest.

References

1. Lorenzo, P.; Gonzalez, L.O.; Reigosa, M.J. The genus Acacia as invader: The characteristic case of *Acacia dealbata* Link in Europe. *Ann. For. Sci.* **2010**, *67*, 101. [CrossRef]
2. Kull, C.A.; Shackleton, C.M.; Cunningham, P.J.; Ducatillon, C.; Dufour-Dror, J.-M.; Esler, K.J.; Friday, J.B.; Gouveia, A.C.; Griffin, A.R.; Marchante, E.; et al. Adoption, use and perception of Australian acacias around the world. *Divers. Distrib.* **2011**, *17*, 822–836. [CrossRef]
3. Le Maitre, D.C.; Gaertner, M.; Marchante, E.; Ens, E.-J.; Holmes, P.M.; Pauchard, A.; O'Farrell, P.J.; Rogers, A.M.; Blanchard, R.; Blignaut, J.; et al. Impacts of invasive Australian acacias: Implications for management and restoration. *Divers. Distrib.* **2011**, *17*, 1015–1029. [CrossRef]
4. Instituto da Conservação da Natureza e das Florestas. *Inventário Florestal Nacional (IFN6)–Principais Resultados*; 2019.
5. European Commission Directive (EU) 2018/2002 of the European Parliament and of the Council of 11 December 2018 amending Directive 2012/27/EU on energy efficiency. *Off. J. Eur. Union* **2018**, *61*, 1–230.
6. Oliveira, C.S.D.; Moreira, P.; Resende, J.; Cruz, M.T.; Pereira, C.M.F.; Silva, A.M.S.; Santos, S.A.O.; Silvestre, A.J.D. Characterization and Cytotoxicity Assessment of the Lipophilic Fractions of Different Morphological Parts of *Acacia dealbata*. *Int. J. Mol. Sci.* **2020**, *21*, 1814. [CrossRef]
7. Freire, C.; Coelho, D.S.C.; Santos, N.M.; Silvestre, A.J.D.; Neto, C. Identification of Δ7 phytosterols and phytosteryl glucosides in the wood and bark of several *Acacia* species. *Lipids* **2005**, *40*, 317–322. [CrossRef]
8. Luís, A.; Gil, N.; Amaral, M.E.; Duarte, A.P. Antioxidant activities of extracts from *Acacia melanoxylon*, *Acacia dealbata* and *Olea europaea* and alkaloids estimation. *Int. J. Pharm. Pharm. Sci.* **2012**, *4*, 225–231.
9. Yáñez, R.; Gómez, B.; Martínez, M.; Gullón, B.; Alonso, J.L.; Estévez, B.G. Valorization of an invasive woody species, *Acacia dealbata*, by means of Ionic liquid pretreatment and enzymatic hydrolysis. *J. Chem. Technol. Biotechnol.* **2014**, *89*, 1337–1343. [CrossRef]
10. Yildiz, S.; Gürgen, A.; Can, Z.; Tabbouche, S.A.; Kiliç, A.O. Some bioactive properties of Acacia dealbata extracts and their potential utilization in wood protection. *Drewno* **2018**, *61*, 81–97. [CrossRef]
11. Devi, S.R.; Prasad, M.N.V. Tannins and related polyphenols from ten common *Acacia* species of India. *Bioresour. Technol.* **1991**, *36*, 189–192. [CrossRef]

12. Freire, C.S.R.; Silvestre, A.J.D.; Neto, C.P. Demonstration of long-chain n-alkyl caffeates and Δ7-steryl glucosides in the bark of *Acacia* species by gas chromatography-mass spectrometry. *Phytochem. Anal.* **2007**, *18*, 151–156. [CrossRef]
13. Lisperguer, J.; Saraiva, Y.; Vergara, E. Structure and thermal behavior of tannins from *Acacia dealbata* bark and their reactivity toward formaldehyde. *J. Chil. Chem. Soc.* **2016**, *61*, 3188–3190. [CrossRef]
14. Neiva, D.M.; Luís, Â.; Gominho, J.; Domingues, F.; Duarte, A.P.; Pereira, H. Bark residues valorization potential regarding antioxidant and antimicrobial extracts. *Wood Sci. Technol.* **2020**, *54*, 559–585. [CrossRef]
15. Casas, M.P.; Conde, E.; Ribeiro, D.; Fernandes, E.; Domínguez, H.; Torres, M.D. Bioactive properties of *Acacia dealbata* flowers extracts. *Waste Biomass Valoriz.* **2020**, *11*, 2549–2557. [CrossRef]
16. López-Hortas, L.; Falqué, E.; Domínguez, H.; Torres, M.D. Microwave hydrodiffusion and gravity versus conventional distillation for *Acacia dealbata* flowers. Recovery of bioactive extracts for cosmetic purposes. *J. Clean. Prod.* **2020**, *274*. [CrossRef]
17. Pereira, F.B.M.; Domingues, F.M.J.; Silva, A.M.S. Triterpenes from *Acacia dealbata*. *Nat. Prod. Lett.* **1996**, *8*, 97–103. [CrossRef]
18. Imperato, F. A chalcone glycoside from *Acacia dealbata*. *Phytochemistry* **1982**, *21*, 480–481. [CrossRef]
19. Rodolphe, P.; Katharina, B.; Meierhenrich, U.J.; Elise, C.; Georges, F.; Nicolas, B. Chemical composition of french mimosa absolute oil. *J. Agric. Food Chem.* **2010**, *58*, 1844–1849. [CrossRef]
20. Soto, M.L.; Parada, M.; Falqué, E.; Domínguez, H. Personal-care products formulated with natural antioxidant extracts. *Cosmetics* **2018**, *5*, 13. [CrossRef]
21. Silva, E.; Fernandes, S.; Bacelar, E.; Sampaio, A. Antimicrobial activity of aqueous, ethanolic and methanolic leaf extracts from *Acacia* spp. and *Eucalyptus nicholii*. *Afr. J. Tradit. Complement. Altern. Med.* **2016**, *13*, 130–134. [CrossRef]
22. Borges, A.; José, H.; Homem, V.; Simões, M. Comparison of Techniques and Solvents on the Antimicrobial and Antioxidant Potential of Extracts from *Acacia dealbata* and *Olea europaea*. *Antibiotics* **2020**, *9*, 48. [CrossRef]
23. de Melo, M.M.R.; Oliveira, E.L.G.; Silvestre, A.J.D.; Silva, C.M. Supercritical fluid extraction of vegetable matrices: Applications, trends and future perspectives of a convincing green technology. *J. Supercrit. Fluids* **2014**, *92*, 115–176. [CrossRef]
24. Reverchon, E.; Marco, I. De Supercritical fluid extraction and fractionation of natural matter. *J. Supercrit. Fluids* **2006**, *38*, 146–166. [CrossRef]
25. Rodrigues, V.H.; de Melo, M.M.R.; Portugal, I.; Silva, C.M. Extraction of *Eucalyptus* leaves using solvents of distinct polarity. Cluster analysis and extracts characterization. *J. Supercrit. Fluids* **2018**, *135*, 263–274. [CrossRef]
26. Rodrigues, V.H.; de Melo, M.M.R.; Portugal, I.; Silva, C.M. Supercritical fluid extraction of *Eucalyptus globulus* leaves. Experimental and modelling studies of the influence of operating conditions and biomass pretreatment upon yields and kinetics. *Sep. Purif. Technol.* **2018**, *191*, 173–181. [CrossRef]
27. Domingues, R.M.A.; de Melo, M.M.R.; Oliveira, E.L.G.; Neto, C.P.; Silvestre, A.J.D.; Silva, C.M. Optimization of the supercritical fluid extraction of triterpenic acids from *Eucalyptus globulus* bark using experimental design. *J. Supercrit. Fluids* **2013**, *74*, 105–114. [CrossRef]
28. de Melo, M.M.R.; Oliveira, E.L.G.; Silvestre, A.J.D.; Silva, C.M. Supercritical fluid extraction of triterpenic acids from *Eucalyptus globulus* bark. *J. Supercrit. Fluids* **2012**, *70*, 137–145. [CrossRef]
29. Domingues, R.M.A.; Oliveira, E.L.G.; Freire, C.S.R.; Couto, R.M.; Simões, P.C.; Neto, C.P.; Silvestre, A.J.D.; Silva, C.M. Supercritical fluid extraction of *Eucalyptus globulus* bark-A promising approach for triterpenoid production. *Int. J. Mol. Sci.* **2012**, *13*, 7648–7662. [CrossRef]
30. de Melo, M.M.R.; Carius, B.; Simões, M.M.Q.; Portugal, I.; Saraiva, J.; Silva, C.M. Supercritical CO2 extraction of *V. vinifera* leaves: Influence of cosolvents and particle size on removal kinetics and selectivity to target compounds. *J. Supercrit. Fluids* **2020**, *165*, 104959. [CrossRef]
31. de Melo, M.M.R.; Vieira, P.G.; Şen, A.; Pereira, H.; Portugal, I.; Silva, C.M. Optimization of the supercritical fluid extraction of *Quercus cerris* cork towards extraction yield and selectivity to friedelin. *Sep. Purif. Technol.* **2020**, *238*, 116395. [CrossRef]
32. Martins, P.F.; de Melo, M.M.R.; Sarmento, P.; Silva, C.M. Supercritical fluid extraction of sterols from *Eichhornia crassipes* biomass using pure and modified carbon dioxide. Enhancement of stigmasterol yield and extract concentration. *J. Supercrit. Fluids* **2016**, *107*, 441–449. [CrossRef]
33. Evans, C.S.; Qureshi, M.Y.; Bell, E.A. Free amino acids in the seeds of *Acacia* species. *Phytochemistry* **1977**, *16*, 565–570. [CrossRef]
34. Domingues, R.; Guerra, A.; Duarte, M.; Freire, C.; Neto, C.; Silva, C.; Silvestre, A. Bioactive Triterpenic Acids: From Agroforestry Biomass Residues to Promising Therapeutic Tools. *Mini Rev. Org. Chem.* **2014**, *11*, 382–399. [CrossRef]
35. de Melo, M.M.R.; Domingues, R.M.A.; Silvestre, A.J.D.; Silva, C.M. Extraction and purification of triterpenoids using supercritical fluids: From lab to exploitation. *Mini Rev. Org. Chem.* **2014**, 362–381. [CrossRef]
36. Xu, F.; Huang, X.; Wu, H.; Wang, X. Beneficial health effects of lupenone triterpene: A review. *Biomed. Pharmacother.* **2018**, *103*, 198–203. [CrossRef]
37. Saleem, M. Lupeol, a novel anti-inflammatory and anti-cancer dietary triterpene. *Cancer Lett.* **2009**, *285*, 109–115. [CrossRef]
38. Meneses-Sagrero, S.E.; Navarro-Navarro, M.; Ruiz-Bustos, E.; Del-Toro-Sánchez, C.L.; Jiménez-Estrada, M.; Robles-Zepeda, R.E. Antiproliferative activity of spinasterol isolated of *Stegnosperma halimifolium* (Benth, 1844). *Saudi Pharm. J.* **2017**, *25*, 1137–1143. [CrossRef]
39. Lou-Bonafonte, J.M.; Martínez-Beamonte, R.; Sanclemente, T.; Surra, J.C.; Herrera-Marcos, L.V.; Sanchez-Marco, J.; Arnal, C.; Osada, J. Current Insights into the Biological Action of Squalene. *Mol. Nutr. Food Res.* **2018**, *62*, 1–16. [CrossRef]

40. Aragão, G.F.; Carneiro, L.M.V.; Junior, A.P.F.; Vieira, L.C.; Bandeira, P.N.; Lemos, T.L.G.; de B. Viana, G.S. A possible mechanism for anxiolytic and antidepressant effects of alpha- and beta-amyrin from Protium heptaphyllum (Aubl.) March. *Pharmacol. Biochem. Behav.* **2006**, *85*, 827–834. [CrossRef]
41. Aragão, G.F.; Cunha Pinheiro, M.C.; Nogueira Bandeira, P.; Gomes Lemos, T.L.; de Barros Viana, G.S. Analgesic and anti-inflammatory activities of the isomeric mixture of alpha- and beta-amyrin from protium heptaphyllum (Aubl.) March. *J. Herb. Pharmacother.* **2007**, *7*, 31–47. [CrossRef] [PubMed]
42. da Silva Júnior, W.F.; Lima Bezerra de Menezes, D.; Calvarho de Oliveira, L.; Scherer Koester, L.; Olveira de Almeida, P.D.; Lima, E.S.; Pereira de Azevedo, E.; da Veiga Júnior, V.F.; Neves de Lima, Á.A. Inclusion complexes of β and HPβ-cyclodextrin with α, β amyrin and in vitro anti-inflammatory activity. *Biomolecules* **2019**, *9*, 241. [CrossRef]
43. Holanda Pinto, S.A.; Pinto, L.M.S.; Cunha, G.M.A.; Chaves, M.H.; Santos, F.A.; Rao, V.S. Anti-inflammatory effect of α, β-Amyrin, a pentacyclic triterpene from Protium heptaphyllum in rat model of acute periodontitis. *Inflammopharmacology* **2008**, *16*, 48–52. [CrossRef]
44. Oliveira, F.A.; Vieira-Júnior, G.M.; Chaves, M.H.; Almeida, F.R.C.; Florêncio, M.G.; Lima, R.C.P., Jr.; Silva, R.M.; Santos, F.A.; Rao, V.S.N. Gastroprotective and anti-inflammatory effects of resin from Protium heptaphyllum in mice and rats. *Pharmacol. Res.* **2004**, *49*, 105–111. [CrossRef]
45. Okoye, N.N.; Ajaghaku, D.L.; Okeke, H.N.; Ilodigwe, E.E.; Nworu, C.S.; Okoye, F.B.C. Beta-Amyrin and alpha-Amyrin acetate isolated from the stem bark of Alstonia boonei display profound anti-inflammatory activity. *Pharm. Biol.* **2014**, *52*, 1478–1486. [CrossRef] [PubMed]
46. Jæger, D.; O'Leary, M.C.; Weinstein, P.; Møller, B.L.; Semple, S.J. Phytochemistry and bioactivity of *Acacia sensu stricto* (Fabaceae: Mimosoideae). *Phytochem. Rev.* **2019**, *18*, 129–172. [CrossRef]
47. Subhan, N.; Burrows, G.E.; Kerr, P.G.; Obied, H.K. Phytochemistry, Ethnomedicine, and Pharmacology of *Acacia*. *Stud. Nat. Prod. Chem.* **2018**, *57*, 247–326. [CrossRef]
48. Pitzer, K.S.; Schreiber, D.R. Improving equation-of-state accuracy in the critical region; equations for carbon dioxide and neopentane as examples. *Fluid Phase Equilib.* **1988**, *41*, 1–17. [CrossRef]
49. Pöhler, H.; Kiran, E. Volumetric properties of carbon dioxide + ethanol at high pressures. *J. Chem. Eng. Data* **1997**, *42*, 384–388. [CrossRef]
50. Falco, N.; Kiran, E. Volumetric properties of ethyl acetate + carbon dioxide binary fluid mixtures at high pressures. *J. Supercrit. Fluids* **2012**, *61*, 9–24. [CrossRef]
51. Gujar, A.; Anderson, T.; Cavagnino, D.; Patel, A. Comparative analysis of mass spectral matching for confident compound identification using the Advanced Electron Ionization source for GC-MS. *Thermoscientific* **2018**, *10598*, 1–7.
52. Zuluaga, A.M.; Mena-García, A.; Soria Monzón, A.C.; Rada-Mendoza, M.; Chito, D.M.; Ruiz-Matute, A.I.; Sanz, M.L. Microwave assisted extraction of inositols for the valorization of legume by-products. *LWT* **2020**, *133*, 109971. [CrossRef]
53. Zuluaga, A.M.; Mena-García, A.; Chito-Trujillo, D.; Rada-Mendoza, M.; Sanz, M.L.; Ruiz-Matute, A.I. Development of a microwave-assisted extraction method for the recovery of bioactive inositols from lettuce (*Lactuca sativa)* byproducts. *Electrophoresis* **2020**, *41*, 1804–1811. [CrossRef] [PubMed]
54. Mena-García, A.; Rodríguez-Sánchez, S.; Ruiz-Matute, A.I.; Sanz, M.L. Exploitation of artichoke byproducts to obtain bioactive extracts enriched in inositols and caffeoylquinic acids by Microwave Assisted Extraction. *J. Chromatogr. A* **2020**, *1613*, 460703. [CrossRef]
55. Ruiz-Aceituno, L.; García-Sarrió, M.J.; Alonso-Rodriguez, B.; Ramos, L.; Sanz, M.L. Extraction of bioactive carbohydrates from artichoke (*Cynara scolymus* L.) external bracts using microwave assisted extraction and pressurized liquid extraction. *Food Chem.* **2016**, *196*, 1156–1162. [CrossRef] [PubMed]
56. Chóez-Guaranda, I.; Ruíz-Barzola, O.; Ruales, J.; Manzano, P. Antioxidant activity optimization and GC-MS profile of aqueous extracts of *Vernonanthura patens* (Kunth) H. Rob. leaves. *Nat. Prod. Res.* **2020**, *34*, 2505–2509. [CrossRef]
57. Somsub, W.; Kongkachuichai, R.; Sungpuag, P.; Charoensiri, R. Effects of three conventional cooking methods on vitamin C, tannin, myo-inositol phosphates contents in selected Thai vegetables. *J. Food Compos. Anal.* **2008**, *21*, 187–197. [CrossRef]
58. Warren, C.R.; Aranda, I.; Cano, F.J. Responses to water stress of gas exchange and metabolites in *Eucalyptus* and *Acacia* spp. *Plant Cell Environ.* **2011**, *34*, 1609–1629. [CrossRef]
59. Croze, M.L.; Soulage, C.O. Potential role and therapeutic interests of myo-inositol in metabolic diseases. *Biochimie* **2013**, *95*, 1811–1827. [CrossRef] [PubMed]
60. Pütter, K.M.; van Deenen, N.; Müller, B.; Fuchs, L.; Vorwerk, K.; Unland, K.; Bröker, J.N.; Scherer, E.; Huber, C.; Eisenreich, W.; et al. The enzymes OSC1 and CYP716A263 produce a high variety of triterpenoids in the latex of *Taraxacum koksaghyz*. *Sci. Rep.* **2019**, *9*, 5942. [CrossRef] [PubMed]
61. Yendo, A.C.A.; De Costa, F.; Gosmann, G.; Fett-Neto, A.G. Production of plant bioactive Triterpenoid saponins: Elicitation strategies and target genes to improve yields. *Mol. Biotechnol.* **2010**, *46*, 94–104. [CrossRef]
62. Fukushima, E.O.; Seki, H.; Ohyama, K.; Ono, E.; Umemoto, N.; Mizutani, M.; Saito, K.; Muranaka, T. CYP716A Subfamily Members are Multifunctional Oxidases in Triterpenoid Biosynthesis. *Plant Cell Physiol.* **2011**, *52*, 2050–2061. [CrossRef] [PubMed]
63. Brendolise, C.; Yauk, Y.K.; Eberhard, E.D.; Wang, M.; Chagne, D.; Andre, C.; Greenwood, D.R.; Beuning, L.L. An unusual plant triterpene synthase with predominant α-amyrin—Producing activity identified by characterizing oxidosqualene cyclases from *Malus × domestica*. *FEBS J.* **2011**, *278*, 2485–2499. [CrossRef] [PubMed]

processes MDPI

Article

Improvement of Enzymatic Saccharification of Cellulose-Containing Raw Materials Using *Aspergillus niger*

Ekaterina Budenkova [1], Stanislav Sukhikh [1], Svetlana Ivanova [2,3,*], Olga Babich [1], Vyacheslav Dolganyuk [1,4], Philippe Michaud [5] and Olga Kriger [1]

1 Institute of Living Systems, Immanuel Kant Baltic Federal University, A. Nevskogo Street 14, 236016 Kaliningrad, Russia; abudenkova@kantiana.ru (E.B.); stas-asp@mail.ru (S.S.); olich.43@mail.ru (O.B.); dolganuk_vf@mail.ru (V.D.); olgakriger58@mail.ru (O.K.)
2 Natural Nutraceutical Biotesting Laboratory, Kemerovo State University, Krasnaya Street 6, 650043 Kemerovo, Russia
3 Department of General Mathematics and Informatics, Kemerovo State University, Krasnaya Street 6, 650043 Kemerovo, Russia
4 Department of Bionanotechnology, Kemerovo State University, Krasnaya Street 6, 650043 Kemerovo, Russia
5 CNRS, SIGMA Clermont, Institut Pascal, Université Clermont Auvergne, F-63000 Clermont-Ferrand, France; philippe.michaud@uca.fr
* Correspondence: pavvm2000@mail.ru; Tel.: +7-384-239-6832

Abstract: Enzymatic hydrolysis of cellulose-containing raw materials, using *Aspergillus niger*, were studied. Filter paper, secondary cellulose-containing or starch-containing raw materials, miscanthus cellulose after alkaline or acid pretreatment, and wood chip cellulose, were used as substrates. The study focused on a wild *A. niger* strain, treated, or not (control), by ultraviolet (UV) irradiations for 45, 60, or 120 min (UV45, UV60, or UV120), or by UV irradiation for 120 min followed by a chemical treatment with NaN_3 + ItBr for 30 min or 80 min (UV120 + CH30 or UV120 + CH80). A mixture of all the *A. niger* strains (MIX) was also tested. A citrate buffer, at 50 mM, wasthe most suitable for enzymatic hydrolysis. As the UV exposure time increased to 2 h, the cellulase activity of the surviving culturewas increased ($r = 0.706$; $p < 0.05$). The enzymatic activities of the obtained strains, towards miscanthus cellulose, wood chips, and filter paper, were inferior to those obtained with commercial enzymes (8.6 versus 9.1 IU), in some cases. Under stationary hydrolysis at 37 °C, pH = 4.7, the enzymatic activity of *A. niger* UV120 + CH30 was 24.9 IU. The enzymatic hydrolysis of secondary raw materials, using treated *A. niger* strains, was themost effective at 37 °C. Similarly, the most effective treatment of miscanthus cellulose and wood chips occurred at 50 °C. The maximum conversion of cellulose to glucose was observed using miscanthus cellulose (with alkaline pretreatment), and the minimum conversion was observed when using wood chips. The greatest value of cellulase activity was evidenced in the starch-containing raw materials, indicating that *A. niger* can ferment not only through cellulase activity, but also via an amylolytic one.

Keywords: cellulose-containing raw materials; enzymatic hydrolysis; *Aspergillus niger*; bioethanol; cellulase activity; ultraviolet treatment

1. Introduction

The development of efficient methods for the production of biofuels is becoming increasingly important, due to the depletion of fossil fuels and the environmental conditions that are associated with their extraction. Possessing environmental and economic advantages, bioethanol is able to compete in the global fuel market. It can be used alone or as an additive to gasoline [1].

Bioenergy production has significantly increased in the recent decades, as part of the transition to a carbon-neutral energy system, and its growth is expected to continue, particularly in the form of biofuels such as bioethanol. The global market is forecasted to increase the share of bioethanol production from the current 2% to 27% by 2050 [2].

Bioethanol production has been strongly supported by the United States and Brazil, which are the leading producing countries [3]. Several patents that are devoted to methods of converting cellulosic material into ethanol by fermentation (RU 2432368, RU 2456394, RU 2529371, and RU 2593724) and GOST R 33872-2016 [4–6], are registered in Russia. However, Russian biofuel production is only inthe pilot stage [7].

Cellulosic biomass is attractive as a source of raw materials, due to its availability and renewability [3,8–11]. The study [12] showed that microorganisms with visual differences in the morphology and growth of colonies (including bacteria, yeast, and filamentous fungi), were found in mangroves in Brazil.Metagenomic sequencing data showed the predominance of the following bacterial types: *Proteobacteria* (57.8%), *Firmicutes* (12.3%), and *Actinobacteria* (8.4%). It was found that these microorganisms produce a large number of enzymes that are involved in the degradation of polycyclic aromatic compounds. Specific sequences that are involved in cellulolytic degradation, belonging to cellulases, hemicellulases, carbohydrate-binding domains, dockerins, and cohesins, were identified. It became possible to isolate cultivated fungi and bacteria that are associated with biomass decomposition and the potential for use in biofuel production. These results showed that cellulosolytic microorganisms that are present in mangroves possess all the basic molecular tools for building the cellulosome, which is necessary for the efficient degradation of cellulosic material and the release of sugar.

To provide an economically viable process for its bioconversion into bioethanol, it is necessary to achieve effective splitting of polymeric lignocellulosic raw materials into monomeric sugars, which are further fermented by microorganisms (yeast or bacteria). The composition and concentration of sugars in the lignocellulose hydrolysate vary depending on the composition of the raw material and the processing method. Other important factors of alcoholic fermentation include the selection and preparation of the substrate, the selection and adaptation of the microorganisms, and the process conditions [1].

Enzymatic hydrolysis is conducted using the cellulase enzyme complex. Endoglucanases rapidly and randomly break down the internal cellulose sites. The products of this reaction serve as a substrate for the action of exoglucanases, which mainly release cellobiose by breaking down oligosaccharides. Enzymatic hydrolysis occurs under the synergistic action of the following three main classes of cellulases: endoglucanase, exoglucanase, and β-glucosidase [13]. The main advantages of enzymatic hydrolysis include technological accessibility, the high specificity of enzymes, and reduced generation of by-products and inhibitory products. The disadvantages include the high cost of enzyme preparations and the duration of the hydrolysis process (48–72 h). It is important to consider the concentration and structure of the substrate (reaction rate) and enzyme concentration (price/reaction rate ratio) in enzymatic hydrolysis [14].

The main constraints of the lignocellulosic biomass fermentation are the crystalline structure of cellulose and the formation of a lignin–cellulose complex. High pressure and temperature lead to the destruction of glucose and the formation of reaction by-products [3].

Fungal amylases are used to hydrolyze carbohydrates, proteins, and other components of soybeans and wheat, into peptides, amino acids, sugars, and other low-molecular-weight compounds [13,15]. The study [13] was carried out to assess the amylase activity of the amylolytic fungi *Aspergillus niger*, using cassava waste as a fodder substrate. Fungi were isolated from soil samples and identified as *Aspergillus niger*, based on lactophenol cotton blue staining and plating on an appropriate fungal growth medium. The production of *Aspergillus niger* amylase was detected by the disappearance of the blue color in the agar medium, with starch around the microbial colonies after incubation. Cassava was used as a substrate for amylase production. Solid-phase fermentation was performed using *Aspergillus niger* to obtain amylase. Amylase activity was determined by the following four methods: the DNSA method, dextrinization method, reduction in starch–iodine color intensity, and assay on plates.

Solid-phase fermentation has great potential for the amylase enzyme production by *Aspergillus niger* [16].Hard substrates that are rich in starch, such as rice bran, wheat bran, rye bran, coconut oil cake, ginger cake, and peanut cake, are used.These agro-industrial residues are cheap raw materials for amylase production.It was found that *Aspergillus niger* is the best amylase producer.The addition of carbohydrates and nitrogen increased the yield of amylase [16].

Membrane enzymatic hydrolysis, in comparison with chemical hydrolysis, has significant application potential [17,18], due to the higher yield of a low-toxic product, environmental friendliness, complete decomposition of lignocellulosic biomass [16], high degree of conversion of carbohydrates, absence of toxic waste, greater yield of the target product, and the ability to modify and cultivate microorganisms to obtain significant quantities of high-quality biofuel. Nonetheless, this method requires modernization and adaptation.

Using membranes to extract and reuse cellulase saves energy, reduces the cost of purchasing an enzyme preparation, and shortens the time that is required by the stages of enzymatic hydrolysis [14]. Membrane filtration is widely used in this process. In this case, the enzyme, while maintaining its relative catalytic activity, recirculates up to 75% of the enzyme and is reused in the process of hydrolysis [11]. The membranes are made from hollow fibers of cellulose acetate, polysulfone, nylon, and polyethersulfone.

The rigid structure of lignocellulosic biomass prevents the enzymatic hydrolysis of polysaccharides, which prevents the conversion of biomass into bioethanol [3]. The combined use of several methods in pretreatment (chemical, physical, electrical, and biological) promotes the degradation of lignocellulose, reduces the impermeability of the biomass, and facilitates the access of enzymes to their substrates; however, the processes of the pretreatment of lignocellulosic materials cause an increased formation of undesirable by-products, originating from lignin, which can inhibit the growth of microbes and, consequently, reduce the fermentation yield [19,20]. The use of microorganisms helps to avoid many of these difficulties and, in combination with the membrane methods of enzymatic purification, helps to intensify this process.

It is known [21] that *Aspergillus niger* secretes endoglucanase, exoglucanase, andβ-glucosidases (EC 3.2.1.21, EC 3.2.1.4, EC 3.2.1.74), which have many biotechnological and industrial applications. It was found that the purified enzyme with a molecular weight of 116 kDa, forms monomers in the solution, according to the data of native gel electrophoresis, and has a pI value of 4.55, as found for most fungal β-glucosidases. Small-angle X-ray experiments showed that these β-glucosidases have a tadpole-like structure, with an N-terminal catalytic domain and a C-terminal fibronect in III-like domain (FnIII) linked by a long linker peptide (~100 amino acid residues) in an extended conformation. This molecular organization resembles the organization of other cellulases (such as, for example, cellobiohydrolases), which often contain a catalytic domain that is associated with a cellulose-binding module, which mediates their binding to insoluble and polymeric cellulose. The reasons why β-glucosidases from *Aspergillus niger* act on small soluble substrates and have the form of a tadpole molecule, are not entirely clear. However, assays with reduced enzymes with various polymeric substrates suggest that these β-glucosidases have little or no ability to bind and adsorb cellulose, xylan, and starch, but have a high lignin affinity. Molecular dynamics simulations have shown that clusters of residues that are located in the C-terminal domain of FnIII, strongly interact with lignin fragments. Modeling has shown that numerous arginine residues, scattered over the surface of FnIII, play an essential role in the interaction with lignin through the π-stacking of cations with lignin aromatic rings. These results indicate that the C-terminal domain of FnIII can act to immobilize the enzyme on the cell wall and prevent the unproductive binding of cellulase to lignin [21].

Aspergillus niger fungus produces a significant amount of cellulase and protease activity in its extracellular environment. The maturation rate of fungi during the takeover of large territories is unique (within three days, intensive sporulation occurs). This study is aimed at experimentally selecting the optimal process for the enzymatic hydrolysis of cellulose-containing raw materials, using *Aspergillus niger* exposed to ultraviolet (UV) radiation and chemical treatments.

2. Materials and Methods

2.1. Microorganisms

Aspergillus niger F-1270 was acquired from the collection of microorganisms of the Federal Institution "State Research Institute of Genetics and Selection of Industrial Microorganisms of the National Research Center" Kurchatov Institute (Moscow, Russia). The recommended cultivation conditions were as follows: temperature of 26 °C, and a malt-peptone agar medium (malt extract 30 g/L, peptone 1 g/L, agar 20 g/L). The culture medium was sterilized at 121 °C for 20 min. The cultivation was carried out by the quadrant or the thinning streak methods on Petri dishes, in test tubes with a cotton and gauze plug. The passage of microscopic fungi was carried out with a bacteriological loop. In addition to the recommended medium, we used the Czapek–Dox and the Mandels–Weber mineral nutrient media to stimulate cellulolytic activity [18].

Potato agar medium (potato extract/dried and milled potato waste 4 g/L, agar-agar 15 g/L, pH 5) was used to stimulate starch-degrading activity. The media were used for modifications with/without added cellulose and with/without added glucose. Malt-peptone agar was used to disperse microorganisms.

At the stage of the exposure to mutagenic factors, these media were used with the addition of glucose at a concentration of 1 g/L. The media were sterilized by autoclaving at 121 °C for 20 min. A KH_2PO_4 solution and solution of trace elements ($FeSO_4{\cdot}7H_2O$ 5.0 mg/L, $MnSO_4{\cdot}7H_2O$ 1.6 mg/L, $ZnSO_4{\cdot}7H_2O$ 1.4 mg/L, $CoCl_2{\cdot}6H_2O$ 20 mg/L) were prepared and sterilized separately, and the combination with the stock solution was carried out after cooling solutions to room temperature.

The medium used for preparing a spore suspension and enzymatic hydrolysis contained the following (g/L): 0.3 urea, 1.4 $(NH_4)_2SO_4$, 2.0 KH_2PO_4, 0.4 $CaCl_2{\cdot}2H_2O$, 0.3 $MgSO_4{\cdot}7H_2O$, 1.0 peptone, 0.2 Tween 80, 0.005 $FeSO_4{\cdot}7H_2O$, 0.0016 $MnSO_4{\cdot}7H_2O$, 0.0014 $ZnSO_4{\cdot}7H_2O$, 0.02 $CoCl_2{\cdot}6H_2O$, 10.0 cellulose; pH 5.

As a control of cellulolytic activity, we used the "Cellulase Ultra" enzyme preparation from *Trichoderma reeseii* (Sibbiopharm, Berdsk, Russia), a widespread cellulase-producing culture [22], which can be used as a reference for studying the cellulase production of *Aspergillus niger*. This preparation hydrolyzes β-1,4-glycosidic bonds with an activity of 2500 U per 1 g of enzyme at working ranges of temperatures and pH between 30 to 65 °C and pH 2–7, respectively (optimum at 50–60 °C and pH 4.0–5.5).To prepare the stock solution of the enzyme preparation (control 3), 10 mL (with an accuracy of 0.2 μL) of the preparation was suspended in a small amount of distilled water (up to 50 cm^3) on a magnetic stirrer for 15 min. The suspension was quantitatively transferred into a volumetric flask with a capacity of 200 cm^3 and the volume was brought up to the mark with distilled water. The resulting suspension was centrifuged at 7000 rpm for 15 min. The supernatant was used for analysis.

2.2. Characterization and Pretreatment of Raw Materials

The cellulose of miscanthus (*Miscanthus sinensis Andersson*) was used as a substrate for enzymatic hydrolysis. The cellulose of alder chips was used as the primary raw material, and starch and cellulose-containing household waste (potato and bread) were used as secondary raw materials. Miscanthus cellulose is a natural polysaccharide built from anhydro-D-glucopyranose units interconnected with a β-glycosidic bond, with a degree of polymerization (the number of elementary units in a cellulose macromolecule) of 15–20 thousand. The chemical composition of cellulose obtained from miscanthus included α-cellulose 89.4%, ash content 0.34%, acid-insoluble lignin 1.45%, pentosans 9.4%; the crystallinity index was 0.6. Alder chip cellulose is a natural cellulose, high-molecular-weight linear polysaccharide, poly-1.4-β-D-glucopyranosyl-D-glucopyranose with a degree of polymerization from 2000 to 26,000. The average degree of polymerization of alder chip cellulose is 5000–10,000. The chemical composition of cellulose obtained from miscanthus included α-cellulose 88.4%, ash content 0.04%, lignin 0.44%, pentosans 6.0%; the crystallinity index was 0.66. Cellobiose disaccharide is the main structural element that is repeated in the polysaccharide macromolecule. Cellulose-containing waste includes the following: α-cellulose 73.7–77.0%, ash content 0.03%, residual amount of lignin 3.7%, pentosans 9–16%, resins and fats 0.04%; the viscosity of cellulose solutions is 31.0 mPa·s, the crystallinity index is 0.56% [23,24].

The secondary raw materials were not additionally pretreated. Starch-containing raw materials were used to advance knowledge of the *Aspergillus niger* activity against potato starch. Filter paper (blue ribbon filter) was used to determine the cellulolytic activity of microorganisms.

The primary raw material underwent alkaline and acid delignifications. Delignification was performed under hydrolysis conditions with sodium hydroxide solution, namely, 10 g of dry biomass of miscanthus was placed in a two-neck round-bottom flask with a magnetic stirrer, reflux condenser, and thermometer. Water was added to the biomass in a ratio of 10:1, and 4 g of NaOH was added with stirring. The mixture was stirred for 4 h at 75 °C. In the case of acid hydrolysis, trifluoroacetic acid (30 mL) and a 30% hydrogen peroxide solution (20 mL) were used per 10 g of dry biomass supplemented with 50 mL of water. In addition, 2 mL of sulfuric acid (H_2SO_4) was added to the reaction mixture as a catalyst. The mixture was heated at 60 °C for an hour with constant stirring. As a result, fibers were obtained, which were separated by filtration and washed with distilled water.

2.3. MALDI-TOFMethod

The process of optimizing the delignification conditions, and the procedure for the physicochemical analysis of the reaction products, was carried out according to the technique described in [22]. For this, the MALDI-TOF method was used. The MALDI-TOF method is based on the procedure of soft ionization of the test material (analyte), which allows, in the presence of a special substance, the so-called matrix, under the influence of a laser to ionize biological macromolecules (peptides, proteins, DNA, oligonucleotides, lipopolysaccharides, and sugars) without their fragmentation and destruction. The most widely used matrix is α-cyano-4-hydroxycinnamic acid (CHCA), sinapinic acid (SA), ferulic acid (FA), and 2,5-dihydroxybenzoic acid (DHB). After the laser beam ionization, the system scans for microbial proteins, which generally fall in the range of 4000 to 20,000 Da (60 to 70% of the dry weight of the bacterial cell). After desorption, predominantly single-charged ionized molecules are accelerated in an electric field, enter the separating part of the device (a tube in the cavity of which a vacuum is maintained), after passing through which the ions reach the detector. The speed of movement and, accordingly, the time it takes to travel from the point of ionization to the detector, is inversely proportional to the mass of the ions. Knowing the path length of ion movement from the ionizer to the detector, as well as the time of this movement, it is possible to calculate the speed of ion movement and, based on its value, calculate the mass of particles present in the analyte, as well as generate a spectrum characterizing the qualitative composition of the object under study [25].

2.4. Induced Mutagenesis of Microorganisms

The studied microorganisms were consistently subjected to physical and chemical mutagenic effects to enhance their cellulolytic properties. The technique used to induce mutagenesis is described hereafter and was created based on those presented in [20,26–30] and adapted to the existing conditions.

Physical mutagenesis was carried out in the microbiological safety cabinet BMB-II-Laminar-S-1.2 (221.120) Class II (Type A2) using a Philips TUV 30W G13 ultraviolet lamp (LAMSYSTEMS, Mias, Russia) with maximum radiation at253.7 nm. A suspension based on sodium phosphate buffer (pH 7.0) at a concentration of 10^8 spores/mL was prepared from a 5-day culture of microorganisms. A quantity of 2 mL of the resulting suspension was placed on a Petri dish with a nutrient medium. The medium contained the following (g/L): 0.3 urea, 1.4 $(NH_4)_2SO_4$, 2.0 KH_2PO_4, 0.4 $CaCl_2 \cdot 2H_2O$, 0.3 $MgSO_4 \cdot 7H_2O$, 1.0 peptone, 0.2 Tween 80, 0.005 $FeSO_4 \cdot 7H_2O$, 0.0016 $MnSO_4 \cdot 7H_2O$, 0.0014 $ZnSO_4 \cdot 7H_2O$, 0.02 $CoCl_2 \cdot 6H_2O$, 10.0 cellulose; pH 5.

Then, the cups were positioned at a distance of 10–13 cm from the UV lamp. The exposure time varied from 45 to 120 min, after which the dishes were incubated in the dark at 30 °C for three days. The spore solution was prepared as follows: A suspension was prepared from a 3-day culture of microorganisms based on sodium phosphate buffer (pH 7.0) and a concentration of 10^8 spores/mL. The resulting suspension was mixed with a solution of sodium azide/ethidium bromide in ratios of 9:1 and 1:1. The concentration of sodium azide in the solution was 5 and 400 mg/mL, ethidium bromide 5 and 150 mg/mL, respectively. After exposure for 30 and 80 min at room temperature, the spores were washed three times with sterile deionized water. The supernatant was removed after 15 min of centrifugation at 6000× *g*. Spores were resuspended in sodium phosphate buffer and transferred to plates with a nutrient medium. They were incubated for 5 days, after which the cellulolytic activity was checked and the most effective colonies were selected for subsequent studies. A spore suspension was washed with 0.05 M citrate buffer (pH = 4.8) from a plate with a 5-day culture.

Spore concentration was determined using a Hirst volumetric impaction sampler (General Electric Company plc, Borehamwood, UK). The spores fixed on the strip were examined using a light microscope. Ten fields of view were selected and the number of spores was counted on the strip segment corresponding to a 3-h log.

Chemical mutagenesis was performed as described above using the sole sodium azide and ethidium bromide treatment. Spores were resuspended in sodium phosphate buffer and transferred onto plates with a nutrient medium. Spores were incubated for 5 days, followed by a verification of cellulolytic activity and the selection of the most effective colonies for subsequent studies.

2.5. CellulolyticActivities Measurement

Cellulase activity was determined using an enzymatic assay.

The Bradford method [31] was used to obtain the protein concentration. This method is based on the shift of the absorption maximum of the optical density of the acid blue 90 (Coomassie Brilliant Blue R-250) dye from 470 to 595 nm, observed due to the binding of the protein to the dye.

The enzymatic activity of the preparation was determined in comparison with a standard sample (SS) of this enzyme under the same experimental conditions. The enzymatic activity (EA) in the appropriate units (IU or U) was calculated by the following formula [32]:

$$EA = P \cdot C \cdot A0/P0, \tag{1}$$

where $A0$—SS enzymatic activity is in units (IU or U) per 1 g of protein or preparation; $P0$—value of the measured parameter for SS; p—value of the measured parameter for the test preparation; C—coefficient equalizing the concentrations of solutions of the test preparation and SS.

Enzyme activity is expressed in international units IU (or U) and is equal to the amount of enzyme that catalyzes the conversion of 1 μM of reducing sugars (glucose equivalent) in 1 min at 50 °C.

2.6. Aspergillus Niger Growth Parameters and Determination of the Concentration of Spores in Suspension

The protocol used for the enzymatic hydrolysis is an adaptation of the hydrolysis techniques described in the literature [32,33]. Enzymatic hydrolysis was carried out in 1 and 0.25 L conical flasks with cotton and gauze plugs with a working volume of 200 and 50 mL, respectively. The flasks were closed with plastic stoppers and periodically stirred in an IKA KS4000I control incubator shaker (DV-Ekspert, Moscow, Russia). Stirring trajectory was orbital. Torque range was from 10 to 500 rpm. The reaction was carried out under static conditions at a temperature of 30–37 °C for 3–7 days. The substrate concentration in all samples was 60 g/L. The inoculum concentration varied from 10^6 to 10^8 spores/mL. The growth curve of *A. niger* during cultivation is shown in Figure 1. The strains were grown at a temperature of 27–30 °C.

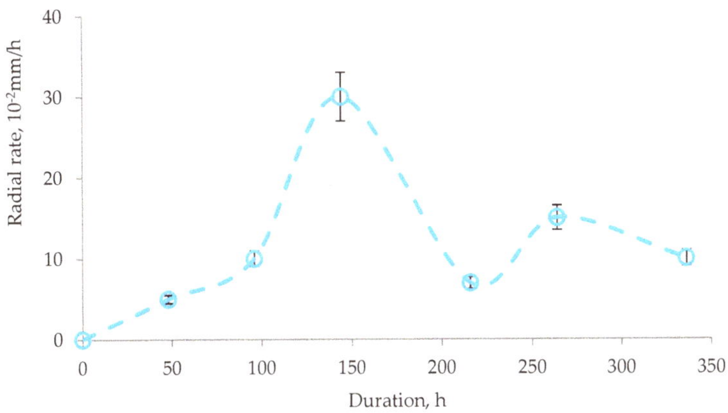

Figure 1. *A. niger* growth curve.

The concentration of the spore suspension was obtained measuring A_{600} Smart Spec Plus spectrophotometer (BioRad Laboratories Inc., Hercules, CA, USA) (r = 0.63). Enzymatic activity was interrupted by heating in an autoclave to 121 °C without standing.

2.7. Optimization of Conditions for Enzymatic Hydrolysis

Original and treated strains were cultivated under the same conditions. To optimize the parameters of enzymatic hydrolysis, the cellulase activity of treated and original strains of *Aspergillus niger* were studied at temperatures of 30, 37, and 50 °C. The enzymatic activity of original and treated microorganisms was compared at these temperatures. The influence of the composition of the nutrient medium, on which the microorganism culture (UV microorganism) was incubated, on the cellulolytic activity during hydrolysis was assessed. For this purpose, microorganisms were inoculated in malt-peptone agar, Czapek–Dox medium, Mandels–Weber saline medium, and potato agar [18]. These media were used because they are standard nutrient media for the cultivation of microorganisms (in particular, fungi), contain all the necessary nutrient components (sources of sodium, potassium, magnesium, iron, agar, peptone, sucrose, and other nutrient components), have optimal pH for the development of microorganisms. They are easy to prepare and convenient to use.

The hydrolysis duration was 1 h and 48 h. To optimize the enzymatic hydrolysis parameters, the total cellulolytic activity level of all studied strains was determined relative to the following different compositions of raw materials: filter paper, recycled cellulose-containing raw materials, secondary starch-containing raw materials (fruit waste, potato starch), miscanthus cellulose after alkaline pretreatment, miscanthus cellulose after acid pretreatment and cellulose. Small substandard fruit products affected by pests (a mixture of apricots and prunes 1:1) were used as substrates (fruit waste) for enzymatic hydrolysis. Fruits were ground using a KT-1321 laboratory grinder (Profit, Moscow, Russia). Potato starch was obtained by processing substandard, non-standard small potatoes, as described in [34].

2.8. Method for the Analysis of Hydrolysis Products

The content of reducing substance in the sample was determined using the reagent 3,5-dinitrosalicylic acid (DNS) according to the method described in GOST R 54905-2012 (Enzyme preparation. Methods of beta-glucanase enzyme activity determination), which allows determining the total amount of reducing sugars (in our studies it was glucose) [35,36]. Glucose is the most important reducing substance in hydrolyzed cellulose samples.

A solution of 1% DNS per 1 L was prepared by dissolving 10 g of DNS in a small amount of water at room temperature and the subsequent gradual addition of 16.05 g of sodium hydroxide and 300 g of potassium–sodium tartrate (in a water bath at 47 ± 1 °C, "soft" cooling that does not destroy the molecules of substances). The reagents were purchased from Pushchinskie Laboratorii (Moscow, Russia). The resulting solution had a bright yellow color. To construct a calibration graph, a stock solution of D-glucose (1 µmol/mL) and a series of its dilutions (0.3–0.6 µmol/mL) were prepared based on citrate or acetate buffer, depending on the medium in which the hydrolysis was carried out. Before analysis, the DNS solution was measured on a spectrophotometer to clarify the absorption maximum. The obtained value (530 nm) was slightly lower than that indicated in GOST (540 nm). To assess the content of the reducing substance, and to build a calibration curve, a 2 mL solution of DNS was thoroughly mixed with 1 mL of the sample and heated in a water bath for 5 min with an accuracy of a few seconds. After cooling, the sample volume was brought to 25 mL with buffer and analyzed on a Shimadzu UV-1800 spectrophotometer (Shimadzu, Kyoto, Japan) at a wavelength of 530 nm in cuvettes with a 10 mm light-absorbing layer.

2.9. Statistical Analysis Methods

Statistical processing was performed using Excel (2019, Microsoft, Redmond, Washington, USA) and IBM SPSS Statistics 22 (2013, SPSS: An IBM Company, Chicago, IL, USA). All experiments were carried out in triplicate. Data are presented as the median ± standard deviation. The Kruskal–Wallis test was used to compare the medians of the samples (significance of differences at $p < 0.05$). For intergroup comparisons, the Mann–Whitney U-test was used with the Bonferroni correction (significance of differences at $p < 0.01$). To check for the presence of a correlation between the UV exposure time and quantitative indicators of cellulolytic activity, the Spearman's rank correlation coefficient was used (significance of differences at $p < 0.05$).

3. Results

3.1. Evaluation of the Original Strain Cellulolytic Activity

The effect of the buffer on the cellulolytic activity of *A. niger* (37 °C, pH = 4.7) is shown in Table 1.

Table 1. Effect of buffer on *A. niger* cellulolytic activity (concentration of reducing substance, mg/mL).

Buffer	C_M	Control 1		Control 2	
		1 h	48 h	1 h	48 h
citrate	1	0.22 ± 0.01	0.23 ± 0.01	0.21 ± 0.03	0.78 ± 0.03
	0.1	0.22 ± 0.01	0.23 ± 0.01	0.19 ± 0.06	0.61 ± 0.09
	0.05	0.23 ± 0.01	0.23 ± 0.02	0.24 ± 0.04	0.45 ± 0.03
acetate	1	0.35 ± 0.01	0.34 ± 0.01	0.38 ± 0.02	0.37 ± 0.03
	0.1	0.35 ± 0.01	0.34 ± 0.02	0.36 ± 0.01	0.42 ± 0.02
	0.05	0.37 ± 0.03	0.47 ± 0.03	0.24 ± 0.04	0.40 ± 0.01

C_M—molar concentration; control 1—without microorganisms; control 2—with the original *A. niger* strain (the concentration of the reducing substance has been reduced). Hydrolysis substrate–filter paper (50 mg/L, according to the commercial enzyme dosage). Data are presented as median ± standard deviation (n = 3).

The cellulolytic activity of the original *A. niger* strain was evaluated, compared to the control without microorganisms (Table 1). The 50 mM citrate buffer was found to be more suitable for enzymatic hydrolysis with *A. niger*, and this advantage was noticeable after 48 h of hydrolysis.

The cellulolytic activity of *A. niger* (37 °C, pH = 4.7) in various substrates is shown in Table 2. In addition to reducing sugars, a protein complex was found.

Table 2. *A. niger* cellulolytic activity (concentration of the reducing substance, mg/mL) in different substrates.

Samples	Substrates					
	1	2	3	4	5	6
Control 1	0.24 ± 0.04	0.35 ± 0.01	0.34 ± 0.01	0.39 ± 0.01	0.34 ± 0.01	0.36 ± 0.01
Control 2	0.37 ± 0.03	0.41 ± 0.01	0.52 ± 0.01	0.75 ± 0.01	0.67 ± 0.01	0.38 ± 0.01

Control 1—without microorganisms; control 2—with the original *A. niger* strain. Hydrolysis substrates: 1—filter paper (50 mg/L, according to the commercial enzyme dosage), 2—secondary cellulose-containing raw materials, 3—secondary starch-containing raw materials, 4—miscanthus cellulose after alkaline pretreatment, 5—miscanthus cellulose after acid pretreatment, 6—wood chip cellulose. Data are presented as median ± standard deviation (n = 3).

When comparing the cellulase activity of *A. niger* with non-model substrates, it was found that the maximum conversion of cellulose into reducing sugars was observed when using miscanthus cellulose, and the minimum when using alder chips (Table 2). Reducing substances were found in insignificant amounts in control 1 (without microorganisms), since reducing sugars were included in the culture media of *A. niger*.

The MALDI-TOF method was used to study the residues of oligosaccharides and lignin in the filtrate. The cellulose yield was 87% with alkaline treatment and 95% with the acid one. The resulting substrate was sterilized by autoclaving for 20 min at 121 °C. The results of studying oligosaccharides and lignin that were extracted in the process of producing cellulose (pulp) from the miscanthus biomass, are shown in Figure 2.

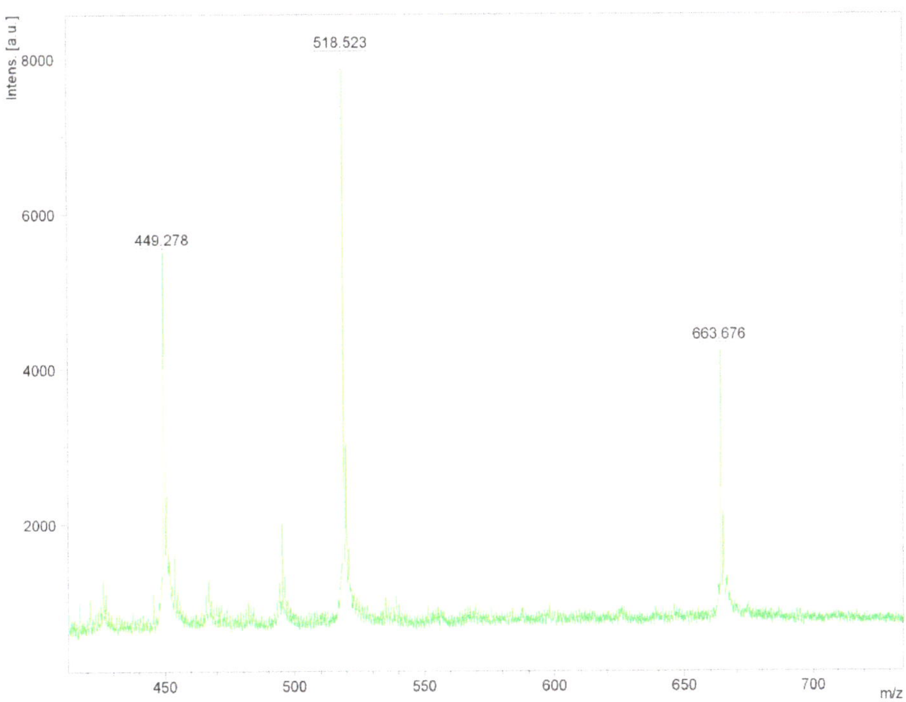

Figure 2. The study of the reaction mixture aqueous phase after delignification by MALDI-TOF spectrometry.

Next, we studied the aqueous phase of the reaction mixture after delignification by MALDI-TOF spectrometry. The analysis of the obtained MALDI-TOF mass spectrometry data resulted in a conclusion on the structure of lignin fragments extracted into an alkaline solution (Figure 3).

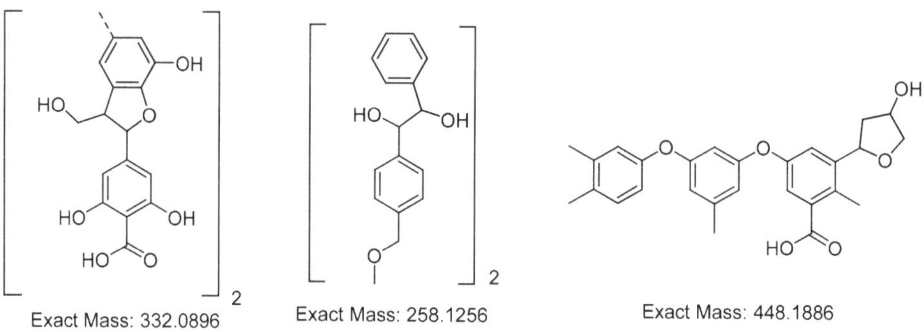

Figure 3. Structure of lignin fragments extracted during the cellulose (pulp) production from miscanthus biomass by hydrotropic delignification in the presence of sodium hydroxide.

The results of the studied oligosaccharides and lignin from the miscanthus biomass, by the method of hydrotropic delignification, presented in Figure 1, show that the signal with a mass of 663.676 *m/z* corresponds to dimeruphenylcumaronic acid, the signal with a mass of 518.523 *m/z* corresponds to the dimer 1-(4-methoxyphenyl)phenyl)-2-phenylethane-1,2-diol, and the signal with a mass of 449.278 *m/z* corresponds to a phenoxy-phenylcumaronic acid fragment.

3.2. *Analysis of the Enzyme Preparation Cellulolytic Activity*

In the second stage of the study, the cellulase activity of a commercial enzyme preparation was studied, in the concentration range of 0.1–10 mg/1.5 mL of citrate buffer, relative to a model cellulose substrate, i.e., filter paper (50 mg/L, according to the commercial enzyme dosage). The results are shown in Table 3. The obtained data are consistent with the dosages and units of the commercial enzyme for all types of cellulose-containing raw materials.

The enzyme preparation with a concentration of 10 mg/mL on the miscanthus cellulose substrate after alkaline pretreatment (Table 3) had the highest cellulase activity.

3.3. *Cellulolytic Activity of A. niger Mutant Strains*

UV radiation damages the overall microbial structure. Numerous kinds of mutations are caused by UV irradiation, as follows: base substitution (transitions and transversions) and reading frame shifts (deletions and insertions). Complex mutations also occurin DNA, when one of its sections is replaced by one of a different length and a different nucleotide composition. Mutations are formed unevenly along the DNA; therefore, chemical mutagenesis is additionally used. As a result of the induced mutagenesis of *A. niger* spores by UV irradiation and/or chemical mutagenesis, mutated strains were obtained, as shown in Figure 4.

A qualitative assessment of the cellulolytic activity of the UV-treated *A. niger* strain is shown in Figure 5.

A qualitative reaction using a Congo red solution (Figure 4) showed an increase in the cellulolytic activity of the treated *A. niger* samples, by increasing the light spot radius around the colony.

Table 3. Characterization of the cellulase activity (concentration of the reducing substance, mg/mL) of the enzyme preparation.

Enzyme Preparation Concentration, mg/mL	Substrates					
	1	2	3	4	5	6
			30 °C			
Control 2	0.24 ± 0.04^{a}	0.35 ± 0.01^{a}	0.34 ± 0.01^{a}	0.39 ± 0.01^{ab}	0.34 ± 0.01^{a}	0.38 ± 0.01^{a}
0.1	0.37 ± 0.01^{ab}	0.46 ± 0.01^{b}	0.61 ± 0.01^{b}	0.37 ± 0.01^{ab}	0.50 ± 0.01^{b}	0.39 ± 0.01^{a}
1	0.50 ± 0.01^{b}	0.52 ± 0.01^{b}	0.67 ± 0.01^{b}	0.50 ± 0.01^{b}	1.22 ± 0.01^{c}	0.73 ± 0.01^{b}
10	0.71 ± 0.01^{c}	0.70 ± 0.01^{c}	0.90 ± 0.01^{c}	0.71 ± 0.01^{c}	2.40 ± 0.01^{d}	1.27 ± 0.02^{c}
			37 °C			
Control 2	0.24 ± 0.04^{a}	0.35 ± 0.01^{a}	0.34 ± 0.01^{a}	0.39 ± 0.01^{a}	0.34 ± 0.01^{a}	0.38 ± 0.01^{a}
0.1	0.39 ± 0.01^{b}	0.55 ± 0.01^{b}	1.00 ± 0.01^{b}	0.54 ± 0.01^{b}	0.90 ± 0.01^{b}	0.43 ± 0.01^{a}
1	1.80 ± 0.01^{c}	1.11 ± 0.01^{c}	2.13 ± 0.01^{c}	2.09 ± 0.01^{c}	2.01 ± 0.01^{c}	0.75 ± 0.01^{b}
10	1.84 ± 0.01^{c}	2.34 ± 0.01^{d}	3.06 ± 0.01^{d}	7.42 ± 0.03^{d}	3.00 ± 0.01^{d}	3.42 ± 0.02^{c}
			50 °C			
Control 2	0.24 ± 0.04^{a}	0.35 ± 0.01^{a}	0.34 ± 0.01^{a}	0.39 ± 0.01^{a}	0.34 ± 0.03^{a}	0.38 ± 0.01^{a}
0.1	0.48 ± 0.01^{b}	0.88 ± 0.01^{b}	0.49 ± 0.01^{b}	0.83 ± 0.01^{b}	1.03 ± 0.01^{b}	0.47 ± 0.01^{a}
1	1.71 ± 0.01^{c}	2.31 ± 0.01^{c}	2.06 ± 0.01^{c}	3.15 ± 0.01^{c}	2.08 ± 0.01^{c}	1.32 ± 0.01^{b}
10	2.19 ± 0.01^{d}	4.59 ± 0.02^{d}	2.53 ± 0.01^{c}	5.14 ± 0.03^{d}	4.23 ± 0.02^{d}	3.76 ± 0.01^{c}

Control 2—with the original *A. niger* strain. Hydrolysis substrates: 1—filter paper, 2—secondary cellulose-containing raw materials, 3—secondary starch-containing raw materials, 4—miscanthus cellulose after alkaline pretreatment, 5—miscanthus cellulose after acid pretreatment, 6—wood chip cellulose. Data are presented as median ± standard deviation ($n = 3$). Values in a column followed by the same letter a, b, c and d do not differ significantly ($p > 0.05$) as assessed by the Kruskal–Wallis test.

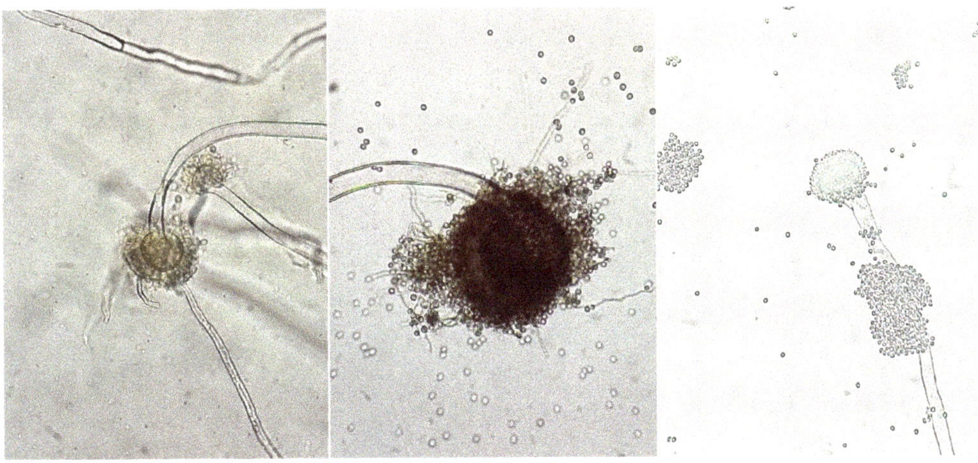

(**a**) (**b**) (**c**)

Figure 4. Morphology of *A. niger* conidiophores: (**a**)—original; (**b**)—UV-treated (2ч); (**c**)—chemically treated, (magnification 400×).

Quantitative assessment of the enzymatic activity of cellulase that was obtained using treated *A. niger*, on Mandels–Weber medium (M-W) and potato agar (PA) media, is shown in Table 4. These media were chosen because they contain the main nutrient components for the microbiological production of cellulases, which are as follows: $(NH_4)_2SO_4$, KH_2PO_4, $CaCl_2 \cdot 2H_2O$, $MgSO_4 \cdot 7H_2O$, $FeSO_4 \cdot 7H_2O$, $MnSO_4 \cdot H_2O$, $ZnSO_4 \cdot 7H_2O$, $CoC_{12} \cdot 6H_2O$, peptone, carbohydrates, and other nutrients. When cultivated on these media, *A. niger* and reference *T. reesei* show the greatest growth [37].

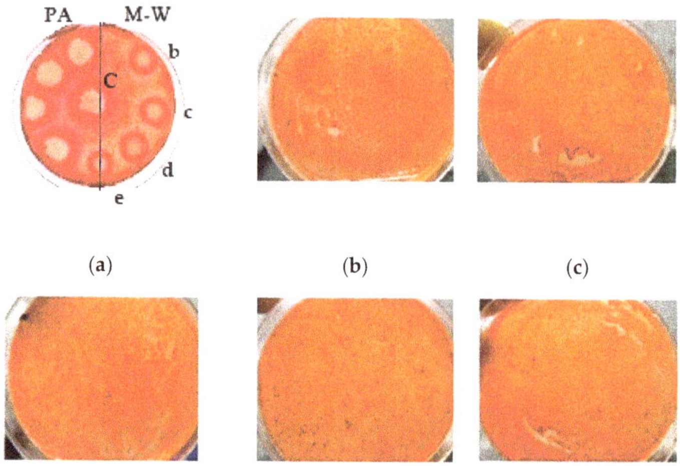

Figure 5. Qualitative assessment of the cellulolytic activity of the UV-treated *A. niger* strain: (**a**) all strains on one plate (left—mutants grown on potato agar (PA); right—grown on Mandels–Weber (M-W) medium; in the center control (C)—original strain; from top to bottom there are strains with different modes of exposure); (**b**) strain obtained by UV irradiation for 45 min; (**c**) strain obtained by UV irradiation for 1 h; (**d**) strain obtained by intermittent UV irradiation for 1 h, (**e**) a strain obtained by UV irradiation for 2 h; (**f**) strain obtained by UV irradiation for 2 h with additional exposure (NaN3 + ItBr).

Table 4. Quantitative assessment (concentration of the reducing substance, mg/mL) of the enzymatic activity cellulase obtained using treated *A. niger*.

Samples	Substrates			
	M-W		PA	
	1	2	1	3
Control 1	0.04 ±0.03 [a]	0.10 ±0.03 [a]	0.04 ± 0.01 [a]	0.06 ± 0.02 [a]
Control 2	0.23 ± 0.01 [b]	0.38 ± 0.03 [b]	0.23 ± 0.01 [b]	0.36 ± 0.08 [b]
UV45	0.64 ± 0.01 [c]	1.45 ± 0.05 [c]	0.70 ± 0.02 [c]	1.61 ± 0.03 [c]
UV60	0.69 ± 0.02 [c]	1.84 ± 0.01 [cd]	0.68 ± 0.06 [c]	15.63 ± 4.41 [d]
UV60-I	0.82 ± 0.09 [c]	1.46 ± 0.01 [c]	0.73 ± 0.02 [c]	1.68 ± 0.21 [c]
UV120	0.36 ± 0.02 [b]	2.41 ± 0.81 [d]	0.36 ± 0.02 [b]	16.99 ± 3.86 [d]
UV120 + CH30	1.78 ± 0.14 [d]	2.41 ± 0.01 [d]	1.10 ± 0.01 [d]	3.76 ± 0.01 [e]
UV120 + CH80	0.68 ± 0.01 [c]	2.10 ± 0.01 [d]	0.69 ± 0.01 [c]	2.05 ± 0.01 [be]
MIX	1.24 ± 0.01 [d]	6.44 ± 0.01 [e]	0.76 ± 0.04 [c]	16.04 ± 0.02 [d]

Control 1—without microorganisms, control 2—original *A. niger* strain, UV45—mutant *A. niger* strain obtained by UV irradiation for 45 min; UV60—mutant *A. niger* strain obtained by UV irradiation for 60 min; UV60-I—mutant *A. niger* strain obtained by interval UV irradiation for 60 min; UV120—mutant *A. niger* strain obtained by UV irradiation for 120 min; UV120 + CH30—mutant *A. niger* strain obtained by UV irradiation for 120 min followed by chemical treatment with NaN_3 + ItBr for 30 min; UV120 + CH80—mutant *A. niger* strain obtained by UV irradiation for 120 min followed by chemical treatment with NaN_3 + ItBr for 80 min; MIX—a mixture of all *A. niger* strains (original, UV, UV + CH). Hydrolysis substrates: 1—filter paper, 2—secondary cellulose-containing raw materials, 3—secondary starch-containing raw materials; M-W—Mandels–Weber medium, PA—potato agar. Data are presented as median ± standard deviation (n = 3). Values in a column followed by the same letter a, b, c, d and e do not differ significantly ($p > 0.05$) as assessed by the Kruskal–Wallis test.

Quantification of the enzymatic activity of the treated strains also showed a positive result (Table 4). The highest enzymatic activity was observed for the MIX sample on PA medium, and it was 16.04 ± 0.02 mg/mL.

3.4. Optimization of Conditions for Enzymatic Hydrolysis

The numerical values of the optimization of the temperature of enzymatic hydrolysis are shown in Table 5. The optimal growth range for *A. niger* was 27–30 °C.

Table 5. Optimization of the temperature factor during enzymatic hydrolysis.

Substrates	T, °C	Reducing Substance, mg/mL			
		Control 1	Control 2	UV120	UV120 + CH30
1	30	0.24 ± 0.04 $^{a/a}$	0.36 ± 0.01 $^{a/a}$	0.39 ± 0.01 $^{a/a}$	0.57 ± 0.01 $^{a/b}$
	37	0.24 ± 0.04 $^{a/a}$	0.37 ± 0.03 $^{a/a}$	0.36 ± 0.02 $^{a/a}$	1.78 ± 0.14 $^{b/b}$
	50	0.24 ± 0.05 $^{a/a}$	0.36 ± 0.06 $^{a/a}$	0.55 ± 0.07 $^{b/b}$	1.16 ± 0.05 $^{c/c}$
2	30	0.35 ± 0.01 $^{a/a}$	0.41 ± 0.01 $^{a/a}$	1.56 ± 0.01 $^{a/b}$	1.92 ± 0.01 $^{a/c}$
	37	0.39 ± 0.03 $^{a/a}$	0.41 ± 0.01 $^{a/a}$	2.41 ± 0.01 $^{b/b}$	2.41 ± 0.01 $^{b/b}$
	50	0.39 ± 0.04 $^{a/a}$	0.43 ± 0.02 $^{a/a}$	2.06 ± 0.52 $^{b/b}$	2.50 ± 0.86 $^{b/c}$
3	30	0.34 ± 0.01 $^{a/a}$	0.37 ± 0.01 $^{a/a}$	10.94 ± 0.11 $^{a/b}$	2.05 ± 0.01 $^{a/c}$
	37	0.37 ± 0.09 $^{a/a}$	0.52 ± 0.01 $^{a/b}$	16.99 ± 3.86 $^{b/c}$	3.76 ± 1.01 $^{b/d}$
	50	0.63 ± 0.02 $^{b/a}$	0.66 ± 0.01 $^{b/a}$	12.34 ± 1.37 $^{c/b}$	2.78 ± 0.52 $^{a/c}$
4	30	0.39 ± 0.01 $^{a/a}$	0.39 ± 0.01 $^{a/a}$	0.87 ± 0.01 $^{a/b}$	1.05 ± 0.01 $^{a/b}$
	37	0.37 ± 0.01 $^{a/a}$	0.75 ± 0.01 $^{b/b}$	1.68 ± 0.41 $^{b/c}$	1.07 ± 0.01 $^{a/b}$
	50	0.36 ± 0.02 $^{a/a}$	1.05 ± 0.09 $^{c/b}$	4.27 ± 0.56 $^{c/c}$	1.08 ± 0.35 $^{a/b}$
5	30	0.34 ± 0.01 $^{a/a}$	0.39 ± 0.01 $^{a/ab}$	0.54 ± 0.01 $^{a/b}$	2.15 ± 0.01 $^{a/c}$
	37	0.34 ± 0.01 $^{a/a}$	0.67 ± 0.01 $^{b/b}$	0.61 ± 0.01 $^{a/b}$	4.83 ± 0.01 $^{b/c}$
	50	0.85 ± 0.03 $^{b/a}$	0.77 ± 0.05 $^{b/b}$	2.64 ± 0.34 $^{b/c}$	1.72 ± 0.88 $^{c/d}$
6	30	0.34 ± 0.01 $^{a/a}$	0.39 ± 0.01 $^{a/a}$	0.36 ± 0.01 $^{a/a}$	1.01 ± 0.01 $^{a/b}$
	37	0.36 ± 0.01 $^{a/a}$	0.38 ± 0.01 $^{a/a}$	0.38 ± 0.01 $^{a/a}$	1.31 ± 0.01 $^{a/b}$
	50	0.36 ± 0.01 $^{a/a}$	0.36 ± 0.03 $^{a/a}$	1.06 ± 0.02 $^{b/b}$	1.09 ± 0.21 $^{a/b}$

Control 1—without microorganisms, control 2—original *A. niger* strain, UV120—mutant *A. niger* strain obtained by UV irradiation for 120 min; UV120 + CH30—mutant *A. niger* strain obtained by UV irradiation for 120 min followed by chemical treatment with NaN3 + ItBr for 30 min. Hydrolysis substrates: 1—filter paper, 2—secondary cellulose-containing raw materials, 3—secondary starch-containing raw materials, 4—miscanthus cellulose after alkaline pretreatment, 5—miscanthus cellulose after acid pretreatment, 6—wood chip cellulose. Data are presented as median ± standard deviation ($n = 3$). Values in a column/row followed by the same letter a, b, c and d do not differ significantly ($p > 0.05$) as assessed by the Kruskal–Wallis test.

The recommended temperature for enzymatic hydrolysis with UV120 + CH30, using filter paper, secondary starch-containing raw materials, miscanthus cellulose after acid pretreatment, and wood chip cellulose, was 37 °C (Table 5).

The amount of protein produced by *A. niger*, determined by the Bradford method, is shown in Table 6.

Table 6. The amount of protein produced by *A. niger* according to the Bradford method.

Substrates	Protein Amount, mg/mL			
	Control 2	Control 3	UV120	UV120 + CH30
1	1008 ± 20 ^{a}	1152 ± 30 ^{b}	1075 ± 20 ^{a}	1196 ± 30 ^{b}
2	1004 ± 20 ^{a}	1049 ± 20 ^{ab}	1063 ± 20 ^{b}	1064 ± 20 ^{b}
3	1022 ± 20 ^{a}	1109 ± 30 ^{a}	1250 ± 30 ^{b}	1150 ± 30 ^{ac}
4	1060 ± 20 ^{a}	1200 ± 30 ^{b}	1180 ± 30 ^{b}	1185 ± 30 ^{b}
5	1039 ± 20 ^{a}	1104 ± 20 ^{ab}	1004 ± 20 ^{a}	1199 ± 30 ^{b}
6	1007 ± 20 ^{a}	1109 ± 20 ^{b}	1004 ± 20 ^{a}	1083 ± 20 ^{ab}

Control 2—original *A. niger* strain, control 3—enzyme preparation, UV120—mutant *A. niger* strain obtained by UV irradiation for 120 min; UV120 + CH30—mutant *A. niger* strain obtained by UV irradiation for 120 min followed by chemical treatment with NaN_3 + ItBr for 30 min. Hydrolysis substrates: 1—filter paper, 2—secondary cellulose-containing raw materials, 3—secondary starch-containing raw materials, 4—miscanthus cellulose after alkaline pretreatment, 5—miscanthus cellulose after acid pretreatment, 6—wood chip cellulose. Data are presented as median ($n = 3$). Values in a row followed by the same letter a, b and c do not differ significantly ($p > 0.05$) as assessed by the Kruskal–Wallis test.

The greatest amount of proteins (Table 5) is produced by strain UV120, on a substrate containing secondary starch-containing raw materials.

According to Formula (1), the activity of the cellulolytic enzymes of *A. niger*, original and subjected to undirected mutagenesis (in addition to a commercial preparation for comparison), during hydrolysis of the considered substrates, was calculated (Table 7).

Table 7. Comparison of the enzymatic activity of microorganisms.

Substrates	T, °C	Activitiesper Unit Time, IU			
		Control 2	Control 3	UV120	UV120 + CH30
1	30	0.7 ^{a}	2.7 ^{b}	0.8 ^{a}	1.8 ^{d}
	37	0.7 ^{a}	9.1 ^{b}	0.7 ^{a}	8.6 ^{b}
	50	0.7 ^{a}	10.8 ^{b}	1.7 ^{a}	5.1 ^{d}
2	30	0.1 ^{a}	2.3 ^{b}	6.7 ^{c}	8.7 ^{cd}
	37	0.1 ^{a}	11.1 ^{b}	11.2 ^{b}	11.2 ^{b}
	50	0.2 ^{a}	24.2 ^{b}	9.3 ^{c}	11.7 ^{d}
3	30	0.2 ^{a}	2.8 ^{b}	58.9 ^{c}	9.5 ^{d}
	37	0.8 ^{a}	15.2 ^{b}	92.3 ^{c}	18.8 ^{b}
	50	0.2 ^{a}	12.1 ^{b}	65.1 ^{c}	11.9 ^{b}
4	30	0.1 ^{a}	2.3 ^{b}	2.7 ^{b}	3.7 ^{d}
	37	2.1 ^{a}	38.8 ^{b}	7.3 ^{c}	3.9 ^{d}
	50	3.8 ^{a}	26.2 ^{b}	21.7 ^{b}	4.0 ^{a}
5	30	0.3 ^{a}	11.1 ^{b}	1.1 ^{c}	10.1 ^{b}
	37	1.8 ^{a}	15.3 ^{b}	1.5 ^{a}	24.9 ^{d}
	50	0.0 ^{a}	18.7 ^{b}	9.9 ^{c}	4.8 ^{d}
6	30	0.3 ^{a}	5.2 ^{b}	0.1 ^{a}	3.7 ^{d}
	37	0.1 ^{a}	17.1 ^{b}	0.1 ^{a}	5.3 ^{d}
	50	0.0 ^{a}	24.2 ^{b}	3.9 ^{c}	4.1 ^{cd}

Control 2—original *A. niger* strain, control 3—enzyme preparation, UV120—mutant *A. niger* strain obtained by UV irradiation for 120 min; UV120 + CH30—mutant *A. niger* strain obtained by UV irradiation for 120 min followed by chemical treatment with NaN_3 + ItBr for 30 min. Hydrolysis substrates: 1—filter paper, 2—secondary cellulose-containing raw materials, 3—secondary starch-containing raw materials, 4—miscanthus cellulose after alkaline pretreatment, 5—miscanthus cellulose after acid pretreatment, 6—wood chip cellulose. Data are presented as median (n = 3). Values in a row followed by the same letter a, b, c and d do not differ significantly ($p > 0.05$) as assessed by the Kruskal–Wallis test.

The most active strain was UV120 when secondary starch-containing raw materials were used as a substrate (Table 7).

The Kruskal–Wallis test showed statistically significant differences ($p < 0.05$) when comparing the cellulolytic activity of four groups (control 2, control 3, UV120, strain UV120 + CH30), relative to the substrates 1, 2, 3, 4, and 5, under all three temperatures (30, 37, and 50 °C). When using the substrate six, significant differences between the groups were found at temperatures of 30 and 37 °C. In all the cases, differences were found with a 1 h hydrolysis. The influence of the composition of the nutrient medium on which the culture of the microorganism (UV) was incubated, on the cellulolytic activity during hydrolysis, was assessed (Figure 6a). To assess the contribution of the pH level during the hydrolysis process, a substrate was used that showed the highest sensitivity to the action of cellulases of the studied microorganism. The UV-treated strain (UV120) was used as an enzyme source, as the most effective one (Figure 6b).

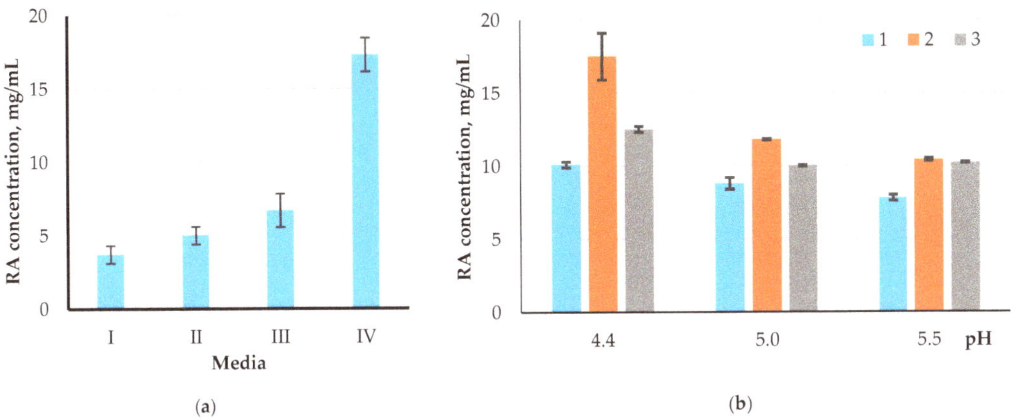

Figure 6. Comparison of the enzymatic activity of UV120 strain when cultivated on different nutrient media (**a**) and different pH values of the medium during hydrolysis (**b**): I—MPA, II—C-D, III—M-W, IV—PA; 1—30 °C, 2—37 °C, 3—50 °C.

To summarize the data, a comparative histogram is presented, which clearly shows the level of total cellulolytic activity of all the studied strains relative to different types of substrate (Figure 7).

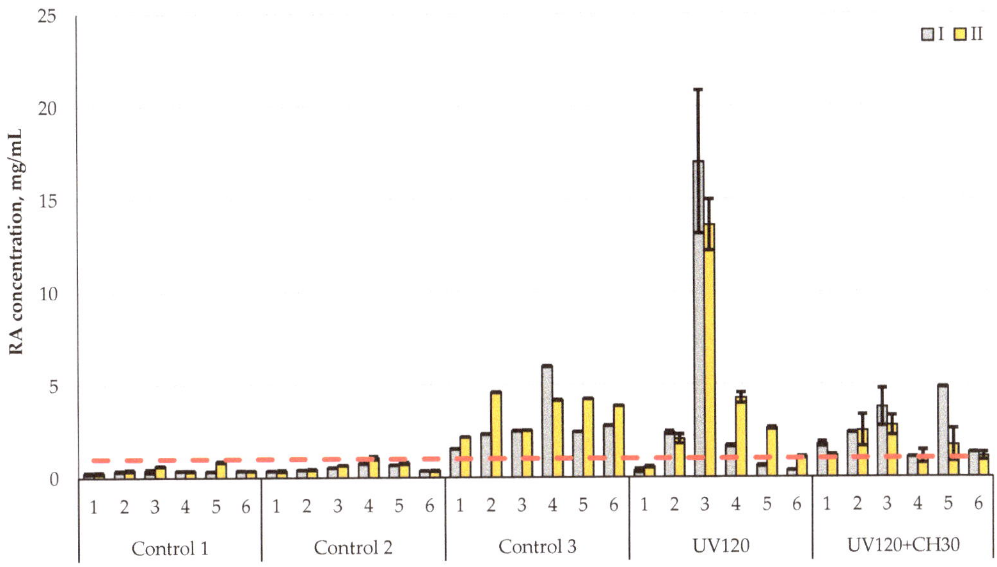

Figure 7. Comparison of cellulolytic activity of original (control 2) and treated *A. niger* (UV120 and UV120 + CH30) with a commercial enzyme preparation (control 3) at the following different process temperatures: I—37 °C; II—50 °C. Control 1—without microorganisms, control 2—original *A. niger* strain, control 3—enzyme preparation, UV120—mutant *A. niger* strain obtained by UV irradiation for 120 min; UV120 + CH30—mutant *A. niger* strain obtained by UV irradiation for 120 min followed by chemical treatment with NaN_3 + ItBr for 30 min. Hydrolysis substrates: 1—filter paper, 2—secondary cellulose-containing raw materials, 3—secondary starch-containing raw materials, 4—miscanthus cellulose after alkaline pretreatment, 5—miscanthus cellulose after acid pretreatment, 6—wood chip cellulose. The red dotted line is the maximum level of enzymatic activity of the original *A. niger* strain. Data are presented as median ± standard deviation (n = 3).

The greatest reducing substanceis evidenced in starch-containing raw materials, which indicates that *A. niger* can ferment not only through cellulase activity, but also via amylolytic activity [16].

4. Discussion

In the first stage of this study, the cellulolytic activity of the original *A. niger* strain was assessed in relation to six insoluble substrates, one of which being the model (filter paper). It was found that citrate buffer, with a concentration of 50 mM, was more suitable for enzymatic hydrolysis, and the advantage was noticeable after 48 h of hydrolysis (Table 1). The hydrolysis conditions were as follows: stationary, 30 °C, and pH = 4.7. The recommended substrate (filter paper) concentration for hydrolysis is 50 mg/L. With a further increase in the concentration of the substrate, the formation of the reducing substance decreased. The relatively low degree of hydrolysis (Table 1) is explained by the endothermicity of this process. The degree of hydrolysis increased concomitantly with the increasein temperature, and the equilibrium shiftedtowards the hydrolysis products formation, as confirmed by the data in Table 3.

When using acetate buffer as a medium for the hydrolysis process, the yield of the reducing substance was slightly higher in comparison with that of the citrate buffer. However, when a suspension of microorganisms was added to the acetate buffer, the yield of the reducing substance dropped. It is likely that the acetate buffer inhibited the functioning of the microorganisms. In addition, it was noted [32] that the acetate buffer is able to suppress the enzymatic activity of microorganisms, which can adversely affect further use of the hydrolysate under alcoholic fermentation.

When comparing the cellulase activity of *A. niger* with non-model substrates (Table 2), it was found that the maximum conversion of cellulose into reducing sugars was observed when using miscanthus cellulose (with alkaline pretreatment), and the minimum one was observed using wood chips. The hydrolysis conditions were as follows: stationary, 37 °C, and pH = 4.7.

As a result of the first stage, it was revealed that the studied original strain exhibited stable, but low, cellulase activity. Our results agree with the data that are available in the literature. Thus, in foreign studies in this area, it is customary to use either directed genetically treated microorganisms or commercial enzyme preparations. In Russian studies, it is customary to use commercial enzyme preparations. Within the framework of this study, a decision was made to activate cellulase activity by physicochemical action, and also to optimize the enzymatic hydrolysis. The expected results were obtained, which confirms the adequacy of the functioning of a commercial enzyme preparation; an increase in cellulolytic activity, with an increase in the dose of the preparation, and with an increase in the temperature of the hydrolysis. Our results allow for the correlation of the level of enzymatic activity in the studied microorganisms, with cultivation conditions. The hydrolysis conditions were as follows: stationary, 37 °C, and pH = 4.7. The filter paper was hydrolyzed according to a commercial enzyme dosage of 50 mg/L. The obtained data are consistent with the dosages and units of the commercial enzyme for all types of cellulose-containing raw materials.

A qualitative reaction, using a Congo red solution (Figure 3), showed an increase in the cellulolytic activity of the treated *A. niger* samples, by increasing the radius of the light spot around the colony (the dye reacts with glucose obtained from the cleavage of cellulose contained in the medium). A variant of the medium without the addition of simple carbohydrates was used for verification purposes.

A quantitative assessment of the enzymatic activity of the mutant strains also showed a positive result (Table 4). The obtained data are consistent with the dosages and units of the commercial enzyme for all types of cellulose-containing raw materials.

After mutagenic action, the spore suspension was separated and partly cultivated on a nutrient medium containing cellulose, and partly on a medium containing potato waste. The total cellulase activity of the first passage was maximal in the cultures that were exposed to UV treatment for two hours; as the UV exposure time increased to two hours, the cellulase activity of the surviving part of the culture increased ($r = 0.706$; $p < 0.05$).

The studied temperature range was limited from 30 to 50 °C. With respect to secondary raw materials, enzymatic hydrolysis using treated strains was more efficient at 37 °C, whereas miscanthus cellulose and wood chips were more efficiently hydrolyzed at 50 °C. The obtained data are consistent with the literature [26,34,38–44]. In these studies, the highest enzymatic activity of the relative secondary raw materials was found in the UV strain (92.3 IU/h or 1.5 IU/min).

Comparatively, similar data were obtained in [26], in which the cellulase activity of *A. niger* during the hydrolysis of fruit wastes was 64.5 IU. Zao et al. [43] showed an increase in enzymatic activity with the combined use of recombinant *A. niger* and *T. reesei*, up to 12 IU. In a study by Abdullah et al. [44], the cellulase activity of UV-treated *Bacillus* sp. increased to 11 IU compared to the control (7.5 IU).

The enzymatic activity of the strains that were obtained in this work, relative to the primary raw material (miscanthus cellulose and wood chips), and filter paper in some cases, was inferior to that of the commercial enzyme preparation (8.6 versus 9.1 IU in the hydrolysis of filter paper, for example). However, it was possible to increase the total cellulase activity in the case of joint use.

During the fermentation of miscanthus cellulose (Table 8) in an acidic medium, *A. niger* UV120 + CH30, obtained by us, exhibited a greater enzymatic activity than the microorganisms that have been presented in other studies [43,44]. However, during the hydrolysis of fruit waste and wood chips, it was possible to achieve a greater enzymatic activity of *A. niger*.

Table 8. Cellulose fermentation indicators.

Hydrolysis Temperature, °C	Hydrolysis pH	Microorganisms	Secondary Raw Materials	Enzymatic Activities, IU	Sources
37	4.7	*A. niger* UV120 + CH30	Miscanthus cellulose	24.9	Our results
40	4.0	*A. niger*	Fruit waste	64.5	[26]
40	5.1	*A. niger* and *T. reesei*	Miscanthus cellulose	12.0	[43]
35	4.8	*Bacillus* sp.	Miscanthus cellulose	11.0	[44]
50	4.7	*A. niger*	Wood chips	92.3	[34,38–44]

The results of the enzymatic activity of the samples vary significantly (Table 8). However, they prove the possibility of obtaining an enzyme that breaks down cellulose of various origins, and the very fact of this enzymatic cleavage. The variation in the parameters is explained by the difference in the microorganisms producing the enzyme, and the conditions of enzymatic hydrolysis (different temperature, time, pH of the process) [26,34,38–44].

The results of the study favor the use of the PA and M-W media, in addition to a more acidic medium (pH = 4.4). A higher enzymatic activity of cellulases inthe mutated *A. niger*, compared to the original strain, was shown. When using the *A. niger* UV strain, the highest yield of the reducing substance was observed when using secondary starch-containing raw materials, at both 37 and 50 °C. It is more expedient to carry out enzymatic hydrolysis of the wood chips and cellulose of miscanthus using a UV strain at a temperature of 50 °C.

In this case, miscanthus cellulose after alkaline pretreatment was more exposed to degradation. The CH strain exhibited a cellulolytic activity with less deviation from the mean relative to the type of substrate used. The activity of this strain was not statistically significant relative to the hydrolysis temperature, except for the case in which the substrate was miscanthus cellulose after acid pretreatment.

The model substrate (filter paper) and wood chips were the least susceptible to enzymatic degradation under the influence of the cellulolytic complex, by both UV and CH strains of *A. niger*.

5. Conclusions

A number of stages were undertaken in this research, as follows: the pretreatment of various cellulose-containing raw materials was carried out, the cellulolytic activity of microorganisms was studied, physicochemical undirected mutagenesis of microorganisms was conducted to intensify the enzymatic activity, and the optimal conditions for the enzymatic hydrolysis of filter paper, secondary cellulose-containing raw materials, secondary starch-containing raw materials, miscanthus cellulose after alkaline pretreatment, miscanthus cellulose after acid pretreatment, and wood chip cellulose, were selected. As a result, the treated strains of *Aspergillus niger*, with an increased ability to break down cellulose and starch, were obtained. The analysis of the hydrolysis parameters was carried out to identify the recommended conditions, as follows: hydromodule 1:20, temperature 37 °C. To obtain a higher yield of RA for the cultivation of microorganisms, a Mandels–Weber salt medium should be used for the further hydrolysis of the cellulose-containing substrate, and a PA medium for the hydrolysis of starch-containing raw materials. A quantity of 0.05 M citrate buffer with pH 4.4 is more suitable as a medium for enzymatic hydrolysis.

Our future research will focus on the intensification of the process of enzymatic hydrolysate production for subsequent alcoholic fermentation. However, there are numerous potential applications of fermented cellulose. Some studies [45,46] showed that fermented celluloses are a vivid example of natural sources of innovative technologies for use in materials science and biotechnology. A variation in the applications exists for nanocellulose (from mass-produced industrial goods to high-quality medical products).

Research teams are actively working to globally expand the range of nanocellulose products, and find entirely new areas of application [8,46–48]. It has been established that microbial cellulose is an extremely versatile biomaterial that can be used in a wide variety of applied scientific purposes, such as paper products, electronics, acoustics, and biomedical devices. Biomedical devices have recently received a large amount of attention due to an increased interest in tissue products for wound care, and the regeneration of damaged or diseased organs. Due to its unique nanostructure and properties, microbial cellulose is a natural candidate for numerous applications in medicine and tissue engineering. The microbial cellulose membrane can be successfully used as a wound-healing device for severely damaged skin, and for replacing small-diameter blood vessels. Microbial cellulose microfibril nonwoven ribbons strongly resemble the structure of native extracellular matrices, suggesting that they can function as a scaffold for the production of many tissue-engineered structures. In addition, microbial cellulose membranes with a unique nanostructure have many other uses in wound healing and regenerative medicine, for example, in guided tissue regeneration, the treatment of periodontitis, or as a replacement for the dura mater (the membrane that surrounds brain tissue). Microbial cellulose can function as a scaffold for the regeneration of a wide variety of tissues, showing that it could ultimately become an excellent technological platform for medicine.

Author Contributions: Conceiving and designing the research—O.B. and O.K.; analyzing and interpreting the data—S.S., S.I. and V.D.; contributing reagents, materials, analysis tools or data—E.B. and V.D.; writing review and editing—V.D., S.I., P.M., O.B. and O.K. All authors have read and agreed to the published version of the manuscript.

Funding: This research was funded by the RUSSIAN FOUNDATION FOR BASIC RESEARCH, grant number 19-416-390001.

Institutional Review Board Statement: Not applicable.

Informed Consent Statement: Not applicable.

Data Availability Statement: Data are contained within the article.

Conflicts of Interest: The authors declare no conflict of interest.

References

1. Djelal, H.; Chniti, S. Identification of strain isolated from dates (*Phœnix dactylifera* L.) for enhancing very high gravity ethanol production. *ESPR* **2017**, *24*, 9886–9894. [CrossRef] [PubMed]
2. Neelakandan, T.; Usharani, G. Optimisation and production of bioethanol from Cashew apple juice using yeast cells by Saccharomyces cerevisiae. *AEJSR* **2009**, *4*, 85–88.
3. Verardi, A.; Lopresto, C.G. Bioconversion of lignocellulosic biomass to bioethanol and biobutanol. In *Lignocellulosic Biomass to Liquid Biofuels*; Yousuf, A., Sannino, F., Eds.; Academic Press, Elsevier Ltd.: Cambridge, UK, 2020; pp. 67–125.
4. Pavlov, I.N.; Denisova, M.N.; Makarova, E.I.; Budaeva, V.V.; Sakovich, G.V. Versatile thermobaric setup and production of hydrotropic cellulose therein. *Cellul. Chem. Technol.* **2015**, *49*, 847–852.
5. Siqueira, J.G.W.; Rodrigues, C. Current advances in on-site cellulase production and application on lignocellulosic biomass conversion to biofuels: A review. *Biomass Bioenergy* **2020**, *132*, 105419. [CrossRef]
6. Zucaro, A.; Fiorentino, G. Constraints, impacts and benefits of lignocellulose conversion pathways to liquid biofuels and biochemical. In *Lignocellulosic Biomass to Liquid Biofuels*; Yousuf, A., Sannino, F., Eds.; Academic Press, Elsevier Ltd.: Cambridge, UK, 2020; pp. 249–282.
7. Santos, F.; Eichler, P. Production of second-generation ethanol from sugarcane. In *Sugarcane Biorefinery, Technology and Perspectives*, 1st ed.; Santos, F., Rabelo, S.C., Eds.; Academic Press, Elsevier Ltd.: Cambridge, UK, 2020; pp. 195–228.
8. Klemm, D.; Heublein, B.; Fink, H.-P.; Bohn, A. Cellulose: Fascinating biopolymer and sustainable raw material. *J. Ger. Chem. Soc.* **2005**, *44*, 3358–3393.
9. Sanchez, O.J.; Cardona, C.A. Trends in biotechnological production of fuel ethanol from different feedstocks. *Bioresour. Technol.* **2008**, *99*, 5270–5295. [CrossRef] [PubMed]
10. Sharma, P.R.; Joshi, R.; Sharma, S.K.; Hsiao, B.S. A simple approach to prepare carboxycellulose nanofibers from untreated biomass. *Biomacromolecules* **2017**, *18*, 2333–2342. [CrossRef]
11. Hafid, H.S.; Rahman, N.A.A. Feasibility of using kitchen waste as future substrate for bioethanol production: A review. *Renew. Sustain. Energy Rev.* **2017**, *74*, 671–686. [CrossRef]
12. Thompson, C.E.; Beys-da-Silva, W.O.; Santi, L.; Berger, M.; Vainstein, M.H.; Guimarães, J.A.; Vasconcelos, A.T.R. A potential source for cellulolytic enzyme discovery and environmental aspects revealed through metagenomics of Brazilian mangroves. *AMB Expr.* **2013**, *3*, 65. [CrossRef]
13. Sakthi, S.S.; Kanchana, D.; Saranraj, P.; Usharani, G. Evaluation of amylase activity of the amylolytic fungi Aspergillus niger using Cassava as substrate. *Int. J. Appl. Microbiol. Sci.* **2012**, *1*, 24–34.
14. Soares, J.F.; Confortin, T.C. Dark fermentative biohydrogen production from lignocellulosic biomass: Technological challenges and future prospects. *Renew. Sustain Energy Rev.* **2020**, *117*, 109484. [CrossRef]
15. Bellaouchi, R.; Abouloifa, H.; Rokni, Y.; Hasnaoui, A.; Ghabbour, N.; Hakkou, A.; Bechchari, A.; Asehraou, A. Characterization and optimization of extracellular enzymes production by Aspergillus niger strains isolated from date by-products. *JGEB* **2021**, *19*, 50. [CrossRef]
16. Ramasamy, S.; Benazir, J.F.; Ramalingam, S.; Kumar, R.; Hari, A.; Raman, N.; Nidhiya, K.A.; Kavitha, G.; Lakshmi, R. Amylase production by Aspergillus Niger under solid state fermentation using agroindustrial wastes. *IJEST* **2011**, *3*, 1756–1763.
17. Kriger, O.; Budenkova, E. The process of producing bioethanol from delignified cellulose isolated from plants of the Miscanthus genus. *Bioengineering* **2020**, *7*, 61. [CrossRef] [PubMed]
18. Cyrus, E.T.; Juwon, A.D. effects of radiation and chemical mutagenesis on expression of aflatoxigenic traits in Aspergillus parasiticus SMS08-C. *Res. J. Microbiol.* **2015**, *10*, 205. [CrossRef]
19. Janusz, G.; Pawlik, A.; Sulej, J.; Burek, U.Ś.; Wilkołazka, A.J.; Paszczyński, A. Lignin degradation: Microorganisms, enzymes involved, genomes analysis and evolution. *FEMS Microbiol. Rev.* **2017**, *41*, 941–962. [CrossRef]
20. Javed, U.; Aman, A. Utilization of corncob xylan as a sole carbon source for the biosynthesis of endo-1, 4-β xylanase from Aspergillus niger KIBGE-IB36. *Bioresour. Bioprocess.* **2017**, *4*, 1–7. [CrossRef]
21. Lima, M.A.; Oliveira-Neto, M.; Kadowaki, M.A.; Rosseto, F.R.; Prates, E.T.; Squina, F.M.; Leme, A.F.; Skaf, M.S.; Polikarpov, I. Aspergillus niger β-glucosidase has a cellulase-like tadpole molecular shape: Insights into glycoside hydrolase family 3 (GH3) β-glucosidase structure and function. *J. Biol. Chem.* **2013**, *288*, 32991–33005.
22. Reddy, G.P.K.; Narasimha, G. Cellulase production by Aspergillus niger on different natural lignocellulosic substrates. *Int. J. Curr. Microbiol. Appl. Sci.* **2015**, *4*, 835–845.
23. Bušić, A.; Marđetko, N.; Kundas, S.; Morzak, G.; Belskaya, H.; Ivančić Šantek, M.; Komes, D.; Novak, S.; Šantek, B. Bioethanol production from renewable raw materials and its separation and purification: A review. *Food Technol. Biotechnol.* **2018**, *56*, 289–311. [CrossRef] [PubMed]
24. Choudhury, A.R.K. Easy-care finishing. In *Textiles, Principles of Textile Finishing*; Choudhury, A.K.R., Ed.; Woodhead Publishing, Elsevier Ltd.: Cambridge, UK, 2017; pp. 245–284. [CrossRef]

25. Il'ina, A.; Kulikova, O.; Maltsev, D.; Krasnov, M.; Rybakova, E.; Skripnikova, V.; Kuznetsova, E.; Buriak, A.; Iamskova, V.; Iamskov, I. MALDI-TOF mass spectrometric identification of novel intercellular space peptides. *Appl. Biochem. Microbiol.* **2011**, *47*, 118–122. [CrossRef]
26. Kumar, A.K.; Parikh, B.S. Use of combined UV and chemical mutagenesis treatment of Aspergillus terreus D34 for hyper-production of cellulose-degrading enzymes and enzymatic hydrolysis of mild-alkali pretreated rice straw. *Bioresour. Bioprocess.* **2015**, *2*, 35. [CrossRef]
27. Meena, S.S.; Vinay, S. In vitro optimisation of fungal cellulase production from fruit waste for handmade paper industries. *Biotechnology* **2018**, *17*, 35–43. [CrossRef]
28. Chandra, M.S.; Viswanath, B. Cellulolytic enzymes on lignocellulosic substrates in solid-state fermentation by *Aspergillus niger*. *Indian J. Microbiol.* **2007**, *47*, 323–328. [CrossRef]
29. Edor, S.P.; Edogbanya, O.P. Cellulase activity of Aspergillus niger in the biodegradation of rice husk. *MOJBM* **2018**, *3*, 49–51. [CrossRef]
30. Zapata, Y.; Galviz-Quezada, A. Cellulases production on paper and sawdust using native Trichoderma asperellum. *Univ. Sci.* **2018**, *23*, 419–436. [CrossRef]
31. Mæhre, H.K.; Dalheim, L.; Edvinsen, G.; Elvevoll, E.O.; Jensen, I.-J. Protein determination—method matters. *Foods* **2018**, *7*, 5. [CrossRef] [PubMed]
32. Brummer, V.; Jurena, T. Enzymatic hydrolysis of pretreated waste paper–Source of raw material for production of liquid biofuels. *Bioresour. Technol.* **2014**, *152*, 543–547. [CrossRef]
33. Zhao, C.; Deng, L. Mixed culture of recombinant Trichoderma reesei and Aspergillus niger for cellulase production to increase the cellulose degrading capability. *Biomass Bioenergy* **2018**, *112*, 93–98. [CrossRef]
34. Omojasola, F.P.; Jilani, O.P. Cellulase production by Trichoderma longi, Aspergillus niger and Saccharomyces cerevisae cultured on waste materials from orange. *Pak. J. Biol. Sci.* **2008**, *11*, 2382–2388. [CrossRef]
35. Eleazu, C.H.; Eleazu, K.; Amarachi, I. In vitro starch digestibility, α-amylase and α-glucosidase inhibitory capacities of raw and processed forms of three varieties of Livingstone potato (Plectranthus esculentus). *Innov. Food Sci. Emerg. Technol.* **2016**, *37*, 37–43. [CrossRef]
36. Hernández-López, A.; Félix, D.A.S.; Sierra, Z.Z.; Bravo, I.G.; Dinkova, T.D.; Avila-Alejandre, A.X. Quantification of reducing sugars based on the qualitative technique of Benedict. *ACS Omega* **2020**, *5*, 32403–32410. [CrossRef] [PubMed]
37. Oberoi, H.; Dhillon, G.; Bansal, S. production of cellulases through solid state fermentation using kinnow pulp as a major substrate. *Food Bioproc. Tech.* **2010**, *3*, 528–536. [CrossRef]
38. Chandra, R.; Raj, A.; Purohit, H.J.; Kapley, A. Characterisation and optimisation of three potential aerobic bacterial strains for kraft lignin degradation from pulp paper waste. *Chemosphere* **2007**, *67*, 839–846. [CrossRef] [PubMed]
39. Sudhakar Reddy, C.; Alekhya, V.V.L.; Saranya, K.; Kakkara, A.; Jha, C.; Diwakar, P.; Dadhwal, V. Monitoring of fire incidences in vegetation types and Protected Areas of India: Implications on carbon emissions. *J. Earth Syst. Sci.* **2017**, *126*, 11. [CrossRef]
40. Kumar, V.; Hart, A.J.; Wimalasena, T.T.; Tucker, G.A.; Greetham, D. Expression of RCK2 MAPKAP (MAPK-activated protein kinase) rescues yeast cells sensitivity to osmotic stress. *Microb. Cell Fact.* **2015**, *14*, 85. [CrossRef] [PubMed]
41. Zapata, L.; Pich, O.; Serrano, L.; Kondrashov, F.A.; Ossowski, S.; Schaefer, M.H. Negative selection in tumor genome evolution acts on essential cellular functions and the immunopeptidome. *Genome Biol.* **2018**, *19*, 67. [CrossRef] [PubMed]
42. Abdullah, A.; Hamid, H.; Christwardana, M.; Hadiyanto, H. Optimization of cellulase production by Aspergillus niger ITBCC L74 with Bagasse as Substrate using Response Surface Methodology. *J. Biosci.* **2018**, *25*, 115–125.
43. Zhao, X.; Wei, C.-M.; Yang, L.; Chou, M. Reply. Physical review letters. *Phys. Rev. Lett.* **2005**, *94*, 219702. [CrossRef]
44. Abdullah, N.H.; Shamsuddin, A.; Wahab, E.; Aziati, A. The relationship between organizational culture and product innovativeness. *Procedia-Soc. Behav. Sci.* **2014**, *129*, 140–147. [CrossRef]
45. Klemm, D.; Cranston, E.D.; Fischer, D.; Gama, M.; Kedzior, S.A.; Kralisch, D.; Kramer, F.; Kondo, T.; Lindström, T.; Nietzsche, S.; et al. Nanocellulose as a natural source for groundbreaking applications in materials science: Today's state (Review). *Mater. Today* **2018**, *21*, 720–748. [CrossRef]
46. Sharma, P.R.; Sharma, S.K.; Antoine, R.; Hsiao, B.S. Efficient removal of arsenic using zinc oxide nanocrystal-decorated regenerated microfibrillated cellulose scaffolds. *ACS Sustain. Chem. Eng.* **2019**, *7*, 6140–6151. [CrossRef]
47. Velusamy, S.; Sakthivel, S.; Sangwai, J.S. Effect of imidazolium-based ionic liquids on the interfacial tension of the alkane–water system and its influence on the wettability alteration of quartz under saline conditions through contact angle measurements. *Ind. Eng. Chem. Res.* **2017**, *56*, 13521–13534. [CrossRef]
48. Geng, L.; Peng, X.; Zhan, C.; Naderi, A.; Sharma, P.R.; Mao, Y.; Hsiao, B.S. Structure characterization of cellulose nanofiber hydrogel as functions of concentration and ionic strength. *Cellulose* **2017**, *24*, 5417–5429. [CrossRef]

processes

MDPI

Article

The Influence of Solvent and Extraction Time on Yield and Chemical Selectivity of Cuticular Waxes from *Quercus suber* Leaves

Rita Simões, Isabel Miranda * and Helena Pereira

Centro de Estudos Florestais (CEF), Associate Laboratory TERRA, Instituto Superior de Agronomia, Universidade de Lisboa, Tapada da Ajuda, 1349-017 Lisbon, Portugal
* Correspondence: imiranda@isa.ulisboa.pt

Abstract: The cuticular lipid compounds, usually named cuticular waxes, present in the cuticular layering of *Quercus suber* adult leaves were extracted with solvents of different polarities (n-hexane, dichloromethane and acetone) and analysed by GC–MS. *Q. suber* leaves have a substantial cuticular wax layer (2.8% of leaf mass and 239 μg/cm^2), composed predominantly by terpenes (43–63% of all compounds), followed by aliphatic long chain molecules, mainly fatty acids, and by smaller amounts of aliphatic alcohols and n-alkanes. The major identified compound was lupeol (1.2% of leaves in n-hexane extract). The recovery and composition of cuticular lipids depended on the solvent and extraction time. The non-polar or weak polar solvents n-hexane and dichloromethane extracted similar lipid yields (77% and 86% of the total extract, respectively) while acetone solubilised other cellular compounds, namely sugars, with the lipid compounds representing 43% of the total extract. For cuticular lipids extraction, solvents with a low polarity such as n-hexane are the more suitable with an adequate extraction duration, e.g., n-hexane with a minimum extraction of 3 h.

Keywords: cork oak; cuticular wax; solvents; terpenes; lupeol

1. Introduction

The extracellular surfaces of plant leaves are covered by a layer consisting of cutin, an insoluble polyester glyceride and of a complex mixture of lipids soluble in organic solvents, commonly called cuticular waxes, which are located within and above the cutin structural matrix and named, respectively, intracuticular and epicuticular waxes [1,2]. This hydrophobic interface confers a resistance to a wide range of biotic and abiotic stresses and is involved in the regulation of non-stomatal water loss and gas exchange [2–4]. The functional barrier against water diffusion through the cuticle is preferentially established by intracuticular lipids [5–7] which consist of long-chain aliphatic molecules and alicyclic waxes, including triterpenoids, while the epicuticular lipids, comprising long-chain aliphatic molecules (e.g., alkanes, primary alcohols, fatty acids), are important for interactions between leaf surface and environment [5–7].

The cuticle, as a boundary between the plant and environment, has an enhanced role under adverse environmental conditions. This is the case for most species growing in the Mediterranean region under high solar irradiances, air temperatures, and vapor pressure deficits with a limited water availability, thereby showing a high sclerophyll character. Under drought stress, plants react namely by an ABA-induced stomatal closure, the accumulation of cuticular lipids, and the formation of a deep root system which improves their drought tolerance. Cork oak (*Quercus suber* L.), an evergreen sclerophyllous tree species distributed in the western Mediterranean basin, is an example of a species with great economic importance due to its production of cork, a material with a very interesting set of properties that feeds a dedicated industrial chain [8].

In leaves of *Q. suber* sampled in spring and summer, Martins et al. [9] described a layer of cuticular wax composed mainly of n-alkyl esters (25–45% of the wax extract)

and alkanols (18–50%) with a significant content variability. A more comprehensive study by our group on the cork oak leaves of trees from six seed provenance sources observed that a substantial wax layer (154–235 μg/cm^2) covered the leaves, with the majority of compounds being pentacyclic triterpenoids and long chain aliphatic compounds, and related the cork oak cuticular characteristics to the species since the trees of different seed origins did not differ [10]. The triterpene fraction contained a large amount of lupeol, a compound that has shown beneficial effects on human health [11–15].

Given the important role of the cuticular waxes in establishing the tree adaptation to adverse environmental conditions, it is of a high interest to clarify the influence of extraction conditions in their content and compositional determinations. In this study, we provide a characterization of the lipophilic compounds present in the leaf cuticle of *Q. suber* extracted with solvents of different polarities and with different extraction times. The results presented here will contribute to the knowledge of the lipophilic compounds in *Q. suber* cuticular leaves and their extraction specificity, allowing to direct the solubilisation process to obtain potential highly valuable phytochemicals of diverse industrial interest.

2. Materials and Methods

2.1. Plant Material

Mature leaves were collected from two mature *Q. suber* L. trees that were never submitted to cork removal, grown on the campus of the School of Agriculture, in the region of Lisbon, Portugal. The leaves were collected randomly from different branches on the south exposed crown side, in the lower part of the canopy up to a height of approximately 2 m, making up a total sample per tree of about 200 leaves. A composite sample was prepared by combining the leaves of the two trees.

The area of 80 leaves randomly selected from the composite sample was measured by digitalizing and it was calculated with Leica Qwin vs. 3.0 Image Analysis Software.

2.2. Cuticular Wax Extraction

The influence of the solvent polarity and extraction time on the yields of solubilized cuticular waxes and on their chemical profile was studied. Conventional Soxhlet extractions were used for the extraction of the cuticular waxes from the whole and intact cork oak leaves taken randomly from the leaf composite sample. Three replicates of leaf samples (without petiole) with a dry mass of around 1.5 g (corresponding to about 10 leaves) were placed in the Soxhlet apparatus and the lipophilic fractions were extracted using organic solvents n-hexane (dielectric constant 2.02), dichloromethane (dielectric constant 9.1), and acetone (dielectric constant 20.7), with different extraction times (5 min, 1, 2, 3, and 6 h with each solvent). The amount of the soluble compounds removed with each solvent and extraction time was determined from the mass difference in the extracted leaves after drying at 105 °C. The results were expressed as a mass percent of the leaf dry mass (g per 100 g leaves) and on a leaf surface area basis (μg per cm^2 leaf area) as the ratio between the extract and the two-sided leaf surface area, obtained by digitalization.

The lipid fraction in the crude extracts was determined from the mass of lipidic compounds estimated by GC–MS (alkanes, alkanols, fatty acids, terpenes, and sterols) and expressed as a percent of the extract mass and as a percent of the leaf dry mass. The estimated quantification of the compounds in the GC–MS chromatograms was done by the calculation of each compound peak area and expressed as a percent of the total chromatogram peak area.

2.3. Cuticular Wax Composition

The extracts were analysed using gas chromatography–mass spectrometry (GC–MS) after trimethylsilylation according to Simões et al. [10]. The extract was treated with BSTFA (N,O-bis(trimethylsilyl)trifluoroacetamide) in pyridine (30 min at 70 °C) and analysed in a gas chromatograph coupled to a mass spectrometer (EMIS, Agilent 5973 MSD, Palo Alto, CA, USA) using a Zebron 7HG-G015-02 column (30 m, 0.25 mm; ID, 0.1 μm film thickness) with an electron ionization at 70 eV, and helium as a carrier gas at 1.0 mL min^{-1} flux. The column temperature was initially held at 50 °C for 1 min, raised to 150 °C at a rate of 10 °C min^{-1}, then to 300 °C at 4 °C min^{-1}, to 370 °C at 5 °C min^{-1}, and at 8 °C min^{-1} until it reached 380 °C, and followed by an isothermal 5 min period. The compounds were identified and quantified as TMS derivatives by comparing their mass spectra with a GC–MS spectral library (Wiley, NIST) and by comparing their fragmentation profiles with the published data, reference compounds, ion fragmentation patterns, or retention times [10]. The results are given as a percent of each compound peak area to the total chromatogram peak areas. The results are to be considered as only semiquantitative since no internal standards were added nor were the response factors of the different compounds calculated.

The proportion of specific chemical classes in the extracts (e.g., lipids, terpenes, aromatics, sugars), expressed as a % of the extract or in mg per g of leaf dry mass, was calculated using the extract amount per dry leaf mass and the compound area proportion in the total GC–MS chromatogram.

3. Results and Discussion

In the present work, the effect of the solvent and extraction time on the yield and composition of the cuticular wax removed from the *Q. suber* leaves surface was studied using three solvents with a different polarity index (by increasing polarity: n-hexane, dichloromethane and acetone) and different extraction times. In the procedure used here, the solubilisation of the compounds refers to both the adaxial and abaxial leaf surfaces. With the relatively long extraction time by Soxhlet (from 1 h to 6 h), the cuticular lipids that are released correspond to a mixture of both the epicuticular and intracuticular lipids [5,16]. Therefore, the resulting extracts reflect the total lipid composition, averaging over the entire depth of the cuticle, rather than assessing only the surface. The potential exception is the 5 min extraction which may target mostly the epicuticular waxes, as discussed later.

Although some techniques have been applied in other works to study separately the chemical composition of epicuticular and intracuticular lipids, e.g., by physically stripping the epicuticular lipids from the leaf surfaces or by using very short dipping times [5,7,17], this was not the objective of the present work. Therefore, the results presented below on the cuticular lipids comprise both epicuticular and intracuticular compounds.

A specific observation should be made on the term cuticular waxes used throughout this work. Although wax is chemically defined as very long-chain esters, the compounds solubilised from the leaf cuticle by the organic solvents have a complex composition comprising long chain fatty acids, primary and secondary alcohols, aldehydes, wax esters, ketones and linear hydrocarbons, that are derived from fatty acid precursors [18]. A more accurate designation should therefore be "extracellular surface lipids" [19]. However, most literature dealing with leaf cuticles uses the term cuticular waxes, and this was maintained here.

3.1. Extraction Yields

Table 1 shows the average extract yields (in g per 1000 g of dry leaves) and leaf surface coverage (in µg per cm^2 leaf surface) obtained with the different solvents and extraction times. A previous study from our group on the cuticular waxes of cork oak leaves from different tree provenances showed that a 6 h Soxhlet extraction with dichloromethane solubilised on average 2.2% of the dry leaf mass, corresponding to a surface coverage of 189 µg/cm^2 [10]. The results obtained here with dichloromethane after 6 h (2.8% of leaf mass and 239 µg/cm^2) are similar to those previously reported.

There was a clear difference between the solvents in their extracting ability: the yields were higher for acetone and hexane (3.6% and 3.4% of the leaf mass, respectively, after 6 h of extraction) in comparison to dichloromethane. The extraction time also had an effect on the extract yield (Table 1). Increasing the extraction time resulted in a higher removal, although the relative extraction power of the three solvents was maintained, i.e., lower for dichloromethane. It should be noted that the extraction yields are related to the solubilization power and selectivity of the solvent to both lipid and non-lipid compounds, as given namely by their polarity, and discussed in detail in the following sections.

Table 1. Yields of material solubilized from the surface of whole *Q. suber* leaves by different solvents and extraction times, in g per 100 g dry leaves, and μg per cm^2 of leaf surface.

	5 min	1 h	2 h	3 h	6 h
n-Hexane					
g/100 g leaves	0.33 ± 0.04	1.92 ± 0.72	1.98 ± 0.25	3.11 ± 0.11	3.39 ± 0.19
μg/cm^2	28.13 ± 3.09	165.67 ± 62.33	170.58 ± 21.52	286.02 ± 9.27	287.15 ± 23.18
Dichloromethane					
g/100 g leaves	0.51 ± 0.03	1.17 ± 0.09	1.23 ± 0.13	1.79 ± 0.33	2.77 ± 0.17
μg/cm^2	43.83 ± 2.88	101.36 ± 7.66	108.30 ± 11.17	154.10 ± 28.52	239.00 ± 14.25
Acetone					
g/100 g leaves	0.35 ± 0.07	1.27 ± 0.25	2.70 ± 0.15	3.16 ± 0.02	3.57 ± 0.49
μg/cm^2	30.29 ± 5.84	109.42 ± 21.81	232.86 ± 12.81	272.24 ± 2.12	308.33 ± 42.32

The short 5 min extraction time led to a very partial removal of the soluble compounds, e.g., hexane, dichloromethane, and acetone solubilised only, respectively, 10%, 18%, and 10% of the extracts obtained with a 6 h extraction.

These results point out that a full, or at least a major, solubilisation of the leaf cuticular compounds is only possible if sufficient extraction time is given. Therefore, protocols that use a short or very short contact between the solvent and the leaf surface will only solubilise a small fraction of the cuticular lipids, namely only the epicuticular waxes, as previously discussed [10]. This is the reason for the difference in the extraction yield found between the present results and a previous study on young cork oak leaves using a quick dipping and shaking in the chloroform [9].

3.2. Effect of Solvent on Lipid Extraction

The extraction yield, i.e., the compounds solubilized by the solvent in relation to the leaf mass, is a quantitative but undifferentiating indicator of the molecules removed from the surface of the leaves, meaning that other non-lipid compounds may be solubilized, and more so with solvents with an increased polarity. Previous reports on the composition of the solubilized compounds from *Q. suber* leaves using dichloromethane showed that 80% of the extract was of a lipid nature, with about 2% corresponding to aromatics and a small proportion of sugars, as calculated from the corresponding peak areas in relation to the total chromatogram peak area [10].

The composition of the extracts is therefore an important aspect when analysing the solvent efficiency to solubilize the cuticular lipids. Considering the total amount of lipidic compounds present in the cuticular extracts, as determined from their GC–MS analysis (which will be reported in detail in the next section), by grouping alkanes, alkanols, fatty acids and glycerides, terpenes, and sterols, a new insight is obtained on the solvent role to attain higher lipid yields. Figure 1 represents the lipid yields obtained with the three solvents and five extraction times. The proportion of the lipid extractives was consistent with the order of the polarity of the solvent.

The 6 h acetone extraction shows the lowest proportion of extracted lipid compounds as estimated by their peak area proportion in the total chromatogram (15.3 mg/g of dry leaves, corresponding to 43% of the total extract), while the non-polar or weak polar solvents n-hexane and dichloromethane extracted a similar lipid proportion estimated as 26.1 and 24.1 mg/g of dry leaves, corresponding to 77% and 86% of the total extract, respectively (Figure 1). The lipid proportion in the extracts increased with the extraction time, e.g., from 1.7% to 2.6% with n-hexane, and from 0.5% to 1.5% with acetone for 1 h and 6 h extractions, respectively.

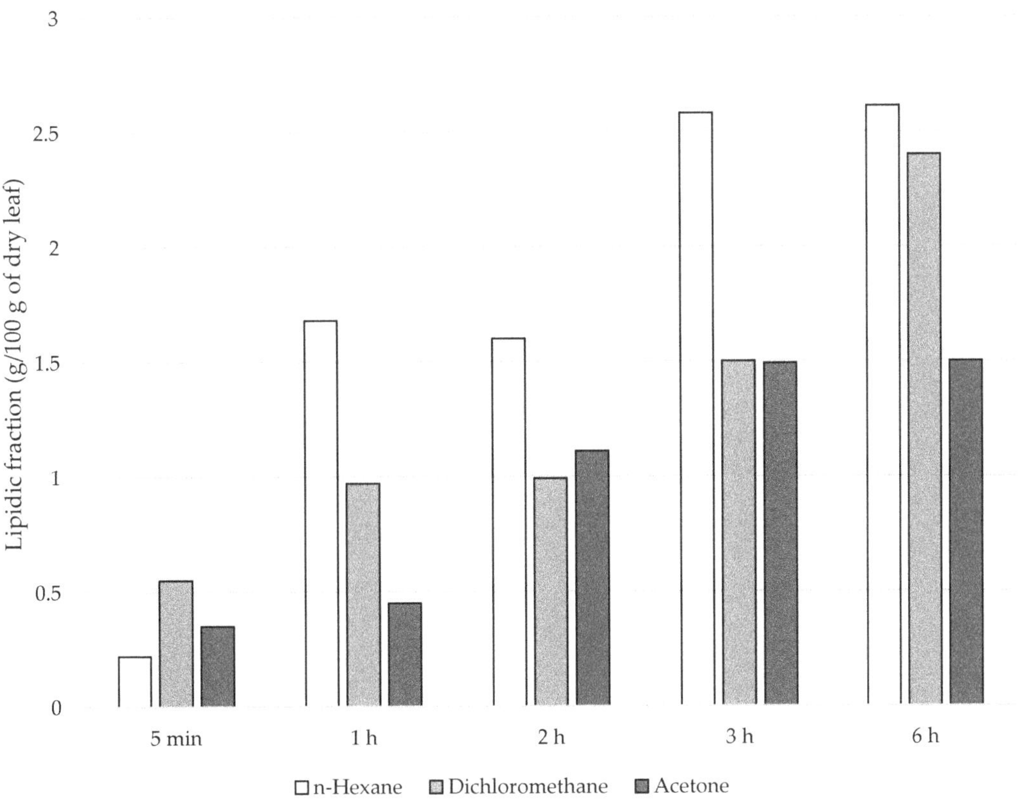

Figure 1. Lipidic fraction obtained from extraction of *Q. suber* leaves with solvents with different polarity indices with different extraction times.

For solubilisation of the cuticular lipids from *Q. suber* leaves (Figure 1), n-hexane was the most effective solvent in providing the highest recovery, which was attained with a 3 h extraction in the experimental conditions used. Dichloromethane, as used in the previous *Q. suber* cuticular studies, also provided a similar lipid recovery with a 6 h extraction [10].

3.3. Compositional Analysis of Extracts by Chemical Family

The composition by the chemical family of the cuticular waxes of cork oak leaves solubilised by different solvents with different extraction times, as a percentage of the total peak area in the GC–MS chromatograms, are presented in Table 2, and detailed for the three solvents in Tables 3–5. The results are a semiquantitative approximation of compound proportion since no internal standards were added or response factors were calculated.

The composition of n-hexane and dichloromethane extracts is quite similar. With the exception of the 5 min extracts, that will be discussed subsequently, the extracts are mainly made up of terpenes (43–63% of all compounds), followed by aliphatic long chain molecules, mainly fatty acids (11–26% of all compounds), and by smaller amounts of aliphatic alcohols, n-alkanes, and with some phenolics. This is in accordance with the reported composition of the cuticular wax from *Q. suber* leaves where triterpenes and aliphatic compounds predominate (61–72% and 17–23% of the total compounds, respectively) [10]. For the short extraction time of 5 min, the compounds solubilised were aliphatic alcohols that represented

98% and 96% of all compounds, respectively, in the n-hexane and dichloromethane extracts. This suggests that the more easily accessible external layer of the cuticle predominantly contains alkanols [20].

Table 2. Composition of the cuticular extracts of leaves from *Q. suber* solubilised by different solvents and extraction times by chemical class as a percentage of total peak area in the GC–MS chromatograms.

	Proportion of Total Compounds, %														
	n-Hexane					**Dichloromethane**					**Acetone**				
Family	**5 min**	**1 h**	**2 h**	**3 h**	**6 h**	**5 min**	**1 h**	**2 h**	**3 h**	**6 h**	**5 min**	**1 h**	**2 h**	**3 h**	**6 h**
Alkanes	0.3	6.9	3.5	6.9	6.6	0.4	3.4	3.1	2.2	3.6	-	0.5	1.8	0.4	0.2
Alkanols	98.4	5.5	4.0	3.7	7.0	95.5	5.1	7.3	5.5	7.3	2.1	3.8	3.7	4.2	2.0
Fatty acids	0.2	11.8	11.0	16.3	16.1	0.9	12.6	25.0	25.3	20.0	39.6	5.4	11.6	11.7	9.0
Glycerides	-	0.1	0.2	0.5	0.3	-	1.0	2.5	0.9	0.2	-	1.1	1.9	1.8	1.2
Sterols	-	0.01	3.2	4.2	2.7	-	2.0	3.0	2.7	3.4	-	1.1	1.5	1.6	1.2
Terpenes	0.1	59.7	63.4	52.0	44.6	2.0	59.9	42.7	48.4	44.8	-	19.5	20.3	20.5	22.2
Aromatics	-	1.7	4.5	5.6	4.1	-	4.7	5.6	5.7	4.5	13.5	2.9	3.2	2.6	2.7
Sugars	-	0.0	0.0	0.0	0.0	-	0.1	0.5	1.4	0.6	39.7	56.8	50.0	41.3	46.8
Others	-	1.6	0.9	0.6	0.3	-	1.1	0.5	1.4	1.4	5.0	6.2	1.7	9.9	8.4
Total Identified	99.0	87.3	90.7	89.8	81.7	99.5	90.0	90.7	93.7	85.5	99.8	93.4	92.0	89.7	91.6

Table 3. Composition of the cuticular waxes of cork oak leaves solubilised by n-hexane with different extraction times, as determined by GC–MS, in % of total peak area (only compounds with over 1% are shown; complete composition in Supplementary Table S1).

	Extraction Time				
Wax Constituent (% Total Peak Area)	**5 min**	**1 h**	**2 h**	**3 h**	**6 h**
Alkanes					
Heptacosane (C_{27})	-	-	-	1.12	1.02
Triacontane (C_{30})	-	5.10	2.47	4.89	4.66
Alkanols					
Hexadecan-1-ol ($C_{16}OH$)	-	1.15	-	-	-
Docosan-1-ol ($C_{22}OH$)	1.24	-	-	-	-
Tretracosan-1-ol ($C_{24}OH$)	96.64	2.91	3.11	3.05	4.34
Fatty acids					
Saturated					
Hexadecanoic acid ($C_{16:0}$)	-	7.57	5.35	5.57	5.32
Hexacosanoic acid ($C_{26:0}$)	-	-	-	1.00	1.05
Octacosanoic acid ($C_{28:0}$)	-	-	-	2.61	3.39
Triacontanoic acid ($C_{30:0}$)	-	-	-	1.08	1.37
Unsaturated					
9,12-Octadecadienoic acid ($C_{18:2}$)	-	-	-	1.12	0.81
9,12,15-Octadecatrienoic acid ($C_{18:3}$)	-	-	2.28	3.43	2.11
Sterols					
β-Systosterol	-	-	2.82	3.50	2.50
Terpenes					
pentacyclic triterpenes					
β-Amyrin	-	3.02	2.83	3.11	2.47
Germanicol	-	11.19	17.31	7.49	6.00
Lupeol	-	43.93	40.43	37.76	33.04
Betulin	-	-	-	1.12	0.97
Aromatics					
Hexadecyl-(E)-p-coumarate	-	-	4.21	5.32	3.66

Table 4. Composition of the cuticular waxes of cork oak leaves solubilised in dichloromethane with different extraction times, as determined by GC–MS in % of total peak area (only compounds with over 1% are shown; complete composition in Supplementary Table S2).

	Extraction Time				
Wax Constituent (% Total Peak Area)	**5 min**	**1 h**	**2 h**	**3 h**	**6 h**
Alkanes					
Nonacosane (C_{29})	-	2.53	2.18	1.47	2.56
Alkanols					
Hexadecan-1-ol ($C_{16}OH$)	0.90	-	-	-	-
Docosan-1-ol ($C_{22}OH$)	2.38	0.77	1.12	1.10	1.85
Tretracosan-1-ol ($C_{24}OH$)	92.12	3.47	4.59	3.31	4.62
Fatty acids					
Saturated					
Hexadecanoic acid ($C_{16:0}$)	-	7.60	9.84	9.71	9.68
Octacosanoic acid ($C_{28:0}$)	-	1.60	2.31	1.14	2.49
Unsaturated					
9,12-Octadecadienoic acid ($C_{18:2}$)	-	0.14	1.73	2.38	0.75
9,12,15-Octadecatrienoic acid ($C_{18:3}$)	-	0.39	6.43	8.04	2.69
Glycerides					
Glycerol	-	0.95	2.14	0.51	-
Sterols					
β-Systosterol	-	1.59	2.55	2.33	3.24
Terpenes					
Pentacyclic triterpenes					
β-Amyrin	-	2.75	2.29	2.56	2.11
Germanicol	-	9.15	5.46	6.46	4.77
Lupeol	1.61	45.44	31.56	35.28	35.37
Aromatics					
Hexadecyl-(E)-p-coumarate	-	4.27	5.31	5.19	4.09
Sugars					
Fructofuranose	-	0.09	0.54	1.43	0.18

Table 5. Composition of the cuticular waxes of cork oak leaves solubilised in acetone with different extraction times, as determined by GC–MS, in % of total peak area (only compounds with over 1% are shown; complete composition in Supplementary Table S3).

	Extraction Time				
Wax Constituent (% Total Peak Area)	**5 min**	**1 h**	**2 h**	**3 h**	**6 h**
Alkanols					
1,4-Butanediol	1.85	-	-	-	-
Tretracosan-1-ol ($C_{24}OH$)	-	3.22	3.21	2.74	1.28
Fatty acids					
Saturated					
Hexadecanoic acid ($C_{16:0}$)	-	2.51	5.54	5.92	4.35
Unsaturated					
9,12-Octadecadienoic acid ($C_{18:2}$)	-	0.03	0.66	1.23	0.48
9,12,15-Octadecatrienoic acid ($C_{18:3}$)	-	0.07	0.61	0.07	1.26
Diacids					
Butanedioic acid ($C_{4:0}$)	-	0.95	0.99	2.45	1.69
Glycerides					
Glycerol	-	1.04	1.72	1.36	0.95
Sterols					
β-Systosterol	-	0.32	0.99	1.43	1.15

Table 5. *Cont.*

Wax Constituent (% Total Peak Area)	Extraction Time				
	5 min	1 h	2 h	3 h	6 h
Terpenes					
pentacyclic triterpenes					
β-Amyrin	-	0.72	-	1.18	1.19
Germanicol	-	2.89	2.05	2.71	2.86
Lupeol	-	14.53	16.25	14.71	16.15
Aromatics					
Catechine	13.18	-	-	-	-
Hexadecyl-(E)-p-coumarate		1.67	1.62	2.35	2.42
Sugars					
Myo Inositol	19.98	-	15.34	-	−0.11
Scyllo Inositol	-	15.44	14.27	12.39	14.88
Deoxinositol	-	6.28	1.81	5.24	6.3
D-Fructose	-	15.47	11.27	10.68	8.61
(α/β) D-Glucopyranose (isomer)	-	17.11	6.28	11.18	15.01
Sucrose	18.62	0.48	0.21	0.47	0.4
Others compounds					
Quininic acid	38.18	4.6	0.94	7.2	7.15
Shikimic acid, 4TMS derivative	5.03	-	-	-	-

The compositional profile of the acetone extract clearly differs, with sugars being the main chemical class (57–47% of all compounds), followed by triterpenes (19–23%), and fatty acids (5–12%) with lower proportions of the other chemical families. The presence of other non-lipid compounds, mainly of sugars, indicates that, in contrast to the other solvents used, acetone penetrates into the leaf interior and solubilises the more polar sugar molecules. The degradation of the cellular membranes by solvents such as chloroform and acetone have been reported, leading to the solubilisation of cellular compounds not related to the cuticular matrix, for instance of indole alkaloids [21], and chlorophyll and cytoplasmic sterols. The effect of acetone on the leaf cells was noticed already for the short 5 min extraction with the solubilisation of fatty acids, sugars, and aromatics. This is in agreement with reports of an instant destruction of cell integrity when using chloroform [21].

The recovery (in g/100 g leaves) of the extracted cuticular lipids by the chemical family under the different conditions is shown in Figure 2. The difference in the acetone extract in comparison with the hexane and dichloromethane extracts is clear with the major extract fraction being sugars representing 1.67 g/100 g leaves (6 h extraction). Hexane and dichloromethane permit the recovery of cuticular terpenes (1.51 and 1.24 g/100 g leaves, respectively) and fatty acids (0.55 and 0.55 g/100 g leaves, respectively).

The extraction time had only a moderate impact on the recovery of the extracellular cuticular lipids with the differences given more by the extraction yield (Table 1) than by the compositional profile, which remains independent of the extraction time (Table 2). For instance, the recovery of terpenes with n-hexane increased from 11.5 mg/g of leaves with 1 h to 15.1 mg/g of leaves with 6 h of contact, while the extraction with dichloromethane was the one most impacted by duration, e.g., terpenes recovery increased from 7.0 mg/g of leaves with 1 h to 12.4 mg/g of leaf with 6 h of contact (Figure 2).

These results should be taken with caution and restricted to the experimental procedure used here (Soxhlet extraction). Martins et al. [9] studied the extraction of extracellular surface lipids in the leaves of young *Q. suber* by immersion for a few seconds in chloroform, thus extracting only the epicuticular layer, and this explains the compositional differences in relation to our results, for example, the extract contained only small amounts of triterpenoids. Loneman et al. [19] applied a quick surface extraction method by dipping maize seedling leaves with chloroform and hexane:diethyl ether for 1 to 10 min, and observed that the recovery and composition depended on the solvent and duration of the extraction.

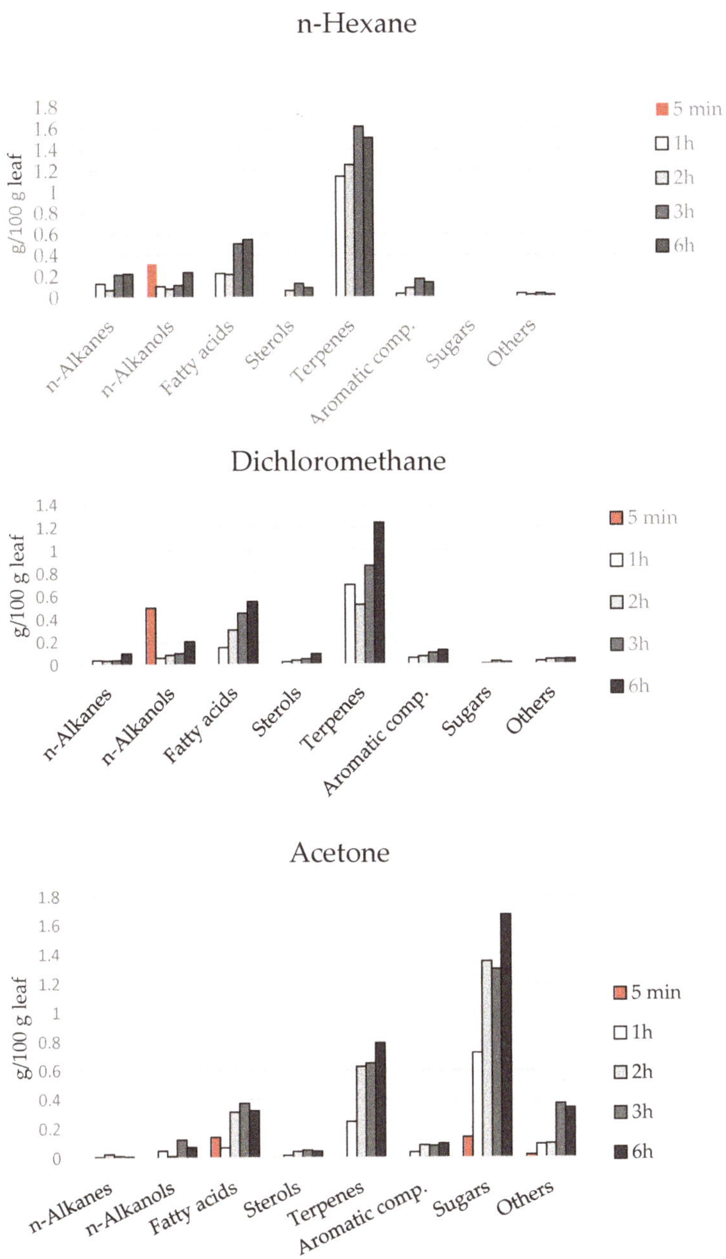

Figure 2. Composition (g/100 g leaves) by chemical family of the cuticular lipids extracted from *Q. suber* leaves with different solvents and extraction time.

3.4. Chemical Composition of Extracts

The composition of n-hexane, dichloromethane, and acetone extracts is summarised in Tables 3–5 that include only the compounds with at least 1.0% of the chromatogram peak area. The complete compositional profile and the GC–MS chromatogram of the derivatised extracts at 5 min and 3 h is given as the supplementary data (Supplementary Tables S1–S3, respectively, for n-hexane, dichloromethane, and acetone extracts) and Supplementary Figures S1–S3, respectively, for n-hexane, dichloromethane, and acetone extracts at 5 min and 3 h, respectively, exemplify the quality of the chromatograms that were obtained for all the cases.

In the terpenic family, the triterpenoid lupeol was among the most prominent individual compound, followed by germanicol and α- and β-amyrin, e.g., lupeol with 33–43% of all compounds (about 75% of the terpenic fraction) and germanicol with 6–17% of all compounds (about 15% of terpenic fraction). Smaller amounts of other triterpenoids such as betulin, betulinic, oleanolic, and ursonic acids were also identified (Supplementary Tables S1–S3). All of the identified triterpenes were previously reported as cuticular wax components of *Q. suber* leaves [10].

Triterpenoids are one of the largest and structurally diverse classes of natural specialized metabolites with various ecological and biological properties, with a broad spectrum of relevant clinical activities, such as antihypertensive, antidiabetic, anti-inflammatory, anticancer, and antimicrobial properties leaves [11–15]. The high proportion of lupeol in the cuticular extracts may lead to considering *Q. suber* leaves as a source of this compound, e.g., a lupeol recovery of 1.18 g/100 g of leaves may be obtained with a 3 h n-hexane extraction.

Four sterols were identified in the extracts (β-systosterol, δ-tocopherol; stigmasterol and 25-hydroxycholesterol, Supplementary Tables S1–S3) with β-systosterol showing the highest abundance with 2.5–3.5% of all compounds.

The hydrocarbon fraction includes fatty acids, n-alkanols, and n-alkanes. With 6 h of contact, fatty acids are mostly saturated fatty acids, accounting to 79% and 59% of the total fatty acids in the n-hexane and dichloromethane extracts, and 59% of the total fatty acids content in the acetone extract. Fatty acids were identified in the homologous series from heptanoic acid (C7:0) to triacontanoic acid (C30:0), with a strong even-over-odd carbon atom predominance. Hexadecanoic acid (palmitic acid, C16:0) was the most abundant saturated fatty acid in the three extracts (5.3% of all compounds in n-hexane extract, 9.7% in dichloromethane extract, and 4.4% in acetone extract). Smaller amounts of octacosanoic acid (C28:0), hexacosanoic acid (C26:0), and triacontanoic acid (C30:0) were also found. Three unsaturated acids were identified in the extracts: 9,12,15-octadecatrienoic (linoleic acid), 9,12-octadecadienoic (linolenic acid), and 9-cis-hexadecenoic (palmitoleic acid) acids.

Fatty alcohols represented only a small fraction of the total amount of lipophilic extractives. A series of n-fatty alcohols ranging from C4 to C28 were present in cork oak wax leaves extracts, with tetracosanol (C24OH) as the most abundant (4.3% of all compounds in the n-hexane wax extract, 4.6% in the dichloromethane extract, and 1.4% in the acetone wax extract)

A series of n-alkanes ranging from nonadecane (C19) to hentriacontane (C31) occurred in the extracts, with a strong odd-over-even carbon atom number predominance. Nonacosane (C29) was the most abundant in the dichloromethane extract (72% of alkanes) and triacontane (C30 in the n-hexane extract (74% of alkanes)).

The composition of the identified aliphatic compounds is similar to that obtained for the cuticular wax of *Q. suber* leaves from different sources extracted with dichloromethane with 6 h of contact, that included fatty acids with a chain length C30, C28, and C16, n-alkanes with chain lengths ranging between C15 and C31 [10].

Low amounts of acylglycerols including mono-glycerides such as monopalmitin, 2-monostearin, 1-linolenoylglycerol, and 3-hydroxypropyl hexadecanoate were identified in all the extracts, together representing 0.45%, 0.57%, and 0.19% of the n-hexane, dichloromethane, and acetone extracts, respectively.

Aromatic compounds were also detected in all the extracts such as benzoic acid and p-coumarates. P-coumarates were the dominant aromatic compound corresponding to 5.3%, 4.1%, and 2.4% of the n-hexane, dichloromethane, and acetone extracts, respectively.

High amounts of sugars were found in the acetone extracts (56% of all compounds) including fructose, glucose, sucrose, the free cyclitols myo-inositol, scyllo-inositol, and deoxinositol. Scyllo-inositol has been considered as a potential therapeutic agent for Aβ amyloid disorders such as Alzheimer's disease, Down's syndrome, and cerebrovascular dementia [22,23].

4. Conclusions

This study confirmed that the extent and selectivity of the solubilisation of non-polar and amphipathic molecules that are part of the cuticular layering of plant leaves, as exemplified for the case of *Q. suber* adult leaves, depends on the solvent and on the extraction duration. Less polar solvents such as n-hexane or dichloromethane solubilize the lipid components and a small number of aromatics, while more polar solvents, such as acetone, remove a large proportion of sugars arising from the solubilisation of cellular material. For the purpose of cuticular lipids extraction, solvents with a low polarity, such as n-hexane, are the more suitable and an adequate extraction duration should be considered for an extensive removal, including both the epicuticular and intracuticular lipids. In the present case, n-hexane with a minimum 3 h extraction is proposed. Although not specifically studied here, the procedural method for the extraction should also play a role in the extent of solubilization. The cuticular lipid compounds, usually named cuticular waxes, include as major components terpenes and a large fraction of long-chain fatty acids and other long-chain aliphatics, with a low proportion of sterols, aromatics, and other compounds. It is therefore probable that the terpenes will establish the transpiration barrier in association with the long chain aliphatic compounds.

Supplementary Materials: The following supporting information can be downloaded at: https://www.mdpi.com/article/10.3390/pr10112270/s1, Table S1: Composition of the cuticular waxes of cork oak leaves solubilised in n-hexane with different extraction times as determined by GC–MS, in % of total peak área; Table S2: Composition of the cuticular waxes of cork oak leaves solubilised in dichloromethane with different extraction times as determined by GC–MS, in % of total peak area; Table S3: Composition of the cuticular waxes of cork oak leaves solubilised in acetone with different extraction times as determined by GC–MS, in % of total peak area. Figure S1: GC–MS chromatogram of the derivatised n-hexane extract from the cuticular waxes of cork oak leaves. Figure S2: GC–MS chromatogram of the derivatised dichloromethane extract from the cuticular waxes of cork oak leaves; Figure S3: GC–MS chromatogram of the derivatised acetone extract from the cuticular waxes of cork oak leaves

Author Contributions: Conceptualization, H.P. and I.M.; methodology, I.M. and R.S.; validation, H.P., I.M. and R.S.; formal analysis, R.S.; investigation, R.S.; data curation, R.S., H.P. and I.M.; writing—original draft preparation, I.M. and R.S.; writing—review and editing, H.P., I.M. and R.S. All authors have read and agreed to the published version of the manuscript.

Funding: This research was funded by Fundação para a Ciência e a Tecnologia (FCT) through funding of the Forest Research Centre (UIDB/00239/2020). Funding for this work was also provided by a doctoral scholarship from FCT SUSFOR Doctoral Programme (PD/BD/128259/2016).

Data Availability Statement: Not applicable.

Acknowledgments: Rita Simões acknowledges a doctoral scholarship from FCT with the SUSFOR Doctoral Programme (PD/BD/128259/2016). We thank Ana Rodrigues for her technical assistance in field sampling.

Conflicts of Interest: The authors declare no conflict of interest.

References

1. Pollard, M.; Beisson, F.; Li, Y.; Ohlrogge, J.B. Building lipid barriers: Biosynthesis of cutin and suberin. *Trends Plant Sci.* **2008**, *13*, 236–246. [CrossRef] [PubMed]
2. Buschhaus, C.; Jetter, R. Composition differences between epicuticular and intracuticular wax substructures: How do plants seal their epidermal surfaces? *J. Exp. Bot.* **2011**, *62*, 841–853. [CrossRef] [PubMed]
3. Jeffree, C.E. The fine structure of the plant cuticle. In *Biology of the Plant Cuticle*; Riederer, M., Müller, C., Eds.; Annual Plant Review; Blackwell: Oxford, UK, 2006; Volume 23, pp. 11–125.
4. Schreiber, L.; Schönherr, J. *Water and Solute Permeability of Plant Cuticles: Measurement and Data Analysis*; Springer: Berlin/Heidelberg, Germany, 2009.
5. Jetter, R.; Riederer, M. Localization of the transpiration barrier in the epi- and intracuticular waxes of eight plant species: Water transport resistances are associated with fatty acyl rather than alicyclic components. *Plant Physiol.* **2016**, *170*, 921–993. [CrossRef] [PubMed]

6. Zeisler-Diehl, V.; Müller, Y.; Schreiber, L. Epicuticular wax on leaf cuticles does not establish the transpiration barrier, which is essentially formed by intracuticular wax. *J. Plant Physiol.* **2018**, *227*, 66–74. [CrossRef] [PubMed]
7. Bourgault, R.; Matschi, S.; Vasquez, M.; Qiao, P.; Sonntag, A.; Charlebois, C.; Mohammadi, M.; Scanlon, M.J.; Smith, L.G.; Molina, I. Constructing functional cuticles: Analysis of relationships between cuticle lipid composition, ultrastructure and water barrier function in developing adult maize leaves. *Ann. Bot.* **2020**, *125*, 79–91. [CrossRef] [PubMed]
8. Pereira, H. *Cork: Biology, Production and Uses*; Elsevier Science: Amsterdam, The Netherlands, 2007.
9. Martins, C.M.C.; Mesquita, S.M.M.; Vaz, W.L.C. Cuticular waxes of the holm (*Quercus ilex* L. subsp. ballota (Desf.) Samp.) and cork (*Q. suber* L.) oaks. *Phytochem. Anal.* **1999**, *10*, 1–5. [CrossRef]
10. Simões, R.; Rodrigues, A.; Ferreira-Dias, S.; Miranda, I.; Perreira, H. Chemical composition of cuticular waxes and pigments and morphology of leaves of *Quercus suber* trees of different provenance. *Plants* **2020**, *9*, 1165. [CrossRef] [PubMed]
11. Fernández, M.A.; Heras, B.; García, M.D.; Sáenz, M.T.; Villar, A. New insights into the mechanism of action of the anti-inflammatory triterpene lupeol. *J. Pharm. Pharmacol.* **2001**, *53*, 1533–1539. [CrossRef] [PubMed]
12. Yamashita, K.; Lu, H.; Lu, J.; Chen, G.; Yokoyama, T.; Sagara, Y.; Manabe, M.; Kodama, H. Effect of three triterpenoids, lupeol, betulin, and betulinic acid on the stimulus-induced superoxide generation and tyrosyl phosphorylation of proteins in human neutrophils. *Clin. Chim. Acta* **2002**, *325*, 91–96. [CrossRef]
13. Prasad, S.; Kumar, Y.V.; Srivastava, S.; Shukla, Y. Protective effects of lupeol against benzo[a]pyrene induced clastogenicity in mouse bone marrow cells. *Mol. Nutr. Food Res.* **2008**, *52*, 1117–1120. [CrossRef] [PubMed]
14. Siddique, H.R.; Saleem, M. Beneficial health effects of lupeol triterpene: A review of preclinical studies. *Life Sci.* **2011**, *88*, 285–293. [CrossRef] [PubMed]
15. Ahmad, R.; Khan, A.; Lee, H.J.; Rehman, I.U.; Khan, I.; Alam, S.I.; Kim, M.O. Lupeol, a plant-derived triterpenoid, protects mice brains against Aβ-induced oxidative stress and neurodegeneration. *Biomedicines* **2020**, *8*, 380. [CrossRef] [PubMed]
16. Jetter, R.; Schäffer, S.; Riederer, M. Leaf cuticular waxes are arranged in chemically and mechanically distinct layers: Evidence from *Prunus laurocerasus* L. *Plant Cell Environ.* **2000**, *23*, 619–628. [CrossRef]
17. Sharma, P.; Kothari, S.L.; Rathore, M.; Gour, V. Properties, variations, roles, and potential applications of cuticular wax: A review. *Turk. J. Bot.* **2018**, *42*, 135–149. [CrossRef]
18. Yeats, T.H.; Rose, J.K. The formation and function of plant cuticles. *Plant Physiol.* **2013**, *163*, 5–20. [CrossRef] [PubMed]
19. Loneman, D.M.; Peddicord, L.; Al-Rashid, A.; Nikolau, B.J.; Lauter, N.; Yandeau-Nelson, M.D. A robust and efficient method for the extraction of plant extra cellular surface lipids as applied to the analysis of silks and seedling leaves of maize. *PLoS ONE* **2017**, *12*, e0180850. [CrossRef] [PubMed]
20. Fernández, V.; Guzmán-Delgado, P.; Graça, J.; Santos, S.; Gil, L. Cuticle structure in relation to chemical composition: Re-assessing the prevailing m7odel. *Front. Plant Sci.* **2016**, *7*, 427. [CrossRef] [PubMed]
21. Abouzeid, S.; Beutling, U.; Selmar, D. Stress-induced modification of indole alkaloids: Phytomodificines as a new category of specialized metabolites. *Phytochemistry* **2019**, *159*, 102–107. [CrossRef] [PubMed]
22. Fenili, D.; Brown, M.; Rappaport, R.; McLaurin, J. Properties of *scyllo*-inositol as a therapeutic treatment of AD-like pathology. *J. Mol. Med.* **2007**, *85*, 603–611. [CrossRef] [PubMed]
23. Lee, D.; Lee, W.S.; Lim, S.; Kim, Y.K.; Jung, H.-Y.; Das, S.; Lee, J.; Luo, W.; Kim, K.-T.; Chung, S.-K. A guanidine-appended *scyllo*-inositol derivative AAD-66 enhances brain delivery and ameliorates Alzheimer's phenotypes. *Sci. Rep.* **2017**, *7*, 14125. [CrossRef] [PubMed]

processes

Article

Evaluation of the Gas Emissions during the Thermochemical Conversion of Eucalyptus Woodchips

João Silva [1,2,*], Carlos Castro [1], Senhorinha Teixeira [2] and José Teixeira [1]

1 MEtRICs Research Centre, University of Minho, 4800-058 Guimarães, Portugal
2 ALGORITMI Research Centre/LASI, University of Minho, 4800-058 Guimarães, Portugal
* Correspondence: js@dem.uminho.pt

Abstract: The combustion of solid biomass in industrial boilers involves a sequence of processes that include heating, drying, devolatilization, and char conversion. To maintain a repeatable and fully controlled environment, and to monitor all the dynamics involved in the phenomena at a real scale, field-scale experiments become necessary to perform investigations. In this way, to evaluate different thermochemical conversion conditions of biomass particles under an oxidative atmosphere, and to quantify the emission of the main gas compounds continuously, a small-scale reactor was developed and presented in this paper. Hence, in this work, larger particles of eucalyptus are burned at 400 and 800 °C under different stoichiometric conditions to understand the differences between different biomass conversion regimes (gasification and combustion). The analysis of the mass loss at the different temperatures was characterized by only two different and consecutive stages for both thermochemical conditions. The first region does not present the influence on the air flow rate; however, there is a significant difference in the second region. This fact highlighted the importance of the diffusion of oxygen during the char conversion. Regarding the quantification of the gas compounds, an increase of around 3 times in the CO and CO_2 emissions when gasification occurs was observed at 400 °C. However, at 800 °C, the same trend was verified, also verifying a considerable amount of CH_4.

Keywords: biomass; combustion; gas emissions; macro thermogravimetric analysis; pyrolysis; woodchips

1. Introduction

Solid biomass fuels, unlike fossil fuels such as coal, do not take millions of years to develop and, every year, a vast amount of biomass grows through the photosynthesis process by absorbing CO_2 from the atmosphere. Solid biomass is thus considered a renewable energy source and an interesting route to diversify energy production and reduce the dependence on fossil fuels [1]. Eucalyptus is the most representative solid biomass species in Portugal [2]. It was reported in 2015 that this species occupied 882,000 ha of the total forest area. It is still growing and is a major resource for paper production as well as fuel for heat production (both in households and in industry) [3]. Although the major source of solid biomass comes from the trunk, the literature emphasizes other streams of solid biomass with high potential. As an example, Roman et al. [4] have studied the potential of forest waste biomass for briquette production. Such works are a major contribution to the area as they present alternatives for the current solid biomass streams.

Regarding the main technological pathways for the production of heat and power, or combined heat and power, through the utilization of solid biomass, combustion and gasification are the main options [5]. Combustion represents one of the oldest technologies of biomass thermochemical conversion and utilization. The combustion process is an exothermic reaction between oxygen and biomass-volatile compounds that produce heat. The heat released is the main source of energy used in this process [6]. In a typical biomass combustion process, three main stages can be observed: drying, devolatilization, and char burning. The drying stage is associated with the water evaporation present in the biomass. During the devolatilization stage, volatile compounds are released and burned with the oxygen present in the atmosphere. For the char combustion stage, the remaining carbon reacts with the oxygen, leaving ashes at the end of the combustion [7,8]. In the operation of biomass boilers, the design of the air supply system, including primary and secondary air, plays an important role in the combustion efficiency of biomass [9,10]. Yin et al. [11] reported that for grate-firing, one of the main technologies in biomass combustion, the overall excess air is usually set to 25% or above.

In its turn, gasification consists of burning biomass with insufficient oxygen under sub-stoichiometric conditions to produce combustible gases, which are collectively referred to as syngas. This is an attractive method of efficient energy extraction, mainly because a considerable amount of CO and CH_4 are obtained in comparison with combustion. Hence, in solid biomass gasification, an air-to-fuel ratio around 1.5:1 to1.8:1 is necessary, while a combustion ratio is around 3.8:1 [12].

Since, in both technologies, the main stages of biomass thermochemical conversion remain the same, the study and comprehension of the gas-release evolution in a practical way can provide knowledge to develop computer models or to design equipment, such as furnaces, stoves, boilers, and gasifiers [13]. Among the three stages, devolatilization is considered one of the most important for heat and power generation. Hence, it is related to the oxidation, under different stoichiometric conditions depending on the conversion technology, of the volatile compounds and has a significant impact on the exothermic reaction. For that reason, it is also important to study and understand the composition of the gas released during the devolatilization stage. Usually, complete combustion yields only CO_2 and H_2O as reaction products. However, other compounds can be formed, such as CH_4, CO, and H_2, increasing pollutant emissions and contaminating the environment [14].

To study the thermochemical conversion behavior of solid biomass, TGA (Thermo Gravimetric Analysis), a well-known thermal analysis technique and one of the most used, is applied [15,16]. Several authors have applied TGA to characterize the conversion of samples with a reduced size and mass in a kinetic way. However, TGA experiments are limited to assess the gas phase, and the lack of information in the literature relative to the gas emissions in this type of work persists [17–20]. Nevertheless, the gaseous release process analysis can be evaluated using the same technique but on a larger scale, commonly known as macro TGA [21]. In this way, the experiments are closer to the real thermochemical conversion processes, either in industrial or domestic equipment. Hence, macro TGA experiments take into account heat and mass transfer effects.

Regarding the literature concerning macro TGA experiments, Becidan et al. [22] presented the application of chromatography and spectrophotometry to study the gases released during the combustion of biomass residues. A fraction of the exhaust gases is collected and analyzed by a Fourier-Transform Infrared Spectroscopy (FTIR) analyzer and a micro-gas chromatograph. The FTIR was used to quantify CO_2, CO, CH_4, C_2H_2, and C_2H_4. The gas samples were also quantified online using a micro-gas chromatograph equipped with two thermal conductivity detectors and a double injector connected to two columns to separate and quantify CO_2, hydrocarbons (CH_4, C_2H_2, C_2H_4, and C_2H_6), and the remaining gases (H_2, O_2, CH_4, CO, and N_2) in another column. Brunner et al. [23] and Gauthier et al. [24] also applied both techniques to analyze the NOx emissions, ash release, and the main gaseous species and tar, respectively. Additionally, Weissinger et al. [25] described the release of nitrogen compounds using FTIR spectroscopy. Both works refer to the importance of the determination of nitrogen gaseous compounds that may serve as input profiles for Computational Fluid Dynamics (CFD) simulations. Bennadji et al. [26] and Nikku et al. [27] measured the fractions of light species from pyrolysis at low temperatures and compared the reactivity of municipal solid wastes with biomass and coal samples through the FTIR technique, respectively. Hu et al. [28] analyzed the influence of different atmospheres in the gaseous conversion using the mass spectrometer.

Although the quantification of the gaseous compounds released during the thermal conversion of biomass is not addressed, there are works in the literature where the conversion of biomass was analyzed separately through the macro TGA technique. Baumgarten et al. [29] and Samuelson et al. [30] analyzed the combustion behavior under typical isothermal conditions in the start-up of furnaces. Orang et al. [31] observed the effect of moisture content on combustion behavior. The author highlighted the higher drying and the ignition times due to the increase of the moisture content.

Thus, macro TGA experiments provide the possibility to control and maintain external heat fluxes in order to better represent, at a small-scale, the conditions expected in industrial boilers.

Gauthier et al. [24] used a purpose-built horizontal lamp tube reactor to perform pyrolysis of centimeter-scale wood particles, for temperatures ranging between 450 and 1050 °C. Yang et al. [32] studied the effect of the particle size (pinewood cubes ranging from 5 to 35 mm) on pinewood combustion in a batch reactor by measuring the mass loss rate and the temperature profile at different bed locations and gas composition in the out-of-bed flue gases. Ryu et al. [33] studied the combustion of four biomass materials with different fuel properties under fuel-rich conditions, measuring temperature, mass loss, and gas composition. Mahmoudi et al. [34] focused on developing a numerical model using the Euler-Lagrange model in which the fluid phase is a continuous phase and each particle is tracked with the Lagrangian approach to understand the combustion phase. This work was performed along with an experimental study, with temperature and mass loss monitoring, to validate the numerical model. Wurzenberger et al. [35] created a combined transient single particle and fuel-bed model of a furnace in order to optimize its efficiency and emissions by acquiring information about all the physical and chemical effects on the process. Markovic et al. [36] studied the combustion of wood waste with pre-heated primary air up to 350 °C and the secondary air distributed via nozzles above the waste layer. The authors measured temperature, gas composition, mass loss, and the influence of primary air speed, fuel moisture, and inert content on the combustion characteristics. Eric et al. [37] focused the study on the kinetics of loose biomass in a vertical tube reactor measuring the fuel mass loss rate, with two biomass combustion models (piston and batch model). Most of the testing conditions resemble the traditional (micro) TGA operation where the sample follows a pre-defined heating curve. Usually, this leads to heating rates well below those expected inside a furnace. Long et al. [38] reported an alternative approach in which the sample was introduced into a reactor set at a predefined temperature. The authors only reported mass loss rates. Lelis et al. [39] measured the mass loss and variation of the elemental composition of pine wood pellets. The authors used a macro TGA to investigate the influence of temperature and time on devolatilization of C, H, and N. The experiments were carried out at a constant temperature, and it was found that the rate of release of N is higher than other compounds. However, no information on the actual species formed was provided.

This work presents an experimental facility developed to study how biomass fuels may behave in industrial power plants. Thus, the mass loss profiles at different thermo-chemical conditions of eucalyptus woodchips are presented together with the composition of the gases released over time. Furthermore, it is important to point out that the motivation for this work is related to the need to understand the composition of the gas compounds released during the conversion of solid biomass particles. It is necessary to know the composition, the amount of pyrolysis products in different reactor thermal conditions, and the reaction rate of the particles. A recent investigation mentioned that numerical prediction inside a grate-fired boiler depends on the devolatilization kinetics mechanism, which can significantly affect the outputs from the bed model [40]. Most of the CFD models usually employed biomass elemental composition and enthalpy conservation equations or models, like that proposed by Thunman et al. [41] and Neves et al. [42], to determine the composition of pyrolysis products (e.g., [43]). However, the results of this type of approach can produce unrealistic results. Therefore, there is a clear need to develop macro TGA experiments to obtain experimental information that may be used to develop mathematical models to describe the devolatilization of biomass. These models can be used as an input to CFD models for grate-type combustors.

2. Materials and Methods

2.1. Samples

The woodchips necessary for the macro TGA experiments were prepared from large eucalyptus trunks by means of a knife chipper. Hence, larger particles were obtained, with dimensions similar to the ones used in biomass power plants, ranging from a few millimeters up to hundreds of millimeters. The particles were then spread in a room to be air-dried and to reduce the moisture content. After this, the particles were then sieved, and the particle size distribution was assessed by horizontal screening according to the standard EN 15149-1:2010 Part 2 using sieves with square hole apertures of 3.15, 8, 16, and 50 mm. The most representative particle class size was between 8 to 16 mm, which is in agreement with the analysis carried out on a biomass power plant [44]. In this way, the particles within this class size were collected and used for the experimental program. At that time, and after being air-dried for approximately one month, the moisture content of the particles was observed to be between 10 to 15% (dry basis). The elemental and proximate composition of the particles was also evaluated considering the standards for solid fuel characterization (CEN/TS 15414:2006 and CEN/TS 15104:2005, respectively), and the results are presented in Table 1.

Table 1. Composition of the eucalyptus woodchips.

Proximate Analysis (wt.%, Dry Basis)		Ultimate Analysis (wt.%, Dry Ash Free)	
Volatile matter	88.90	Carbon	48.68
Ash	1.00	Hydrogen	6.91
Fixed carbon	10.10	Nitrogen	0.23
		Oxygen	44.18

2.2. Experimental Apparatus and Procedure

A lab-scale reactor to represent the different thermochemical conversion conditions of solid biomass particles was designed. During the design stage, some important constraints were addressed. The first issue was related to the possible amount of biomass in each experiment and the ease of access to the interior of the reactor in order to introduce the sample. This was particularly important in order to consider secondary reactions in the fuel bed appropriately and also to consider high heating rates of the fuel comparable to real-scale grate-fired boilers. Secondly, high flexibility regarding analytical equipment connected with the reactor, and easy handling during the experiment without any interference with the sample, was also important. Moreover, online recording of relevant operation data was paramount.

Hence, the reactor can replicate the behavior of a fuel sample used in different combustion or gasification devices and, thereby, evolve through the different reaction stages (drying, devolatilization, and char combustion). The reactor had a cylindrical shape, 200 mm in diameter and 350 mm in height. In the surrounding walls, there was a 2 kW electrical heater and refractory material to avoid heat losses. At the upper part of the reactor, there was a rotating lid with a rip of 10 mm to allow the connection of a small basket with the biomass particles to a digital scale. Furthermore, two type K thermocouples were connected to a digital temperature controller, Eurotherm brand, that controlled its operation. The desired temperature in the equipment can be defined in the set-point temperature, but there is no possibility to monitor and acquire their variation over time. There are other external devices necessary to develop the experiment. An external flowmeter and scale were used to control the gas flow rate supplied to the reactor and to measure the mass loss variation of the sample, respectively.

The weight measurement of the sample was carried out by using a perforated cylindrical basket of 60 mm in diameter and 50 mm in height inside the reactor, suspended from the digital scale by a stainless-steel wire of 2 mm in diameter. Consequently, as there were three different devices, a data acquisition and monitoring program was developed using LabVIEW software to centralize the information of the different parameters and to be able to record the data throughout the experiments. Consequently, the LabVIEW program continuously recorded the weight, air flow rate, and temperature. In addition to these devices, a portable gas analyzer, Rapidox 5100 model, was used to measure the main gas compounds (CO_2, CO, H_2, and CH_4) released during the experiment. All the gases have a $\pm 1\%$ full scale accuracy and a 0.1% resolution. The collection and analysis of the gas samples were only possible through the utilization of a vacuum pump and a particle and moisture filter.

Thus, the mass loss and the gas emissions during each experiment are measured over time. Figure 1 presents a schematic diagram of the apparatus involved in the macro TGA experiments.

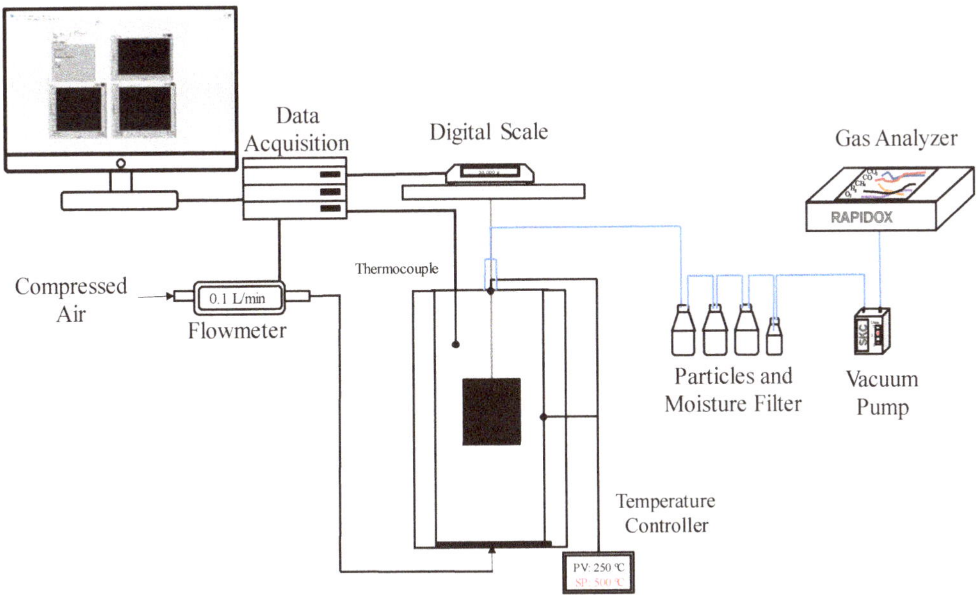

Figure 1. Experimental apparatus.

Regarding the experimental procedure, the small-scale reactor was turned on and preheated to the desired temperature before each experiment. The temperatures considered in this work were 400 and 800 °C in order to present a comprehensive view of biomass conversion over a wide temperature range. Inside the reactor, the walls radiated heat to the surface of the basket, which was in the middle of the reactor. The perforated basket allowed for an air flow to enter and react with the woodchip particles. After reaching a constant temperature in the reactor, the basket was removed, the sample with approximately 20 g was quickly loaded, and the basket was then introduced back into the reactor.

During each experiment, the air flow was controlled at different air flow rates in order to reproduce gasification and combustion regimes at any desired temperature. The air flow necessary for each conversion regime was determined by using Equations (1) and (2), which represent a stoichiometric combustion reaction and the air-to-fuel ratio (λ).

$$C_xH_yO_z + a(O_2 + 3.76\,N_2) \rightarrow b\,CO_2 + cH_2O + d\,N_2 \tag{1}$$

$$\lambda = \frac{\frac{m_{air,real}}{m_{fuel}}}{\frac{a \cdot M_{air}}{x \cdot C + y \cdot H + z \cdot O}} \tag{2}$$

Hence, taking into account the elemental composition of the eucalyptus woodchips presented in Table 1 and the reference value of the air-to-fuel ratio for gasification (0.1 to 0.3) and combustion (1.4), the air flow rate used for gasification experiments was 0.1 L/min, 15.5 for combustion experiments, and 50 L/min for experiments using 400 and 800 °C, respectively. The different air flow rates used for experiments at the highest temperature were necessary because the reaction is considerably faster than at 400 °C. The experiments were run in duplicate for 15 and 10 min for experiments at 400 and 800 °C, respectively. The average values were computed and are reported in the following section. The standard deviation during all experiments was not higher than 0.09%.

3. Results and Discussion

3.1. Thermochemical Conversion: Mass Loss

The mass loss of the samples of eucalyptus woodchips for gasification and combustion regime is presented in Figure 2. Two distinct zones can be identified, each one with a nearly linear variation with time.

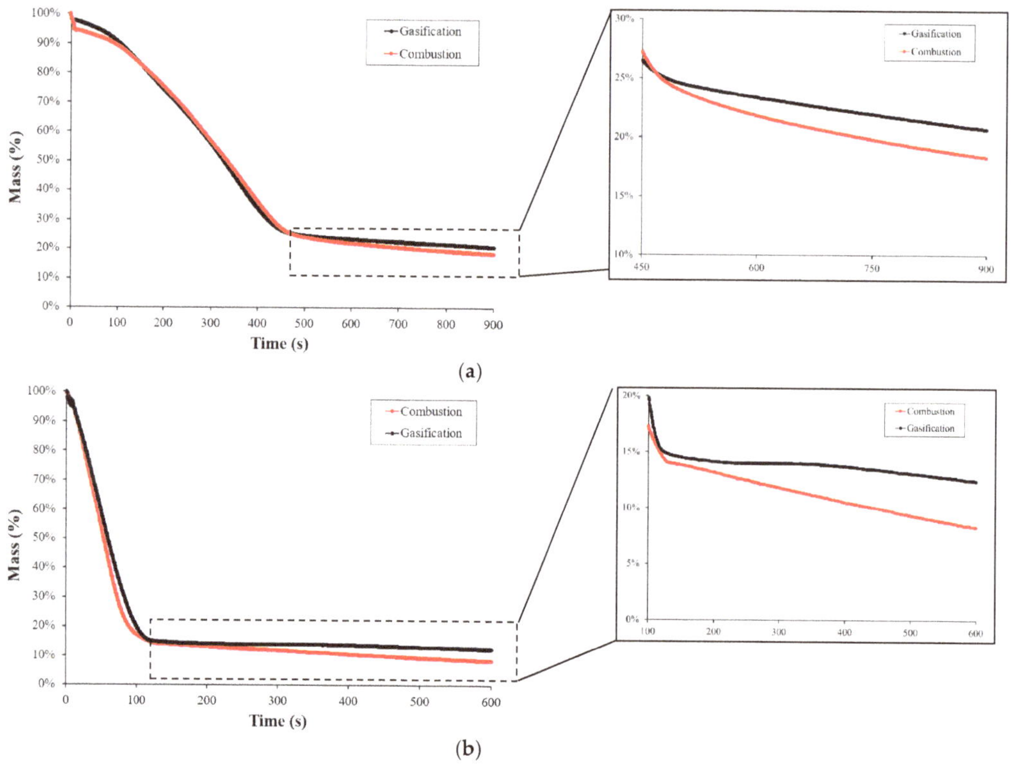

Figure 2. Mass loss curves considering the gasification and combustion regime at: (**a**) 400 °C and (**b**) 800 °C.

In the first region, which includes both the drying and devolatilization stages, there is no significant difference between gasification and combustion conversion regimes. At 400 °C the average mass loss was 75%, while at 800 °C the mass loss was approximately

85%. This suggests that temperature has a small impact on the sample's mass loss. However, temperature has a major impact on the devolatilization rate. The average mass loss rate detected was 0.16 and 0.72%/s in gasification conditions at 400 and 800 °C, respectively. However, in combustion conditions, the devolatilization rate increases from 0.16 to 0.70%/s at 400 and 800 °C, respectively. This means there is an increase of nearly four times the devolatilization rate by increasing the process temperature from 400 to 800 °C.

However, after the first region, the mass loss decreases over time but at a much slower rate. Hence, when the moisture and volatile matter are completely released, char oxidation starts and the mass loss starts to decrease more slowly. Mau et al. [45] showed this phenomena in his work by performing TGA on several char samples.

Hence, in the second region, it is possible to observe that there are differences between the mass loss curves at different thermochemical conditions and at both temperatures. For this stage, the average mass loss rate was 0.016 and 0.006%/s in gasification conditions at 400 and 800 °C, and 0.024 and 0.012%/s in combustion conditions at 400 and 800 °C, respectively. There are two main characteristics that can be noticed. In the first one, the mass loss rate was higher in combustion conditions, indicating that the presence of air enhances thermal loss due to the higher diffusion of oxygen into the solid biomass particles when compared with reduced oxygen in gasification. Simultaneously, at 400 °C the mass loss rate was higher compared with the tests performed at 800 °C for both conditions. This suggests that the remaining volatile matter that could not be devolatilized in the first stage is used to improve the char oxidation. This is in agreement to the review by Li [46] in which the author demonstrated the importance of the volatile-char interactions during the gasification process.

Furthermore, another important issue that can be observed by analyzing the mass loss curves is that the drying and devolatilization stages happen at the same time. One of the reasons might be because the moisture content of the samples was not so significant. However, higher moisture content can introduce some variance on the mass loss on a TGA experiment [47]. As it can be observed, there are no variations of the mass loss in the early stages, which states clearly that the gasification and combustion processes of larger particles do not have the sequence of distinct conversion stages when large particles are burned. In smaller samples, these two steps can be clearly identified as separate events, as the thermal gradients inside the particles are negligible [48]. Table 2 presents some characteristic parameters of the mass loss curves, which corroborate the abovementioned findings. Additionally, the results of the final mass suggest that char conversion also depends on the environment temperature. This fact is highlighted by the difference observed in the result between the experiments at low and high temperatures. Hence, mass transfer (diffusion of oxygen) and kinetics control the second mass region.

Table 2. Characteristic parameters of the mass loss curves at different regimes of thermochemical conversion in the same period of the experiment (600 s).

Temperature	Mass Loss—1st Stage (%)	Devolatilization Time (s)	Final Mass (%)
Gasification			
400 °C	74.49	466	23.35
800 °C	84.60	117	12.44
Combustion			
400 °C	75.19	477	21.80
800 °C	85.80	126	8.30

3.2. Gas Release: Product Distribution

The gases released during all conversion periods were collected. Figure 3 presents the average results from these experiments and shows, in particular, that gaseous emissions are strongly dependent on the operating temperature and the thermochemical conversion regime. The standard deviation during all experiments was not higher than 0.5%. The remaining volume percentage corresponds to the nitrogen and oxygen content.

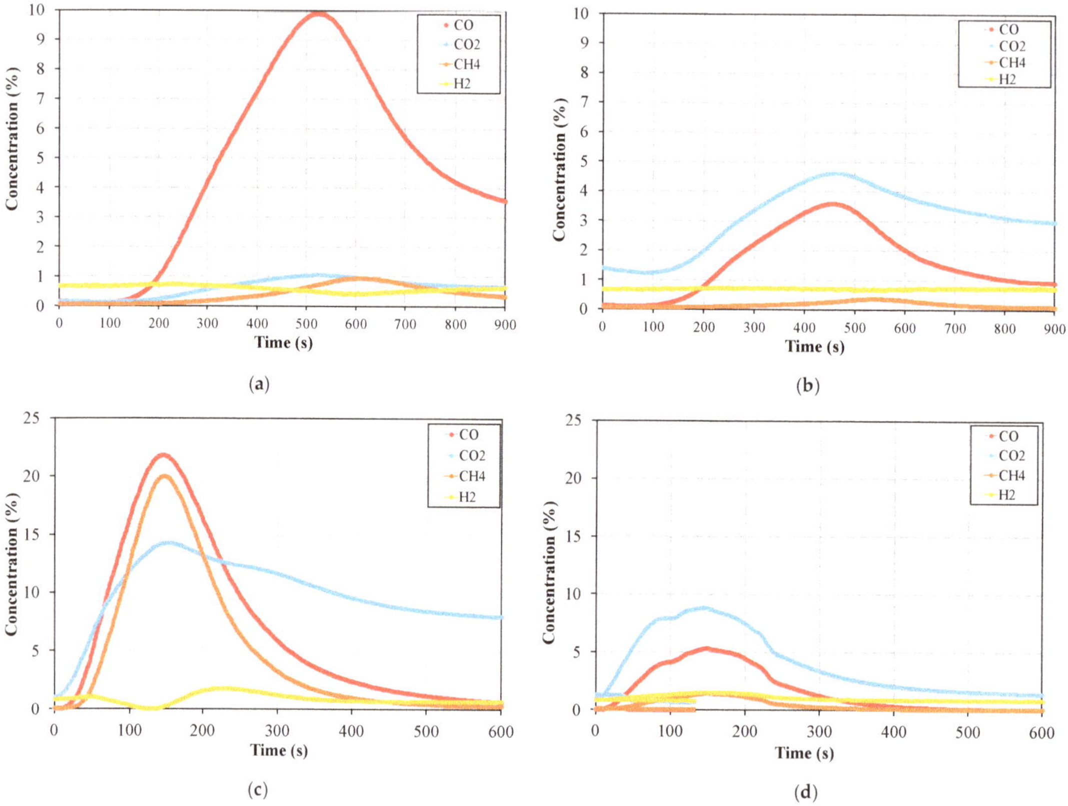

Figure 3. Gas emissions at different reactor temperatures and conversion conditions: (**a**) gasification at 400 °C, (**b**) combustion at 400 °C, (**c**) gasification at 800 °C, (**d**) combustion at 800 °C.

A similar trend was reported by Neves et al. [42], where it was found that at temperatures above 500 °C, the gaseous products strongly become temperature-dependent, leading to a substantial increase in the CO mass fractions. This was considered a result of secondary reactions, which resulted in the decrease in the tar mass fraction, also due to the conversion to CH_4. The same behavior was described by Mehrabian et al. [43] based on dedicated experiments and data collected from the literature. Secondary reactions of the volatiles are negligible at low temperatures, and most of the permanent gases result directly from biomass thermal degradation. Within the low temperature range, gases like CO and CO_2 are the main permanent gas compounds with low quantities of CH_4. As the temperature increases, secondary reactions occur, and an increase in CO and CH_4 are attributed to the decrease of tar. Here, due to higher temperatures, the yields of the volatiles have a strong correlation with the temperature and CO is considered responsible for the conversion of two-thirds of the tar.

Although the tar was not measured during the experiments, the results are in accordance with the theory. As depicted in Figure 3, there is a considerable increment of the CO from 10 to 21%, CH_4 from 1 to 20%, and CO_2 from 1 to almost 15% (peak) in gasification conditions, when increasing the temperature. In combustion conditions, the increase in temperature also enhanced the concentration of CO and CO_2 from 3.5 to 5% and 4.5 to 9%, respectively. Regarding the influence of the air in the process, at 400 °C, increasing the air flow rate enhanced the combustion process by converting the CO into CO_2. At 800 °C, once again, the air revealed a major influence on gas conversion, specifically for the CO and CH_4 that decreased their concentration from 21 to 5% and 20 to 1%, respectively. This fact is in line with the previously mentioned findings reported in the literature. To quantify the differences between both conversion regimes and the influence of the reactor temperature, the volume of each gas compound, as well as their average value, was computed during the experiment. Table 3 presents these values and it is possible to observe that all gas compounds, except H_2, increased when the experiment was developed at 800 °C, particularly in sub-stoichiometric conditions.

Table 3. Normalized gas emissions considering the different regimes of thermochemical conversion, temperatures, and the duration of the experiment (600 s). The numbers in brackets refer to the average percentage during the experiment.

	Gasification		Combustion	
	400 °C	800 °C	400 °C	800 °C
Value (L/g Biomass)				
CO_2	0.32 (0.57)	3.96 (9.99)	1.43 (3.01)	1.28 (4.05)
CO	2.58 (4.50)	2.79 (7.07)	0.86 (1.82)	0.53 (1.73)
CH_4	0.18 (0.30)	2.02 (5.13)	0.08 (0.17)	0.13 (0.44)
H_2	0.32 (0.63)	0.34 (0.87)	0.32 (0.69)	0.36 (1.06)

4. Conclusions

This work presented a purpose-built facility developed to study how solid biomass fuels behave in conditions similar to those found in industrial power plants. This work stands out from the major of macro TGA in literature as it employs a single work temperature in contrast to the usual increase of the temperature on a fixed heating rate [38,49]. From a practical point of view, this information may be valuable for developing mathematical models. Hence, the mass loss profiles at different thermochemical conditions of eucalyptus woodchips are presented together with the composition of the gases released over time at two different temperatures.

The reactor proved to be a very useful equipment to analyze the combustion behavior, study the phenomena, which are relevant in an industrial furnace on a small scale, and demonstrated good reproducibility. Furthermore, the weight loss and the release quantification of the different gas compounds were particularly important to expose the differences between the different thermochemical conversion regimes. The main results and findings led to the following conclusions:

- Mainly due to the low moisture content, only two different stages can characterize the mass loss of eucalyptus woodchips at low and high temperatures with different kinetics for both thermochemical regimes.
- The first region, which corresponds to a mass loss of 75 to 85% (gasification and combustion, respectively), does not present the influence of the air flow rate, which defines the thermochemical conversion condition; however, there is a significant difference between gasification and combustion conversion at different temperatures. Therefore, the kinetics of the reaction in the devolatilization stage is mainly dependent on the temperature.
- The second conversion stage, in its turn, is dependent on the air flow rate and, therefore, dependent on the diffusion of the oxygen supplied to the solid biomass particles. However, it was verified that at the lowest reactor temperature this reaction presented a contribution of the temperature. This might be related to the remaining volatile matter that was not consumed in the devolatilization stage and was used to react with the remaining carbon.
- Regarding the gases released during all conversion periods, a strong dependency on the reactor temperature and the thermochemical conversion regime was observed.

- All gas compounds, except H_2, increased substantially with reactor temperature and, mainly, when gasification occurred. This suggests that the success of obtaining a better combustible gas during the gasification process depends substantially on process temperature [50].
- The yield of the lowest temperature, comparing both thermochemical conversion conditions, the CO and CO_2 emissions are approximately 3 times higher when gasification occurs. However, at 800 °C, the same trend was verified, while a considerable amount of CH_4 was also verified.
- The gas emissions also showed the impact of the air injection in the conversion of combustible gases and in non-combustible gases. By increasing the CO_2, CO, and CH_4, concentrations for both temperatures were reduced.
- The measurement of the gas emissions in the purpose-built facility also demonstrated a useful strategy to further define the correct boundary conditions for CFD simulations.

For future work, data concerning reaction kinetics and gas emissions will be incorporated into a numerical model to describe the thermochemical conversion behavior inside an industrial grate-fired boiler. This type of information is one of the main drawbacks of commercial simulation software. Additionally, this approach extends to other types of fuels whose behavior is less understood.

Author Contributions: Conceptualization, J.S., C.C. and J.T.; Methodology, J.S., C.C. and J.T.; Investigation, J.S.; Writing—original draft preparation, J.S., C.C., J.T. and S.T.; Writing—review and editing, J.S., C.C., J.T. and S.T.; Supervision, J.T. and S.T. All authors have read and agreed to the published version of the manuscript.

Funding: This work was supported by the Portuguese Foundation for Science and Technology (FCT) within the R&D Units Project Scope UIDB/00319/2020 (ALGORITMI), and R&D Units Project Scope UIDP/04077/2020 (MEtRICs).

Institutional Review Board Statement: Not applicable.

Informed Consent Statement: Not applicable.

Acknowledgments: The first author would like to express his gratitude for the support given by FCT through the Ph.D. Grant SFRH/BD/130588/2017.

Conflicts of Interest: The authors declare no conflict of interest.

References

1. Demirbas, A. Potential applications of renewable energy sources, biomass combustion problems in boiler power systems and combustion related environmental issues. *Prog. Energy Combust. Sci.* **2005**, *31*, 171–192. [CrossRef]
2. Ferreira, S.; Monteiro, E.; Brito, P.; Vilarinho, C. Biomass resources in Portugal: Current status and prospects. *Renew. Sustain. Energy Rev.* **2017**, *78*, 1221–1235. [CrossRef]
3. GPP—Gabinete de Planeamento Políticas e Administração Geral. Caderno de Análise e Prospetiva Cultivar. 2018. Available online: https://doi.org/2183-5624 (accessed on 11 September 2022).
4. Roman, K.; Barwicki, J.; Rzodkiewicz, W.; Dawidowski, M. Evaluation of Mechanical and Energetic Properties of the Forest Residues Shredded Chips during Briquetting Process. *Energies* **2021**, *14*, 3270. [CrossRef]
5. Anca-Couce, A.; Hochenauer, C.; Scharler, R. Bioenergy technologies, uses, market and future trends with Austria as a case study. *Renew. Sustain. Energy Rev.* **2020**, *135*, 110237. [CrossRef]
6. Basu, P. *Biomass Gasification, Pyrolysis and Torrefaction: Practical Design and Theory*, 2nd ed.; Elsevier Inc.: Amsterdam, The Netherlands, 2013; p. 530. [CrossRef]
7. Vainio, E. Fate of Fuel-Bound Nitrogen and Sulfur in Biomass-Fired Industrial Boilers. Ph.D. Thesis, Åbo Akademi University, Turku, Finland, 2014.
8. Sadaka, S.; Johnson, D. Biomass Combustion. Agriculture and Natural Resources. 2010. Available online: https://www.researchgate.net/publication/268207461_Biomass_Combustion (accessed on 11 September 2022).
9. Kirch, T.; Birzer, C.; van Eyk, P.; Medwell, P.R. Influence of Primary and Secondary Air Supply on Gaseous Emissions from a Small-Scale Staged Solid Biomass Fuel Combustor. *Energy Fuels* **2017**, *32*, 4212–4220. [CrossRef]
10. Sun, J.; Shen, Z.; Zhang, L.; Zhang, Q.; Lei, Y.; Cao, J.; Huang, Y.; Liu, S.; Zheng, C.; Xu, H.; et al. Impact of primary and secondary air supply intensity in stove on emissions of size-segregated particulate matter and carbonaceous aerosols from apple tree wood burning. *Atmos. Res.* **2017**, *202*, 33–39. [CrossRef]

11. Yin, C.; Rosendahl, L.A.; Kær, S.K. Grate-firing of biomass for heat and power production. *Prog. Energy Combust. Sci.* **2008**, *34*, 725–754. [CrossRef]
12. Perera, S.M.; Wickramasinghe, C.; Samarasiri, B.; Narayana, M. Modeling of thermochemical conversion of waste biomass—A comprehensive review. *Biofuel Res. J.* **2021**, *8*, 1481–1528. [CrossRef]
13. Ragland, K.; Aerts, D.; Baker, A. Properties of wood for combustion analysis. *Bioresour. Technol.* **1991**, *37*, 161–168. [CrossRef]
14. Hellén, H.; Hakola, H.; Haaparanta, S.; Pietarila, H.; Kauhaniemi, M. Influence of residential wood combustion on local air quality. *Sci. Total Environ.* **2008**, *393*, 283–290. [CrossRef]
15. Jia, Y.; Li, Z.; Wang, Y.; Wang, X.; Lou, C.; Xiao, B.; Lim, M. Visualization of Combustion Phases of Biomass Particles: Effects of Fuel Properties. *ACS Omega* **2021**, *6*, 27702–27710. [CrossRef] [PubMed]
16. Grønli, M.G.; Varhegyi, G.; Di Blasi, C. Thermogravimetric Analysis and Devolatilization Kinetics of Wood. *Ind. Eng. Chem. Res.* **2002**, *41*, 4201–4208. [CrossRef]
17. Chen, Z.; Hu, M.; Zhu, X.; Guo, D.; Liu, S.; Hu, Z.; Xiao, B.; Wang, J.; Laghari, M. Characteristics and kinetic study on pyrolysis of five lignocellulosic biomass via thermogravimetric analysis. *Bioresour. Technol.* **2015**, *192*, 441–450. [CrossRef]
18. Mishra, R.K.; Mohanty, K. Pyrolysis kinetics and thermal behavior of waste sawdust biomass using thermogravimetric analysis. *Bioresour. Technol.* **2018**, *251*, 63–74. [CrossRef] [PubMed]
19. Xiao, R.; Yang, W.; Cong, X.; Dong, K.; Xu, J.; Wang, D.; Yang, X. Thermogravimetric analysis and reaction kinetics of lignocellulosic biomass pyrolysis. *Energy* **2020**, *201*, 117537. [CrossRef]
20. Gaitán-Álvarez, J.; Moya, R.; Puente-Urbina, A.; Rodriguez-Zúñiga, A. Thermogravimetric, Devolatilization Rate, and Differential Scanning Calorimetry Analyses of Biomass of Tropical Plantation Species of Costa Rica Torrefied at Different Temperatures and Times. *Energies* **2018**, *11*, 696. [CrossRef]
21. Fernandez, A.; Soria, J.; Rodriguez, R.; Baeyens, J.; Mazza, G. Macro-TGA steam-assisted gasification of lignocellulosic wastes. *J. Environ. Manag.* **2018**, *233*, 626–635. [CrossRef]
22. Becidan, M.; Skreiberg, Ø.; Hustad, J.E. Products distribution and gas release in pyrolysis of thermally thick biomass residues samples. *J. Anal. Appl. Pyrolysis* **2007**, *78*, 207–213. [CrossRef]
23. Brunner, T.; Biedermann, F.; Kanzian, W.; Evic, N.; Obernberger, I. Advanced Biomass Fuel Characterization Based on Tests with a Specially Designed Lab-Scale Reactor. *Energy Fuels* **2013**, *27*, 5691–5698. [CrossRef]
24. Gauthier, G.; Melkior, T.; Grateau, M.; Thiery, S.; Salvador, S. Pyrolysis of centimetre-scale wood particles: New experimental developments and results. *J. Anal. Appl. Pyrolysis* **2013**, *104*, 521–530. [CrossRef]
25. Weissinger, A. In situ FT-IR spectroscopic investigations of species from biomass fuels in a laboratory-scale combustor: The release of nitrogenous species. *Combust. Flame* **2004**, *137*, 403–417. [CrossRef]
26. Bennadji, H.; Smith, K.; Shabangu, S.; Fisher, E. Low-Temperature Pyrolysis of Woody Biomass in the Thermally Thick Regime. *Energy Fuels* **2013**, *27*, 1453–1459. [CrossRef]
27. Nikku, M.; Deb, A.; Sermyagina, E.; Puro, L. Reactivity characterization of municipal solid waste and biomass. *Fuel* **2019**, *254*, 115690. [CrossRef]
28. Hu, Q.; He, X.; Yao, Z.; Dai, Y.; Wang, C.-H. Gaseous production kinetics and solid structure analysis during isothermal conversion of biomass pellet under different atmospheres. *J. Energy Inst.* **2021**, *98*, 53–62. [CrossRef]
29. Baumgarten, B.; Reinhardt, J.; Lepski, C.; Risio, B.; Thorwarth, H. Kinetics of Wood Devolatilization during Start-up. *Energy Fuels* **2019**, *33*, 11285–11291. [CrossRef]
30. Samuelsson, L.N.; Umeki, K.; Babler, M.U. Mass loss rates for wood chips at isothermal pyrolysis conditions: A comparison with low heating rate powder data. *Fuel Process. Technol.* **2017**, *158*, 26–34. [CrossRef]
31. Orang, N.; Tran, H. Effect of feedstock moisture content on biomass boiler operation. *TAPPI J.* **2015**, *14*, 629–637. [CrossRef]
32. Bin Yang, Y.; Ryu, C.; Khor, A.; Sharifi, V.N.; Swithenbank, J. Fuel size effect on pinewood combustion in a packed bed. *Fuel* **2005**, *84*, 2026–2038. [CrossRef]
33. Ryu, C.; Yang, Y.B.; Khor, A.; Yates, N.E.; Sharifi, V.N.; Swithenbank, J. Effect of fuel properties on biomass combustion: Part I. Experiments—Fuel type, equivalence ratio and particle size. *Fuel* **2006**, *85*, 1039–1046. [CrossRef]
34. Mahmoudi, A.H.; Markovic, M.; Peters, B.; Brem, G. An experimental and numerical study of wood combustion in a fixed bed using Euler–Lagrange approach (XDEM). *Fuel* **2015**, *150*, 573–582. [CrossRef]
35. Wurzenberger, J.C.; Wallner, S.; Raupenstrauch, H.; Khinast, J. Thermal conversion of biomass: Comprehensive reactor and particle modeling. *AIChE J.* **2002**, *48*, 2398–2411. [CrossRef]
36. Markovic, M.; Bramer, E.A.; Brem, G. Experimental investigation of wood combustion in a fixed bed with hot air. *Waste Manag.* **2014**, *34*, 49–62. [CrossRef] [PubMed]
37. Erić, A.; Nemoda, S.; Komatina, M.; Dakić, D.; Repić, B. Experimental investigation on the kinetics of biomass combustion in vertical tube reactor. *J. Energy Inst.* **2019**, *92*, 1077–1090. [CrossRef]
38. Long, Y.; Zhou, H.; Meng, A.; Li, Q.; Zhang, Y. Interactions among biomass components during co-pyrolysis in (macro)thermogravimetric analyzers. *Korean J. Chem. Eng.* **2016**, *33*, 2638–2643. [CrossRef]
39. Fraga, L.G.; Silva, J.; Teixeira, J.C.; Ferreira, M.E.C.; Teixeira, S.F.; Vilarinho, C.; Gonçalves, M.M. Study of Mass Loss and Elemental Analysis of Pine Wood Pellets in a Small-Scale Reactor. *Energies* **2022**, *15*, 5253. [CrossRef]
40. Jiang, M.; Lai, A.; Law, A. Solid Waste Incineration Modelling for Advanced Moving Grate Incinerators. *Sustainability* **2020**, *12*, 8007. [CrossRef]

41. Thunman, H.; Niklasson, F.; Johnsson, F.; Leckner, B. Composition of Volatile Gases and Thermochemical Properties of Wood for Modeling of Fixed or Fluidized Beds. *Energy Fuels* **2001**, *15*, 1488–1497. [CrossRef]
42. Neves, D.; Thunman, H.; Matos, A.; Tarelho, L.; Gómez-Barea, A. Characterization and prediction of biomass pyrolysis products. *Prog. Energy Combust. Sci.* **2011**, *37*, 611–630. [CrossRef]
43. Mehrabian, R.; Shiehnejadhesar, A.; Scharler, R.; Obernberger, I. Multi-physics modelling of packed bed biomass combustion. *Fuel* **2014**, *122*, 164–178. [CrossRef]
44. Silva, J.P.; Teixeira, S.; Grilo, É.; Peters, B.; Teixeira, J.C. Analysis and monitoring of the combustion performance in a biomass power plant. *Clean. Eng. Technol.* **2021**, *5*, 100334. [CrossRef]
45. Mau, V.; Gross, A. Energy conversion and gas emissions from production and combustion of poultry-litter-derived hydrochar and biochar. *Appl. Energy* **2018**, *213*, 510–519. [CrossRef]
46. Li, C.-Z. Importance of volatile–char interactions during the pyrolysis and gasification of low-rank fuels—A review. *Fuel* **2013**, *112*, 609–623. [CrossRef]
47. Magdziarz, A.; Werle, S. Analysis of the combustion and pyrolysis of dried sewage sludge by TGA and MS. *Waste Manag.* **2014**, *34*, 174–179. [CrossRef]
48. Fraga, L.G.; Silva, J.; Teixeira, S.; Soares, D.; Ferreira, M.; Teixeira, J. Influence of Operating Conditions on the Thermal Behavior and Kinetics of Pine Wood Particles Using Thermogravimetric Analysis. *Energies* **2020**, *13*, 2756. [CrossRef]
49. Skreiberg, A.; Sandquist, J.; Sørum, L. TGA and macro-TGA characterisation of biomass fuels and fuel mixtures. *Fuel* **2011**, *90*, 2182–2197. [CrossRef]
50. Basu, P. *Biomass Gasification, Pyrolysis and Torrefaction: Pratical Design and Theory*, 3rd ed.; Academic Press: Cambridge, MA, USA, 2018. [CrossRef]

processes

MDPI

Article

Chemical and Functional Characterization of Extracts from Leaves and Twigs of *Acacia dealbata*

Ricardo Correia [1], Maria Paula Duarte [1], Elisabete Muchagato Maurício [2], João Brinco [3], José Carlos Quintela [4], Marco Gomes da Silva [5] and Margarida Gonçalves [1,6,*]

1 MEtRICs/NOVA School of Science and Technology, Universidade NOVA de Lisboa, Campus de Caparica, 2829-516 Caparica, Portugal
2 Faculty of Engineering/CBIOS, Universidade Lusófona, 1749-024 Lisboa, Portugal
3 CENSE-Center for Environmental and Sustainability Research & CHANGE—Global Change and Sustainability Institute, NOVA School of Science and Technology, Universidade NOVA de Lisboa, Campus de Caparica, 2829-516 Caparica, Portugal
4 Natac Biotech, C/Electrónica 7, 28923 Alcorcón, Madrid, Spain
5 LAQV/REQUIMTE, Department of Chemistry, NOVA School of Science and Technology, Universidade NOVA de Lisboa, Campus de Caparica, 2829-516 Caparica, Portugal
6 VALORIZA-Research Center for Endogenous Resource Valorization, Polytechnic Institute of Portalegre, 7300-555 Portalegre, Portugal
* Correspondence: mmpg@fct.unl.pt

Abstract: The purpose of this work was to evaluate the recovery of bioactive extracts from *Acacia dealbata* leaves and twigs and to characterize their chemical composition and functional properties. Fresh and air-dried samples were extracted by maceration at room temperature and by hot extraction at 60 °C using aqueous solutions of acetone, ethanol, and methanol. The highest extraction yields (14.8 and 12.0% for dried leaves and twigs, respectively) were obtained with 70% acetone, for both extraction procedures. Extracts were characterized for total phenolics content (TPC), total flavonoid content (TFC) and total proanthocyanidin content (TPrAC). Bioactive extracts with high TPC (526.4 mg GAE/g extract), TFC (198.4 mg CatE/g extract), and TPrAC (631.3 mg PycE/g extract) were obtained using maceration, a technically simple and low-energy process. The non-polar fraction of selected extracts was characterized using gas chromatography and time of flight mass spectrometry (GC-TOFMS). The main components detected were phytol, squalene, α-tocopherol, lupenone, and lupeol. The antioxidant activity of the extracts was characterized through DPPH and FRAP assays. Antimicrobial activity of the extracts against different bacteria was also determined. The highest DPPH and FRAP activities were obtained from dried twigs from Alcobaça (1068.3 mg TE/g extract and 9194.6 mmol Fe^{2+}/g extract, respectively). Extracts from both leaves and twigs showed antimicrobial properties against *Staphylococcus aureus*, *Staphylococcus epidermidis*, methicillin resistant *Staphylococcus aureus* (MRSA), *Enterococcus faecalis*, *Bacillus cereus*, *Streptococcus mutans*, and *Streptococcus mitis*. The results obtained demonstrate the feasibility of recovering valuable components from these biomass fractions that may be further valorized for energy production in a biorefinery concept.

Keywords: *Acacia dealbata*; leaves; twigs; antioxidant; antimicrobial; invasive species

1. Introduction

The genus *Acacia* comprises more than 1350 species distributed throughout tropical and warm temperate areas of the world [1]. Most of those species are native to Australia but spread all over the world due to a wide variety of useful applications such as sand and dunes stabilization, extraction of tannins, essences or gums, valorization as timber or fodder crop, and production of biofuels [2–4]. The ability to easily adapt to changing environments, the large seed production and accumulation of massive seed banks for long periods, the high capacity for reproduction after fires and cuttings, and the allelopathic properties are some of the characteristics that contribute to the success of *Acacia* spp. as invaders, leading to negative impacts on ecosystem biodiversity [5–7].

A. dealbata is one of the most widespread *Acacia* species [7,8] and is classified as invasive in Portuguese territory [9]. Periodic removal operations to minimize its proliferation generate high amounts of biomass that is usually burned for energy production or landfilled, but these solutions are not economically sustainable due to the high costs of biomass collection and transportation. However, different fractions of *Acacia* spp. biomass have been evaluated as raw materials to produce functional extracts that can be used in the nutraceutical, cosmetic, or food industries. In fact, it is widely documented that bark, wood, leaves, flowers, pods, seeds or roots of *Acacia* spp. are rich in bioactive secondary metabolites (e.g., amines and alkaloids, cyanogenic glycosides, cyclitols, fatty acids and seed oils, gums, non-protein amino acids, terpenes, tannins and other flavonoids, and simple phenolics) [4,10] and have been used in traditional medicine for a wide range of ailments, such as diabetes, worm infection, dysmenorrhea, eczema, malaria, gout, jaundice, abdominal pain, kidney problems, constipation, leprosy, piles, pneumonia, rheumatism, fever, and cancer [11].

Plant extracts are known for their ability to act as antioxidants and reduce oxidative stress [12], a physiologic condition considered to play a key role in the pathogenesis of several degenerative diseases, such as cardiovascular diseases, diabetes neurodegeneration, or cancer [13]. Additionally, it has been reported that plant secondary metabolites also possess antimicrobial properties, which is important in the development of alternatives to antibiotics due to the increasing resistance to conventional antimicrobial agents [14]. Therefore, the production of bioactive extracts from *A. dealbata* biomass is an additional pathway for the valorization of these biomass materials, complementing energy applications and contributing to the sustainability of the forest cleaning and management actions, reducing the risk of fire, and improving the social-economic development of rural areas.

Extraction of value-added components from *Acacia* spp. has been focused mainly on the bark, flower, wood, and leaves as reviewed by Correia et al. [15]. Concerning the leaves fraction, antioxidant or antimicrobial activities have been determined for *A. farnesiana* [16], *A. karroo* [17–20], *A. longifolia* [21], *A. pycnantha* [22,23], *A. saligna* [24–27], or *A. nilotica* [28]. The antioxidant and antimicrobial activities of *A. dealbata* leaves were evaluated by Borges et al. [29], for acetonic and ethanolic extracts of fresh *A. dealbata* leaves, obtained with different extraction methods. Ethanolic extracts of dried *A. dealbata* leaves were also found to have antimicrobial activity against the food poisoning agent *Bacillus cereus* [30].

Research on extraction of functional components from twigs of *Acacia* spp. is scarce. Extracts of *A. nilotica* twigs were characterized for antimicrobial activity against oral pathogens [31–34], while extracts of *A. pennata* twigs were described as having some potential application in the prevention of Alzheimer's disease [35].

The extraction methods explored so far included solid-liquid extraction for a predetermined period under agitation at room temperature (1 h [29], 8 h [34], or 2 days [30,33]) ultrasound-assisted extraction [29], Soxhlet [29,35], microwave-assisted extraction [29], cold percolation [31], and supercritical fluid extraction [32,36]. Most of these methods are energy-consuming or involve high investment costs and may promote thermal degradation of the extract components [37].

This study aimed to investigate the production of bioactive extracts from leaves and twigs of *A. dealbata*, using maceration at room temperature, a method with low energy requirements and easy to implement on an industrial scale. Extraction with different aqueous solvents (acetone, ethanol, and methanol) was applied to fresh and dried biomass, collected in two different locations to evaluate the influence of solvent polarity, biomass water content, and geographic origin on the characteristics of the extracts, namely their yield, composition, and properties. To the best of our knowledge, this is also the first study on the production and characterization of bioactive extracts from *A. dealbata* twigs. The extracts were characterized for mass yield, total phenolic content, total flavonoid content, total proanthocyanidins content, in vitro antioxidant activity tests, and antimicrobial activity against several bacteria and yeasts. To identify non-polar components that might contribute to the antioxidant or antimicrobial properties of the extracts, a representative set of leaves and twigs extracts were fractionated and characterized by GC-TOFMS.

2. Materials and Methods

2.1. Plant Material

Branches of at least ten different *A. dealbata* trees of Caparica (CAP) and Alcobaça (ALC) regions were collected. Samples of around 500 g of branches were collected from each tree and combined to obtain a composite sample from that geographical origin. For each composite sample, leaves were separated from the twigs and both fractions were analyzed separately. Fresh and dried leaves and twigs were manually cut into small pieces about 1 cm in length before impregnation with the solvent.

2.2. Chemicals and Reagents

Acetone, ethanol 96°, methanol, and petroleum ether used in the extraction processes were purchased from CarloErba Reagents (Val de Reuil, France). Folin-Ciocalteau reagent (Panreac, Barcelona, Spain), sodium carbonate (Labkem, Barcelona, Spain), and gallic acid (Merck, Darmstadt, Germany) were used to determine phenolics content. The reagents sodium nitrite ($NaNO_2$), sodium hydroxide (NaOH), and aluminium chloride ($AlCl_3$) used in the determination of flavonoids were all purchased from Merck (Darmstadt, Germany). Butanol (Panreac, Barcelona, Spain), ferrous sulfate heptahydrate ($FeSO_4 \cdot 7H_2O$) (Panreac, Barcelona, Spain), hydrochloric acid (Chem-Lab NV, Zedelgem, Belgium), and Pycnognol® (generously provided by Horphag Research Ltd, Geneva, Switzerland) were used in the acid-butanol assay. The reagent 2,2-diphenyl-1-picrylhydrazyl (DPPH) (Sigma-Aldrich, St. Louis, MO, USA) was used in the DPPH assay. The reagents 2,4,6-tris(2-pyridyl)-s-triazine (TPTZ), acetate buffer 3.6, iron (III) chloride ($FeCl_3$), and iron (II) sulfate ($FeSO_4$) used in the ferric reducing antioxidant power (FRAP) assay were also purchased from Merck (Darmstadt, Germany). Ketoconazole, ofloxacin and vancomycin were purchase from Sigma-Aldrich (St. Louis, MO, USA). Brain heart infusion agar (BHIA) was purchased from Merck (Darmstadt, Germany). Mueller-Hinton agar (MHA), Sabouraud dextrose agar (SDA) and Trypto casein-soy agar (TSA) were purchase from Biokar diagnostics (Allonne, France).

2.3. Extraction Procedure

A portion of leaves and twigs was processed in a fresh state, during the first 24 h from harvest, while the remaining biomass was left to dry in air, at room temperature, until a constant weight was attained. Two extraction methods were compared with the fresh and dried raw material: maceration at room temperature for 48 h in the case of the leaves and 24 h for the twigs; extraction times were selected for leaves and twigs as the maceration time necessary to achieve constant values of extract yield and extract composition according to a previous study (Supplementary Materials; Table S2). Hot extraction was performed at 60 °C for 1 h, for both leaves and twigs, to evaluate the effect of temperature in extract yield and properties. Higher temperatures were not selected to avoid thermal degradation of the extracts. Maceration and hot extraction were performed in a single step, at a biomass:solvent ratio of 1:6 (w:v), using 70% of aqueous acetone (70% ACE), 70% aqueous ethanol (70% ET), and 70% of aqueous methanol (70% MET) as extraction solvents. After the extraction period, the liquid extract was separated from the biomass by filtration and stored at −20 °C until analysis. An adequate aliquot was evaporated to dryness at reduced pressure to determine the mass yield (expressed as percentage, on a wet basis). The extraction process was made in duplicate for each one of the four biomass samples of leaves and twigs (fresh and dried material, from Caparica and Alcobaça).

Extraction of spent raw material was also assessed after the removal of lipophilic components by petroleum ether. Dried leaves and twigs were extracted with petroleum ether by maceration at room temperature using a biomass:solvent ratio of 1:5, and the lipophilic extracts were stored for subsequent analysis and not included in this study. The spent biomass was left to dry in air at room temperature and then submitted to extraction of polar components by maceration at room temperature with 70% ACE at the same conditions that were used for the raw leaves and twigs. Mass yields were determined as described above, and the extracts were stored at −20 °C until analysis.

2.4. Extracts Characterization

Total phenolic content (TPC) of the extracts was measured by the Folin-Ciocalteu method, following an adaptation of the method described by Singleton et al. [38]. Briefly, 0.5 mL of extract, 2 mL of distilled water, and 0.5 mL of Folin-Ciocalteu reagent were added to a test tube and allowed to stand for 5 min. Then, 2 mL of aqueous sodium bicarbonate (10% m/V) were added, and the mixture was incubated for 1 h in the dark. Absorbance was then measured at 760 nm and total phenolic content was determined using a calibration curve constructed with gallic acid standards. Results were expressed as gallic acid equivalents (mg GAE/g of dry extract, and mg GAE/g of fresh biomass).

Total flavonoid content was determined as described by Barros et al. [39]. Briefly, 0.5 mL of extract was mixed with 2 mL of distilled water and subsequently with 0.15 mL of $NaNO_2$ solution (5%). After 6 min, 0.15 mL of $AlCl_3$ solution (10%) was added and allowed to stand further 6 min. After this period 2 mL of NaOH solution (4%) was added to the mixture, and distilled water was then added to bring the final volume to 5 mL. Then, the mixture was completely mixed and allowed to stand for 15 min. Finally, absorbance was measured at 510 nm (Biochrom Libra S4). The results were expressed in catechin equivalents (mg CatE/g of dry extract, and mg CatE/g of fresh biomass). A calibration curve was also made using rutin as a standard for comparison of the results with the ones found in the literature expressed in rutin equivalents (mg RE/g of dry extract, and mg RE/g of fresh biomass).

Total proanthocyanidin content (TPrAC) was determined by the butanol-HCl assay, exactly as described by Skerget et al. [40]. Briefly, 5 mL of an acidic ferrous solution (77 mg $FeSO_4 \cdot 7H_2O$ in 500 mL HCl/BuOH (2/3)) was added to 0.5 mL of the extract. The tubes were covered and put in a water bath at 95 °C for 15 min. The absorbance was read at 540 nm (Biochrom Libra S4). The same procedure was made using a proanthocyanidin commercial extract (Pycnogenol®) to build a calibration curve, and the results were expressed as Pycnogenol® equivalents (mg PycE/g of dry extract, and mg PycE/g of fresh biomass).

Antioxidant activity was evaluated by the DPPH free radical scavenging assay and by the ferric reducing antioxidant power assay. For the DPPH assay [41], 0.5 mL of extract were mixed with 4 mL of DPPH solution (2,2-diphenyl-1-picrylhydrazyl at 45 mg/L in methanol), and then incubated at room temperature for 30 min. After incubation, absorbance was measured at 517 nm and the results were expressed in Trolox equivalents by using a calibration curve obtained with Trolox as a standard (mg TE/g of dry extract, and mg TE/g of fresh biomass).

The FRAP assay was carried out using the procedure of Benzie and Strain with slight modifications [42]. The FRAP reagent was prepared from 0.3 M acetate buffer (pH 3.6), 10 mM TPTZ solution in 40 mM HCl, and 20 mM iron (III) chloride solution in proportions of 10:1:1 (*v*/*v*), respectively. This reagent was freshly prepared before analysis and warmed to 37 °C in a water bath prior to use. One hundred microliters of sample were added to 3 mL of the FRAP reagent and the mixture was kept at 37 °C for 30 min. After this period, absorbance was recorded at 593 nm. A standard curve was made using iron (II) sulfate solution, and the results were expressed as mmol Fe(II)/g of dry extract and as mmol Fe(II)/g of fresh biomass.

In all these characterization assays, two replicates of each sample were analyzed in triplicate.

2.5. Antimicrobial Activity

Antimicrobial activity was assayed against Gram-negative bacteria (*Escherichia coli* ATCC® 8739™ and *Pseudomonas aeruginosa* ATCC® 9027™), Gram-positive bacteria (*Staphylococcus aureus* ATCC® 6538™, *Staphylococcus epidermidis* ATCC®12228™, methicillin resistant *Staphylococcus aureus* (MRSA) ATCC® 33591™, *Enterococcus faecalis* ATCC® 29212™, *Bacillus cereus* ATCC®11778™, *Streptococcus mutans* ATCC®25175™ and *Streptococcus mitis* NCIMB®13770) and a yeast (*Candida albicans* ATCC® 10231™). All microbial strains were kept at −70 °C in broth with glycerol (15% *v*/*v*). The antimicrobial activity was assessed by the well-diffusion assay, according to Pereira et al. [43]. Briefly, microorganisms were subcultured on TSA (*E. coli*, *P. aeruginosa*, *S. Choleraesuis*, *S. aureus*, MRSA, *S. epidermidis*, *E. faecalis* and *B. cereus*), BHIA (*S. mutans* and *S. mitis*) or SDA (*C. albicans*) and incubated at 30 ± 2 °C (*B. cereus* and *C. albicans*) or 35 ± 2 °C (remaining bacteria). Isolated colonies were suspended in saline medium (NaCl, 0.85% *w*/*v*), and the turbidity was adjusted to 0.5 on the McFarland scale (approx. 1–2 × 10^8 CFU/mL for bacteria and 1–5 × 10^6 CFU/mL for yeasts) (DEN-1B McFarland Densitometer, Grant-bio). Subsequently, microbial suspensions were spread on BHIA (*S. mutans* and *S. mitis*) or MHA (remaining microorganisms) Petri dishes, wells (6 mm in diameter) were aseptically punched, and 50 µL of extracts (10 mg/mL) were poured into the wells. The plates were incubated, in the dark, for 24 h at 30 ± 2 °C (*B. cereus* and *C. albicans*) or 35 ± 2 °C (remaining bacteria). Antimicrobial activity was evaluated by measuring the diameter of the growth inhibition zone (mm) around the well. Paper disks impregnated with vancomycin (Gram-positive bacteria), ofloxacin (Gram-negative bacteria) and ketoconazole (*C. albicans*) were used as positive controls. Ethanol (70% *w*/*v*) was used as a negative control. Data are presented as the mean ± standard deviation. All the determinations were performed in triplicates.

2.6. GC-TOFMS Analysis

The extracts selected for analysis of non-polar components were derivatized using a procedure adapted from Popova et al. [44]. Briefly, 2 mL of each extract was evaporated to dryness and the residue was diluted in 100 µL of pyridine and 100 µL of the derivatizing agent N,O-bis(trimethylsilyl)trifluoroacetamide (BSTFA); the mixture was incubated at room temperature for 12 h, diluted in petroleum ether (1.5 mL), dried with anhydrous sodium sulphate and kept at −5 °C until analysis. Analyses were performed with an Agilent 7890B (Palo Alto, CA, USA) gas chromatograph equipped with a multi-mode inlet. An Agilent HP-5MS UI fused silica capillary column (30 m × 0.25 mm i.d., 0.25 µm d f—film thickness) was used in all separations. An aliquot of the derivatized sample (1 µL) was injected via a LECO L-PAL3 autosampler fitted with a 10 µL syringe. The injector was operated in solvent vent mode, with the split valve opened at 100 mL/min for 25 s, and the injector at 70 °C. Then, the split valve was closed for 120 s, and the inlet temperature was raised to 300 °C at 700 °C/min. Finally, the split valve was opened again at 20 mL/min until the end of the run. The oven program was as follows: 50 °C for 1 min, then 7 °C/min until 100 °C, then 3 °C/min until 220 °C, and finally 10 °C/min until 295 °C, held for 8 min. The transfer line to the MS was kept at 300 °C. Detection was performed with a LECO Pegasus BT Time-of-Flight mass spectrometer (Saint Joseph, MI, USA). The MS was operated with the ion source at 250 °C, electron ionization at 70 eV, acquisition from m/z 40 to 550, 10 spectra per second and an acquisition delay of 480 s for liquid samples. Data acquisition, system control and spectra deconvolution were performed using LECO ChromaTOF version 5.40. NIST MS Search Program Version 2.3 g was used for spectra matching (NIST, 2015). Linear retention index (LRIs) values for sample peaks were calculated by analyzing the commercial alkane standard solution C8–C40, using the aforementioned chromatographic conditions [45].

2.7. Statistical Analysis

Statistical analysis of the results was performed using the SPSS software, version 23 (IBM Analytics, Armonk, NY, USA). Analysis of variance (ANOVA) was performed to evaluate significant differences between averages, and the Tukey post hoc test ($p < 0.05$) was used to perform multiple comparisons between averages.

3. Results and Discussion

3.1. Extract Yield and Composition

The recovery of bioactive extracts from *A. dealbata* leaves and twigs was evaluated for different extraction methods (maceration at room temperature and hot extraction), different solvents (70% ACE, 70% ET, and 70% MET), different geographical origins (Caparica and Alcobaça) and for different biomass water contents (fresh biomass and dried biomass).

Leaves and twigs were collected in Alcobaça (39°36′36.5″ N; 8°59′45.0″ W, at 32–40 m of altitude) and Caparica (38°37′22.6″ N; 9°10′38.5″ W, at 44–48 m of altitude). Both locations are classified as Csb (Warm-summer Mediterranean climate) according to Köppen-Geiger's classification system for climate but with some differences between them. Monthly average daily temperatures are in similar ranges (10.6–20.4 °C for Alcobaça and 11.6–21.9 °C for Caparica), but Alcobaça is characterized by a higher level of annual rainfall (696 mm against 591 mm) and relative humidity (73–81% against 68–80%) [46,47]. The fresh leaves and twigs that were processed in the first 24 h from the collection had moisture contents in the ranges of 47.7–58.3% and 43.4–51.9%, respectively. After air-drying at room temperature, moisture contents were in the range of 8.1–9.3% for leaves and 8.8–11.2% for twigs.

Extraction yields and compositions (TPC, TFC, and TPrAC contents) for the extracts obtained leaves and twigs, and the different extraction conditions are presented in Table 1.

Making an overall analysis of this set of data, significantly higher extraction yields were obtained for dried raw material, and significantly higher TPC, TFC, and TPrAC contents were found in the extracts produced with 70% ACE (except for hot extraction from twigs). The TPC, TFC, and TPrAC contents of twigs extracts with 70% ACE were also higher than those contents in the leaves extracts obtained with the same solvent.

The extraction yield obtained for leaves with 70% ACE (10.5–15.8%, on a wet basis) was higher than the ones reported by Borges et al. for *A. dalbata* leaves, extracted with different solvents (water, methanol, ethanol, acetone, dichloromethane, and hexane) and extraction methods (maceration under agitation, Soxhlet, and the use of ultrasound and microwaves), and solvent-to biomass ratio of 1:10, that were in the range of 2.8–12.0% of fresh weight [29]. Extraction yields were also higher than those observed by Luís et al. [48] for ethanolic, hydroethanolic (1:1 in volume basis), methanolic, and acetonic extracts of aerial parts (wood, bark, and leaves) of *A. dealbata* (3.26–9.51% dry weight) using Soxhlet extraction and were in the same range of the described for *A. melanoxylon* aerial parts (11.88–15.41% dry weight) by the same authors. Concerning twigs, Lomarat et al. found an extraction yield of dried *A. pennata* twigs of 15.1%, obtained by Soxhlet extraction with methanol [35].

For the best extraction solvent (70% ACE), TPC in leaves and twigs extracts varied in the range 345.8–478.4 mg GAE/g extract and 367.1–526.5 mg GAE/g extract, respectively. These contents are higher than those found by Luís et al. [48] in extracts of aerial parts of *A. dealbata* (203.10–290.65 mg GAE/g extract) and *A. melanoxylon* (100.10–138.76 mg GAE/g extract), and higher than the reported for *A. farnesiana* leaves extracts (63.2–247.9 mg GAE/g extract) [49], similar to the TPC contents found for *A. mearnsii* (163.9–646.6 mg GAE/g extract) [50], *A. tortilis* (260.7–512.4 mg GAE/g extract) [49], *A. nilotica* (42.18–116.60 mg GAE/g extract) [51], and *A. ataxacantha* (63.26–115.57 mg GAE/g extract) [51] leaves extracts, but lower than the observed for *A. longifolia* leaves extracts (524.9–858.8 mg GAE/g extract) [49]. Leaves and twigs are important sources of phenolics, with contents of 36.2–71.2 mg GAE/g of fresh leaves and 30.1–59.9 mg GAE/g of fresh twigs, respectively (see full results in Supplementary Material, Table S1). These are in a similar range as what was observed by Ferreira-Santos et al. [52] with *Pinus pinaster* bark, a known source of phenolic compounds, which gave 65.1 and 68.2 mg GAE/g of bark by extraction with 70% and 50% aqueous ethanol, respectively.

Table 1. Extraction yields (%, wet basis), TPC (mg GAE/g extract), TFC (mg CatE/g extract), and TPrAC (mg PycE/g extract) of extracts from fresh and dried leaves and twigs from Alcobaça (ALC) and Caparica (CAP), using different extraction solvents and extraction methods; TPC, TFC, and TPrAC expressed relative to the mass of biomass are presented in the Supplementary Material, Table S1.

	Samples	Maceration at Room Temperature			Hot Extraction		
		70% ACE	70% ET	70% MET	70% ACE	70% ET	70% MET
Leaves							
Yield	Fresh (ALC)	12.2 ± 0.1 ^{h}	10.8 ± 0.1 ^{j}	11.6 ± 0.1 ^{i}	12.2 ± 0.1 ^{h}	11.0 ± 0.1 ^{j}	9.9 ± 0.1 ^{l}
	Fresh (CAP)	10.5 ± 0.2 ^{k}	10.0 ± 0.1 ^{l}	9.8 ± 0.1 ^{l}	12.3 ± 0.1 ^{i}	9.8 ± 0.1 ^{l}	9.9 ± 0.1 ^{l}
	Dried (ALC)	14.8 ± 0.1 ^{c}	14.4 ± 0.1 ^{de}	14.1 ± 0.1 ^{de}	15.8 ± 0.1 ^{a}	13.5 ± 0.1 ^{f}	13.0 ± 0.1 ^{g}
	Dried (CAP)	14.0 ± 0.1 ^{e}	14.3 ± 0.1 ^{de}	13.5 ± 0.1 ^{f}	15.6 ± 0.1 ^{a}	15.8 ± 0.1 ^{a}	15.2 ± 0.1 ^{b}
TPC	Fresh (ALC)	449.1 ± 8.0 ^{b}	338.9 ± 6.6 ^{ghi}	329.0 ± 7.6 ^{ijk}	428.4 ± 3.7 ^{c}	361.0 ± 5.9 ^{de}	336.7 ± 6.9 ^{hj}
	Fresh (CAP)	345.8 ± 4.0 ^{fgh}	295.8 ± 6.4 ^{m}	288.3 ± 3.4 ^{mn}	352.6 ± 5.8 ^{def}	324.5 ± 9.5 ^{kl}	314.7 ± 7.6 ^{l}
	Dried (ALC)	478.5 ± 4.5 ^{a}	324.9 ± 3.3 ^{jkl}	336.1 ± 3.2 ^{hijk}	450.6 ± 3.7 ^{b}	362.2 ± 3.6 ^{d}	351.3 ± 5.4 ^{def}
	Dried (CAP)	350.2 ± 4.4 ^{efg}	292.3 ± 4.3 ^{mn}	255.2 ± 5.3 ^{p}	352.6 ± 5.4 ^{def}	272.4 ± 4.4 ^{o}	282.4 ± 4.7 ^{no}
TFC	Fresh (ALC)	98.7 ± 2.5 ^{cde}	81.4 ± 2.5 ^{kl}	75.3 ± 2.6 ^{m}	90.7 ± 3.1 ^{gh}	83.1 ± 3.4 ^{ijkl}	77.7 ± 3.3 ^{lm}
	Fresh (CAP)	94.9 ± 2.4 ^{efg}	97.7 ± 3.4 ^{cde}	89.3 ± 3.7 ^{h}	106.5 ± 2.8 ^{b}	96.6 ± 3.9 ^{def}	95.7 ± 2.8 ^{defg}
	Dried (ALC)	114.8 ± 2.0 ^{a}	82.7 ± 1.8 ^{ijkl}	81.7 ± 1.3 ^{jkl}	103.2 ± 1.5 ^{bc}	88.2 ± 1.8 ^{hi}	87.0 ± 1.9 ^{hij}
	Dried (CAP)	105.4 ± 1.7 ^{b}	92.1 ± 1.8 ^{fgh}	78.9 ± 2.3 ^{klm}	101.2 ± 2.8 ^{bcd}	83.6 ± 2.3 ^{ijk}	82.1 ± 1.9 ^{jkl}
TPrAC	Fresh (ALC)	330.7 ± 7.9 ^{b}	192.4 ± 6.2 ^{h}	203.7 ± 9.3 ^{gh}	292.7 ± 9.4 ^{c}	191.0 ± 6.1 ^{h}	211.6 ± 8.4 ^{g}
	Fresh (CAP)	255.4 ± 5.5 ^{de}	140.2 ± 6.0 ^{j}	120.0 ± 5.1 ^{k}	263.6 ± 8.1 ^{d}	142.2 ± 8.3 ^{j}	146.4 ± 5.3 ^{j}
	Dried (ALC)	357.0 ± 8.0 ^{a}	201.5 ± 5.5 ^{gh}	189.4 ± 6.3 ^{hi}	331.5 ± 5.1 ^{b}	241.9 ± 8.8 ^{ef}	238.1 ± 7.8 ^{f}
	Dried (CAP)	279.3 ± 4.5 ^{c}	211.1 ± 4.9 ^{g}	133.5 ± 5.1 ^{jk}	280.0 ± 7.6 ^{c}	174.9 ± 8.5 ^{i}	198.2 ± 5.8 ^{gh}
Twigs							
Yield	Fresh (ALC)	7.9 ± 0.1 ^{h}	6.9 ± 0.1 ^{kl}	6.9 ± 0.1 ^{kl}	9.0 ± 0.1 ^{e}	8,3 ± 0.1 ^{g}	7.7 ± 0.1 ^{hi}
	Fresh (CAP)	7.3 ± 0.1 ^{j}	6.7 ± 0.1 ^{l}	7.0 ± 0.1 ^{k}	8.0 ± 0.2 ^{g}	7.6 ± 0.1 ^{i}	7.0 ± 0.1 ^{k}
	Dried (ALC)	12.0 ± 0.1 ^{a}	10.7 ± 0.1 ^{b}	10.6 ± 0.1 ^{b}	8.8 ± 0.1 ^{ef}	7.1 ± 0.1 ^{k}	10.1 ± 0.1 ^{c}
	Dried (CAP)	9.0 ± 0.1 ^{e}	8.6 ± 0.1 ^{f}	10.1 ± 0.1 ^{c}	8.8 ± 0.1 ^{ef}	8.3 ± 0.1 ^{g}	9.5 ± 0.1 ^{d}
TPC	Fresh (ALC)	526.5 ± 4.5 ^{a}	406.6 ± 4.3 ^{g}	348.3 ± 6.3 ^{j}	514.6 ± 2.2 ^{b}	420.9 ± 3.3 ^{ef}	419.1 ± 3.0 ^{ef}
	Fresh (CAP)	412.6 ± 6.7 ^{fg}	294.2 ± 4.9 ^{n}	273.7 ± 7.9 ^{o}	367.1 ± 3.8 ^{i}	344.8 ± 5.3 ^{jk}	324.0 ± 6.1 ^{l}
	Dried (ALC)	499.0 ± 3.9 ^{c}	389.5 ± 7.3 ^{h}	329.0 ± 8.3 ^{l}	485.6 ± 3.5 ^{d}	412.5 ± 5.5 ^{fg}	402.6 ± 6.2 ^{g}
	Dried (CAP)	419.1 ± 3.7 ^{ef}	311.0 ± 6.5 ^{m}	251.6 ± 3.4 ^{p}	425.1 ± 5.8 ^{e}	341.5 ± 6.1 ^{jk}	334.9 ± 3.3 ^{kl}
TFC	Fresh (ALC)	166.2 ± 4.3 ^{cd}	139.1 ± 4.0 ^{ijkl}	130.9 ± 2.1 ^{m}	166.3 ± 3.9 ^{c}	137.6 ± 3.3 ^{jkl}	135.3 ± 2.6 ^{klm}
	Fresh (CAP)	178.7 ± 4.7 ^{b}	147.6 ± 2.3 ^{fgh}	130.0 ± 3.8 ^{mn}	153.0 ± 2.8 ^{f}	153.2 ± 2.4 ^{ef}	144.7 ± 4.5 ^{ghi}
	Dried (ALC)	166.3 ± 3.0 ^{c}	143.0 ± 3.4 ^{hij}	133.8 ± 2.0 ^{lm}	163.8 ± 1.7 ^{cd}	139.7 ± 2.8 ^{ijkl}	141.0 ± 2.9 ^{hijk}
	Dried (CAP)	198.4 ± 2.7 ^{a}	149.8 ± 3.3 ^{fg}	123.8 ± 2.4 ^{n}	201.0 ± 1.9 ^{a}	159.6 ± 1.4 ^{de}	151.5 ± 2.7 ^{f}
TprAC	Fresh (ALC)	585.8 ± 10.3 ^{b}	307.5 ± 7.1 ^{i}	286.4 ± 11.6 ^{j}	553.3 ± 7.0 ^{c}	375.2 ± 6.8 ^{g}	387.7 ± 9.0 ^{g}
	Fresh (CAP)	440.3 ± 9.1 ^{e}	94.2 ± 6.9 ^{m}	72.3 ± 6.3 ^{n}	388.3 ± 9.2 ^{g}	209.1 ± 7.8 ^{k}	142.1 ± 7.3 ^{l}
	Dried (ALC)	576.4 ± 5.5 ^{b}	340.3 ± 9.5 ^{h}	305.1 ± 6.9 ^{i}	516.9 ± 8.2 ^{d}	351.0 ± 8.1 ^{h}	384.6 ± 6.5 ^{g}
	Dried (CAP)	631.3 ± 7.9 ^{a}	379.9 ± 9.8 ^{g}	306.0 ± 7.2 ^{i}	614.3 ± 8.4 ^{a}	417.6 ± 6.1 ^{f}	432.4 ± 6.4 ^{ef}

Values (mean ± standard deviation) are average of two replicates of each sample, analyzed individually in triplicate (n = 2 × 3); different letters indicate significant differences of means within samples, for each determined parameter (one-way ANOVA, Tukey test at $p < 0.05$).

For 70% ACE, TFC in leaves and twigs extracts lied between 90.7 and 114.8 mg CatE/g extract and 153.0 to 198.4 mg CatE/g extract, respectively. A calibration curve was made using rutin as a standard for comparison purposes, giving 228.8–348.0 mg RE/g extract for leaves, and 375.5–601.4 mg RE/g extract (Supplementary Material, Table S3). TFC in leaves extracts were in the range observed by Tung et al. [53] with leaves of *A. nilotica* (2.3–355.3 mg RE/g extract), while TFC in twigs extracts were higher. The TFC values found in the extracts of *A. dealbata* twigs with 70% ACE were higher than those observed in bark extracts of various *Acacia* spp. including *A. dealbata* (74.0–134.0 mg RE/g extract) [54], in leaves and bark extracts of *A. nilotica* and *A. ataxacantha* (0.34–49.90 mg RE/g extract) [51], or in *Pinus pinaster* bark extracts, a known source of flavonoids (77–161 mg CatE/g extract) [52]. Relatively to the mass of raw material, TFC contents varied in the range 9.9–17.0 mg CatE/g of leaves (30.2–51.4 mg RE/g of leaves), and 12.6–20.0 mg CatE/g of twig (38.2–60.5 mg RE/g of twig) (see full results in Supplementary Material, Table S4), which are much superior to the reported by Sowndhararajan et al. for the bark of *A. dealbata* and other *Acacia* spp. (2.4–15.5 mg RE/g of bark), and also other known sources of flavonoids such as yellow, green, and black teas (0.2–3.8 mg CatE/g of tea) [55], and red fruits (0.2–0.8 mg CatE/g of tea) [56].

Regarding TPrAC, the results were expressed relative to a standard source of proanthocyanidins, the commercial extract Pycnogenol®, a mixture of proanthocyanidins extracted from bark of Atlantic Pinus trees. As observed for TPC and TFC, the extracts with higher TPrAC contents were obtained with 70% ACE, with TPrAC values in the range of 255.4–357.0 mg PycE/g extract for leaves, and in the range of 388.3–631.3 mg PycE/g extract for twigs.

The results of this study showed that the leaves and twigs of *A. dealbata* from Caparica and from Alcobaça regions can be an abundant source of extracts rich in phenolics, especially flavonoids and proanthocyanidins, with average contents similar or higher than comparable extratcs described in the literature.

For a better interpretation of the impact of the extraction method (regardless of the drying status or the origin of the plant), the average of the measured variables for all extracts produced by maceration was compared with the average of all extracts produced by hot extraction for each solvent. The same approach was followed to evaluate the average values of the measured variables for all extracts obtained from fresh and dried leaves (regardless of other extraction parameters). The average values of the measured variables for the extracts obtained from a specific origin (Caparica or Alcobaça) were also determined and compared. These results were then subject to analysis of variance (ANOVA) followed by Tukey's test at $p < 0.05$ and are shown in Supplementary Material, Table S5.

For the extraction yield, statistically significant differences were only found between fresh and dried raw materials, as shown in the Figure 1. Yields in the range 13.9–15.0%, and 10.3–11.6% were found for dried and fresh leaves, respectively and yields in the range of 8.7–10.1%, and 7.2–8.1% were found for dried and fresh twigs, respectively.

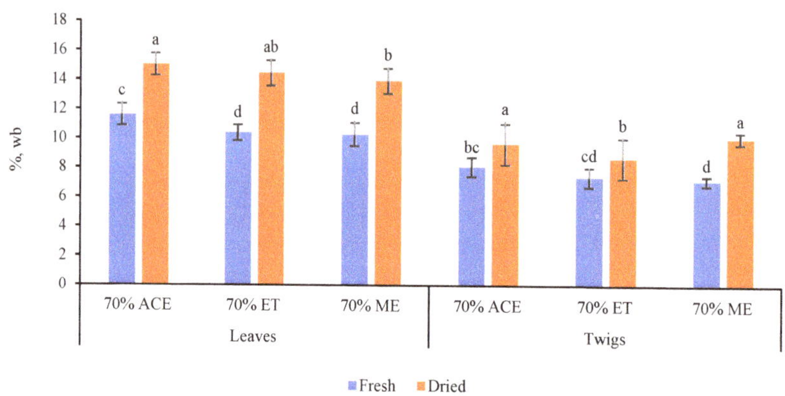

Figure 1. Comparison of the average extraction yields for fresh and dried leaves and twigs (regardless the extraction method and the provenience of the raw material); different letters indicate significant differences of means (one-way ANOVA, Tukey test at $p < 0.05$).

In what concerns the TPC, TFC, and TPrAC of the extracts, no significant differences were found between maceration and hot extraction, except for methanolic extracts that presented slightly higher contents for hot extraction (Supplementary Material, Table S5). Figure 2 highlights differences coming from the drying status and origin of the raw material, for all determined composition parameters.

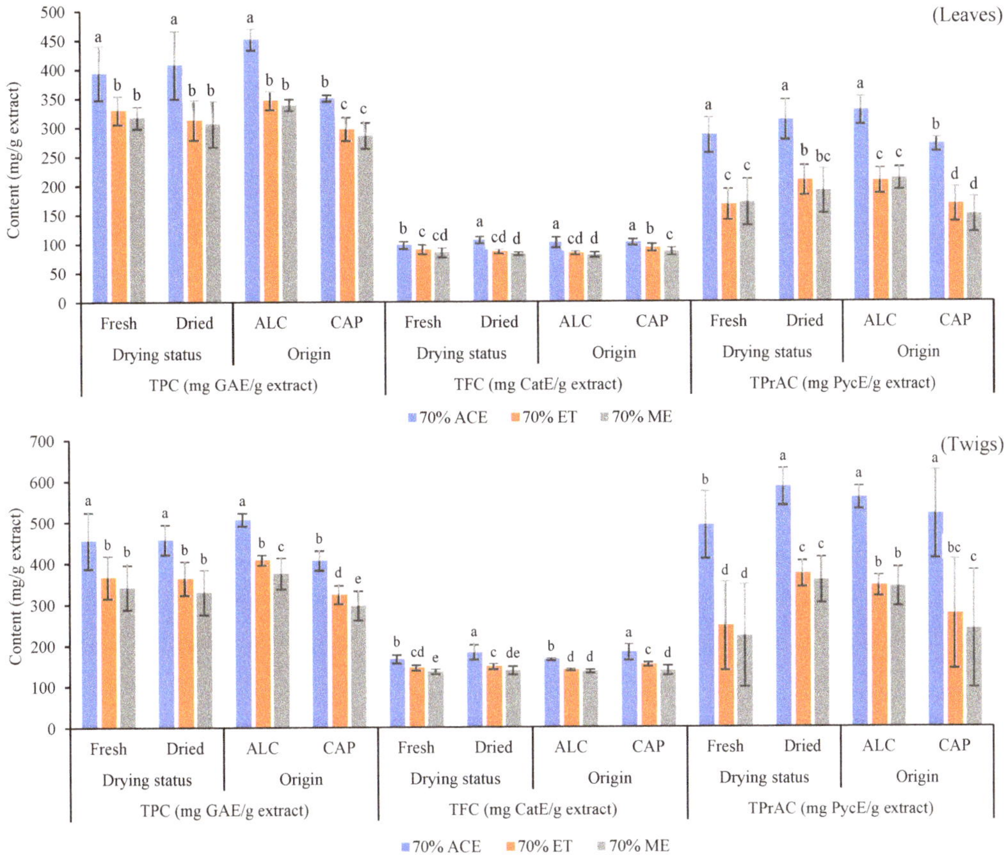

Figure 2. Comparison of the average values of TPC (mg GAE/g extract), TFC (mg CatE/g extract), and TPrAC (mg PycE/g extract) for fresh or dried material (regardless the extraction method and the provenience of the raw material), and for plants from Alcobaça (ALC) and Caparica (CAP) (regardless the extraction method and the drying status); different letters indicate significant differences of means (one-way ANOVA, Tukey test at $p < 0.05$).

Overall, it was confirmed that the higher average values for all the parameters considered were higher for the extracts obtained with 70% ACE. Regarding leaves, the extracts from the Alcobaça region had higher average values of TPC and TPrAC and higher average values of TFC (only for 70% ET and 70% ME extracts) than those from Caparica region. In the case of twigs, extracts from dried biomass had higher average values of TFC and TPrAC than those obtained from wet biomass with the same solvents. Extracts of twigs from Alcobaça region showed higher average values of TPC than those from the Caparica region. Differences in the average values comparing fresh and dried plants were most noticed in the extraction yield, TFC, and TPrAC, with dried material originating better results. The differences of extraction yield or extract composition between fresh or dried biomass may be explained by the change of the extraction solvent characteristics, since the presence of biomass moisture corresponds to a reduction of the concentration of the organic component (acetone, ethanol, or methanol). Differences coming from the plant origin can be explained by the fact that secondary metabolites are synthesized by plants to protect them in adverse conditions from pathogens and environmental stresses, being influenced by several biotic and abiotic factors, such as water, drought, salinity, temperature, radiation, or chemical stresses [57]. Thus, differences in edaphoclimatic conditions between Alcobaça and Caparica can explain the differences observed.

3.2. Antioxidant Activity

The antioxidant activity of the extracts obtained from *A. dealbata* leaves and twigs was assessed by DPPH and FRAP assays, and the results are presented in Table 2.

Table 2. Results from DPPH (mg TE/g extract) and FRAP (mmol Fe^{2+}/g extract) of extracts from fresh and dried leaves and twigs from Alcobaça (ALC) and Caparica (CAP), using different extraction solvents and extraction methods; DPPH and FRAP results expressed relative to the mass of biomass are presented in Supplementary Material, Table S1.

	Samples	Maceration at Room Temperature			Hot Extraction		
		70% ACE	70% ET	70% MET	70% ACE	70% ET	70% MET
Leaves							
DPPH	Fresh (ALC)	740.8 ± 10.1 ^{b}	597.9 ± 8.3 ^{e}	535.8 ± 8.5 ^{i}	740.2 ± 6.1 ^{b}	636.7 ± 6.0 ^{d}	526.8 ± 5.3 ^{i}
	Fresh (CAP)	560.0 ± 7.2 ^{h}	348.5 ± 5.7 ^{n}	335.4 ± 5.9 ^{n}	540.0 ± 8.4 ^{gh}	428.4 ± 9.0 ^{k}	393.6 ± 7.2 ^{m}
	Dried (ALC)	893.4 ± 10.8 ^{a}	578.2 ± 6.5 ^{fg}	576.4 ± 8.2 ^{fg}	886.5 ± 7.5 ^{a}	634.3 ± 9.1 ^{d}	685.4 ± 5.5 ^{c}
	Dried (CAP)	446.3 ± 7.0 ^{j}	412.2 ± 5.6 ^{l}	337.0 ± 8.3 ^{n}	588.9 ± 5.5 ^{ef}	392.6 ± 7.3 ^{m}	406.6 ± 6.0 ^{lm}
FRAP	Fresh (ALC)	7664.1 ± 167.6 ^{b}	6285.3 ± 95.4 ^{ef}	5620.3 ± 118.5 ^{gh}	7702.8 ± 164.7 ^{b}	6501.3 ± 132.2 ^{de}	5423.2 ± 142.5 ^{hi}
	Fresh (CAP)	6612.1 ± 167.8 ^{d}	4386.9 ± 93.0 ^{l}	4403.9 ± 129.7 ^{l}	6015.2 ± 177.6 ^{def}	5072.4 ± 225.1 ^{jk}	4835.8 ± 178.6 ^{k}
	Dried (ALC)	8885.4 ± 93.1 ^{a}	6299.9 ± 71.9 ^{ef}	5334.0 ± 79.0 ^{ij}	6962.5 ± 63.4 ^{c}	6175.8 ± 138.3 ^{f}	3300.7 ± 98.7 ^{n}
	Dried (CAP)	5846.6 ± 90.7 ^{g}	4855.0 ± 58.6 ^{k}	3934.8 ± 103.5 ^{m}	5766.3 ± 81.2 ^{g}	4097.2 ± 124.8 ^{m}	4449.8 ± 82.2 ^{l}
Twigs							
DPPH	Fresh (ALC)	996.8 ± 6.7 ^{b}	757.2 ± 11.6 ^{e}	511.1 ± 11.3 ^{j}	994.7 ± 6.2 ^{bc}	778.8 ± 14.0 ^{d}	782.7 ± 8.1 ^{d}
	Fresh (CAP)	675.5 ± 11.0 ^{h}	254.2 ± 6.6 ^{q}	230.2 ± 9.2 ^{r}	596.3 ± 11.5 ^{i}	379.7 ± 11.8 ^{n}	358.4 ± 11.4 ^{o}
	Dried (ALC)	1068.3 ± 8.5 ^{a}	665.5 ± 11.2 ^{h}	514.2 ± 11.8 ^{j}	974.2 ± 6.4 ^{c}	791.2 ± 14.3 ^{d}	782.1 ± 7.1 ^{d}
	Dried (CAP)	731.3 ± 6.1 ^{f}	418.5 ± 7.9 ^{m}	321.3 ± 8.3 ^{p}	710.2 ± 5.2 ^{g}	474.5 ± 9.9 ^{k}	452.6 ± 8.5 ^{l}
FRAP	Fresh (ALC)	6852.4 ± 208.0 ^{de}	6546.5 ± 221.0 ^{e}	3989.4 ± 61.6 ^{j}	7829.9 ± 140.9 ^{c}	6851.8 ± 225.2 ^{de}	6554.3 ± 112.4 ^{e}
	Fresh (CAP)	6951.8 ± 146.0 ^{d}	3306.8 ± 109.3 ^{k}	3075.8 ± 75.0 ^{k}	6192.0 ± 53.1 ^{f}	4626.8 ± 255.7 ^{hi}	4461.1 ± 137.7 ^{i}
	Dried (ALC)	9194.6 ± 165.8 ^{a}	4770.3 ± 86.1 ^{h}	3783.0 ± 84.8 ^{j}	8804.2 ± 112.4 ^{b}	6987.7 ± 163.4 ^{d}	7098.5 ± 87.1 ^{d}
	Dried (CAP)	7754.4 ± 119.0 ^{c}	5354.9 ± 141.1 ^{g}	3969.0 ± 85.0 ^{j}	8003.8 ± 171.7 ^{c}	5585.2 ± 140.0 ^{g}	5429.1 ± 83.3 ^{g}

Values (mean ± standard deviation) are average of two replicates of each sample, analyzed individually in triplicate (n = 2 × 3); different letters indicate significant differences of means within samples, for each antioxidant determination method (one-way ANOVA, Tukey test at $p < 0.05$).

The highest antioxidant activity of extracts from leaves or twigs was observed when using 70% ACE as extraction solvent, regardless of the extraction method, drying status, or the geographical origin of the plant. When using this solvent, the antioxidant activity of twigs extracts was higher than the one of corresponding leaves extracts.

The DPPH assay measures the ability of extract components to act as free radical scavengers or hydrogen donors antioxidants [58]. For 70% ACE extracts, the values lied between 446.3–893.4 mg TE/g extract (58.6–132.1 mg TE/g of leaves) for leaves extract and between 596.3–1068.3 mg TE/g extract (49.0–128.3 mg TE/g of twigs). These values are higher than those reported by Zheleva-Dimitrova et al. for extracts from leaves (31.94–493.90 mg TE/g extract) and stem bark (8.51–349.17 mg TE/g extract) of *A. nilotica* and *A. ataxacantha* [51]. These results are also higher than the ones reported for known antioxidant sources such as yellow, green, and black teas (11.930–26.521 µmol TE/100 g, which corresponds to 0.030–0.066 mg TE/g of tea leaves) [55], red fruits (64.14–177.11 µmol TE/g, corresponding to 16.1–44.3 mg TE/g of fruit) [56], or pine bark (approximately between 50 and 350 mg TE/g of extract) [59].

The reducing power of the obtained extracts was measured through the FRAP assay, which evaluates the reduction of Fe^{3+} in the complex (Fe^{3+}-TPTZ) to a ferrous form

(Fe^{2+}-TPTZ), by donation of electrons from the antioxidants, resulting in stopping the free radical production chain [13,60,61]. For 70% ACE the FRAP values were in the range 5766.3–8885.4 μmol Fe^{2+}/g extract and 6192.0–9194.6 μmol Fe^{2+}/g extract, for leaves and twigs extracts, respectively. These values higher than those found in leaves (233.17 μmol Fe^{2+}/g extract), pods (254.42 μmol Fe^{2+}/g extract), and seed (178.14 μmol Fe^{2+}/g extract) extracts of *A. leucophloea* [62]. Values of 152 mg Fe^{2+}/g (corresponding to 2721.2 μmol Fe^{2+}/g extract) and 575 mg Fe^{2+}/g (corresponding to 10296.4 μmol Fe^{2+}/g extract) were found for *Pinus radiata* bark extract and standardized *Pinus pinaster* bark extract Pycnogenol®, respectively [63].

The extraction solvent significantly affected the antioxidant activity, with the best results being obtained when using 70% ACE, regardless of the origin and the drying status of the raw material. Higher antioxidant activities of extracts obtained with aqueous acetone solutions were also observed with stem bark extracts of *Butea monosperma* (Lam.) Kuntze [64], bark extracts of *A. leucophloea*, *A. ferruginea*, *A. dealbata* and *A. pennata* [54], and flower extracts of safflower (*Carthamus tinctorius* L.) [65].

A better understanding of the influence of the extraction method, drying status, and provenience of the plant on the antioxidant activity can be derived from the analysis of variance applied using the same approach as for the impact of these factors on the composition and extract yield. The results are presented in Table 3. Besides the confirmation of 70% ACE as the best extraction solvent for the removal of antioxidant compounds, it can be seen that the geographical origin of the plant had the major impact on the antioxidant activity of the produced extracts, since the extracts of leaves and twigs from Alcobaça showed significantly higher antioxidant activity than the corresponding extracts from biomass collected in the Caparica region. The extraction method only gave significant differences on methanolic and ethanolic extracts of twigs, with hot extraction giving the highest antioxidant activity by both DPPH and FRAP assays. The effect of using fresh or dried material was most reflected in the antioxidant activity of twigs given by the FRAP assay, with dried twigs extracted with 70% ACE originating a significantly higher value than the one obtained using fresh leaves.

Table 3. Average values of DPPH (mg TE/g extract) and FRAP (mmol Fe^{2+}/g extract) assays for extracts obtained by maceration and hot extraction (regardless drying status and origin), from fresh and dried material (regardless extraction method and origin), and from plants from Alcobaça (ALC) and Caparica (CAP) (regardless extraction process and drying status).

Samples	Leaves			Twigs		
	70% ACE	70% ET	70% MET	70% ACE	70% ET	70% MET
DPPH						
Maceration	660.1 ± 174.7 [a]	484.2 ± 109.0 [b]	446.1 ± 113.5 [b]	868.0 ± 171.5 [a]	523.8 ± 203.4 [bc]	394.2 ± 125.7 [c]
Hot extraction	696.2 ± 131.2 [a]	523.0 ± 115.9 [b]	503.1 ± 120.0 [b]	818.9 ± 174.4 [a]	606.0 ± 186.4 [b]	593.9 ± 195.7 [b]
Fresh	652.5 ± 90.2 [a]	502.9 ± 121.4 [b]	447.9 ± 88.0 [b]	815.8 ± 186.2 [a]	542.5 ± 235.1 [b]	470.6 ± 210.4 [b]
Dried	703.8 ± 197.2 [a]	504.3 ± 106.5 [b]	501.3 ± 140.5 [b]	871.0 ± 157.5 [a]	587.4 ± 152.6 [b]	517.5 ± 171.7 [b]
Alcobaça	815.2 ± 76.8 [a]	611.8 ± 26.2 [b]	581.1 ± 64.7 [bc]	1008.5 ± 37.0 [a]	748.2 ± 51.7 [b]	647.5 ± 138.1 [c]
Caparica	541.1 ± 57.3 [c]	395.4 ± 31.3 [d]	368.1 ± 33.6 [d]	678.3 ± 53.1 [c]	381.7 ± 83.2 [d]	340.6 ± 81.9 [d]
FRAP						
Maceration	7252.1 ± 1174.0 [a]	5456.8 ± 873.7 [b]	4823.3 ± 746.5 [bc]	7688.3 ± 969.5 [a]	4994.6 ± 1199.0 [c]	3704.3 ± 386.4 [d]
Hot extraction	6693.0 ± 746.5 [a]	5461.7 ± 981.1 [b]	4502.4 ± 801.6 [c]	7707.5 ± 976.6 [a]	6012.9 ± 1007.8 [b]	5885.8 ± 1046.0 [b]
Fresh	7079.9 ± 644.8 [a]	5561.5 ± 898.9 [b]	5070.8 ± 509.5 [b]	6956.5 ± 610.9 [b]	5333.0 ± 1491.8 [cd]	4520.2 ± 1306.5 [d]
Dried	6865.2 ± 1287.9 [a]	5357.0 ± 946.6 [b]	4254.8 ± 765.0 [c]	8439.3 ± 611.2 [a]	5674.5 ± 841.4 [c]	5069.9 ± 1364.5 [cd]
Alcobaça	7803.7 ± 715.8 [a]	6315.6 ± 159.6 [b]	4919.5 ± 966.2 [c]	8170.3 ± 940.4 [a]	6289.1 ± 926.3 [c]	5356.3 ± 1518.6 [d]
Caparica	6141.4 ± 380.2 [b]	4602.9 ± 412.3 [cd]	4406.1 ± 347.9 [d]	7225.5 ± 737.5 [b]	4718.4 ± 921.4 [de]	4233.7 ± 873.4 [e]

Different letters indicate significant differences of means within samples for each set of parameters (one-way ANOVA, Tukey test at $p < 0.05$).

3.3. Extraction of Spent Raw Material

Leaves and twigs of *Acacia* spp. contain lipophilic compounds such as terpenes, long-chain alcohols, fatty acids, sterols, aromatic compounds, among others [66]. These lipophilic components may be valorized in different application fields, and their presence might limit the access of polar solvents to the matrix, affecting extraction of the bioactive components. To test this hypothesis the dried leaves and twigs were treated with petroleum ether, to remove non-polar components and extraction with 70% ACE was performed on the spent leaves and twigs using equivalent conditions to those used with the raw biomass. The characterization of the lipophilic extracts will be addressed elsewhere.

The comparison of composition and antioxidant activity of extracts obtained from dried material macerated at room temperature with 70% ACE, and those of extracts obtained from spent material in similar conditions is shown in Table 4.

Table 4. Extract yield, composition, and antioxidant activity of extracts obtained from raw and spent leaves and twigs, using maceration at room temperature with 70% ACE as extraction solvent.

Sample		Yield	TPC	TFC	TPrAC	DPPH	FRAP
Leaves	Alcobaça-spent	15.5 ± 0.1	436.4 ± 11.1	116.1 ± 2.8	441.0 ± 6.8	796.0 ± 9.9	8807.8 ± 225.1
	Alcobaça-raw	14.8 ± 0.1	478.5 ± 4.5	114.8 ± 2.0	357.0 ± 8.0	893.4 ± 10.8	8885.4 ± 93.1
	Caparica-spent	14.5 ± 0.1	350.5 ± 2.4	97.0 ± 1.8	356.9 ± 1.6	661.7 ± 3.5	6431.1 ± 85.8
	Caparica-raw	14.0 ± 0.1	350.2 ± 4.4	105.4 ± 1.7	279.3 ± 4.5	446.3 ± 7.0	5846.6 ± 90.7
Twigs	Alcobaça-spent	12.9 ± 0.1	448.3 ± 8.0	152.8 ± 3.6	610.1 ± 5.8	838.6 ± 8.6	8171.1 ± 153.0
	Alcobaça-raw	12.0 ± 0.1	499.0 ± 3.9	166.3 ± 3.0	576.4 ± 5.5	1068.3 ± 8.5	9194.6 ± 165.8
	Caparica-spent	10.2 ± 0.1	389.0 ± 5.3	147.6 ± 2.2	544.2 ± 4.7	644.8 ± 9.6	7151.6 ± 160.4
	Caparica-raw	9.0 ± 0.1	419.1 ± 3.7	198.4 ± 2.7	631.3 ± 7.9	731.3 ± 6.1	7754.4 ± 119.0

The extracts produced from spent leaves and twigs had a slightly higher yield and slightly higher proanthocyanidin contents than those obtained from the raw dried materials. However, TPC and TFC, as well as the antioxidant activity, were slightly lower when using spent raw material. The removal of lipophilic compounds from the plant matrix may facilitate the penetration of the solvent, thus allowing obtaining higher extract yields. The slight decrease in the antioxidant activity may be explained by the previous removal of lipophilic components with antioxidant activity, such as terpenes or sterols. Despite this, it should be noted that the values are quite close and that this option allows more efficient use of this resource by obtaining a wider range of products from the same amount of raw material.

3.4. Correlations between Composition and Antioxidant Activity

The correlation analysis between TPC, TFC, TPrAC, and antioxidant activities by DPPH and FRAP assays for the extracts of fresh and spent flowers are given in Table 5.

All the correlations found in this work were statistically significant at $p < 0.01$, with the exception made for the correlation between TFC and antioxidant activity by the DPPH assay for leaves (non-significant) and twigs (significant at $p < 0.05$) and for the correlation between TFC and FRAP assay in the case of leaves (significant at $p < 0.05$). The Pearson's correlation coefficients found indicate very strong correlations between TPC and antioxidant activity given by DPPH and FRAP assays, and strong correlations between TPrAC and the same assays [67], for both leaves and twigs. Moderate correlation was found between TFC of leaves and twigs and both antioxidant assays. Ref. [68] also found strong correlations between TPC, TFC and DPPH values ($r = 0.771$ and $r = 0.815$), for extracts of *A. nilotica* leaves. However, [69] found non-significant correlation between TPC of extracts of different components (leaves, flowers, and pods) of three *Acacia* species (*A. nilotica*, *A. seyal*, and *A. laeta*) and DPPH results, having found a significant correlation between TFC and DPPH values instead. Significant and strong correlations between TPC and the DPPH and

FRAP results were found in *Pinus densiflora* bark extract [59] and in medicinal herbs and spices [70].

Table 5. Correlation analysis between total phenolics content (TPC), total flavonoid content (TFC), total proanthocyanidins content (TPrAC) and antioxidant activities given by DPPH and FRAP assays, for leaves extracts; significance of the correlations assessed at $p < 0.05$ and at $p < 0.01$ (two-tailed).

		TPC	TFC	TPrAC	DPPH	FRAP
Leaves	TPC	1				
	TFC	r = 0.563 ** $p = 0.003$	1			
	TPrAC	r = 0.801 ** $p = 0.000$	r = 0.553 ** $p = 0.003$	1		
	DPPH	r = 0.922 ** $p = 0.000$	r = 0.373 $p = 0.061$	r = 0.766 ** $p = 0.000$	1	
	FRAP	r = 0.864 ** $p = 0.000$	r = 0.490 * $p = 0.011$	r = 0.742 ** $p = 0.000$	r = 0.778 ** $p = 0.000$	1
Twigs	TPC	1				
	TFC	r = 0.511 ** $p = 0.008$	1			
	TPrAC	r = 0.754 ** $p = 0.000$	r = 0.671 ** $p = 0.000$	1		
	DPPH	r = 0.967 ** $p = 0.000$	r = 0.396 * $p = 0.045$	r = 0.765 ** $p = 0.000$	1	
	FRAP	r = 0.861 ** $p = 0.000$	r = 0.603 ** $p = 0.011$	r = 0.856 ** $p = 0.000$	r = 0.884 ** $p = 0.000$	1

** correlation is significant at $p < 0.01$ (two-tailed); * correlation is significant at $p < 0.05$ (two-tailed).

These findings demonstrate that phenolic compounds have an important role in the observed antioxidant activity given by DPPH and FRAP assays. Moreover, the abundance of these compounds in the produced extracts suggests that leaves and twigs have a strong potential to be explored for obtaining antioxidants for diverse applications, such as in nutraceutical or cosmetic industries.

3.5. Antimicrobial Activity

A representative group of extract samples was selected to assess antimicrobial activity against several microorganisms responsible for foodborne disease, skin infections, caries, and oral infections. Some of these microorganisms, namely *S. aureus*, *Enterococcus* spp., *P. aeruginosa*, *E. coli*, and *C. albicans*, are usually involved in hospital-acquired infections and have developed resistance to antibiotics, and virulence factors can exacerbate microbial drug resistances [71,72]. Antibiotic resistance is emerging as a serious worldwide problem and natural extracts with antimicrobial activity have been suggested as a solution to this problem [73]. Natural products from plants could also target microbial virulence factors and thus play an important role to combat microbial infections and overcoming antibiotic resistances [74]. Microbial virulence factors encompass a wide range of molecules produced by pathogens, such as toxins, enzymes, exopolysaccharides, cell surface structures such as capsules, lipopolysaccharides, glyco- and lipoproteins [75]. Moreover, natural extracts with antimicrobial activity could be an less polluting and more safe alternative to synthetic antimicrobial substances applied in food industry and in oral, cosmetics, and pharmacological formulations [43].

Both leaves and twigs extracts were able to inhibit the growth of Gram-positive bacteria but were inactive against Gram-negative bacteria and *C. albicans*. The inhibition zones against bacteria *S. aureus*, MRSA, *S. epidermis*, *E. faecalis*, *B. cereus*, *S. mutans and S. mitis* are presented in Table 6. Overall, extracts were more effective against *B. cereus* (inhibition zones in the range 11.0–12.1 mm for leaves extracts and 9.3–13.0 mm for twigs extracts), while *E. faecalis* showed a sensitivity lower than the remaining microorganisms (inhibition zones in the range 8.3–8.8 mm for leaves extracts and 9.3–9.8 mm for twigs extracts).

Table 6. Inhibition zones (mm) of selected extracts of *A. dealbata* leaves and twigs against *S. aureus*, MRSA, *S. epidermis*, *E. faecalis*, *B. cereus*, *S. mutans* and *S. mitis*.

Samples (0.5 mg/well)	*S. aureus*	*MRSA*	*S. epidermidis*	*E. faecalis*	*B. cereus*	*S. mutans*	*S. mitis*
Leaves							
Dried-70%ACE-maceration (ALC)	10.3 ± 0.1	10.3 ± 0.0	9.8 ± 0.1	8.5 ± 0.1	12.1 ± 0.1	11.0 ± 0.0	11.3 ± 0.6
Dried-70%ET-maceration (ALC)	8.8 ± 0.0	9.5 ± 0.1	8.5 ± 0.1	8.0 ± 0.0	11.5 ± 0.0	10.0 ± 0.0	11.0 ± 0.0
Dried-70%MET-maceration (ALC)	8.5 ± 0.1	9.3 ± 0.1	8.0 ± 0.0	nd	11.0 ± 0.0	10.0 ± 0.0	nd
Dried-70%ACE-maceration (CAP)	9.3 ± 0.0	9.0 ± 0.0	8.5 ± 0.1	8.8 ± 0.0	11.0 ± 0.0	10.0 ± 0.0	9.0 ± 1.7
Dried-70%ACE-hot extraction (ALC)	9.5 ± 0.1	10.5 ± 0.1	9.5 ± 0.1	8.8 ± 0.1	12.0 ± 0.0	10.0 ± 0.0	nd
Fresh-70%ACE-maceration (ALC)	9.5 ± 0.1	9.8 ± 0.0	9.3 ± 0.1	8.3 ± 0.1	11.3 ± 0.0	10.0 ± 0.0	9.3 ± 0.6
Spent leaves-70%ACE-maceration (ALC)	10.5 ± 0.1	10.5 ± 0.1	9.3 ± 0.0	8.5 ± 0.1	11.8 ± 0.0	11.3 ± 0.0	nd
Twigs							
Dried-70%ACE-maceration (ALC)	10.8 ± 0.0	10.5 ± 0.0	10.3 ± 0.1	9.5 ± 0.1	12.5 ± 0.1	9.0 ± 0.0	12.0 ± 0.0
Dried-70%ET-maceration (ALC)	nd	7.8 ± 0.0	nd	nd	10.0 ± 0.0	9.0 ± 0.0	11.0 ± 0.0
Dried-70%MET-maceration (ALC)	nd	nd	nd	nd	9.3 ± 0.0	10.0 ± 0.0	10.0 ± 0.0
Dried-70%ACE-maceration (CAP)	11.0 ± 0.0	11.0 ± 0.0	10.0 ± 0.0	9.5 ± 0.1	13.0 ± 0.0	12.0 ± 0.0	10.0 ± 0.0
Dried-70%ACE-hot extraction (ALC)	10.0 ± 0.0	10.8 ± 0.1	9.8 ± 0.0	9.5 ± 0.1	13.0 ± 0.0	9.0 ± 0.0	11.0 ± 0.0
Fresh-70%ACE-maceration (ALC)	11.0 ± 0.0	10.7 ± 0.0	9.5 ± 0.1	9.3 ± 0.0	12.8 ± 0.1	11.0 ± 0.0	12.0 ± 0.0
Spent twigs-70%ACE-maceration (ALC)	11.0 ± 0.0	11.0 ± 0.0	9.5 ± 0.1	9.8 ± 0.0	12.1 ± 0.0	9.0 ± 0.0	10.0 ± 0.0
Positive control	12.5 ± 0.5 [a]	13.9 ± 0.7 [a]	15.7 ± 1.0 [a]	12.5 ± 0.5 [a]	16.4 ± 1.1 [a]	19.3 ± 0.6 [b]	25.7 ± 0.6 [b]

[a] Vancomycin (5.0 mg); [b] Vancomycin (50.0 mg); nd: not detected.

Comparing leaves and twigs, twigs originated slightly higher inhibition diameters than leaves extracts, whenever acetone was used as extraction solvent. 70% ACE was the best extraction solvent regarding antimicrobial activity, for both leaves and twigs extracts, having produced significantly higher inhibition diameters than 70% ET or 70% MET extracts.

The origin of the raw material had a marked influence in the case of leaves extracts, with extracts obtained with leaves from Alcobaça having better results with all microorganisms, except for *E. faecalis*. In the case of twigs extracts, the differences were not so visible but extracts from Caparica twigs gave better results, exception made against *S. epidermidis*, *E. faecalis*, and *S. mitis*.

For both leaves and twigs extracts, the extraction process (maceration versus hot extraction), as well as the drying state of the raw material (dried versus fresh raw material), and the use of dried versus spent raw material did not significantly affect the antimicrobial activity, with extracts producing very similar inhibition zones against the microorganisms. Regarding the obtention of antibacterial extracts, these are important findings, once it supports the suggestion of using a more sustainable process (maceration at room temperature), with facilitated logistics due to not having the necessity of using fresh raw material. Moreover, the results show the possibility of retrieving a broader range of products from the same amount of raw material, since the previous extraction with petroleum ether to get lipophilic compounds did not affect the antimicrobial activity of the extracts that were then obtained with 70% aqueous acetone.

Considering the concentration of 10 mg/mL used in the antimicrobial assays for comparison purposes, it can be observed that the leaves extracts obtained in this study produced similar effects on the microorganisms to the ones shown by other extracts from leaves of *Acacia* spp. reported in the literature with concentrations ranging from 5 mg/mL to 200 mg/mL. Antimicrobial effects of leaves extracts expressed by inhibition zones were already described against *S. aureus* with *A. dealbata* aqueous and ethanolic extract (10.0–10.2 mm of inhibition, at 5 mg/mL) [29], with *A. saligna* (15 mm by an ethyl acetate extract at 10 mg/mL [24], and 24.7 mm by an ethanolic extract at 200 mg/mL [26]), with extracts of *A. etbaica*, *A. laeta*, and *A. origena* (7.3–21.0 mm at 500 mg/mL) [23], with *A.*

pycnantha (10.2–21.0 mm at 500 mg/mL [23], and 9–12 mm by aqueous and ethanolic extracts, no concentration reported [22]), and with *A. rigidula* and *A. berlandieri* acetonic, methanolic, and acetic acid extracts (6–11.4 mm at 500 mg/mL); against MRSA with *A. saligna* ethanolic extract (22.7 mm at 200 mg/mL) [26]; against *S. epidermis* (6–10 mm by aqueous and ethanolic extracts, no concentration reported) [22]; against *E. faecalis* with *A. rigidula* and *A. berlandieri* acetonic, methanolic, and acetic acid extracts (6–12.1 mm at 500 mg/mL); and against *B. cereus* with *A. dealbata*, *A. melanoxylon*, *A. baileyana*, and *A. nicholli* aqueous, ethanolic, and methanolic extracts (6–19 mm at 10 mg/mL) [30], and with *A. saligna* ethyl acetate extract (inhibition zone of 16 mm at 10 mg/mL) [24].

Concerning twigs in particular, just a few studies on antimicrobial effects involving *Acacia* spp. were found in the literature, most of them involving *A. nilotica* twigs [31–34]. Antimicrobial activity of twigs extracts produced in this work was in a lower range than the reported by Kumari et al. for *A. nilotica* twigs against *S. aureus* (10.8–40.2 mm), *E. faecalis* (16.2–38.0 mm), and *C. albicans* (14.7–27.0 mm) [31]. However, these results were obtained by applying 2 mg of extract/well [31], while in this work only 0.5 mg of extract/well was applied, which can at least partially explain the differences observed. To the best of our knowledge, no results of antimicrobial effects against MRSA, *S. epidermidis*, or *B. cereus* for twigs extracts of other *Acacia* spp. have been reported in the literature.

It is known that Gram-negative bacteria are more resistant to plant extracts than Gram-positive bacteria, due to the protection of the lipopolysaccharide layer in the outer membrane [73]. *Candida albicans* was included in this study as a model to determine the antifungal activity of the extracts. Despite being a common commensal yeast fungus of the human oral, gastrointestinal, and genital mucosal surfaces and skin, under specific circumstances, such as perturbation of barrier integrity or host immune responses, *C. albicans* causes opportunistic infections that range from superficial infections of the skin to life-threatening systemic infections [76,77]. All tested samples revealed ineffective against Gram-negative *P. aeruginosa* and *E. coli*, which was equally reported by El-Toumy et al. with aqueous, methanolic, and ethyl acetate extracts of *A. saligna* leaves [24].The tested samples were also ineffective against *C. albicans* yeast, as observed by Silva et al. with aqueous and methanolic extracts of *A. dealbata* leaves [30], by Ramli et al. with ethanolic extract of *A. farnesiana* leaves [16], or by Cock and van Vuuren with aqueous and methanolic extracts of *A. karroo* leaves [17]. Antibacterial effects against Gram-negative *E. coli* bacteria [22,25,26,29,30] and *C. albicans* yeast [24,26,30,31] were detected in leaves extracts of *Acacia* spp., but, in some cases, using extract concentrations that were more than double [31] or even twenty times higher [26] than the ones used in this work. In general, a positive correlation was observed between the antimicrobial activity of extracts from leaves and twigs of *A.dealbata*, and their total phenolic content, but this association should be confirmed by additional quantitative studies.

3.6. Characterisation of the Non-Polar Components of the Extracts

To understand if the phenolic extracts also contained non-polar components with antioxidant or antimicrobial activities that might influence the properties of the extracts, a group of selected extracts were derivatized and fractionated to isolate these less polar components and analysed by GC-TOFMS. A total of 32 compounds were tentatively identified using the NIST mass spectra library matching (Supplementary Material, Table S6). Among them, phytol, squalene, α-tocopherol, lupeol, and lupenone (lup-20(29)-en-3-one) (Figure 3) are the most predominant, showing the highest peak areas. Phytol, squalene, lupeol, lupenone, and α-tocopherol have been previously identified in *A. dealbata* leaves [36,66,78].

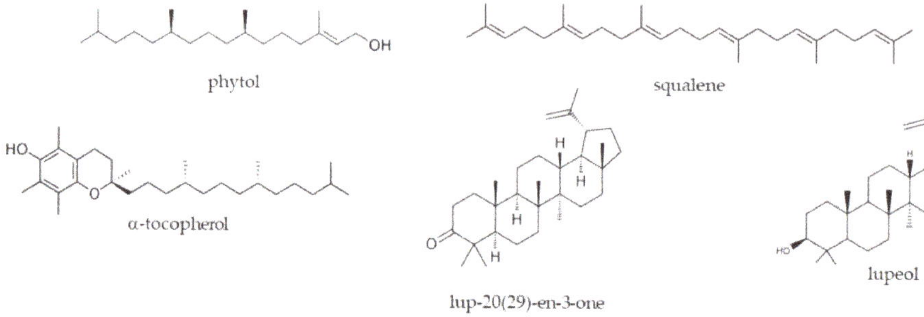

Figure 3. Chemical structures of phytol, squalene, α-tocopherol, lupenone (lup-20(29)-en-3-one), and lupeol.

Phytol is a diterpene known for its wide range of biological activities, such as antimicrobial, antioxidant, cytotoxic, anxiolytic, anticonvulsant, immunomodulatory, antinociceptive, and anti-inflammatory activities, besides its role in the induction of apoptosis and protective autophagy and in the treatment of metabolic disorders [79]. Phytol is the most abundant compound in the analyzed extracts, especially in the leaves fraction, with the extracts obtained from dried leaves showing higher content than the one obtained from fresh leaves (Figure 4).

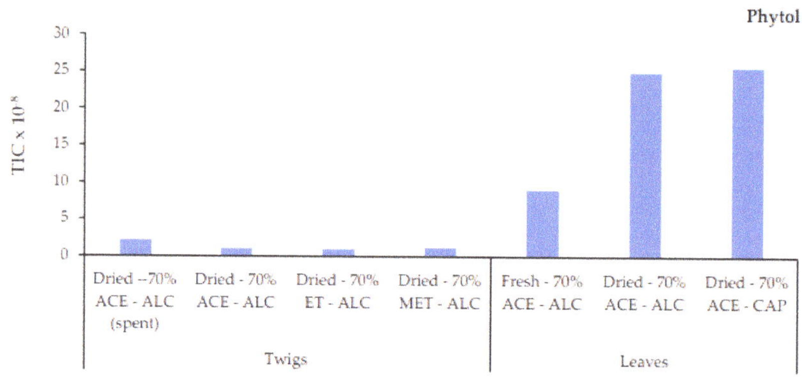

Figure 4. Absolute peak areas of phytol in the analyzed extracts; TIC: Total Ion Current.

Figure 5 shows the absolute peak areas of squalene, α-tocopherol, lupenone, and lupeol for the selected extracts.

Squalene is a triterpene with reported antioxidant, anti-inflammatory, and anti-atherosclerotic properties [80], as well as presenting the potential for cosmetic dermatology applications due to its emollient and antioxidant properties, and for hydration and its antitumor activities [81]. Squalene is more abundant in 70% ACE extracts than in 70% ET or 70% MET ones, with the highest amounts being observed in the dried leaves extracts.

α-Tocopherol is the main form of vitamin E, which has antioxidant, anti-inflammatory, and antitumor properties already described in the literature [82]. α-Tocopherol is not present in the 70% ET and 70% MET extracts, being particularly abundant in the dried leaves extracts.

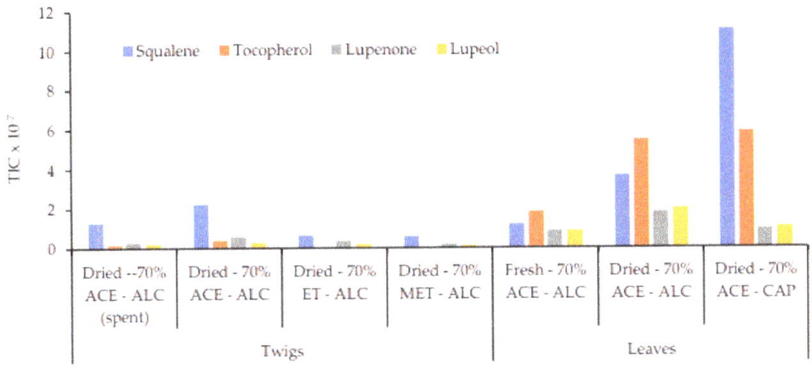

Figure 5. Absolute peak areas of squalene, α-tocopherol, lupenone (lup-20(29)-en-3-one), and lupeol in the analyzed extracts; TIC: Total Ion Current.

Lupenone and lupeol are pentacyclic triterpenes showing a broad spectrum of biological activities. For the former, its importance as a therapeutic and chemo preventive agent for the treatment of inflammation, virus infection, diabetes, cancer, and Chagas disease was already described in the literature [83]. For lupeol, antioxidant, antiprotozoal, anti-inflammatory, antitumor, and antimicrobial activities were previously reported, in addition to its cardio- and hepatoprotective effects and as a cancer chemo preventive agent [84]. Both compounds are more abundant in leaves extracts than in twigs extracts. These results show the ability of acetone to extract both polar and non-polar compounds, which occurred to a lesser extent when using aqueous ethanol or methanol, given the higher polarity of these solvents. The use of dry material favored the extraction of this non-polar fraction as well. The previous extraction of the raw material with petroleum ether naturally decreased the amount of these non-polar compounds in the extract obtained from spent material. The presence of these bioactive compounds in a higher amount in the leaves extracts does not agree with the highest antioxidant and antimicrobial effects observed in the twig's extracts, which suggests that these effects are predominantly due to the action of the more polar compounds present in the extracts. In fact, it was observed a strong positive correlation between TPC and antioxidant activity.

4. Conclusions

This investigation demonstrated the possibility of producing extracts rich in phenolic compounds with antioxidant and antimicrobial properties using leaves and twigs of Portuguese *A. dealbata* by a simple maceration process at room temperature. The highest values of extraction yield, TPC, TFC, and TPrAC as well as antioxidant activity measured by the DPPH and FRAP assays were obtained with 70% ACE using leaves and twigs from Alcobaça. The drying status of the raw material also affected some characteristics of the extracts, improving extraction yield, TFC, and TPrAC for dried leaves and twigs.

As a rule, better results were observed in extracts from the twigs than in the corresponding extracts from the leaves, except for extraction yield. This observation shows the possibility of using the entire fraction of the branches, where the leaves represent by far the highest weight fraction, without the necessity of separating the twigs from the leaves once they add quality to the produced extracts.

Moreover, this work showed the possibility of obtaining a wider range of products from the same amount of raw material via sequential extraction with petroleum ether to remove lipophilic compounds followed by the best extraction solvent to remove phenolic compounds, all without significantly decreasing the quality of the extract.

Significant and strong correlations between TPC and TPrAC and antioxidant activity by DPPH and FRAP assays shows that those phenolic components are the main responsible for the antioxidant and antimicrobial properties of the extracts. Nevertheless, bioactive lipophilic components, such as phytol, squalene, α-tocopherol, lupenone, and lupeol were also detected, especially in the extracts of leaves. Therefore, this work demonstrates the possibility of obtaining high-value extracts with antioxidant and antimicrobial properties from fractions of *A. dealbata* biomass that can be collected regularly to reduce widespread dissemination of this species.

Supplementary Materials: The following supporting information can be downloaded at: https://www.mdpi.com/article/10.3390/pr10112429/s1, Table S1: TPC (mg GAE/g of biomass, wet basis), TFC (mg CatE/g of biomass, wet basis), and antioxidant activity by DPPH (mg TE/g of biomass) and FRAP (mmol Fe^{2+}/g of biomass) of extracts from fresh and dried leaves and twigs from Alcobaça and Caparica, using different extraction solvents and extraction methods; Table S2: Extraction yield, TPC and DPPH values for different times of maceration at room temperature, of plants collected in Caparica using 70% ACE as extraction solvent; Table S3: TFC (mg RE/g extract, wet basis) of extracts from fresh and dried leaves and twigs from Alcobaça and Caparica, using different extraction solvents and extraction methods; Table S4: TFC (mg RutE/g of raw material, wet basis) of extracts from fresh and dried leaves and twigs from Alcobaça and Caparica, using different extraction solvents and extraction methods; Table S5: Average values of extraction yield (%, wet basis), TPC (mg GAE/g extract), TFC (mg CatE/g extract) and TPrAC (mg PycE/g extract) for extracts obtained by maceration and hot extraction (regardless drying status and origin), from fresh and dried material (regardless extraction method and origin), and from plants from Alcobaça and Caparica (regardless extraction process and drying status); Table S6: Chromatographic peak areas of non-polar components co-extracted with phenolic compounds for a group of selected extracts obtained by maceration at room temperature; Figure S1: Chromatographic profile of the non-polar fraction of the extract obtained with 70% acetone by maceration of dried Acacia leaves collected in Alcobaça region. Part 1—Low retention time components; Figure S2: Chromatographic profile of the non-polar fraction of the extract obtained with 70% acetone by maceration of dried Acacia leaves collected in Alcobaça region. Part 2—High retention time components; Figure S3: Chromatographic profile of the non-polar fraction of the extract obtained with 70% methanol by maceration of dried Acacia twigs collected in Alcobaça region. Part 1—Low retention time components.

Author Contributions: Conceptualization, R.C. and M.G.; methodology, R.C. and M.G.; investigation, R.C., E.M.M. and J.B.; writing—original draft preparation, R.C.; writing—review and editing, M.G., M.P.D., J.C.Q., M.G.d.S. and R.C.; supervision, M.G., M.P.D., M.G.d.S. and J.C.Q. All authors have read and agreed to the published version of the manuscript.

Funding: This work was supported by national funds through the Fundação para a Ciência e a Tecnologia, I.P. (Portuguese Foundation for Science and Technology) by the project (UIDB/04077/2020-2023 and UIDP/04077/2020-2023) of Mechanical Engineering and Resource Sustainability Center—MEtRICs, by the project UIDB/05064/2020 of VALORIZA—Research Centre for Endogenous Resource Valorization), by the project (PTDC/CTA-AMB/6587/2020) of CENSE "Center for Environmental and Sustainability Research" and by the project (UID/QUI/50006/2020) of LAQV/REQUIMTE; human resources were also supported by the PhD grants SFRH/BD/133300/2017 and UI/BD/150867/2021.

Data Availability Statement: Not applicable.

Acknowledgments: The authors acknowledge Fundação para a Ciência e a Tecnologia, I.P. (Portuguese Foundation for Science and Technology) for financing the projects (UIDB/04077/2020-2023 and UIDP/04077/2020-2023), UIDB/05064/2020, (PTDC/CTA-AMB/6587/2020) and (UID/QUI/50006/2020) that support the activities of the research units that participated in this work. The authors also acknowledge Fundação para a Ciência e a Tecnologia, I.P. for the PhD grants SFRH/BD/133300/2017 and UI/BD/150867/2021.

Conflicts of Interest: The authors declare no conflict of interest.

References

1. Maslin, B.R. Synoptic overview of *Acacia* sensu lato (Leguminosae: Mimosoideae) in East and Southeast Asia. *Gard. Bull. Singap.* **2015**, *67*, 231. [CrossRef]
2. Breton, C.; Guerin, J.; Ducatillion, C.; Médail, F.; Kull, C.A.; Bervillé, A. Taming the wild and "wilding" the tame: Tree breeding and dispersal in Australia and the Mediterranean. *Plant Sci.* **2008**, *175*, 197–205. [CrossRef]
3. Kull, C.A.; Shackleton, C.M.; Cunningham, P.J.; Ducatillon, C.; Dufour-Dror, J.-M.; Esler, K.J.; Friday, J.B.; Gouveia, A.C.; Griffin, A.R.; Marchante, E.; et al. Adoption, use and perception of Australian acacias around the world. *Divers. Distrib.* **2011**, *17*, 822–836. [CrossRef]
4. Seigler, D.S. Phytochemistry of *Acacia—Sensu lato*. *Biochem. Syst. Ecol.* **2003**, *31*, 845–873. [CrossRef]
5. Fuentes-Ramírez, A.; Pauchard, A.; Cavieres, L.A.; García, R.A. Survival and growth of *Acacia dealbata* vs. native trees across an invasion front in south-central Chile. *For. Ecol. Manag.* **2011**, *261*, 1003–1009. [CrossRef]
6. Le Maitre, D.C.; Gaertner, M.; Marchante, E.; Ens, E.J.; Holmes, P.M.; Pauchard, A.; O'Farrell, P.J.; Rogers, A.M.; Blanchard, R.; Blignaut, J.; et al. Impacts of invasive Australian acacias: Implications for management and restoration. *Divers. Distrib.* **2011**, *17*, 1015–1029. [CrossRef]
7. Lorenzo, P.; González, L.; Reigosa, M.J. The genus *Acacia* as invader: The characteristic case of *Acacia dealbata* Link in Europe. *Ann. For. Sci.* **2010**, *67*, 101. [CrossRef]
8. Richardson, D.M.; Rejmánek, M. Trees and shrubs as invasive alien species—A global review. *Divers. Distrib.* **2011**, *17*, 788–809. [CrossRef]
9. Ministério do Ambiente. *Ministério do Ambiente Decreto-Lei n.º 565/99 de 21 de Dezembro*; Série-A. N.º 295—31-12-1999; Diário da República: Lisboa, Portugal, 1999; pp. 9100–9115.
10. Jæger, D.; O'Leary, M.C.; Weinstein, P.; Møller, B.L.; Semple, S.J. Phytochemistry and bioactivity of *Acacia sensu stricto* (Fabaceae: Mimosoideae). *Phytochem. Rev.* **2019**, *18*, 129–172. [CrossRef]
11. Subhan, N.; Burrows, G.E.; Kerr, P.G.; Obied, H.K. Phytochemistry, Ethnomedicine, and Pharmacology of *Acacia*. In *Studies in Natural Products Chemistry*; Elsevier B.V.: Amsterdam, The Netherlands, 2018; Volume 57, pp. 247–326. ISBN 9780444640574.
12. Miguel-Chávez, R.S. Phenolic Antioxidant Capacity: A Review of the State of the Art. In *Phenolic Compounds—Biological Activity*; IntechOpen: London, UK, 2017; pp. 59–74.
13. Santos-Sánchez, N.F.; Salas-Coronado, R.; Villanueva-Cañongo, C.; Hernández-Carlos, B. Antioxidant Compounds and Their Antioxidant Mechanism. In *Antioxidants*; IntechOpen: London, UK, 2019; pp. 1–28.
14. Ghannoum, M.A.; Rice, L.B. Antifungal Agents: Mode of Action, Mechanisms of Resistance, and Correlation of These Mechanisms with Bacterial Resistance. *Clin. Microbiol. Rev.* **1999**, *12*, 501–517. [CrossRef]
15. Correia, R.; Quintela, J.C.; Duarte, M.P.; Gonçalves, M. Insights for the valorization of biomass from portuguese invasive *Acacia* spp. in a biorefinery perspective. *Forests* **2020**, *11*, 1342. [CrossRef]
16. Ramli, S.; Harada, K.I.; Ruangrungsi, N. Antioxidant, antimicrobial and cytotoxicity activities of *Acacia farnesiana* (L.) Willd. Leaves ethanolic extract. *Pharmacogn. J.* **2011**, *3*, 50–58. [CrossRef]
17. Cock, I.E.; van Vuuren, S.F. South African food and medicinal plant extracts as potential antimicrobial food agents. *J. Food Sci. Technol.* **2015**, *52*, 6879–6899. [CrossRef]
18. Nielsen, T.R.H.; Kuete, V.; Jäger, A.K.; Meyer, J.J.M.; Lall, N. Antimicrobial activity of selected South African medicinal plants. *BMC Complement. Altern. Med.* **2012**, *12*, 74. [CrossRef]
19. Nyila, M.A.; Leonard, C.M.; Hussein, A.A.; Lall, N. Activity of South African medicinal plants against *Listeria monocytogenes* biofilms, and isolation of active compounds from *Acacia karroo*. *S. Afr. J. Bot.* **2012**, *78*, 220–227. [CrossRef]
20. Priyanka, C.; Kumar, P.; Bankar, S.P.; Karthik, L. In vitro antibacterial activity and gas chromatography–mass spectroscopy analysis of *Acacia karoo* and Ziziphus mauritiana extracts. *J. Taibah Univ. Sci.* **2015**, *9*, 13–19. [CrossRef]
21. Lima, C.P.; Cunico, M.M.; Auer, C.G.; Miguel, O.G.; Miguel, M.D.; da Silva, C.B.; Andrade, C.A.; Kerber, V.A. Potencial alelopático e antifúngico do extrato das folhas de *Acacia longifolia* (Andr.) Willd. *Visão Acadêmica* **2013**, *14*, 16–25. [CrossRef]
22. Edrah, S.M.; Alafid, F.; Shmeala, H.; Abobaker, D.M. Phytochemical analysis and antibacterial activity of *Acacia pycnantha* from Alkhums Libya. In Proceedings of the 2nd Annual Conference on Theories and Applications of Basic and Biosciences, Bali, Indonesia, 11–12 April 2019; pp. 704–712.
23. Mahmoud, M.F.; Alrumman, S.A.; Hesham, A.E.L. Biological activities of some *Acacia* spp. (Fabaceae) against new clinical isolates identified by ribosomal RNA gene-based phylogenetic analysis. *Pak. J. Pharm. Sci.* **2016**, *29*, 221–229.
24. El-Toumy, S.A.; Salib, J.Y.; Mohamed, W.M.; Morsy, F.A. Phytochemical and antimicrobial studies on *Acacia saligna* leaves. *Egypt. J. Chem.* **2010**, *53*, 705–717. [CrossRef]
25. Noreen, I.; Iqbal, A.; Rabbi, F.-; Muhammad, A.; Shah, Z.; Rahman, Z.U. Antimicrobial activity of different solvents extracts of *Acacia cyanophylla*. *Pak. J. Weed Sci. Res.* **2017**, *23*, 79–90.
26. Gumgumjee, N.M.; Hajar, A.S. Antimicrobial efficacy of *Acacia saligna* (Labill.) HL Wendl. and *Cordia sinensis* Lam. leaves extracts against some pathogenic microorganisms. *Int. J. Microbiol. Immunol. Res* **2015**, *3*, 51–57.
27. Gedara, S.R.; Galala, A.A. New cytotoxic spirostane saponin and biflavonoid glycoside from the leaves of *Acacia saligna* (Labill.) H.L. Wendl. *Nat. Prod. Res.* **2014**, *28*, 324–329. [CrossRef] [PubMed]
28. Fowora, M.A.; Onyeaghasiri, F.U.; Olanlege, A.L.O.; Edu-Muyideen, I.O.; Adebesin, O.O. In Vitro Susceptibility of Dermatophytes to Anti-Fungal Drugs and Aqueous *Acacia nilotica* Leaf Extract in Lagos, Nigeria. *J. Biomed. Sci. Eng.* **2021**, *14*, 74–82. [CrossRef]

29. Borges, A.; José, H.; Homem, V.; Simões, M. Comparison of techniques and solvents on the antimicrobial and antioxidant potential of extracts from *Acacia dealbata* and *Olea europaea*. *Antibiotics* **2020**, *9*, 48. [CrossRef]
30. Silva, E.; Fernandes, S.; Bacelar, E.; Sampaio, A. Antimicrobial activity of aqueous, ethanolic and methanolic leaf extracts from *Acacia* spp. and Eucalyptus nicholii. *Afr. J. Tradit. Complement. Altern. Med.* **2016**, *13*, 130–134. [CrossRef] [PubMed]
31. Kumari, R.; Mishra, R.C.; Yadav, A.; Yadav, J.P. Screening of traditionally used medicinal plants for their antimicrobial efficacy against oral pathogens and GC-MS analysis of acacia nilotica extract. *Indian J. Tradit. Knowl.* **2019**, *18*, 162–168.
32. Kumari, R.; Mishra, R.C.; Yadav, J.P. Preparation and in vitro antimicrobial activity of supercritical fluid extracts of selected Indian plants against oral pathogens and their phytochemicals and statistical analysis. *Int. J. Green Pharm.* **2020**, *14*, 146–154. [CrossRef]
33. Kumari, R.; Mishra, R.C.; Sheoran, R.; Yadav, J.P. Fractionation of antimicrobial compounds from *Acacia nilotica* twig extract against oral pathogens. *Biointerface Res. Appl. Chem.* **2020**, *10*, 7097–7105. [CrossRef]
34. Arshad, M.S.; Hussain, I.; Mahmood, M.S.; Khan, M.N. Evaluation of antimicrobial potential of *Acacia nilotica* (Kikar) against oral pathogens associated with caries and periodontitis. *Pak. J. Agric. Sci.* **2017**, *54*, 423–430. [CrossRef]
35. Lomarat, P.; Chancharunee, S.; Anantachoke, N.; Kitphati, W.; Sripha, K.; Bunyapraphatsara, N. Bioactivity-guided separation of the active compounds in *Acacia pennata* responsible for the prevention of Alzheimer's disease. *Nat. Prod. Commun.* **2015**, *10*, 1431–1434. [CrossRef]
36. Rodrigues, V.H.; De Melo, M.M.R.; Portugal, I.; Silva, C.M. Extraction of added-value triterpenoids from *Acacia dealbata* leaves using supercritical fluid extraction. *Processes* **2021**, *9*, 1159. [CrossRef]
37. Azwanida, N.N. A Review on the Extraction Methods Use in Medicinal Plants, Principle, Strength and Limitation. *Med. Aromat. Plants* **2015**, *4*, 3–8. [CrossRef]
38. Singleton, V.L.; Orthofer, R.; Lamuela-Raventós, R.M. Analysis of total phenols and other oxidation substrates and antioxidants by means of Folin-Ciocalteu reagent. *Methods Enzymol.* **1999**, *299*, 152–178. [CrossRef]
39. Barros, L.; Carvalho, A.M.; Morais, J.S.; Ferreira, I.C.F.R. Strawberry-tree, blackthorn and rose fruits: Detailed characterisation in nutrients and phytochemicals with antioxidant properties. *Food Chem.* **2010**, *120*, 247–254. [CrossRef]
40. Škerget, M.; Kotnik, P.; Hadolin, M.; Hraš, A.R.; Simonič, M.; Knez, Ž. Phenols, proanthocyanidins, flavones and flavonols in some plant materials and their antioxidant activities. *Food Chem.* **2005**, *89*, 191–198. [CrossRef]
41. Blois, M.S. Antioxidant Determinations by the Use of a Stable Free Radical. *Nature* **1958**, *181*, 1199–1200. [CrossRef]
42. Benzie, I.F.F.; Strain, J.J. Hormonelle Beeinflussung des experimentellen Portiocarcinoms. *Anal. Biochem.* **1996**, *239*, 70–76. [CrossRef]
43. Pereira, P.; Mauricio, E.M.; Duarte, M.P.; Lima, K.; Fernandes, A.S.; Bernardo-Gil, G.; Cebola, M.-J. Potential of supercritical fluid myrtle extracts as an active ingredient and co-preservative for cosmetic and topical pharmaceutical applications. *Sustain. Chem. Pharm.* **2022**, *28*, 100739. [CrossRef]
44. Popova, V.; Ivanova, T.; Stoyanova, A.; Nikolova, V.; Hristeva, T.; Zheljazkov, V.D. GC-MS composition and olfactory profile of concretes from the flowers of four *Nicotiana* species. *Molecules* **2020**, *25*, 2617. [CrossRef]
45. van Den Dool, H.; Kratz, P.D. A generalization of the retention index system including linear temperature programmed gas—Liquid partition chromatography. *J. Chromatogr. A* **1963**, *11*, 463–471. [CrossRef]
46. Climate-Data.org. Clima Alcobaça. Available online: https://pt.climate-data.org/europa/portugal/alcobaca/alcobaca-882351/ (accessed on 1 April 2021).
47. Climate-Data.org. Clima Almada. Available online: https://pt.climate-data.org/europa/portugal/almada/almada-7111/ (accessed on 1 April 2021).
48. Luís, A.; Gil, N.; Amaral, M.E.; Duarte, A.P. Antioxidant activities of extracts from *Acacia melanoxylon*, *Acacia dealbata* and *Olea europaea* and alkaloids estimation. *Int. J. Pharm. Pharm. Sci.* **2012**, *4*, 225–231.
49. Gabr, S.; Nikles, S.; Maria, E.; Wenzig, P.; Ardjomand-woelkart, K.; Hathout, R.M.; El-ahmady, S.; Abdel, A.; Singab, A.; Bauer, R. Characterization and optimization of phenolics extracts from *Acacia* species in relevance to their anti-inflammatory activity. *Biochem. Syst. Ecol.* **2018**, *78*, 21–30. [CrossRef]
50. Xiong, J.; Grace, M.H.; Esposito, D.; Wang, F.; Lila, M.A. Phytochemical characterization and anti-inflammatory properties of *Acacia mearnsii* leaves. *Nat. Prod. Commun.* **2016**, *11*, 649–653. [CrossRef]
51. Zheleva-Dimitrova, D.; Sinan, K.I.; Etienne, O.K.; Ak, G.; Sharmeen, J.B.; Dervisoglu, G.; Ozdemir, F.A.; Mahomoodally, M.F.; Zengin, G. Comprehensive chemical characterization and biological evaluation of two *Acacia* species: *A. nilotica* and *A. ataxacantha*. *Food Chem. Toxicol.* **2021**, *156*, 112446. [CrossRef] [PubMed]
52. Ferreira-Santos, P.; Genisheva, Z.; Botelho, C.; Santos, J.; Ramos, C.; Teixeira, J.A.; Rocha, C.M.R. Unravelling the biological potential of *Pinus pinaster* bark extracts. *Antioxidants* **2020**, *9*, 334. [CrossRef] [PubMed]
53. Tung, Y.T.; Wu, J.H.; Hsieh, C.Y.; Chen, P.S.; Chang, S.T. Free radical-scavenging phytochemicals of hot water extracts of *Acacia confusa* leaves detected by an on-line screening method. *Food Chem.* **2009**, *115*, 1019–1024. [CrossRef]
54. Sowndhararajan, K.; Joseph, J.M.; Manian, S. Antioxidant and free radical scavenging activities of Indian acacias: *Acacia leucophloea* (Roxb.) Willd., *Acacia ferruginea* Dc., *Acacia dealbata* Link. and *Acacia pennata* (l.) Willd. *Int. J. Food Prop.* **2013**, *16*, 1717–1729. [CrossRef]
55. Kopjar, M.; Tadić, M.; Piližota, V. Phenol content and antioxidant activity of green, yellow and black tea leaves. *Chem. Biol. Technol. Agric.* **2015**, *2*, 1. [CrossRef]

56. Sariburun, E.; Şahin, S.; Demir, C.; Türkben, C.; Uylaşer, V. Phenolic content and antioxidant activity of raspberry and blackberry cultivars. *J. Food Sci.* **2010**, *75*, 328–335. [CrossRef]
57. Verma, N.; Shukla, S. Impact of various factors responsible for fluctuation in plant secondary metabolites. *J. Appl. Res. Med. Aromat. Plants* **2015**, *2*, 105–113. [CrossRef]
58. Kedare, S.B.; Singh, R.P. Genesis and development of DPPH method of antioxidant assay. *J. Food Sci. Technol.* **2011**, *48*, 412–422. [CrossRef] [PubMed]
59. Venkatesan, T.; Choi, Y.-W.; Kim, Y.-K. Impact of Different Extraction Solvents on Phenolic Content and Antioxidant Potential of *Pinus densiflora* Bark Extract. *Biomed Res. Int.* **2019**, *2019*, 3520675. [CrossRef] [PubMed]
60. Sadiq, M.B.; Tharaphan, P.; Chotivanich, K.; Tarning, J.; Anal, A.K. In vitro antioxidant and antimalarial activities of leaves, pods and bark extracts of *Acacia nilotica* (L.) Del. *BMC Complement. Altern. Med.* **2017**, *17*, 372. [CrossRef] [PubMed]
61. Afsar, T.; Razak, S.; Shabbir, M.; Khan, M.R. Antioxidant activity of polyphenolic compounds isolated from ethyl—Acetate fraction of *Acacia hydaspica* R. Parker. *Chem. Cent. J.* **2018**, *12*, 5. [CrossRef]
62. Zia-Ul-Haq, M.; Cavar, S.; Qayum, M.; Khan, I.; Ahmad, S. Chemical composition and antioxidant potential of *Acacia leucophloea* Roxb. *Acta Bot. Croat.* **2013**, *72*, 133–144. [CrossRef]
63. Astuya, A.; Ziehe, J.; Rivera, A.; Ortiz, S.; Ulloa, V.; Roeckel, M.; Aspé, E.; Fernández, K. Antioxidant and anti-inflammatory activities of *Pinus radiata* bark extract in salmonid cell lines. *Aquac. Res.* **2016**, *48*, 3568–3578. [CrossRef]
64. Salar, R.K.; Seasotiya, L. Free radical scavenging activity, phenolic contents and phytochemical evaluation of different extracts of stem bark of *Butea monosperma* (Lam.) Kuntze. *Front. Life Sci.* **2011**, *5*, 107–116. [CrossRef]
65. Salem, N.; Msaada, K.; Hamdaoui, G.; Limam, F.; Marzouk, B. Variation in Phenolic Composition and Antioxidant Activity during Flower Development of Safflower (*Carthamus tinctorius* L.). *J. Agric. Food Chem.* **2011**, *59*, 4455–4463. [CrossRef]
66. Oliveira, C.S.D.; Moreira, P.; Resende, J.; Cruz, M.T.; Pereira, C.M.F.; Silva, A.M.S.; Santos, S.A.O.; Silvestre, A.J.D. Characterization and cytotoxicity assessment of the lipophilic fractions of different morphological parts of *Acacia dealbata*. *Int. J. Mol. Sci.* **2020**, *21*, 1814. [CrossRef]
67. Akoglu, H. User's guide to correlation coefficients. *Turk. J. Emerg. Med.* **2018**, *18*, 91–93. [CrossRef]
68. Yadav, A.; Yadav, M.; Kumar, S.; Sharma, D.; Yadav, J.P. In vitro antioxidant activities and GC-MS analysis of different solvent extracts of *Acacia nilotica* leaves. *Indian J. Pharm. Sci.* **2018**, *80*, 892–902. [CrossRef]
69. Abdel-Farid, I.B.; Sheded, M.G.; Mohamed, E.A. Metabolomic profiling and antioxidant activity of some *Acacia* species. *Saudi J. Biol. Sci.* **2014**, *21*, 400–408. [CrossRef] [PubMed]
70. Ulewicz-Magulska, B.; Wesolowski, M. Total Phenolic Contents and Antioxidant Potential of Herbs Used for Medical and Culinary Purposes. *Plant Foods Hum. Nutr.* **2019**, *74*, 61–67. [CrossRef] [PubMed]
71. Zeng, X.; Su, W.; Zheng, Y.; Liu, H.; Li, P.; Zhang, W.; Liang, Y.; Bai, Y.; Peng, W.; Yao, H. UFLC-Q-TOF-MS/MS-based screening and identification of flavonoids and derived metabolites in human urine after oral administration of exocarpium citri grandis extract. *Molecules* **2018**, *23*, 895. [CrossRef]
72. Jenkins, D.R. Nosocomial infections and infection control. *Medicine* **2017**, *45*, 629–633. [CrossRef]
73. Ganaie, H.A.; Ali, M.N.; Ganai, B.A.; Meraj, M.; Ahmad, M. Antibacterial activity of 14, 15-dihydroajugapitin and 8-o-acetylharpagide isolated from *Ajuga bracteosa* Wall ex. Benth against human pathogenic bacteria. *Microb. Pathog.* **2017**, *103*, 114–118. [CrossRef]
74. Wu, S.-C.; Liu, F.; Zhu, K.; Shen, J.-Z. Natural Products That Target Virulence Factors in Antibiotic-Resistant *Staphylococcus aureus*. *J. Agric. Food Chem.* **2019**, *67*, 13195–13211. [CrossRef]
75. Leitão, J.H. Microbial Virulence Factors. *Int. J. Mol. Sci.* **2020**, *21*, 5320. [CrossRef]
76. Mayer, F.L.; Wilson, D.; Hube, B. *Candida albicans* pathogenicity mechanisms. *Virulence* **2013**, *4*, 119–128. [CrossRef]
77. Lopes, J.P.; Lionakis, M.S. Pathogenesis and virulence of *Candida albicans*. *Virulence* **2022**, *13*, 89–121. [CrossRef]
78. Pereira, F.B.M.; Domingues, F.M.J.; Silva, A.M.S. Triterpenes from *Acacia dealbata*. *Nat. Prod. Lett.* **1996**, *8*, 97–103. [CrossRef]
79. Islam, M.T.; Ali, E.S.; Uddin, S.J.; Shaw, S.; Islam, M.A.; Ahmed, M.I.; Chandra Shill, M.; Karmakar, U.K.; Yarla, N.S.; Khan, I.N.; et al. Phytol: A review of biomedical activities. *Food Chem. Toxicol.* **2018**, *121*, 82–94. [CrossRef] [PubMed]
80. Lou-Bonafonte, J.M.; Martínez-Beamonte, R.; Sanclemente, T.; Surra, J.C.; Herrera-Marcos, L.V.; Sanchez-Marco, J.; Arnal, C.; Osada, J. Current Insights into the Biological Action of Squalene. *Mol. Nutr. Food Res.* **2018**, *62*, 1800136. [CrossRef] [PubMed]
81. Huang, Z.R.; Lin, Y.K.; Fang, J.Y. Biological and pharmacological activities of squalene and related compounds: Potential uses in cosmetic dermatology. *Molecules* **2009**, *14*, 540–554. [CrossRef] [PubMed]
82. Szewczyk, K.; Chojnacka, A.; Górnicka, M. Tocopherols and tocotrienols—Bioactive dietary compounds; what is certain, what is doubt? *Int. J. Mol. Sci.* **2021**, *22*, 6222. [CrossRef]
83. Xu, F.; Huang, X.; Wu, H.; Wang, X. Beneficial health effects of lupenone triterpene: A review. *Biomed. Pharmacother.* **2018**, *103*, 198–203. [CrossRef]
84. Wal, P.; Wal, A.; Sharma, G.; Rai, A.K. Biological activities of lupeol. *Syst. Rev. Pharm.* **2011**, *2*, 96–103. [CrossRef]

processes
MDPI

Article

Poplar Wood Pretreatment Using Deep Eutectic Solvents for Promoting Enzymatic Hydrolysis

Xuyang Zhao, Ying Zhan, Lihua Han, Xiaoran Sun, Tianyu Zhang and Zheng Zhao *

College of Chemical Engineering, North China University of Science and Technology, Tangshan 063210, China; a15736973753@163.com (X.Z.); zhanying0326@163.com (Y.Z.); tshlh@ncst.edu.cn (L.H.); xiaoranhb@yeah.net (X.S.); a13786654504@163.com (T.Z.)
* Correspondence: buctzhaozheng@163.com; Tel.: +86-315-8805-460

Abstract: Bioethanol produced from lignocellulose is a renewable energy substitute for traditional fossil fuels. Poplar wood as forest waste is popular in bioethanol production. Nonetheless, the complex structure of lignocellulose leads to low reducing sugar and ethanol yields. Thus, lignocellulose pretreatment is necessary to promote enzymatic hydrolysis. Deep eutectic solvents (DESs) have good dissolution capacity, low vapor pressure, a simple synthesis procedure, low synthesis cost and low toxicity. More and more researchers have begun paying attention to the application of DESs in lignocellulose pretreatment. In this work, poplar wood was pretreated using a series of basic DESs based on diol. The effects of the DES species, the basicity of the solvents, the pretreatment temperature and the pretreatment time on the effectiveness of pretreatment and enzyme hydrolysis for poplar wood were investigated, and characterization analysis (Fourier transform infrared spectroscopy, X-ray diffraction and scanning electron microscopy) of poplar wood was carried out to reveal the pretreatment mechanism. The best pretreatment effect was obtained from K: 1, 2-PG, which removed 89.2% and 71.6% of the lignin and hemicellulose, respectively, while preserving 97.5% of the cellulose at 130 °C for 7 h. This enhanced the reducing sugar yield to 82.5% relative to that of the raw sample (3.3%) after 72 h of hydrolysis. The results of the characterization analysis demonstrated that lignin and hemicellulose were removed. Therefore, the DES based on K: 1, 2-PG is a promising solvent for poplar wood pretreatment, and could improve the industrial production of reducing sugar and bioethanol.

Keywords: deep eutectic solvents; diol; poplar wood; pretreatment; enzymatic hydrolysis

1. Introduction

Bioethanol, a source of renewable energy, can replace traditional fossil fuels and, thus, resolve two energy problems. The main sources of lignocellulosic biomass include agricultural and forestry wastes. In particular, a huge amount of forest waste originates from deforestation, timber processing and forest pruning, i.e., about 400 million tons of standard coal annually in China [1]. Nevertheless, the complex structure of lignocellulose hinders the transformation of cellulose into bioethanol. Lignocellulosic biomass contains lignin, hemicellulose and cellulose. Hemicelluloses adhere to the surface of cellulose through van der Waals forces and hydrogen bonds, and connect lignin through ether bonds and glycoside bonds to form a stable network structure [2]. Therefore, it is necessary to pretreat lignocellulosic biomass by removing lignin and hemicellulose in order to improve the reducing sugar and ethanol yields.

Many common traditional pretreatment methods, such as dilute acid, dilute alkali, organic solvent and biological methods [3–7], have one or several disadvantages, including high energy consumption, high toxicity, harsh conditions, corrosion equipment or a long pretreatment time. Therefore, the majority of scientific research workers have begun to look at green solvent ionic liquids (ILs). ILs have the characteristics of high thermal stability, low steam pressure, non-flammability and non-volatilization [8], making them suitable for lignocellulose pretreatment. Nonetheless, the greenness of ILs has been questioned and the industrial application of IL pretreatment technology has been hindered because they possess several drawbacks, such as high synthetic cost, complex synthetic steps and high toxicity in ILs.

Recently, the application of deep eutectic solvents (DESs) in lignocellulosic biomass pretreatment has attracted extensive attention. DESs usually consist of two or more hydrogen bond donors (HBDs) and hydrogen bond acceptors (HBAs). Otherwise, the melting points of DESs are lower than those of individual components. The most important point is that DESs have a few similar properties to ILs and possess merits such as a low cost of synthesis, a simple synthesis procedure, low toxicity and good biocompatibility, which compensate for the shortcomings of ILs. Lee et al. [9] used choline chloride: glycerol, choline chloride: lactic acid and choline chloride: urea to pretreat oil palm empty fruit bundles under ultrasound with 210 W at 50 °C for 15 min. As a result, reducing sugar yields of 35.3%, 36.7% and 35.8%, respectively, were obtained after the pretreated residues were enzymatically hydrolyzed for 72 h. Sai et al. [10] used chloride: lactic acid (molar ratio was 1:10) to pretreat empty fruit bunches at 100 °C for 1 h. After residue hydrolysis for 48 h, 51.1% of the reducing sugar yield was obtained. Li et al. [11] compared the pretreatment capacity of choline chloride: lactic acid and choline chloride: glycine. The research results showed that choline chloride: lactic acid removed 90.4% of the lignin, but choline chloride: glycine only removed 58.4% of the lignin during poplar wood pretreatment at suitable pretreatment conditions (temperature: 120 °C, time: 12 h). Wu et al. [12] reported that the continuous pretreatment of 0.75% (w/w) NaOH and choline chloride: lactic acid for sorghum straw exhibited high pretreatment efficiency. Firstly, 0.75% (w/w) NaOH pretreated sorghum straw at 121 °C for 1 h. Then, choline chloride: lactic acid was used to pretreat sorghum straw at 140 °C for 40 min. As a result, 78.4% lignin and 67.6% hemicellulose were removed, respectively. After the enzymatic hydrolysis of pretreated residue for 72 h, 94.9% of the reducing sugar yield was obtained. However, several problems, such as the larger proportions of cellulose loss and low reducing sugar yield, have yet to be resolved in the current DES pretreatment method. Our research group [13,14] has reported that basic DESs have the capacity to selectively remove the lignin and hemicellulose while retaining the majority of the cellulose in wheat straw. Otherwise, the basic DESs pretreatment method is less studied; in addition, the studies are not extensive and the pretreatment mechanism lacks depth, so it is necessary to conduct systematic research on basic DES pretreatment.

In this paper, a series of basic DESs based on diol (i.e., sodium hydroxide: ethylene, sodium hydroxide: polyethylene glycol-200, potassium hydroxide: polyethylene glycol-300, lithium hydroxide: polyethylene glycol-400, sodium hydroxide: polyethylene glycol-600, sodium hydroxide: 1,4-butyl glycol and potassium hydroxide: 1,2-propylene glycol) were screened to pretreat poplar wood. The effects of the DES species, the basicity of the solvents, the pretreatment temperature (i.e., 70 °C, 90 °C, 110 °C, 130 °C, 150 °C) and the pretreatment time (i.e., 7 h, 9 h, 11 h, 13 h, 15 h, 17 h) on the composition and reducing sugar yield of poplar wood were investigated. Moreover, the crystal structure, surface morphology and chemical structure of poplar biomass were also analyzed via Fourier transform infrared spectroscopy (FT-IR), X-ray diffraction (XRD) and scanning electron microscopy (SEM) to clarify the mechanism of basic DESs based on diol pretreating poplar wood.

2. Materials and Methods

2.1. Materials

Sodium hydroxide, potassium hydroxide (purity: 95%), ethylene glycol, lithium hydroxide (purity: 98%), 1,2-propanediol, 1,4-butanediol (purity: 99%), polyethylene glycol-600, polyethylene glycol-400, polyethylene glycol-300, polyethylene: glycol-200 and cellulase (enzymatic activity: 10,000 $U \cdot g^{-1}$, C805042) were utilized. All of the aforementioned reagents were purchased from Maclean. Poplar wood was obtained from

Shijiazhuang Farm. The poplar wood was milled using a pulverizer and then screened to the desirable particles with a 50–80 mesh. After Soxhlet extraction was performed using ethanol and water for 24 h, the poplar wood particles was airdried and stored at 4 °C.

2.2. Preparation of DESs

The HBAs and HBDs were mixed with a specific molar ratio. Subsequently, the mixture was magnetically stirred at 80 °C until transparent and homogeneous solutions formed [15,16]. Table 1 lists the compositions of prepared DESs and the molar ratios of HBA to HBD in DESs.

Table 1. The preparation of DESs.

HBA	HBD	Molar Ratio (HBA:HBD)	Abbreviation
Sodium hydroxide	Ethylene glycol	1:4	Na: EG
Sodium hydroxide	Polyethylene glycol-200	1:2	Na: PEG-200
Potassium hydroxide	Polyethylene glycol-300	1:4	K: PEG-300
Lithium hydroxide	Polyethylene glycol-400	1:10	Li: PEG-400
Sodium hydroxide	Polyethylene glycol-600	1:4	Na: PEG-600
Sodium hydroxide	1,4-Butyl glycol	1:5	Na: 1,4-BDO
Potassium hydroxide	1,2-Propylene glycol	1:2	K: 1,2-PG

2.3. Pretreatment Process

Firstly, a 50–80 mesh of poplar wood was mixed with DESs at 1:20 (*w*/*w*), then stirred magnetically at a certain temperature for a set length of time. Next, the pretreatment liquid was diluted using hot water, then centrifuged (3000 $r{\cdot}s^{-1}$) and filtered to separate the solid from the liquid. When the separation step was completed, the solid was washed further using deionized water. Centrifuging and washing were repeated until the scrubbing solution was neutral. In the end, the solid residue was freeze-dried for 72 h, then weighed and stored at 25 °C.

2.4. Enzymatic Hydrolysis

Firstly, the dipotassium hydrogen phosphate–citrate buffer solution was adjusted to pH = 4.8. Then, a mixture of 52.5 mL of ready-prepared buffer solution, 7.5 mL of cellulase solution and 0.4 g of the poplar wood sample was shaken in 72 h at 120 r/min and 45 °C. Additionally, 0.6 mL of hydrolysate was withdrawn at regular intervals, and the enzymatic hydrolysis reaction was terminated in a boiling water bath. According to the DNS method, the reducing sugar concentration was determined using the standard curve for reducing sugar ($y = 1.68x - 0.075$, $R^2 = 0.9998$). The reducing sugar yield was calculated based on the reducing sugar concentration [17].

2.5. Components Analysis

2.5.1. Determination of Lignin Content

Firstly, the 72% (*w*/*w*) sulfuric acid (10 mL) hydrolyzed a 0.3 g poplar wood sample for 1.5 h. Afterwards, the filtration of the suspension liquid was carried out to obtain liquids and solids (hydrolysis residue). The deionized water was used to wash the solid until the scrubbing solution was neutral. After filtering, the solid containing acid-insoluble lignin and ash was dried and calcined to determine the acid-insoluble lignin content [18]. The absorbance of the liquid was measured at 205 nm to determine the acid-soluble lignin content. The total lignin content in the poplar wood sample consisted of the acid-soluble lignin content and the acid-insoluble lignin content. According to Equation (1), the lignin removal was calculated.

$$R_l = [1 - (R \times w_{rl})/w_l] \quad (1)$$

R_l—lignin removal (%), R—residue recovery (%), w_{rl}—lignin content of pretreated poplar wood (%) and w_l—lignin content of untreated poplar wood (%).

2.5.2. Determination of Cellulose and Hemicellulose Content

First, 5 mL of 4 mg·mL^{-1} KOH solution (containing 1 mg·mL^{-1} $NaBH_4$) was used to treat 0.1 g of the poplar wood sample for 1 h in a water bath shaker at 25 °C. Next, 10 mL of deionized water and an equal volume of 4 mg·mL^{-1} KOH solution were added to the treatment liquid for washing and diluting. The supernatant and the absorbances were measured at 660 nm (xylose) and 620 nm (glucose), respectively. Then, the concentrations of xylose and glucose were calculated based on the absorbances and the standard curves of xylose ($y = 300.97x - 0.016$, $R^2 = 0.9994$) and glucose ($y = 102.75x + 0.0075$, $R^2 = 0.9992$) [19].

Afterwards, the 72% (w/w) sulfuric acid (4 mL) treated the residue, which was obtained from the above alkali treatment process, in a water bath shaker at 25 °C for 1 h. Then, the treatment liquid was filled to 30 mL by adding deionized water. The absorbances of the supernatant were also measured in order to calculate the concentrations of xylose and glucose, as described above.

Lastly, the cellulose content of the poplar wood sample was calculated according to the dehydration coefficient of glucose (0.90) and the glucose concentration obtained from the acid treatment process. The hemicellulose content in the poplar wood sample was determined according to the dehydration coefficient of glucose (0.90) and xylose (0.88), while the glucose and xylose concentrations were determined during the base treatment process and the xylose concentration was determined during the acid treatment process. According to Equations (2) and (3), the hemicellulose removal and cellulose reservation were calculated, respectively.

$$R_{\mathrm{h}} = [1 - (R \times w_{rh})/w_h] \quad (2)$$

R_{h}—hemicellulose removal (%), R—residue recovery (%), w_{rh} —hemicellulose content in poplar wood residue (%) and w_h —hemicellulose content in untreated poplar wood (%).

$$R_{\mathrm{c}} = [1 - (R \times w_{rc})/w_c] \quad (3)$$

R_{c}—cellulose reservation (%), R—residue recovery (%), w_{rc}—cellulose content in pretreated poplar wood (%) and w_c —lignin content in untreated poplar wood (%).

2.6. SEM Analysis

The crushed poplar wood sample was evenly dispersed on a disc coating with the conductive glue, and then spray-gold treatment was performed. Afterwards, the surface morphology of poplar wood samples before and after pretreatment by DESs was observed using a JSM-IT100 (A) type SEM [20].

2.7. FT-IR Analysis

Firstly, the mixture of the poplar wood sample and potassium bromide, with a mass ratio of 100:1, was pressed into a tablet. Then, a Vertex 80 V spectrometer was used to scan the tablet for 32 s (resolution: 2 cm^{-1}) in order to carry out the chemical structural analysis of the poplar wood sample. The wavenumber ranged from 4000 cm^{-1} to 400 cm^{-1} [21].

2.8. XRD Analysis

The poplar wood sample was scanned from 10° to 50° at a rate of 5°/min, using a Bruker D8 Advanced diffractometer to analyze the crystallinity in the poplar wood. The intensity of the spectrum peaks was used to calculate the crystallinity index, as shown in Equation (4) [21].

$$CrI = \frac{I_{002} - I_{\mathrm{am}}}{I_{002}} \times 100 \quad (4)$$

CrI—crystallinity index, I_{002}—peak intensity of crystalline region ($2\theta = 22.5°$) and I_{am}—peak intensity of amorphous region ($2\theta = 18°$).

3. Results and Analysis

3.1. Effect of DES Type on Pretreatment of Poplar Wood

Figure 1 shows the results of different DESs pretreating poplar wood at 90 °C for 11 h. As depicted in Figure 1, the pretreatment effects of diol monomer-based DESs were better than those of PEG-based DESs. The lignin and hemicellulose removal of Na: EG, Na: 1,4-BDO and K: 1,2-PG (diol monomer based DESs) pretreated residues ranged from 56.6% to 76.4% and from 54.1% to 69.5%, respectively, while Na: PEG-200, K: PEG-300, Li: PEG-400 and Na: PEG-600 (PEG based DESs) only removed 37.6–56.2% of the lignin and 24.3–60.9% of the hemicellulose. It is worth noting that K: 1,2-PG exhibited the best pretreatment capacity for poplar wood; for example, 76.4% lignin and 69.5% hemicellulose were removed, but 97.3% cellulose was preserved during K: 1,2-PG pretreatment of poplar wood.

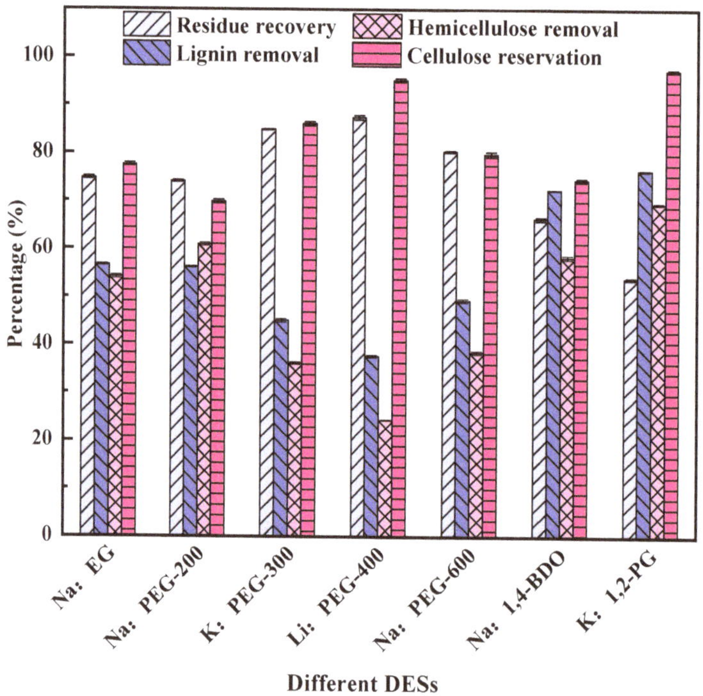

Figure 1. Effect of DES type on pretreatment of poplar wood (pretreatment conditions: 90 °C, 11 h).

The pH values of different DESs were measured as depicted in Figure 2. From Figure 2, we can observe a phenomenon: the pH values of the seven DESs were all greater than 11, which suggests that all of the DESs were basic solvents. In particular, K: 1,2-PG had the strongest basicity among these DESs, with the highest pH value being 13.3. Our previous studies [13,22] have proven that basic solvents can dissolve lignin and hemicellulose, but little cellulose. This explains the phenomenon by which a larger amount of lignin and hemicellulose than cellulose was removed by K: 1,2-PG.

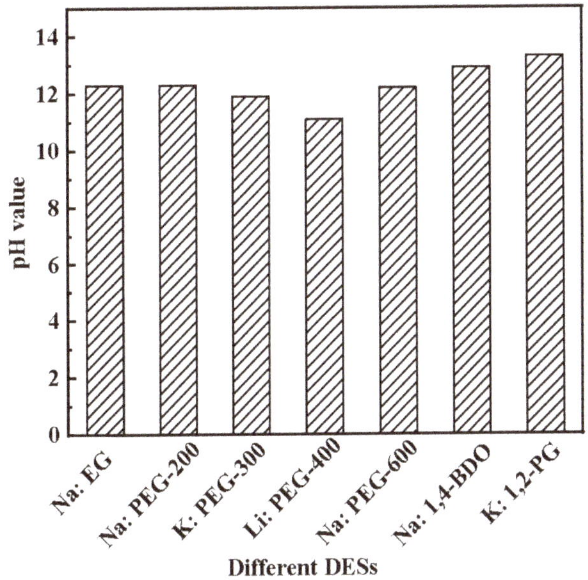

Figure 2. pH values of different DESs (the pH values were obtained by measuring 0.1 mol·L^{-1} DES aqueous solutions).

In addition, the pH values of DESs influence the pretreatment effects. Figure 3 illustrates the relationship between the pH values of DESs and the results of lignin or hemicellulose removal. As shown in Figure 3, the higher the pH values of DESs, the greater the amount of lignin and hemicellulose that is removed. Therefore, K: 1,2-PG can remove more lignin and hemicellulose than other DESs, which can be explained by our previous study, in which the lignin–carbohydrate complex and deacetylation of hemicellulose were destroyed by removing the carbonyl group in hemicellulose and cutting off the ester bond between lignin and hemicellulose during basic DES pretreatment [13].

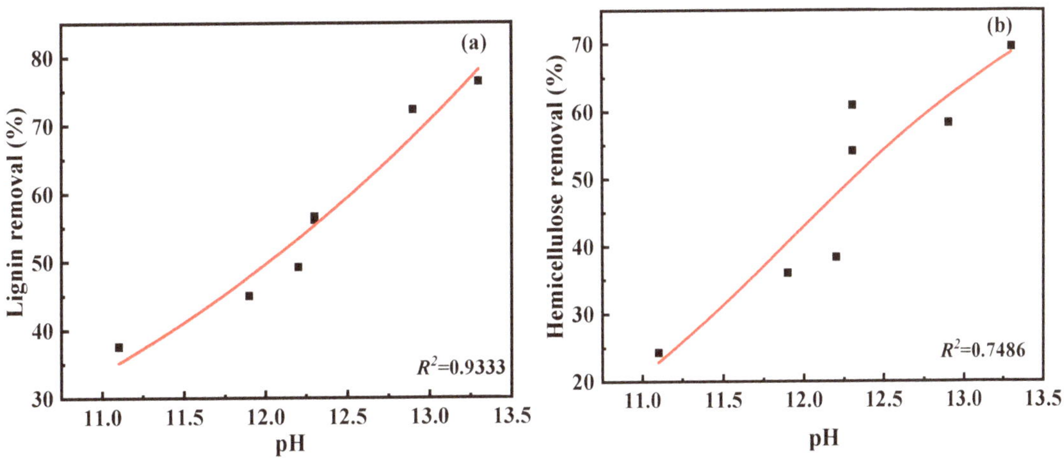

Figure 3. The relationship of lignin removal (**a**) and hemicellulose removal (**b**) with the pH values of different DESs.

Figure 4 shows the influence of DES species on the initial saccharification rate and the reducing sugar yield during the enzyme-catalyzed hydrolysis of poplar wood. According to Figure 4a, the initial saccharification rates of Na: EG-, Na: PEG-200-, K: PEG-300-, Li: PEG-400- and Na: PEG-600-pretreated residues were similar to that of untreated poplar wood (2.7–3.6 mg·mL^{-1}·h^{-1} vs. 2.6 mg·mL^{-1}·h^{-1}) during enzymatic hydrolysis. However, the Na: 1,4-BDO and K: 1,2-PG pretreatment process for poplar wood can significantly increase the initial saccharification rate, i.e., 5.7 mg·mL^{-1}·h^{-1} and 7.5 mg·mL^{-1}·h^{-1}. In particular, the initial saccharification rate of K: 1,2-PG-pretreated residues was its maximum, which is in agreement with the results obtained from pretreatment in Figure 1.

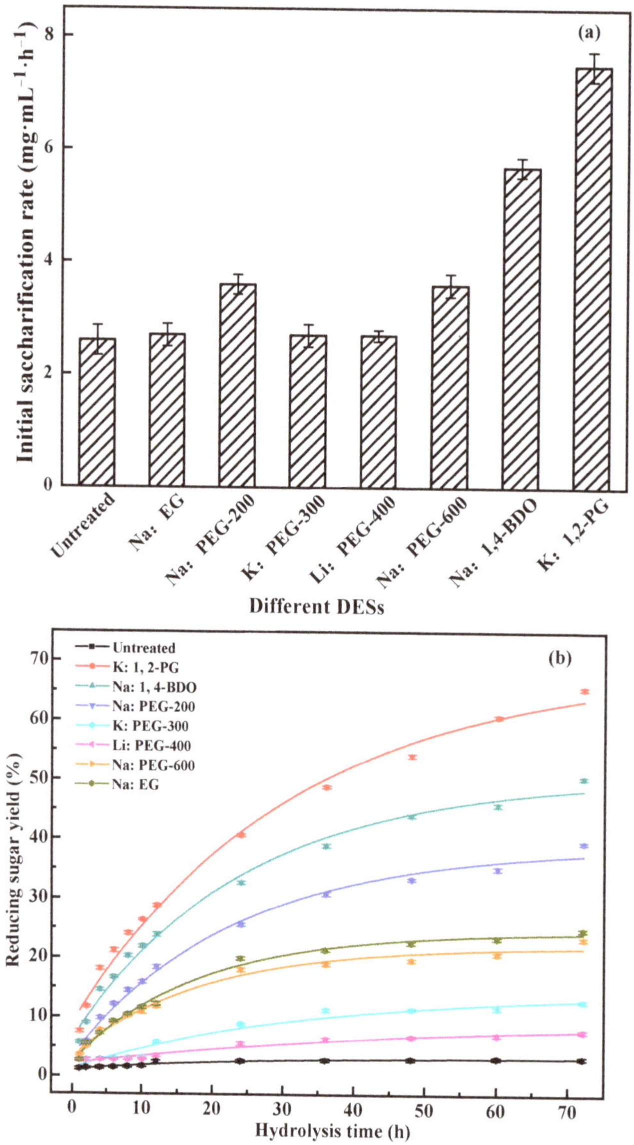

Figure 4. Effect of DES types on enzymatic hydrolysis: (**a**) initial saccharification rate; (**b**) reducing sugar yield.

As observed from Figure 4b, except for Na: EG, the other two diol-based DESs more effectively promoted the production of reducing sugar than PEG-based DESs did. For instance, compared with untreated poplar wood, the reducing sugar yield of Na: PEG-200-, K: PEG-300-, Li: PEG-400- and Na: PEG-600-pretreated residues did not increase significantly (3.3% vs. 7.8–39.6%), while the reducing sugar yields of Na: 1,4-BDO- and K: 1,2-PG-pretreated residues after hydrolysis for 72 h were 15.3 and 19.9 times higher, respectively, than that of untreated poplar wood (i.e., 3.3% vs. 50.5% and 65.6%). This is because the removal of both lignin and hemicellulose has a positive relationship with the reducing sugar yield, as shown in Figure 5. The more lignin and hemicellulose was removed, the higher the obtained reducing sugar yield was, indicating that the removal of lignin and hemicellulose is beneficial to enzyme hydrolysis. From Figures 1 and 4, the pretreatment and hydrolysis results of K: 1,2-PG-pretreated residue were the best of the seven DES-pretreated residues. The lignin and hemicellulose removal reached 76.4% and 69.5%, respectively. About 65.6% of reducing sugar was produced during hydrolysis for 72 h. As a consequence, K: 1,2-PG was determined to be the most effective solvent promoting enzymatic hydrolysis among these seven DESs.

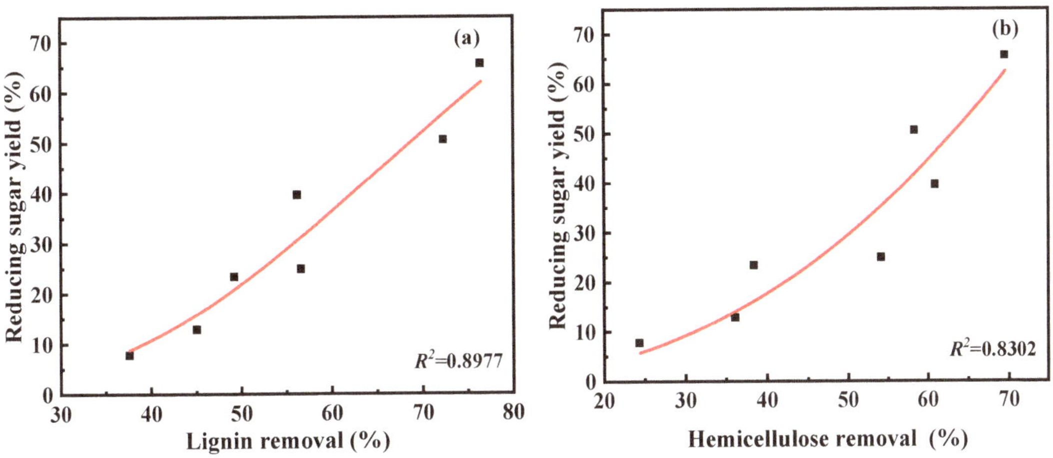

Figure 5. The relationship of reducing sugar yield (72 h) with lignin removal (**a**) or hemicellulose removal (**b**).

3.2. Influence of Pretreatment Temperature on Pretreatment of Poplar Wood

Figure 6 illustrates the effect of the pretreatment temperature on the pretreatment of poplar wood. It can be observed from Figure 6 that, as the temperature increased, the lignin and hemicellulose removal gradually increased from 48.1% to 97.4% and from 48.3% to 91.2%, respectively. This indicates that a high pretreatment temperature contributes to the removal of lignin and hemicellulose. The reason may be that the viscosity of K: 1,2-PG decreases with increasing temperature, and the lower the viscosity, the more diffusible the solute is in the solvent, increasing the contact opportunities between poplar wood and K: 1,2-PG. In addition, there was no obvious change in the lignin removal when the pretreatment temperature was elevated from 130 °C to 170 °C (e.g., 91.1–97.4%). However, cellulose preservation gradually decreased with the increase in the pretreatment temperature. Notably, there was a significant decrease in cellulose preservation from 93.8% to 39.2% when the pretreatment temperature increased from 130 °C to 170 °C, which indicates that an excessive pretreatment temperature causes the degradation of cellulose, resulting in a significant loss of cellulose. Therefore, the balance between the lignin and hemicellulose removal and the cellulose loss should be considered comprehensively. From the perspective of the pretreatment effect, 130 °C can be selected as the suitable pretreatment temperature.

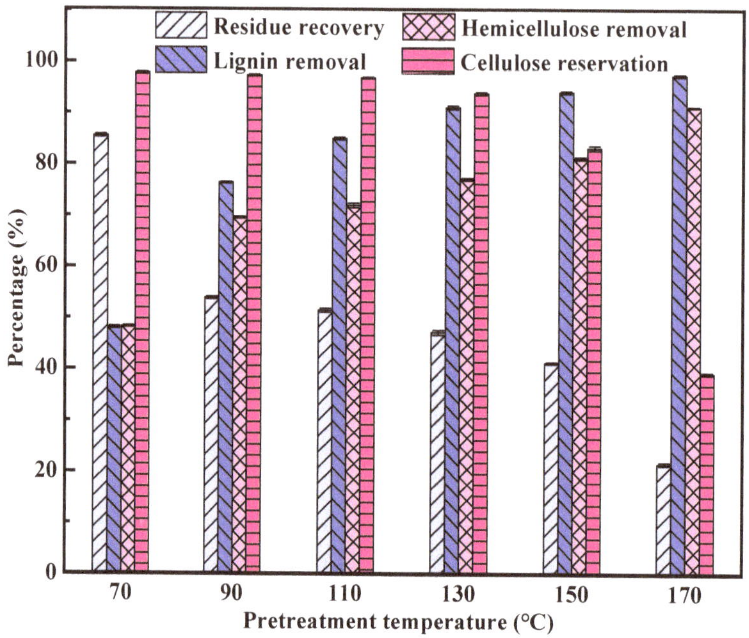

Figure 6. Influence of pretreatment temperature on pretreatment of poplar wood.

Figure 7 exhibits the influence of pretreatment temperature on the hydrolysis results. As shown in Figure 7, the reducing sugar yields and initial saccharification rates of residues obtained from pretreatment at different temperatures all increased in comparison with untreated poplar wood. With the increase in the pretreatment temperature, the reducing sugar yield and initial saccharification rate after hydrolysis for 72 h first gradually increased and then gradually decreased. When the pretreatment temperature was elevated from 70 °C to 130 °C, the reducing sugar yield (hydrolysis for 72 h) and initial saccharification rate increased from 34.1% to 82.6% and from 3.2 $mg \cdot mL^{-1} \cdot h^{-1}$ to 11.7 $mg \cdot mL^{-1} \cdot h^{-1}$, respectively. Nonetheless, when the pretreatment temperature increased from 130 °C to 170 °C, the reducing sugar yield and initial saccharification rate began to decrease gradually from 82.6% to 81.6% and from 11.7 $mg \cdot mL^{-1} \cdot h^{-1}$ to 4.9 $mg \cdot mL^{-1} \cdot h^{-1}$, respectively. This is because degradation products such as organic acids, furfural and phenolic compounds, which were produced during the pretreatment of poplar wood at an excessive pretreatment temperature, may inhibit the cellulase activity, resulting in a low reducing sugar yield and initial saccharification rate.

Consequentially, 130 °C was chosen as the best pretreatment temperature for further investigation. As a result, 91.1% of lignin and 77.1% of hemicellulose were removed, and 93.8% of cellulose was preserved during the pretreatment of poplar wood using K: 1,2-PG at 130 °C for 11 h. The reducing sugar yield reached 82.6% after 72 h of enzymatic hydrolysis.

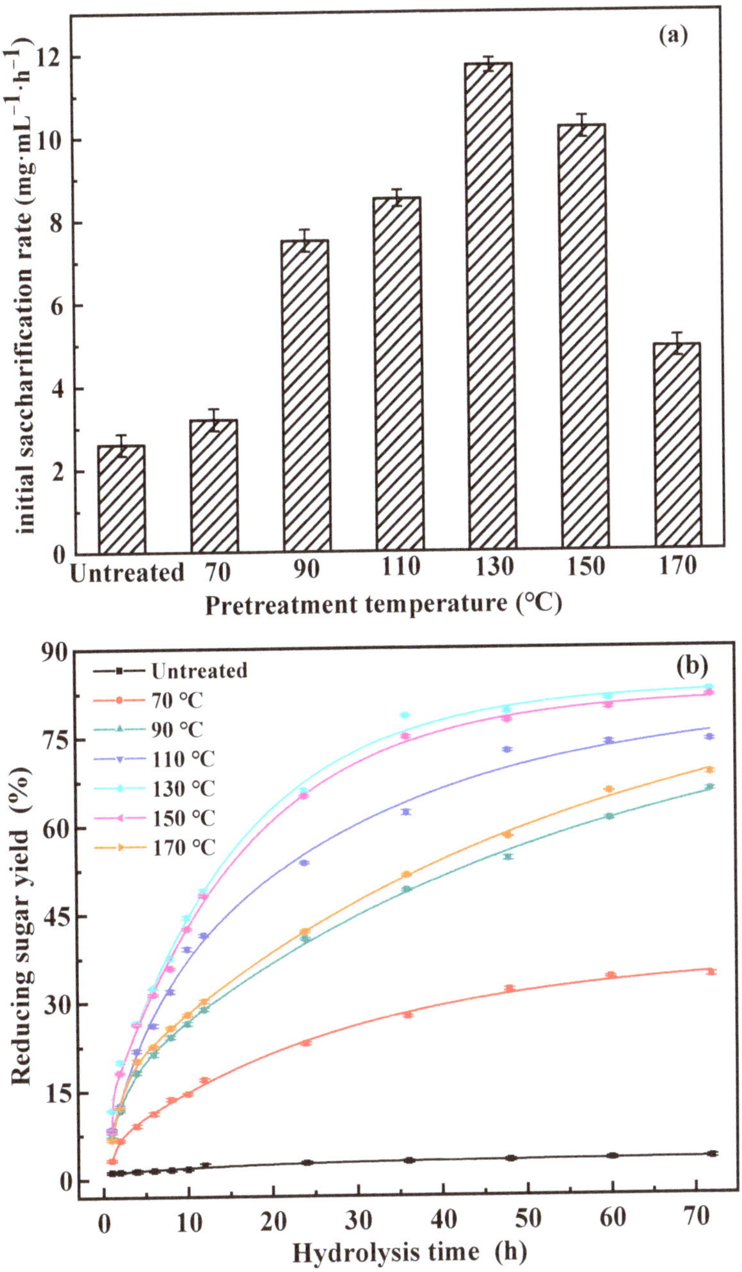

Figure 7. Influence of pretreatment temperature on enzymatic hydrolysis: (**a**) initial saccharification rate; (**b**) reducing sugar yield.

3.3. Influence of Pretreatment Time on Pretreatment of Poplar Wood

Figure 8 depicted the influence of pretreatment time on the residue recovery, lignin removal, hemicellulose removal and cellulose preservation of poplar wood. Figure 8 shows that the residue recovery and cellulose reservation slowly decreased from 50.6% to 46.3%

and from 97.5% to 92.7% with the increase in pretreatment time, respectively, while when the operation time increased, the lignin removal slowly increased from 89.2% to 93.9% and the hemicellulose removal increased from 71.6% to 77.1%. These results indicate that pretreatment time has little effect on poplar wood.

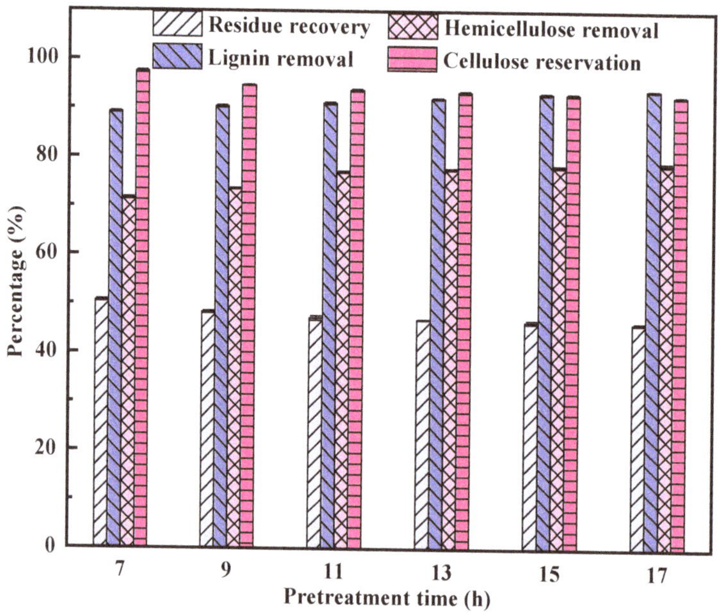

Figure 8. Influence of pretreatment time on the pretreatment of poplar wood.

Figure 9 shows the influence of pretreatment time on the hydrolysis efficiency of poplar wood. As observed from Figure 9, both the reducing sugar yield (72 h) and initial saccharification rate of poplar wood residue for enzymatic hydrolysis were higher than those of untreated poplar wood (82.5~87.3% vs. 3.3% and 10.2–13.2 mg·mL^{-1}·h^{-1} vs. 2.6 mg·mL^{-1}·h^{-1}). These phenomena demonstrate that the removal of lignin and hemicellulose promoted the hydrolysis of poplar wood. Furthermore, the reducing sugar yield (72 h) and initial saccharification rate slowly increased from 82.5% to 87.3% and from 10.2 mg·mL^{-1}·h^{-1} to 13.2 mg·mL^{-1}·h^{-1} as the pretreatment time increased from 7 h to 17 h. The lower hydrolysis results indicate that enzymatic hydrolysis efficiency is not sensitive to pretreatment time.

Therefore, 7 h was chosen as the best pretreatment time. The lignin and hemicellulose removal rates were 89.2% and 71.6%, and the cellulose preservation rate was 97.5% when the poplar wood was pretreated with K: 1,2-PG at 130 °C for 7 h. While the pretreated poplar wood was hydrolyzed by enzymes for 72 h, the reducing sugar yield reached 82.5%.

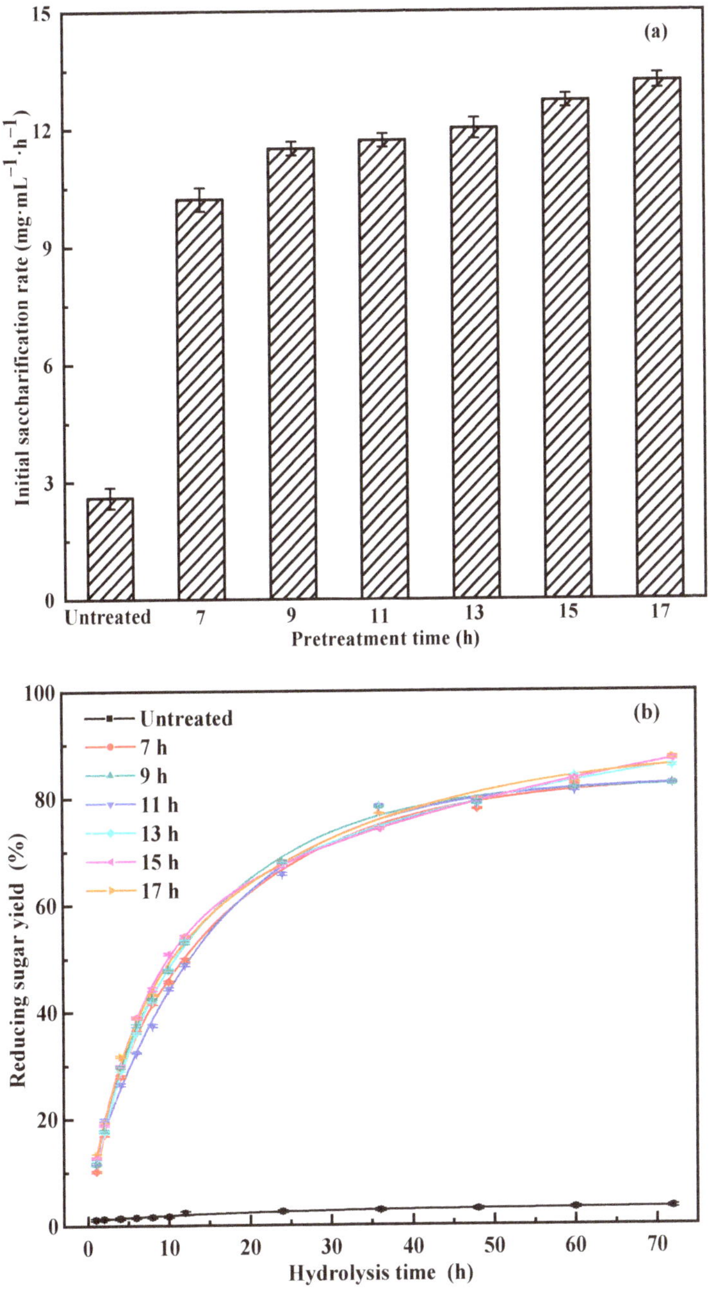

Figure 9. Influence of pretreatment time on enzymatic hydrolysis: (**a**) initial saccharification rate; (**b**) reducing sugar yield.

3.4. Comparison of Different Pretreatment Methods

Figure 10 compares the different methods of poplar wood pretreatment. The pretreatment efficiency of K: 1,2-PG was superior to that of choline chloride: guaiacol: aluminum chloride (molar ratio 25: 50:1, C:G:A) [23], choline chloride:lactic acid (molar ratio 1:8, C:L) [24] and 1-ethyl-3-methyl-imidazolium acetate ($[C_2mim][OAc]$) [25]. For example, K: 1,2-PG removed the greatest amount of lignin and the lowest amount of cellulose of the four pretreatment solvents (89.2% vs. 58.0–88.0% and 97.5% vs. 87.5–94.1%). More significantly, the reducing sugar yield of poplar wood pretreated with K: 1,2-PG exhibited values higher than or similar to those of poplar wood pretreated with the other two DESs, or $[C_2mim][OAc]$, after enzymatic hydrolysis for 72 h (82.5% vs. 32.6–88.2%).

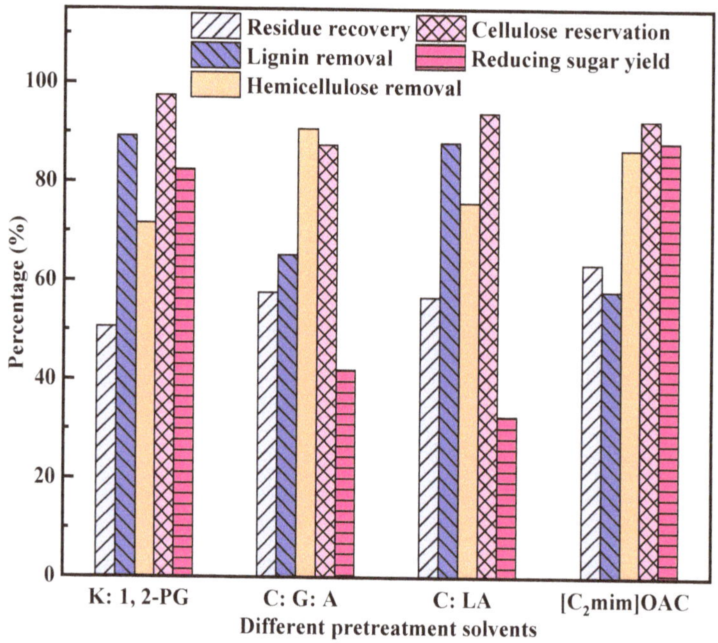

Figure 10. Comparison of different methods of poplar wood pretreatment.

3.5. Structural Characterization of Poplar Wood

3.5.1. XRD

Figure 11 shows the XRD pattern of poplar wood samples before and after pretreatment. From Figure 11, we can observe that the change in the cellulose crystal structure (CrI value) of poplar wood before and after pretreatment with K: 1,2-PG was investigated. The CrI value of K: 1,2-PG-pretreated poplar wood was significantly higher than that of untreated poplar wood (e.g., 57.1 vs. 48.3). This indicates that the crystallinity of pretreated poplar wood significantly increased compared with untreated poplar wood. The increased crystallinity was due to K: 1,2-PG, which has a stronger capacity to remove amorphous structures, i.e., lignin and hemicellulose, than that of structurally ordered cellulose. For example, 89.2% of lignin and 71.6% of hemicellulose were removed, while only 2.5% of cellulose was lost during the pretreatment process with K: 1,2-PG at 130 °C for 7 h.

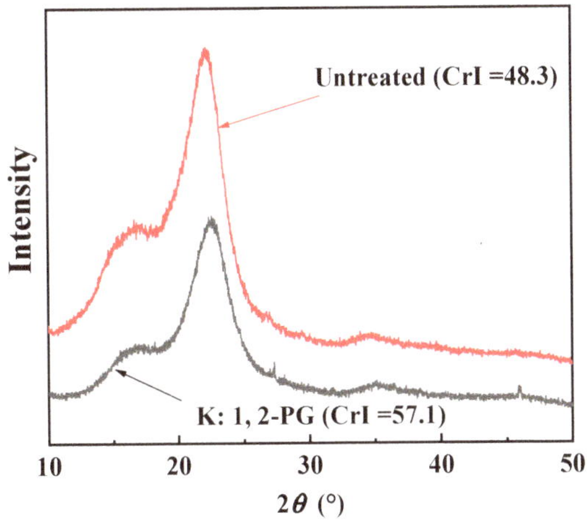

Figure 11. XRD pattern of poplar wood samples before and after K: 1,2-PG pretreatment (pretreatment conditions: 130 °C, 7 h).

3.5.2. FT-IR

Figure 12 depicts the FT-IR spectra of the poplar wood sample before and after K: 1,2-PG pretreatment. As shown in Figure 12, the peaks of K: 1,2-PG-pretreated poplar wood at 1737 cm^{-1}, 1602 cm^{-1}, 1510 cm^{-1} and 1245 cm^{-1} almost disappeared compared with untreated poplar wood. The peak at 1737 cm^{-1} represents the C=O stretching vibration peak in the carbonyl or ester groups. The disappearance of this peak demonstrates that K: 1,2-PG removed the carbonyl group in hemicellulose and cut off the ester bond between lignin and hemicellulose. The peaks at 1602 cm^{-1}, 1510 cm^{-1} and 1245 cm^{-1} represent the aromatic ring skeleton vibration peaks of lignin. The near-disappearance of these peaks illustrates that K: 1,2-PG destroyed the aromatic ring structure of lignin [26–29]. The results obtained from the FT-IR spectra, which were in accordance with the results of pretreatment, confirmed the removal of lignin and hemicellulose.

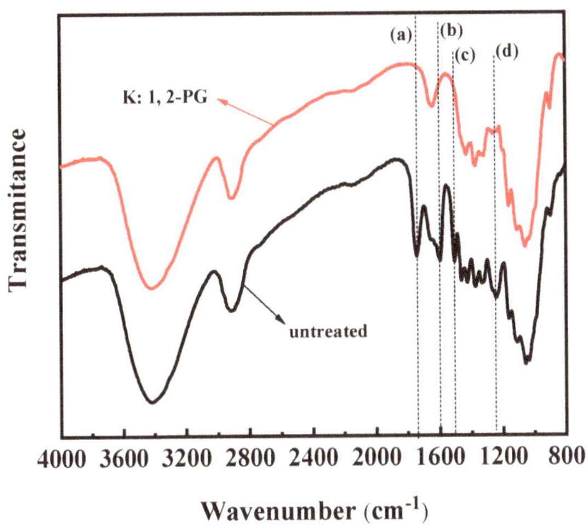

Figure 12. FT-IR spectra of poplar wood samples before and after K: 1,2-PG pretreatment: (**a**) 1737 cm^{-1}, (**b**) 1602 cm^{-1}, (**c**) 1510 cm^{-1} and (**d**) 1245 cm^{-1}.

3.5.3. SEM

Figure 13 illustrates the surface morphology of poplar wood before and after K: 1,2-PG pretreatment. As observed from Figure 13a,b, the intact surface of the untreated poplar wood was displayed, and as a result, it was difficult to see the internal cellulose structure. Figure 13c,d show that many pore structures appeared on the surface of K: 1,2-PG-pretreated residue, and a large number of fiber bundles were distributed in a disorganized manner, which greatly improved the contact area between the enzyme and the cellulose. The broken structure was caused by the removal of lignin and hemicellulose. Hence, the variations in the surface morphology between the untreated and pretreated poplar wood confirm that the removal of lignin and hemicellulose resulted from K: 1,2-PG pretreatment of poplar wood.

Figure 13. Surface morphology of poplar wood before and after K: 1,2-PG pretreatment: (**a**) untreated poplar wood × 37, (**b**) untreated poplar wood × 400, (**c**) K: 1,2-PG-pretreated poplar wood × 37 and (**d**) K: 1,2-PG-pretreated poplar wood × 400.

4. Conclusions

Seven basic DESs based on diol were screened for their ability to improve enzymatic hydrolysis in poplar wood pretreatment. Among these DESs, the pretreatment effects of diol monomer-based DESs were better than those of PEG-based DESs. The best pretreatment and hydrolysis results were obtained using a K: 1,2-PG pretreatment process (pretreatment conditions: 130 °C, 7 h), which furthered the enzymatic hydrolysis process (hydrolysis time: 72 h). As a result, the higher the pH of the DES, the greater the amounts of lignin and hemicellulose that were removed. For instance, the lignin and hemicellulose removal rates were 89.2% and 71.6%, the cellulose preservation rate was 97.5% and the reducing sugar yield reached 82.5%. Otherwise, the results of the FT-IR, XRD and SEM analyses all confirmed the removal of lignin and hemicellulose and revealed the pretreatment mechanism by which K: 1,2-PG removes lignin and hemicellulose by cutting the ester bond between lignin and hemicellulose, breaking the carbonyl group of hemicellulose and the aromatic ring structure of lignin. Thus, the establishment of this new pretreatment system will promote enzymatic hydrolysis in poplar wood.

Author Contributions: Conceptualization, X.Z. and Z.Z.; formal analysis, X.Z., Z.Z. and Y.Z.; investigation, X.Z., Z.Z., Y.Z. and T.Z.; project administration, Z.Z.; resources, L.H., X.S. and Z.Z.; supervision, L.H., X.S. and Z.Z.; writing—original draft, X.Z., Z.Z., Y.Z. and T.Z.; writing—review and editing, Y.Z. and Z.Z. All authors have read and agreed to the published version of the manuscript.

Funding: This research was funded by the Basic Scientific Research Project of Colleges and Universities of the Hebei Province, grant number JQN2022011, and the Key Program of Hebei Province Department of Science and Technology, grant number 20373805D.

Data Availability Statement: The data are listed in this paper.

Conflicts of Interest: The authors declare no conflict of interest.

References

1. Shang, Z.J. *Risk Management Study of HNJY Biomass Cogeneration Project*; Shandong University: Jinan, China, 2021.
2. Mosier, N.; Wyman, C.; Dale, B.; Elander, R.; Lee, Y.Y.; Holtzapple, M.; Ladisch, M. Features of promising technologies for pretreatment of lignocellulosic biomass. *Bioresour. Technol.* **2005**, *96*, 673–686. [CrossRef] [PubMed]
3. Zhang, L.; Xu, C.C.; Champagne, P. Overview of recent advances in thermo-chemical conversion of biomass. *Energ. Convers. Manag.* **2010**, *51*, 969–982. [CrossRef]
4. Prasad, S.; Malav, M.K.; Kumar, S.; Singh, A.; Pant, D.; Radhakrishnan, S. Enhancement of bio-ethanol production potential of wheat straw by reducing furfural and 5-hydroxymethylfurfural. *Bioresour. Technol.* **2018**, *4*, 50–56. [CrossRef]
5. Haque, M.A.; Barman, D.N.; Kang, T.H.; Kim, M.K.; Kim, J.; Kim, H.; Yun, H.D. Effect of dilute alkali on structural features and enzymatic hydrolysis of barley straw (Hordeum vulgare) at boiling temperature with low residence time. *J. Microbiol. Biotech.* **2012**, *22*, 1681–1691. [CrossRef] [PubMed]
6. Te, M.T.D.; Kersten, S.R.; Lange, J.P.; Ruiz, M.P. Cellulosic glycols: An integrated process concept for lignocellulose pretreatment and hydrogenolysis. *Biofuels Bioprod. Biorefining* **2021**, *15*, 1725–1736.
7. Luo, J.; Xu, Y. Comparison of biological and chemical pretreatment on coproduction of pectin and fermentable sugars from apple pomace. *Appl. Biochem. Biotech.* **2020**, *190*, 129–137. [CrossRef]
8. Peleteiro, S.; Rivas, S.; Alonso, J.L.; Santos, V.; Parajo, J.C. Utilization of ionic liquids in lignocellulose biorefineries as agents for separation, derivatization, fractionation, or pretreatment. *J. Agric. Food Chem.* **2015**, *63*, 8093–8102. [CrossRef]
9. Lee, K.M.; Hong, J.Y.; Tey, W.Y. Combination of ultrasonication and deep eutectic solvent in pretreatment of lignocellulosic biomass for enhanced enzymatic saccharification. *Cellulose* **2021**, *28*, 1513–1526. [CrossRef]
10. Sai, Y.W.; Lee, K.M. Enhanced cellulase accessibility using acid-based deep eutectic solvent in pretreatment of empty fruit bunches. *Cellulose* **2019**, *26*, 9517–9528. [CrossRef]
11. Li, L.; Yu, L.; Wu, Z.; Hu, Y. Delignification of poplar wood with lactic acid-based deep eutectic solvents. *Wood Res.* **2019**, *64*, 499–514.
12. Wu, M.; Gong, L.; Ma, C.; He, Y.C. Enhanced enzymatic saccharification of sorghum straw by effective delignification via combined pretreatment with alkali extraction and deep eutectic solvent soaking. *Bioresour. Technol.* **2021**, *340*, 125695. [CrossRef] [PubMed]
13. Zhao, Z.; Chen, X.; Ali, M.F.; Abdeltawab, A.A.; Yakout, S.M.; Yu, G. Pretreatment of wheat straw using basic ethanolamine-based deep eutectic solvents for improving enzymatic hydrolysis. *Bioresour. Technol.* **2018**, *263*, 325–333. [CrossRef] [PubMed]
14. Zhao, X.; Han, L.; Ma, X.; Sun, X.; Zhao, Z. Enhanced Enzymatic Hydrolysis of Wheat Straw to Improve Reducing Sugar Yield by Novel Method under Mild Conditions. *Processes* **2023**, *11*, 898. [CrossRef]
15. Chen, J.; Ali, M.C.; Liu, R.; Munyemana, J.C.; Li, Z.; Zhai, H.; Qiu, H. Basic deep eutectic solvents as reactant, template and solvents for ultra-fast preparation of transition metal oxide nanomaterials. *Chin. Chem. Lett.* **2020**, *31*, 1584–1587. [CrossRef]
16. Ali, M.C.; Liu, R.; Chen, J.; Cai, T.; Zhang, H.; Li, Z.; Qiu, H. New deep eutectic solvents composed of crown ether, hydroxide and polyethylene glycol for extraction of non-basic N-compounds. *Chin. Chem. Lett.* **2019**, *30*, 871–874. [CrossRef]
17. Li, Y.; Zhuo, J.; Liu, P.; Chen, P.; Hu, H.; Wang, Y.; Zhou, S.; Tu, Y.; Peng, L.; Wang, Y. Distinct wall polymer deconstruction for high biomass digestibility under chemical pretreatment in Miscanthus and rice. *Carbohydr. Polym.* **2018**, *192*, 273–281. [CrossRef]
18. Sluiter, A.; Hames, B.; Ruiz, R.; Scarlata, C.; Sluiter, J.; Templeton, D.; Crocker, D.L.A.P. Determination of structural carbohydrates and lignin in biomass. *Lab. Anal. Proced.* **2008**, *1617*, 1–16.
19. Fry, S.C. *The Growing Plant Cell Wall: Chemical and Metabolic Analysis*; The Blackburn Press: Caldwell, NJ, USA, 1988.
20. Hossain, M.A.; Rahaman, M.S.; Yelle, D.; Shang, H.; Sun, Z.; Renneckar, S.; Dong, J.; Tulaphol, S.; Sathitsuksanoh, N. Effects of polyol-based deep eutectic solvents on the efficiency of rice straw enzymatic hydrolysis. *Ind. Crops Prod.* **2021**, *167*, 113480. [CrossRef]
21. Mamilla, J.L.K.; Novak, U.; Grilc, M.; Likozar, B. Natural deep eutectic solvents (DES) for fractionation of waste lignocellulosic biomass and its cascade conversion to value-added bio-based chemicals. *Biomass Bioenergy* **2019**, *120*, 417–425. [CrossRef]
22. Zhao, Z.; Yang, Y.; Abdeltawab, A.A.; Yakout, S.M.; Chen, X.; Yu, G. Cholinium amino acids-glycerol mixtures: New class of solvents for pretreating wheat straw to facilitate enzymatic hydrolysis. *Bioresour. Technol.* **2017**, *245*, 625–632. [CrossRef]

23. Huang, C.; Zhan, Y.; Cheng, J.; Wang, J.; Meng, X.; Zhou, X.; Ragauskas, A.J. Facilitating enzymatic hydrolysis with a novel guaiacol-based deep eutectic solvent pretreatment. *Bioresour. Technol.* **2021**, *326*, 124696. [CrossRef] [PubMed]
24. Su, Y.; Huang, C.; Lai, C.; Yong, Q. Green solvent pretreatment for enhanced production of sugars and antioxidative lignin from poplar. *Bioresour. Technol.* **2021**, *321*, 124471. [CrossRef] [PubMed]
25. Yuan, T.Q.; Wang, W.; Xu, F.; Sun, R.C. Synergistic benefits of ionic liquid and alkaline pretreatments of poplar wood. Part 1: Effect of integrated pretreatment on enzymatic hydrolysis. *Bioresour. Technol.* **2013**, *144*, 429–434. [CrossRef] [PubMed]
26. Francisco, M.; Van, D.B.A.; Kroon, M.C. New natural and renewable low transition temperature mixtures (LTTMs): Screening as solvents for lignocellulosic biomass processing. *Green Chem.* **2012**, *14*, 2153–2157. [CrossRef]
27. Nor, N.A.M.; Mustapha, W.A.W.; Hassan, O. Deep eutectic solvent (DES) as a pretreatment for oil palm empty fruit bunch (OPEFB) in sugar production. *Procedia Chem.* **2016**, *18*, 147–154. [CrossRef]
28. Han, Y.; Bai, Y.; Zhang, J.; Liu, D.; Zhao, X. A comparison of different oxidative pretreatments on polysaccharide hydrolyzability and cell wall structure for interpreting the greatly improved enzymatic digestibility of sugarcane bagasse by delignification. *Bioresour. Bioprocess.* **2020**, *7*, 24. [CrossRef]
29. Sun, F.F.; Wang, L.; Hong, J.; Ren, J.; Du, F.; Hu, J.; Zhou, B. The impact of glycerol organosolv pretreatment on the chemistry and enzymatic hydrolyzability of wheat straw. *Bioresour. Technol.* **2015**, *187*, 354–361. [CrossRef]

MDPI AG
Grosspeteranlage 5
4052 Basel
Switzerland
Tel.: +41 61 683 77 34

Processes Editorial Office
E-mail: processes@mdpi.com
www.mdpi.com/journal/processes

www.ingramcontent.com/pod-product-compliance
Lightning Source LLC
Chambersburg PA
CBHW062050220926
51850CB00010B/2725
* 9 7 8 3 7 2 5 8 6 0 0 3 6 *

Application of Marine Chitin and Chitosan, 3rd Edition

Guest Editors

Jessica Amber Jennings
Tippabattini Jayaramudu

MDPI

Basel • Beijing • Wuhan • Barcelona • Belgrade • Novi Sad • Cluj • Manchester

Guest Editors
Jessica Amber Jennings
Department of Biomedical Engineering
University of Memphis
Memphis
USA

Tippabattini Jayaramudu
Nanoscience and Biomedical Engineering Department
South Dakota School of Mines
Rapid City
USA

Editorial Office
MDPI AG
Grosspeteranlage 5
4052 Basel, Switzerland

This is a reprint of the Special Issue, published open access by the journal *Marine Drugs* (ISSN 1660-3397), freely accessible at: https://www.mdpi.com/journal/marinedrugs/special_issues/V2705R8KY3.

For citation purposes, cite each article independently as indicated on the article page online and as indicated below:

Lastname, A.A.; Lastname, B.B. Article Title. *Journal Name* **Year**, *Volume Number*, Page Range.

ISBN 978-3-7258-5373-1 (Hbk)
ISBN 978-3-7258-5374-8 (PDF)
https://doi.org/10.3390/books978-3-7258-5374-8

Contents

marine drugs

Article

Enhancing Electrospinnability of Chitosan Membranes in Low-Humidity Environments by Sodium Chloride Addition

Hengjie Su [1], Xiaoqi Chen [2], Linna Mao [1] and Ting Li [1,*]

[1] Institute of Biomedical Engineering, Chinese Academy of Medical Sciences & Peking Union Medical College, Tianjin 300192, China
[2] Department of Biomedical Engineering, Tiangong, Tianjin 300387, China
* Correspondence: liting@bme.pumc.edu.cn

Abstract: The electrospinning of pure chitosan nanofibers is highly sensitive to environmental humidity, which limits their production consistency and applicability. This study investigates the addition of sodium chloride (NaCl) to chitosan solutions to enhance spinnability and mitigate the effigurefects of low humidity. NaCl was incorporated into the electrospun chitosan solution, leading to increased conductivity and decreased viscosity. These modifications improved the electrospinning process. Comparative analyses between chitosan membranes (CM) and sodium-chloride-added chitosan membranes (SCM) revealed no significant differences in chemical structure, mechanical strength, or in vitro cell proliferation. This indicates that the addition of 1% (w/v) NaCl does not adversely affect the fundamental properties of the chitosan membranes. The findings demonstrate that NaCl addition is a viable strategy for producing electrospun chitosan nanofibers in low-humidity environments, maintaining their physicochemical properties while enhancing spinnability.

Keywords: chitosan; salt; electrospinning; environmental humidity; conductivity

1. Introduction

Electrospun chitosan membranes are being explored as biocompatible and biodegradable scaffolds for guided bone regeneration (GBR), with potential applications in periodontal treatment and craniofacial regeneration [1–3]. Chitosan, a nature polysaccharide derived from the shell of crustacean, offers advantages such as low cost, biocompatibility, biodegradability, and non-toxicity [4,5]. Consequently, it has been widely used in biomedical fields, including guided bone regeneration, vascular stents, and wound dressings [6,7]. Electrospinning is a method that uses an electrostatic field to create a jet from a polymer solution to form nanofibers [8]. Key factors for the successful spinning of nanofibers include viscosity, surface tension, and electrical conductivity of the solution [5]. These factors vary with changes in the solution concentration and solvent system.

Electrospinning pure chitosan nanofibers is challenging due to several inherent properties of chitosan. The solubility of chitosan in most common solvents is not ideal [9], and the solution viscosity, caused by strong intermolecular and intramolecular hydrogen bonds, makes it difficult to overcome the surface tension during electrospinning [10]. Acetic acid and trifluoroacetic acid (TFA) are widely used solvents for dissolving chitosan. Achieving suitable surface tension for electrospinning typically requires at least 30% acetic acid for chitosan with 56–65% deacetylation (DDA) [11]. TFA has been explored as a solvent because it can block the positive charges of amino groups on chitosan and decrease the electrostatic forces, enabling the production of homogeneous chitosan nanofibers [5]. When comparing the two solvents, TFA produces smoother and more continuous nanofibers than acetic acid [5]. Additionally, the addition of methylene chloride to TFA can result in more uniform fibers produced by electrospinning [12]. Although there have been concerns about the toxicity of TFA [13], studies have proved that electrospun chitosan membranes produced using TFA/DCM solvents are biocompatible both in vitro and in vivo [14,15]. Researchers continue to use TFA/DCM as the solvent for electrospinning chitosan due to its ability to produce a more controlled and stable electrospinning processes [16–20].

However, the success of electrospinning pure chitosan in TFA/DCM depends significantly on environmental moisture [13,21]. It has been observed that the ambient humidity suitable for chitosan electrospinning is between 40–60% [14,21]. The impact of environmental humidity on electrospinning is not fully understood, making pure chitosan electrospinning highly restricted to environmental humidity. It is suspected that lower humidity in the environment results in decreased electrostatic discharge on the spinning jet, as fewer water molecules are available for charge transfer [22]. Consequently, the charge density on the jet is higher at low humidity, causing fiber breakage. If the voltage applied to the electrostatic field is not high enough, or if the solution has low conductivity, the electrostatic force cannot overcome the surface tension, and fibers cannot be produced [23]. To address this, creating a humid atmosphere within the electrospinning apparatus has been explored. However, the dense water mist from the humidifier may cause short circuits in the high voltage electrostatic field, making it difficult to control the mist distribution and posing significant safety risks.

Another approach to improve spinnability is by enhancing the electrical conductivity of the electrospinning solution. In one study, adding tetraethylammonium bromide (TEAB) salt to a polyimides (PI) solution improved its conductivity, allowing successful spinning even at low PI concentrations [8]. Similarly, adding TEAB salt to a polymers of intrinsic microporosity (PIM-1) solution significantly increased its conductivity, enabling the formation of electrospun fibers at a low PIM-1 concentration (10% (w/v)) [24]. Salt ions act as charge carriers in the electric field and affect the motion of the spinning jet [25–27]. Although adding salt to electrospinning solutions to improve spinnability has been explored, no studies have focused on adding salt specifically to pure chitosan solutions for electrospinning. Additionally, the relationship between environmental humidity and the addition of salt to chitosan solutions has not been explored.

This study aims to address the issue of humidity constraints encountered during the electrospinning of chitosan using the TFA/DCM solvent combination. In this study, adding sodium chloride to the electrospun chitosan solution was explored to reduce the impact of environmental moisture. To further investigate the effect of salt on electrospun chitosan membranes, the physical and chemical properties of the electrospun chitosan solution and chitosan membrane (CM) were examined by comparing them with the electrospun chitosan solution with salt and chitosan membrane with salt (SCM).

2. Results and Discussion

2.1. Solution Conductivity and Viscosity

To prepare the chitosan solution with NaCl, 5.5% (w/v) chitosan was dissolved in 10 mL TFA/DCM (7:3 v/v) solution overnight [14,28]. Just before electrospinning, 100 mg of NaCl was dissolved in 300 μL of water and added to the chitosan solution. Given that the water solubility of NaCl is 360 g/L and NaCl is hardly soluble in TFA and DCM, the NaCl solution was added to the chitosan solution in this specified proportion. NaCl was selected to increase the solution's conductivity due to its biocompatibility as a natural component of the human body. Due to the immiscibility of water and DCM, the system separates into two phases after adding the NaCl solution to the electrospun chitosan solution: the TFA/water phase and the TFA/DCM phase. However, the TFA/water phase is very small and nearly invisible to the naked eye due to the tiny volume of water relative to the total volume. The NaCl dissolves in the TFA/water phase but not in the TFA/DCM phase. Given the small amount of water (300 μL) in a much larger volume of TFA/DCM (10 mL), the dissolution of NaCl is limited. This was confirmed in Figure 1a. After three days of deposition of the electrospun chitosan solutions with varying concentrations of NaCl, salt crystals were observed at the bottom of the solutions, indicating that the NaCl was relatively isolated from the electrospun solution (Figure 1a). Additionally, observations on the first day revealed that the mixed state of NaCl in the chitosan solution remained stable during the electrospinning process.

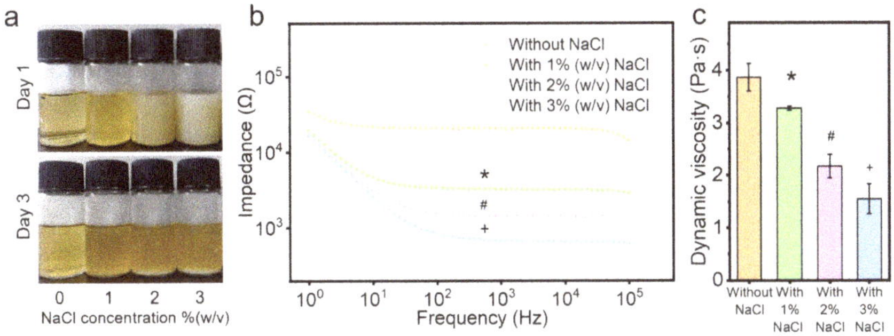

Figure 1. Properties of chitosan solutions with and without NaCl: (**a**) images of electrospun chitosan solutions on day 1 and day 3, (**b**) impedance, and (**c**) dynamic viscosity. *, #, and + denotes significant differences between groups.

An imperative step in electrospinning involves balancing the charge behavior and viscosity of the solution [28]. The charge behavior impacts the spinnability of the solution, as higher solution conductivity enables greater splaying ability to eject the solution into filaments from the Taylor cone [29,30]. Adding an NaCl water solution can enhance the conductivity of the solution, improving the electrospinning process by reducing the impedance and potentially increasing spinnability [31,32]. This is consistent with the results of this study, where the impedance of the chitosan solution significantly decreased ($p < 0.05$) after adding salt (Figure 1b). Chitosan solution with higher salt concentrations showed lower impedance, indicating higher conductivity and potentially greater spinnability.

The viscosity measures the resistance or internal friction of a fluid to flow, greatly influencing the electrospinning process and resulting fiber diameters [28,33]. After adding the NaCl solution, the dynamic viscosity of electrospun chitosan solutions decreased significantly ($p < 0.05$, Figure 1c). These results are consistent with Varnaitė-Žuravliova's research, which showed that conductivity increased and viscosity decreased after adding NaCl to the chitosan/PEO electrospun solution [25]. Previous studies demonstrated that adding NaCl to chitosan solution can lead to the breakdown of the hydrogen bond network of chitosan [25,31,34,35]. The hydroxyl group of chitosan is usually involved in the formation of intermolecular hydrogen bonds, and the destruction of these hydrogen bonds promotes the molecular motion of chitosan, increases the charge density, and decreases the solution viscosity, and thus, enhances spinnability [31,34]. The optimal viscosity range can vary depending on the specific polymer and solvent system used [11,36–40]. When the viscosity is lower than the optimal range, it is difficult to draw continuous nanofibers, resulting in the appearance of beads in the membrane [11]. The viscosity of 10 mL electrospun chitosan solution with 300 mg (3% (w/v)) NaCl decreased below 2 Pa·s (Figure 1c), making it challenging to perform the electrospinning at this viscosity. In lab experiments, it was demonstrated that adding 100 mg (1% (w/v)) and 200 mg (2% (w/v)) NaCl to 10 mL of electrospun chitosan solution allowed successful electrospinning under the environmental moisture levels of 25–35%. However, it was difficult to collect chitosan membranes with 2% (w/v) NaCl at the same volume solution, possibly because the increased amount of water melted the fibers during the process, thereby decreasing the final fiber volume. Hence, the most suitable concentration of NaCl to add in the chitosan solution is 1% (w/v).

2.2. Scanning Electron Microscopy (SEM)

After collecting the electrospun membranes, a post electrospinning treatment involving triethylamine (TEA) and di-tert-butyl dicarbonate (tBOC) was performed to neutralize the membranes, enhancing their aqueous stability and biocompatibility. SEM images showed that both SCM and CM maintained intact fiber structures before and after TEA/tBOC treatment, suggesting that the addition of the NaCl solution did not affect the uniformity of the spinning process, even with the formation of liquid phases (Figure 2). The diameter of SCM (with 2% (*w*/*v*) NaCl) fiber was significantly smaller ($p < 0.05$) than that of CM and SCM (with 1% (*w*/*v*) NaCl) before and after TEA/tBOC treatment (Figure 3a). The fiber diameter distribution showed that most fiber diameters of CM and SCM with 1% (*w*/*v*) NaCl were around 400 nm, whereas the majority of fiber diameters of SCM with 2% (*w*/*v*) NaCl were around 200 nm (Figure 2). Although there was no significant difference in fiber diameters between CM and SCM (with 1% (*w*/*v*) NaCl), a decreasing trend from CM to SCM (with 1% (*w*/*v*) NaCl) was observed. These results indicate that the addition of NaCl to the chitosan electrospinning solution reduces fiber diameter, but small amounts of salt do not significantly affect the fiber diameter. This aligns with findings that fiber diameter is influenced by the viscosity of the electrospun solution [40–42]. In this study, the addition of NaCl significantly decreased the viscosity of the electrospun chitosan solution, resulting in smaller diameters of electrospun chitosan nanofibers. Concurrently, the conductivity of the electrospun chitosan solution increased with the addition of NaCl, indicating an increase in charge density on the jet. This increased the extension level of the jet under the electric field [26]. Therefore, under the same applied voltage and spinning distance, higher solution conductivity enhances the jet's axial stretch, resulting in smaller electrospun fiber diameters [43], see more in Supplementary Materials.

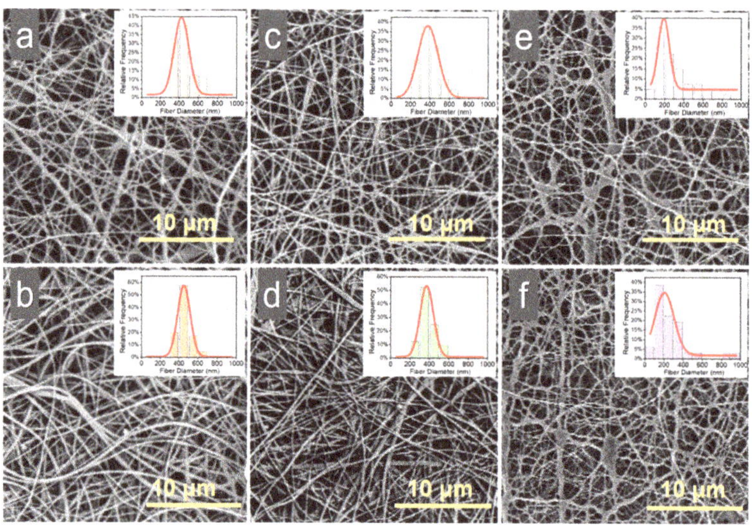

Figure 2. SEM images of electrospun chitosan membranes and the corresponding fiber diameter distribution: (**a**) CM without post-treatment, (**b**) CM with TEA/tBOC treatment, (**c**) SCM (1% NaCl) without post-treatment, (**d**) SCM (1% NaCl) with TEA/tBOC treatment, (**e**) SCM (2% NaCl) without post-treatment, and (**f**) SCM (2% NaCl) with TEA/tBOC treatment. "CM" denotes "chitosan membrane", while "SCM" denotes "chitosan membrane with salt".

After post-electrospinning treatment, there was no significant change ($p > 0.05$) in fiber diameter of both CM and SCM membranes. This is consistent with previous studies showing that post-electrospinning treatment with TEA/tBOC does not cause fiber expansion or structural loss [14].

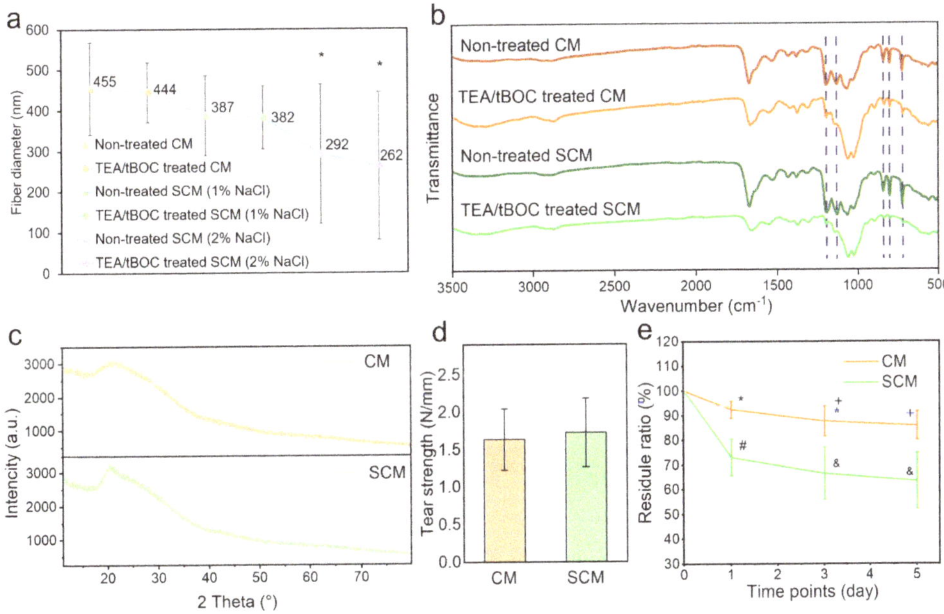

Figure 3. Physicochemical properties of electrospun chitosan membranes: (a) fiber diameter, (b) FTIR spectra, (c) XRD spectra, (d) tear strength, and (e) degradation ratio results. *, +, #, and & denote significant differences between groups.

2.3. Fourier Transform Infrared Spectroscopy (FTIR)

The FTIR spectra of CM and SCM (with 1% (*w*/*v*) NaCl) before and after TEA/tBOC treatment are compared in Figure 3b. The peaks associated with TFA salt at 723 cm^{-1}, 801 cm^{-1}, and 840 cm^{-1} disappeared after the TEA/tBOC treatment of both CM and SCM membranes, which is consistent with the previous study [14]. Additionally, two peaks around 1100–1200 cm^{-1} disappeared after the TEA/tBOC treatment, indicating the removal of trifluoromethyl groups.

The presence of NaCl in the FTIR spectrum was not clearly evident. There were hardly any characteristic bands related to NaCl salt near 1310 cm^{-1}, 1110 cm^{-1}, and 1640 cm^{-1} in the SCM spectra, which is similar to the results of Varnaitė-Žuravliova's study [25]. The spectra of SCM showed the same peaks as those of CM, indicating that the addition of salt to the electrospinning solution of chitosan did not change the chemical composition of the electrospun chitosan film.

2.4. X-Ray Diffraction (XRD)

Figure 3c displays the XRD spectra of both CM and SCM (with 1% (*w*/*v*) NaCl). In both spectra, a broad peak was observed at $2\theta = 20°$, which is characteristic of partially crystalline forms of chitosan, including both anhydrous and hydrated crystalline structures [14]. When comparing with the XRD spectrum of sodium chloride crystals, no distinct characteristic peaks of NaCl are present in the SCM spectrum, indicating the absence of salt crystals in the SCM [44]. This result further supports the FTIR analysis, confirming that the addition of salt to the spinning solution did not alter the chemical composition of the chitosan nanofibers.

2.5. Water Retention

The water retention behavior of the membranes reflects their ability to absorb exudates from wounds. The abundant micropores within the membranes enhance water absorption by increasing surface contact with water, while also providing favorable conditions for cell adhesion and spreading [45]. As shown in the results, the water retention rate of SCM (736.81 ± 37.89%) was significantly higher than that of CM (372.97 ± 11.22%) after 0.5 h and remained stable thereafter (Table 1). Notably, both membranes exhibited minimal expansion in length, 102.79 ± 1.17% for CM and 97.24 ± 0.83% for SCM, indicating that the shape of both membranes remained largely unchanged after water absorption. SCM exhibited initial swelling and thickening during the early stages of water absorption, leading to a slight reduction in side length. FTIR and XRD analysis confirmed no significant differences in the chemical structures of the two membranes. Therefore, the observed variation in water retention may be attributed to differences in fiber diameter. As the fiber diameter decreases, the surface area-to-volume ratio increases, leading to a larger nanoporous space. This allows SCM to have more extensive contact with water, resulting in greater water absorption and swelling [46], see more in Supplementary Materials.

Table 1. CM and SCM membrane swelling extent and water retention content.

Hours (h)	Membrane Swelling Length (%)		Water Retention Content (%)	
	CM (Mean ± std)	SCM (Mean ± std)	CM (Mean ± std)	SCM (Mean ± std)
0	100 ± 0	100 ± 0	100 ± 0	100 ± 0
0.5	102.79 ± 1.17 *	97.24 ± 0.83 +	372.97 ± 11.22 *	736.81 ± 37.89 +
2	102.2 ± 1.35 *	97.61 ± 0.95 +	383.91 ± 2 *	743.39 ± 39.6 +

* and + denote significant differences ($p < 0.05$).

2.6. Tear Strength

The suture pull-out test was used to evaluate the tear strength, indicating the handleability of the membrane in clinical operations. The results showed that the tear strength of CM and SCM (with 1% (w/v) NaCl) was 1.63 N/mm and 1.72 N/mm, respectively (Figure 3d). There was no significant difference ($p > 0.05$), indicating that adding NaCl solution before chitosan electrospinning did not affect the mechanical properties of the chitosan membrane, see more in Supplementary Materials.

2.7. Degradation

The degradation of the membranes was evaluated to assess their breakdown rate when implanted in the human body. In tissue engineering applications, the degradability of electrospun chitosan membranes is advantageous, as it eliminates the need for a second surgery and reduces unnecessary tissue damage. The results showed that both CM and SCM degraded significantly after 5 days, with substantial weight loss (Figure 3e). The degradation rate of CM was consistent with the previous study [47], while SCM exhibited a significantly faster degradation rate. This accelerated degradation could be attributed to the thinner fibers in SCM, which may be more susceptible to breakdown.

2.8. Cell Culture

The cell culture experiment was conducted to evaluate the biocompatibility of CM and SCM (with 1% (w/v) NaCl) after TEA/tBOC treatment. After 5 days of culture, both CM and SCM were proved to be biocompatible with fibroblasts (NIH3T3, Figure 4a) and osteoblasts (MC3T3E1, Figure 4b). Both cell types showed significant proliferation ($p < 0.05$) on CM and SCM from day 1 to day 5, similar to previous studies [14,17]. There was no significant difference ($p > 0.05$) between the two types of membranes at each time point in both the fibroblast and osteoblast cell culture experiments. Live and dead staining graphs showed no obvious dead cells on the 5th day for both CM and SCM, indicating a high cell survival rate. These results demonstrate that adding 1% (w/v) NaCl did not affect the biocompatibility of electrospun chitosan membranes, see more in Supplementary Materials.

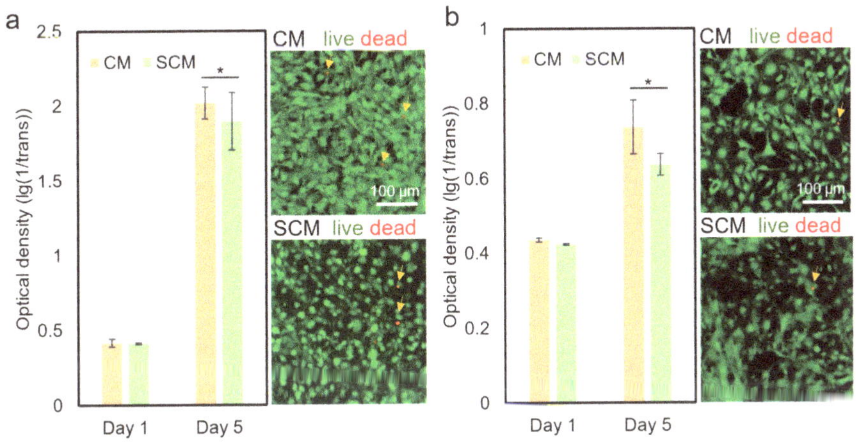

Figure 4. Cell culture results of CM and SCM (with 1% (w/v) NaCl). (**a**) Proliferation of fibroblasts (NIH3T3 cells) on CM and SCM (left), with live/dead staining images on the 5th day (right). (**b**) Proliferation of osteoblasts (MC3T3E1 cells) on CM and SCM (left), with live/dead staining images on the 5th day (right). * denotes a significant difference ($p < 0.05$). "CM" represents "chitosan membrane" and "SCM" represents "chitosan membrane with salt". Arrows indicate dead cells.

In previous studies, NaCl was also a commonly used additive in polymer electrospinning solutions and was proven to be biocompatible with cells [48,49]. MG63 cells proliferated significantly after 11 days on a chitosan/PEO scaffold that was electrospun with 0.2% NaCl. Additionally, smooth muscle cells showed significantly proliferation after 14 days on collagen/elastin membranes that were electrospun with 0.5% (w/v) PEO and 42.5 mM NaCl [45]. These studies demonstrate that the addition of a small amount of sodium chloride does not affect the biocompatibility of electrospun membranes.

3. Materials and Methods

3.1. Chitosan Solution and Membrane Preparation

The electrospun chitosan solution was prepared by dissolving 550 mg of chitosan (70% DDA, MW = 1,000,000, Meryer (Shanghai) chemical technology company, Shanghai, China) in 7 mL of TFA (Shanghai Macklin biochemical technology company, Shanghai, China) and 3 mL of dichloromethane (DCM, Shanghai Macklin biochemical technology company, China) overnight. The salt solution was prepared by dissolving 100 mg of sodium chloride (NaCl, 99% purity, Shanghai Aladdin biochemical technology company, Shanghai, China) in 300 μL of deionized (DI) water. The chitosan solution with NaCl was prepared just before the electrospinning by adding the required amount of salt solution to the prepared electrospun chitosan solution.

Electrospinning of the CM was based on a previously reported method [14,17]. After filling the electrospun solution into a 10 mL syringe, a high voltage of 26 kV was applied to the needle tip. An aluminum foil-covered round plate, rotating at a speed of 8.4 rpm, was placed 15 cm from the needle tip.

After collecting electrospun membranes, a post-electrospinning treatment involving TEA and tBOC, innovated in a previous study, was used to neutralize the membranes [15]. Membranes were immersed in a 10% (w/v) solution of TEA (Shanghai Macklin biochemical technology company, China)/tetrahydrofuran (THF, Shanghai Macklin biochemical technology company, China) for 24 h under mild magnetic stirring to remove the TFA salt. After rinsing the membranes with pure THF twice, the membranes were immersed in a 0.1 g/mL solution of tBOC (Shanghai Aladdin biochemical technology company, China)/THF for 24 h. After washing the membranes with pure THF three times, they were dried between two pieces of nylon mesh in the air to keep them flat.

3.2. Solution Conductivity

The conductivity of 30 mL chitosan solutions (n = 3) was evaluated by electrochemical impedance spectroscopy (EIS) with the electrochemical workstation (CHI 660D, Chenhua, China). The testing electrodes (Wuhu keying new material corporation, Wuhu, China) were selected following the three-electrode system, which included a glassy carbon electrode for the working electrode, a platinum electrode for the counter electrode, and a mercurous sulphate electrode for the reference electrode. The frequency range of EIS measurements was set to 10 to 10^5 Hz, with the voltage of 5 mV.

3.3. Solution Viscosity

The viscosity of chitosan solutions (n = 3) was tested by a glass capillary viscometer (Shanghai Baoshan Qihang glass instrument factory, China) at room temperature. After recording the flowing time between the two markers (t), the dynamic viscosity of the tested solution was calculated using Equation (1):

$$\eta = C \times t \times \rho \quad (1)$$

where η (mPa·s) represents the dynamic viscosity, C (mm^2/S^2) denotes the viscosity number, and ρ (g/cm^3) indicates the solution density.

3.4. Scanning Electron Microscopy (SEM)

The nanostructure of the two types of membranes was examined using an SEM (MIRA LMS, TESCAN, Tempe, AZ, USA). The membranes were attached to an SEM stub and coated with an 8 nm layer of gold-palladium. They were then examined at 2500× magnification. In each sample, more than 20 fiber diameters were measured.

3.5. Fourier Transform Infrared Spectroscopy (FTIR)

FTIR spectra were collected to evaluate the chemical structure of SCM and CM using a Nicolet™ iS20 FTIR spectrometer (Thermo Fisher Scientific, Lenexa, KS, USA). Spectra were collected by scanning samples from 500 cm^{-1} to 4000 cm^{-1} for 32 scans each.

3.6. X-Ray Diffraction (XRD)

XRD analysis was conducted to assess the crystallinity of the chitosan membranes. Both CM and SCM were first immersed in liquid nitrogen and then ground into fine powders using a mortar and pestle. The samples were subsequently analyzed using an automated multipurpose X-ray diffractometer (Ultima IV, Rigaku, Tokyo, Japan) operating in grazing incidence mode with a wavelength of 1.54 Å. Data collection was performed over a 2θ range of 10° to 80°.

3.7. Water Retention

Water absorption was evaluated to determine the swelling extent and water retention capacity of the membranes. Both CM and SCM were cut into 3 cm × 2 cm rectangle. The samples were then immersed in DI water, and the edge length and weight were measured at specific time intervals. The initial edge length and weight were recorded as "l_0" and "w_0", respectively. At each time point, the samples were taken out and the excess water was removed using filter paper. The edge length of samples was recorded as "l_t" and the weight as "w_t". The membrane swelling length was calculated using Equation (2), and the water retention content was calculated using Equation (3):

$$membrane\ swelling\ length\ (\%) = \frac{l_t}{l_0}, \quad (2)$$

$$water\ retention\ content\ (\%) = \frac{m_t}{m_0} \quad (3)$$

3.8. Tear Strength

The tear strength of CM and SCM (n = 3/membrane type) treated with TEA/tBOC was evaluated using an electronic universal testing machine (STS20K, Xiamen Yishite Instruments, Xiamen, China) as an indicator of clinical operability.

Each sample was cut into a 10 × 40 mm^2 rectangle, and surgical sutures were passed through one end at a point 5 mm from the wide edge of the sample and 5 mm from the long edge. The upper clamp of the tension testing machine was used to secure the surgical sutures, while the lower clamp was used to secure the other end of the sample. The load cell used was 50 N, and the extension rate was 10 mm/min. The maximum load was recorded in Newtons (N) and then normalized to film thickness. Three samples of each membrane were tested.

3.9. Degradation

The degradation of the membranes was assessed by measuring mass loss over time. Membranes were cut into 3 cm × 2 cm rectangle and immersed in phosphate-buffered saline (PBS) solution (Beijing Solarbio Science & Technology Co., Beijing, China) containing 100 μg/mL lysozyme (Beijing Solarbio Science & Technology Co., China), and a 5% penicillin–streptomycin–gentamicin solution (Beijing Solarbio Science & Technology Co., China). At each time point, the membranes were removed from the solution, rinsed with DI water, dried overnight at 60 °C, and weighed to determine the change in mass. After weighing, the membranes were returned to the solution to continue the incubation period. A higher concentration of lysozyme was used in this experiment compared to the physiological levels found in human plasma (3–8 μg/mL) to accelerate the degradation process.

3.10. Cell Culture

After high-pressure steam sterilization, round CM and SCM samples (n = 3/membrane type, diameter = 1 cm) were placed into the 48-well plate. Fibroblasts (NIH3T3 cells) and osteoblasts (MC3T3E1 cells) were seeded on the membrane, respectively. The CM and SCM samples were rinsed in culture media and then seeded with cells at 2 × 10^4 cells/well. Cells were grown in DMEM media supplemented with 10% FBS and 1% penicillin/streptomycin. At the required time point, the media was removed, and each well was cleaned with PBS three times. Subsequently, 500 μL of media containing 10% CCK-8 were added to each well, and the plate was cultured in a constant temperature incubator at 37 °C with 5% CO_2 for 2 h. Afterward, 100 μL of the media was transferred into a new 96-well plate, and the absorbance at 450 nm was measured with a microplate reader (SPARK 10M, TECAN, Tempe, AZ, USA). The absorbance value is proportional to the number of cells. The survival rate and morphology of the cells on the fifth day were observed by inverted fluorescence microscopy and staining with Calcein acetoxymethyl ester (Calcein-AM)/propidium iodide (PI).

3.11. Statistical Analysis

A single factor analysis of variance (ANOVA) was used to analyze the results of solution viscosity and fiber diameter. An independent sample t-test was used to analyze the results of mechanical strength, while a two-way ANOVA was employed for analyzing solution conductivity, water retention, degradation, and cell proliferation.

4. Conclusions

In this study, the addition of NaCl to chitosan solutions during electrospinning significantly improved spinnability and reduced the limitations posed by environmental humidity. The incorporation of NaCl led to an increase in solution conductivity and a decrease in viscosity, facilitating the electrospinning process. Comparative analysis of the physicochemical properties of CM and SCM revealed that the addition of 1% (w/v) NaCl did not alter the chemical groups, mechanical strength, or in vitro cell proliferation characteristics of the membranes. However, the reduced fiber diameter in SCM may contribute to its higher water retention capacity and faster degradation rate. These results suggest that 1% (*w*/*v*) NaCl serves as an effective additive to enhance the electrospinning of chitosan in low-humidity environments, optimizing fiber formation while maintaining key physicochemical properties. This study addresses the challenge of humidity limitations in electrospinning chitosan using the TFA/DCM solvent system, bridging the gap in current advancements within the field.

Supplementary Materials: The following supporting information can be downloaded at: https://www.mdpi.com/article/10.3390/md22100443/s1.

Author Contributions: H.S. conceived and designed the study, analyzed the data, and contributed to writing and revising the paper. X.C. carried out significant laboratory work, analyzed the data, and contributed to partial writing. L.M. contributed to paper revision. T.L. supervised the laboratory work and provided funding. All authors have read and agreed to the published version of the manuscript.

Funding: This work was supported by the Chinese Academy of Medical Science Health Innovation project, grant number 2021-I2M-1-058 and 2021-I2M-1-042, Tianjin Outstanding Youth Fund Project, grant number 20JCJQIC00230, and Fundamental Research Funds for the Central Universities, grant number 3332023070.

Data Availability Statement: The data presented in this study are available on request from the corresponding author.

Conflicts of Interest: The authors declare no conflicts of interest.

References

1. Qasim, S.B.; Najeeb, S.; Delaine-Smith, R.M.; Rawlinson, A.; Ur Rehman, I. Potential of Electrospun Chitosan Fibers as a Surface Layer in Functionally Graded GTR Membrane for Periodontal Regeneration. *Dent. Mater.* **2017**, *33*, 71–83. [CrossRef] [PubMed]
2. Zhang, Y.; Wang, T.; Li, J.; Cui, X.; Jiang, M.; Zhang, M.; Wang, X.; Zhang, W.; Liu, Z. Bilayer Membrane Composed of Mineralized Collagen and Chitosan Cast Film Coated with Berberine-Loaded PCL/PVP Electrospun Nanofiber Promotes Bone Regeneration. *Front. Bioeng. Biotechnol.* **2021**, *9*, 684335. [CrossRef]
3. Rodríguez-Méndez, I.; Fernández-Gutiérrez, M.; Rodríguez-Navarrete, A.; Rosales-Ibáñez, R.; Benito-Garzón, L.; Vázquez-Lasa, B.; San Román, J. Bioactive Sr(II)/Chitosan/Poly(ε-Caprolactone) Scaffolds for Craniofacial Tissue Regeneration. In Vitro and In Vivo Behavior. *Polymers* **2018**, *10*, 279. [CrossRef] [PubMed]
4. Toskas, G.; Cherif, C.; Hund, R.-D.; Laourine, E.; Mahltig, B.; Fahmi, A.; Heinemann, C.; Hanke, T. Chitosan(PEO)/Silica Hybrid Nanofibers as a Potential Biomaterial for Bone Regeneration. *Carbohydr. Polym.* **2013**, *94*, 713–722. [CrossRef]
5. Ohkawa, K.; Cha, D.; Kim, H.; Nishida, A.; Yamamoto, H. Electrospinning of Chitosan. *Macromol. Rapid Commun.* **2004**, *25*, 1600–1605. [CrossRef]
6. Yang, W.; Fu, J.; Wang, D.; Wang, T.; Wang, H.; Jin, S.; He, N. Study on Chitosan/Polycaprolactone Blending Vascular Scaffolds by Electrospinning. *J. Biomed. Nanotechnol.* **2010**, *6*, 254–259. [CrossRef]
7. Valachová, K.; El Meligy, M.A.; Šoltés, L. Hyaluronic Acid and Chitosan-Based Electrospun Wound Dressings: Problems and Solutions. *Int. J. Biol. Macromol.* **2022**, *206*, 74–91. [CrossRef]
8. Topuz, F.; Abdulhamid, M.A.; Holtzl, T.; Szekely, G. Nanofiber Engineering of Microporous Polyimides through Electrospinning: Influence of Electrospinning Parameters and Salt Addition. *Mater. Des.* **2021**, *198*, 109280. [CrossRef]
9. Pérez-Nava, A.; Reyes-Mercado, E.; González-Campos, J.B. Production of Chitosan Nanofibers Using the HFIP/Acetic Acid Mixture as Electrospinning Solvent. *Chem. Eng. Process. Process Intensif.* **2022**, *173*, 108849. [CrossRef]
10. Alam, M.R.; Alimuzzaman, S.; Shahid, M.A.; Fahmida-E-Karim; Hoque, M.E. Collagen/Nigella Sativa/Chitosan Inscribed Electrospun Hybrid Bio-Nanocomposites for Skin Tissue Engineering. *J. Biomater. Sci. Polym. Ed.* **2023**, *34*, 1517–1538. [CrossRef]
11. Geng, X.; Kwon, O.-H.; Jang, J. Electrospinning of Chitosan Dissolved in Concentrated Acetic Acid Solution. *Biomaterials* **2005**, *26*, 5427–5432. [CrossRef] [PubMed]
12. Sohofi, N.; Tavanai, H.; Morshed, M.; Abdolmaleki, A. Electrospinning of 100% Carboxymethyl Chitosan Nanofibers. *J. Eng. Fibers Fabr.* **2014**, *9*, 155892501400900110. [CrossRef]
13. Anisiei, A.; Oancea, F.; Marin, L. Electrospinning of Chitosan-Based Nanofibers: From Design to Prospective Applications. *Rev. Chem. Eng.* **2023**, *39*, 31–70. [CrossRef]
14. Su, H.; Liu, K.-Y.; Karydis, A.; Abebe, D.G.; Wu, C.; Anderson, K.M.; Ghadri, N.; Adatrow, P.; Fujiwara, T.; Bumgardner, J.D. In Vitro and in Vivo Evaluations of a Novel Post-Electrospinning Treatment to Improve the Fibrous Structure of Chitosan Membranes for Guided Bone Regeneration. *Biomed. Mater.* **2016**, *12*, 015003. [CrossRef]
15. Su, H.; Fujiwara, T.; Anderson, K.M.; Karydis, A.; Ghadri, M.N.; Bumgardner, J.D. A Comparison of Two Types of Electrospun Chitosan Membranes and a Collagen Membrane in Vivo. *Dent. Mater.* **2021**, *37*, 60–70. [CrossRef]

16. Totito, T.C.; Laatikainen, K.; Bode-Aluko, C.; Pereao, O.; Petrik, L. Fabrication and Characterization of Electrospun Waste Polyethylene Terephthalate Blended with Chitosan: A Potential Single-Use Material. *Polymers* **2023**, *15*, 442. [CrossRef]
17. Su, H.; Fujiwara, T.; Skalli, O.; Selders, G.S.; Li, T.; Mao, L.; Bumgardner, J.D. Porous Nano-Fiber Structure of Modified Electrospun Chitosan GBR Membranes Improve Osteoblast Calcium Phosphate Deposition in Osteoblast-Fibroblast Co-Cultures. *Mar. Drugs* **2024**, *22*, 160. [CrossRef]
18. Korniienko, V.; Husak, Y.; Radwan-Pragłowska, J.; Holubnycha, V.; Samokhin, Y.; Yanovska, A.; Varava, J.; Diedkova, K.; Janus, Ł.; Pogorielov, M. Impact of Electrospinning Parameters and Post-Treatment Method on Antibacterial and Antibiofilm Activity of Chitosan Nanofibers. *Molecules* **2022**, *27*, 3343. [CrossRef]
19. Korniienko, V.; Husak, Y.; Yanovska, A.; Banasiuk, R.; Yusupova, A.; Savchenko, A.; Holubnycha, V.; Pogorielov, M. Functional and Biological Characterization of Chitosan Electrospun Nanofibrous Membrane Nucleated with Silver Nanoparticles. *Appl. Nanosci.* **2022**, *12*, 1061–1070. [CrossRef]
20. Varava, Y.; Samokhin, Y.; Savchenko, A.; Diedkova, K.; Kyrylenko, S.; Korniienko, V. Antimicrobial Electrospun Chitosan Nanofibrous Membranes Functionalized with Silver Nanoparticles. In Proceedings of the 2021 IEEE 11th International Conference Nanomaterials: Applications & Properties (NAP), Odessa, Ukraine, 5–11 September 2021; pp. 1–4.
21. Schiffman, J.D.; Schauer, C.L. Cross-Linking Chitosan Nanofibers. *Biomacromolecules* **2007**, *8*, 594–601. [CrossRef]
22. Nezarati, R.M.; Eifert, M.B.; Cosgriff-Hernandez, E. Effects of Humidity and Solution Viscosity on Electrospun Fiber Morphology. *Tissue Eng. Part C Methods* **2013**, *19*, 810–819. [CrossRef] [PubMed]
23. Qin, X.-H.; Yang, E.-L.; Li, N.; Wang, S.-Y. Effect of Different Salts on Electrospinning of Polyacrylonitrile (PAN) Polymer Solution. *J. Appl. Polym. Sci.* **2007**, *103*, 3865–3870. [CrossRef]
24. Topuz, F.; Satilmis, B.; Uyar, T. Electrospinning of Uniform Nanofibers of Polymers of Intrinsic Microporosity (PIM-1): The Influence of Solution Conductivity and Relative Humidity. *Polymer* **2019**, *178*, 121610. [CrossRef]
25. Varnaitė-Žuravliova, S.; Savest, N.; Baltušnikaitė-Guzaitienė, J.; Abraitienė, A.; Krumme, A. The Investigation of the Production of Salt-Added Polyethylene Oxide/Chitosan Nanofibers. *Materials* **2024**, *17*, 132. [CrossRef]
26. Angammana, C.J.; Jayaram, S.H. Analysis of the Effects of Solution Conductivity on Electrospinning Process and Fiber Morphology. *IEEE Trans. Ind. Appl.* **2011**, *47*, 1109–1117. [CrossRef]
27. Fong, H.; Chun, I.; Reneker, D.H. Beaded Nanofibers Formed during Electrospinning. *Polymer* **1999**, *40*, 4585–4592. [CrossRef]
28. Drew, C.; Wang, X.; Samuelson, L.A.; Kumar, J. The Effect of Viscosity and Filler on Electrospun Fiber Morphology. *J. Macromol. Sci. Part A* **2003**, *40*, 1415–1422. [CrossRef]
29. Choudhury, R.R.; Gohil, J.M.; Dutta, K. Poly(Vinyl Alcohol)-Based Membranes for Fuel Cell and Water Treatment Applications: A Review on Recent Advancements. *Polym. Adv. Technol.* **2021**, *32*, 4175–4203. [CrossRef]
30. Shin, Y.M.; Hohman, M.M.; Brenner, M.P.; Rutledge, G.C. Experimental Characterization of Electrospinning: The Electrically Forced Jet and Instabilities. *Polymer* **2001**, *42*, 09955–09967. [CrossRef]
31. Zahra, F.T.; Zhang, Y.; Ajayi, A.O.; Quick, Q.; Mu, R. Optimization of Electrospinning Parameters for Lower Molecular Weight Polymers: A Case Study on Polyvinylpyrrolidone. *Polymers* **2024**, *16*, 1217. [CrossRef]
32. Ahmadi Bonakdar, M.; Rodrigue, D. Electrospinning: Processes, Structures, and Materials. *Macromol* **2024**, *4*, 58–103. [CrossRef]
33. Nada, A.A.; James, R.; Shelke, N.B.; Harmon, M.D.; Awad, H.M.; Nagarale, R.K.; Kumbar, S.G. A Smart Methodology to Fabricate Electrospun Chitosan Nanofiber Matrices for Regenerative Engineering Applications. *Polym. Adv. Technol.* **2014**, *25*, 507–515. [CrossRef]
34. Su, P.; Wang, C.; Yang, X.; Chen, X.; Gao, C.; Feng, X.-X.; Chen, J.-Y.; Ye, J.; Gou, Z. Electrospinning of Chitosan Nanofibers: The Favorable Effect of Metal Ions. *Carbohydr. Polym.* **2011**, *84*, 239–246. [CrossRef]
35. Bizarria, M.T.M.; d'Ávila, M.A.; Mei, L.H.I. Non-Woven Nanofiber Chitosan/Peo Membranes Obtained by Electrospinning. *Braz. J. Chem. Eng.* **2014**, *31*, 57–68. [CrossRef]
36. Cheng, T.; Hund, R.-D.; Aibibu, D.; Horakova, J.; Cherif, C. Pure Chitosan and Chitsoan/Chitosan Lactate Blended Nanofibres Made by Single Step Electrospinning. *Autex Res. J.* **2013**, *13*, 128–133. [CrossRef]
37. Rieger, K.A.; Birch, N.P.; Schiffman, J.D. Electrospinning Chitosan/Poly(Ethylene Oxide) Solutions with Essential Oils: Correlating Solution Rheology to Nanofiber Formation. *Carbohydr. Polym.* **2016**, *139*, 131–138. [CrossRef]
38. Padmini Tamilarasi, G.; Sabarees, G.; Manikandan, K.; Gouthaman, S.; Alagarsamy, V.; Raja Solomon, V. Advances in Electrospun Chitosan Nanofiber Biomaterials for Biomedical Applications. *Mater. Adv.* **2023**, *4*, 3114–3139. [CrossRef]
39. Mohammad Ali Zadeh, M.; Keyanpour-Rad, M.; Ebadzadeh, T. Effect of Viscosity of Polyvinyl Alcohol Solution on Morphology of the Electrospun Mullite Nanofibres. *Ceram. Int.* **2014**, *40*, 5461–5466. [CrossRef]
40. Geltmeyer, J.; Van der Schueren, L.; Goethals, F.; De Buysser, K.; De Clerck, K. Optimum Sol Viscosity for Stable Electrospinning of Silica Nanofibres. *J. Sol-Gel Sci. Technol.* **2013**, *67*, 188–195. [CrossRef]
41. Fridrikh, S.V.; Yu, J.H.; Brenner, M.P.; Rutledge, G.C. Controlling the Fiber Diameter during Electrospinning. *Phys. Rev. Lett.* **2003**, *90*, 144502. [CrossRef]
42. Kim, B.; Park, H.; Lee, S.-H.; Sigmund, W.M. Poly(Acrylic Acid) Nanofibers by Electrospinning. *Mater. Lett.* **2005**, *59*, 829–832. [CrossRef]
43. Tan, S.-H.; Inai, R.; Kotaki, M.; Ramakrishna, S. Systematic Parameter Study for Ultra-Fine Fiber Fabrication via Electrospinning Process. *Polymer* **2005**, *46*, 6128–6134. [CrossRef]

44. Piyasin, P.; Yensano, R.; Pinitsoontorn, S. Size-Controllable Melt-Electrospun Polycaprolactone (PCL) Fibers with a Sodium Chloride Additive. *Polymers* **2019**, *11*, 1768. [CrossRef]
45. Park, Y.R.; Ju, H.W.; Lee, J.M.; Kim, D.-K.; Lee, O.J.; Moon, B.M.; Park, H.J.; Jeong, J.Y.; Yeon, Y.K.; Park, C.H. Three-Dimensional Electrospun Silk-Fibroin Nanofiber for Skin Tissue Engineering. *Int. J. Biol. Macromol.* **2016**, *93*, 1567–1574. [CrossRef]
46. Sazegar, M.; Bazgir, S.; Katbab, A.A. Preparation and Characterization of Water-Absorbing Gas-Assisted Electrospun Nanofibers Based on Poly(Vinyl Alcohol)/Chitosan. *Mater. Today Commun.* **2020**, *25*, 101489. [CrossRef]
47. Su, H.; Fujiwara, T.; Bumgardner, J.D. A Study of Combining Elastin in the Chitosan Electrospinning to Increase the Mechanical Strength and Bioactivity. *Mar. Drugs* **2021**, *19*, 169. [CrossRef]
48. Singh, Y.P.; Dasgupta, S.; Nayar, S.; Bhaskar, R. Optimization of Electrospinning Process & Parameters for Producing Defect-Free Chitosan/Polyethylene Oxide Nanofibers for Bone Tissue Engineering. *J. Biomater. Sci. Polym. Ed.* **2020**, *31*, 781–803. [CrossRef]
49. Buttafoco, L.; Kolkman, N.G.; Engbers-Buijtenhuijs, P.; Poot, A.A.; Dijkstra, P.J.; Vermes, I.; Feijen, J. Electrospinning of Collagen and Elastin for Tissue Engineering Applications. *Biomaterials* **2006**, *27*, 724–734. [CrossRef]

marine drugs

MDPI

Article

Synthesis and Characterization of a Novel Chitosan-Based Nanoparticle–Hydrogel Composite System Promising for Skin Wound Drug Delivery

Yueying Huang [1], Shuting Hao [1], Jiayu Chen [1], Mengyuan Wang [1], Ziheng Lin [1] and Yanan Liu [1,2,3,*]

[1] Department of Food Science and Engineering, Ningbo University, Ningbo 315800, China; huangyueyingnbu@163.com (Y.H.); 17857395807@163.com (S.H.); 18857475416@163.com (J.C.); w17858875621@163.com (M.W.); anrimoce@outlook.com (Z.L.)

[2] Ningbo Key Laboratory of Detection, Control and Early Warning of Key Hazardous Materials in Food, Ningbo Academy of Product and Food Quality Inspection, Ningbo Fibre Inspection Institute, Ningbo 315048, China

[3] College of Food Science and Technology, Nanjing Agricultural University, Nanjing 210095, China

* Correspondence: yliu98@126.com

Abstract: As a natural preservative, nisin is widely used in the food industry, while its application in biomedicine is limited due to its susceptibility to interference from external conditions. In this study, a nanoparticle–hydrogel composite system was designed to encapsulate and release nisin. Nisin nanoparticles were identified with a smooth, spherical visual morphology, particle size of 122.72 ± 4.88 nm, polydispersity coefficient of 0.473 ± 0.063, and zeta potential of 23.89 ± 0.37 mV. Based on the sample state and critical properties, three temperature-sensitive hydrogels based on chitosan were ultimately chosen with a rapid gelation time of 112 s, outstanding reticular structure, and optimal swelling ratio of $239.05 \pm 7.15\%$. The composite system exhibited the same antibacterial properties as nisin, demonstrated by the composite system's inhibition zone diameter of 17.06 ± 0.83 mm, compared to 20.20 ± 0.58 mm for nisin, which was attributed to the prolonged release effect of the hydrogel at the appropriate temperature. The composite system also demonstrated good biocompatibility and safety, making it suitable for application as short-term wound dressings in biomedicine due to its low hemolysis rate of less than 2%. In summary, our nanoparticle-based hydrogel composite system offers a novel application form of nisin while ensuring its stability, thereby deepening and broadening the employment of nisin.

Keywords: nisin; nanoparticles; thermo-sensitive hydrogels; composite drug delivery system; bacteriostasis

1. Introduction

Nisin is the only bacteriocin in the world that is permitted as a food additive, owing to its GRAS (Generally Regarded as Safe) status approved by the WHO and FDA [1,2]. Nisin, one of the metabolites of *Lactococcus lactis*, is a natural antimicrobial with broad-spectrum antibacterial activity, which is commonly employed in food preservation [3]. In the process of its extensive application, some characteristics of nisin that hinder its effectiveness have gradually emerged, including susceptibility to external environment interference leading to loss of antibacterial activity, susceptibility to hydrolysis by proteolytic enzymes, and poor stability under certain pH conditions [4,5]. Furthermore, nisin has been reported to play a role in immunological modulation, but due to the aforementioned characteristics, little information is available regarding its application in biomedicine [6,7]. Many efforts have been made to ensure the stability of nisin and extend its application. Many methods have been developed to modify and encapsulate it, such as hybridization with active substances or grafting of active groups [8,9], conjugation with polymers [1,10], preparation of encapsulated sustained-release systems [6,11,12], etc.

Nisin is often utilized in the food sector as films or nanofibers for incorporation into food packaging [13,14]. Although nisin currently has fewer biomedical applications than food applications, existing data show that it is mostly utilized in biological dressings and drug delivery systems [15]. At present, there is significant interest in the encapsulation and delivery system of active substances as a research topic, primarily due to its advantages of local drug delivery, ensuring the activity of substances, and high safety [16,17]. The inhibitory mechanism of nisin in destroying the bacterial cell wall demonstrates that moderate contact with the bacteria is essential to suppress its activity, which means that a delivery system capable of stabilizing and releasing it as needed is required [6]. In addition, biomedical applications necessitate delivery mechanisms with high safety and biocompatibility. The application of nanotechnology has achieved remarkable success in enhancing the antibacterial property and stability of nisin as well as the ability to achieve prolonged release, which has been widely used in the food industry [18,19]. However, its use in biomedicine is currently limited due to a scarcity of closed carriers that can make contact with humans. Another typical application is the direct use of nisin to produce hydrogels, which has stringent criteria for the many components involved [20,21]. All elements that influence the activity of nisin must be considered while designing the product. Thus, the nanoparticle–hydrogel composite system is an appropriate form of nisin for biomedical purposes. To date, many studies have been carried out on nanoparticle–hydrogel composite systems, but the encapsulation of nisin in this system has not been documented [22–24].

Chitosan (CS), a product derived from the deacetylation of chitin from marine arthropods, is a natural cationic polymer with good degradability and biocompatibility and is widely used in delivery systems such as hydrogels and nanoparticles [25]. γ-Polyglutamic acid (γ-PGA) is a natural polymer with excellent hydrophilicity, biocompatibility and safety. Due to the presence of carboxyl groups, the negatively charged γ-PGA is able to provide multiple cation-binding sites in aqueous solution, a phenomenon that has been used in nanoencapsulation, self-assembled hydrogels, and other applications [26,27]. There are many cases of synthesizing nanoparticle drug delivery systems using ionic interactions between polymer CS and γ-PGA, mainly relying on the interaction between -$NH4^+$ in CS and -COO^- in γ-PGA, which leads to spontaneous cross-linking to form nanoparticles [28,29]. In addition to this, CS is able to cross-link with other substances to form hydrogel encapsulation systems. Nisin Z, one of the variants of nisin, is an amphiphilic small molecule peptide with strong inhibitory effects on Gram-positive bacteria, and has also been reported to inhibit certain Gram-negative bacteria [30–32]. It has been reported that nisin Z can be extracted from a wide range of lactic acid bacteria strains, e.g., wild strains, strains isolated from marine organisms, etc. [33,34]. Nisin Z also has a wide range of applications, and is capable of assisting aquatic organisms in resisting pathogenic bacteria, in addition to food preservation and preservation [35,36]. Nisin Z is positively charged in aqueous solution and thus is able to bind to negatively charged γ-PGA via electrostatic interactions, while -COO^- on the surface of the γ-PGA is able to further cross-link with CS to form nanoparticles [37]. The nanoparticles are positively charged by chitosan encapsulation, and can be loaded onto supportive hydrogels in order to realize multi-environmental applications. Nanoparticles synthesized using nisin Z and pectin, as reported previously, exhibited a low PDI and high potential value, indicating a strong interaction [38]. However, CS can only be solubilized in an acidic environment, which have been reported to also promote the dissolution and encapsulation of nisin Z [18,39,40]. Therefore, we selected CS as the main constituent of the composite material.

In this study, we used a thermo-sensitive hydrogel based on CS to encapsulate nanoparticles (NPs) made of CS, γ-PGA, and nisin Z. Following this approach, we created an encapsulating, sustained-release, and biocompatible nanoparticle–hydrogel composite system. The surface properties of the nanoparticles are important factors in the formation of composite systems [41]. Particle characteristic parameters, such as the particle size, potential, and polydispersity coefficient, reflected the appropriate particle volume and agglomeration degree of nanoparticles. SEM was used to examine the appearance and conditions of the particles, corresponding to the particle properties. Accordingly, a series of characterizations were performed to assess the properties of the NPs. In addition, the composition of the hydrogel was also optimized to make the hydrogel more compatible with the NPs. The groups with better performance were comprehensively selected as carriers according to a series of characterization results, particularly microstructure, sustained release effect, and swelling performance. This study focused on the antimicrobial properties and biocompatibility of the composite system, which will be evaluated and predicted for applications in biomedicine and other fields.

2. Results and Discussion

2.1. Characterization of Nanoparticles

2.1.1. Properties of Particles

NPs with concentrations of 3, 4, and 5 mg/mL of nisin Z in the final NPs solution were prepared to determine the particle characteristics. As seen in Figure 1a, the nanoparticle sizes at the three concentrations of nisin Z were concentrated between 100 and 200 nm, with nanoparticles with concentrations of 3 and 4 mg/mL of nisin Z showing a small amount of concentrated distribution between 10 and 100 nm. As shown in Table S1, the distribution of nanoparticles with nisin Z concentrations of 5 mg/mL was more uniform for minimum PDI; therefore, it was selected as the loaded particles for the composite system. The average hydrodynamic diameter of nanoparticles with nisin Z concentrations of 5 mg/mL was 122.72 $\pm$ 4.88 nm (mean $\pm$ SD) and the PDI was 0.473 $\pm$ 0.063, demonstrating a minor state of aggregation. The obtained zeta potential of 23.89 $\pm$ 0.37 mV indicates that the surface of the nanoparticles was positively charged and had a certain level of stability. The surface conditions of the NPs were observed and photographed using SEM. Figure 1b shows the smooth surface and spherical shape of the NPs. As shown in Figure 1c, the shape of the NPs changed slightly after being loaded into the hydrogel, showing signs of extrusion or entanglement. In summary, our NPs have good surface properties and particle properties, providing an excellent basis for the formation of composites, though further design and optimization are still needed for both NPs and hydrogels to maintain the same excellent properties.

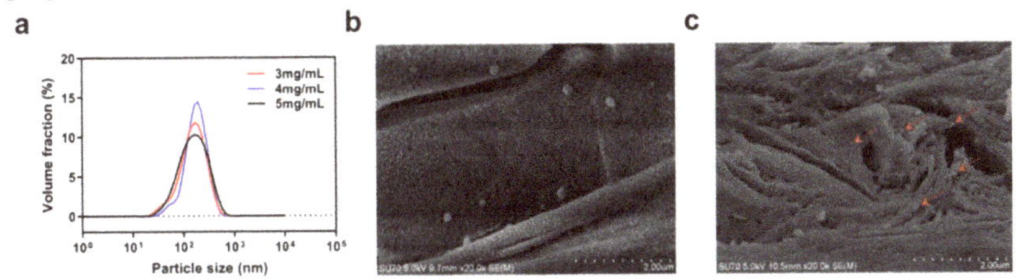

Figure 1. Properties of particles. (**a**) the particle size of nanoparticles with nisin Z concentration of 3, 4, and 5 mg/mL in the final solution, (**b**) SEM of NPs, (**c**) SEM of NPs in freeze-dried nanoparticle–hydrogel composite system. The NPs that were loaded on the hydrogel are indicated by the red arrows in the figure. Scale bar: 2 μm.

2.1.2. Entrapment Efficiency of Nisin Z

Previous reports have shown that the encapsulation efficiency of nisin Z in nanoparticles gradually decreases with an increasing nisin Z concentration. The measured encapsulation efficiency of nanoparticles with nisin Z concentrations of 3, 4, and 5 mg/mL was approximately 80%, and it increased with the increase of nisin Z concentration in our experiment. The encapsulation efficiency of the nanoparticles containing 5 mg/mL nisin Z was 86.18%, indicating a good encapsulation effect. The reason for this phenomenon in this experiment is related to the acidic environment to which nisin Z is exposed. It has been reported that the solubility of nisin Z is related to the pH [39]. Studies have shown that the solubility of nisin Z is higher under acidic conditions, and the corresponding encapsulation efficiency is also higher [18,40]. In the preparation of nanoparticles, the solution is acidic due to the presence of acetic acid, with a pH of approximately 3.40, thus providing suitable environmental conditions for the maximum encapsulation of nisin Z. The encapsulation efficiency of a previously reported nisin Z nanoparticle for beef preservation was 86.6%, demonstrating an excellent preservation effect [42]. In terms of the encapsulation effect, our nanoparticles remain at the same gradient as those in previous studies, which is critical for subsequent activation. The loading capacity of NPs with 5 mg/mL of nisin Z in the final solution was 59.35%, and the loading capacity of NPs with 3 mg/mL and 4 mg/mL of nisin Z in the final solution was 37.16% and 44.08%, respectively. The degree of superiority in the loading capacity results for each NP sample is consistent with the encapsulation rate. Accordingly, the weight percentage of other substances in each NPs can be analyzed. The mass percentage of CS and γ-PGA in the NPs with a concentration of 5 mg/mL of nisin Z in the final solution was 40.65%. In the NPs with 3 mg/mL and 4 mg/mL of nisin Z in the final solution, it was 62.84% and 55.92%, respectively.

2.2. Characterization of Hydrogel and Composite

2.2.1. Choosing the Best Formula

In this study, the effects of time, temperature, and raw material ratio were investigated to select better hydrogels. We also compared the effects of two groups of CS with different molecular weights. Figure 2 shows the state change and final state diagram of each group of samples as the temperature rises from 5 °C to 55 °C. The color (from gray to black) represents the gradual weakening of the fluidity of the samples; that is, gray represents the solution state, and black represents the gel state without fluidity. The two types of CS with different molecular weights showed similarities in terms of the influence of the raw material ratio. When the content of β-Glycerol phosphate disodium salt (β-GP) was low, the fluidity did not change with the increase in temperature. When the β-GP content was high, the fluidity of the sample decreased slightly with an increase in temperature and then remained stable, but no gel change occurred. When the ratios of CS with molecular weight of 100 kDa (CS_{100}) to β-GP were 8:2 and 7:3, the temperature-sensitive gel phenomenon occurred; the same occurred when the ratio of CS with molecular weight of 500 kDa (CS_{500}) to β-GP was 8:2. According to the data on the time required for gelation at 37 °C, the above three groups completed gelation within 5 min; additionally, the time required for gelation gradually decreased with the increase in temperature. Similarly, studies have investigated the effect of CS and β-GP concentrations on gelation, with the results showing that when the concentration of β-GP was 3% (*w*/*v*), the gelation phenomenon occurred within 0.5–1% (*w*/*v*) of CS concentrations. This result is consistent with our study, and gelation of CS and β-GP can only occur within the appropriate dosage range [43]. Three groups of samples (Gel_{100-73}, Gel_{100-82}, and Gel_{500-82}) capable of temperature-sensitive gelation were preliminarily investigated, and subsequent studies were conducted on the basis of these samples to further compare the performance differences.

2.2.2. Rheological Characterization

The time and temperature of gelation were redetermined through time and temperature scanning using rheology. The storage modulus and loss modulus exhibited variations with the passage of time or increase in temperature, while the point of intersection between these two curves indicated the respective gelation time and gelation temperature (Figure 3). In the time-scanning process, Gel_{500-82} first underwent gelation for approximately 1 min and 52 s, and its storage modulus reached a high, stable value within 5 min of gelation, relating to the gel strength. The gelation times of Gel_{100-73} and Gel_{100-82} were relatively close, as can be seen from the temporal scan detail map; Gel_{100-73} gelation occurred at approximately 4 min and 5 s, and Gel_{100-82} gelation occurred at approximately 3 min and 42 s (Figure 3b). Similar to the results of previous experiments, all of our gel samples were able to undergo gelation within a short time [44]. Gel_{500-82} was also the first to undergo gelation transition at approximately 28.24 °C in the temperature scanning process. The gelation temperature point of Gel_{100-82} was 31.56 °C, while that of Gel_{100-73} was 38.56 °C

(Figure 3d). Under the influence of temperature change, the modulus change curve of the sample shows a small fluctuation, likely associated with the environment or equipment, especially for $Gel_{100\text{-}73}$. Small fluctuations had little effect on the measurement of the experimental results, but they were sufficient to reflect the significant influence of temperature change on the sample state, further verifying the temperature responsiveness of the hydrogel. The results obtained from the rheological experiments were consistent with those obtained when the gel effect was investigated. Additionally, our samples exhibited a more sensitive temperature response with the same gelation time as previous hydrogels of the same type [44,45]. The rapid gelation of the above samples at approximately the physiological temperature of the human body (37 °C) provides a prerequisite for their wide application in biomedicine. The time scans of each composite (Figure 3e,f) showed that the addition of NPs did not have significant effect on the gelation time, and each sample was still able to gel within 5 min. However, due to the more complex composition, the gel temperature of the composites in Figure 3g,h showed an increase, and the temperature scanning curves showed small fluctuations. The gel temperature of $Gel_{100\text{-}73}$/NPs even exceeded 40 °C, which is still considered to be suitable for biological excipients.

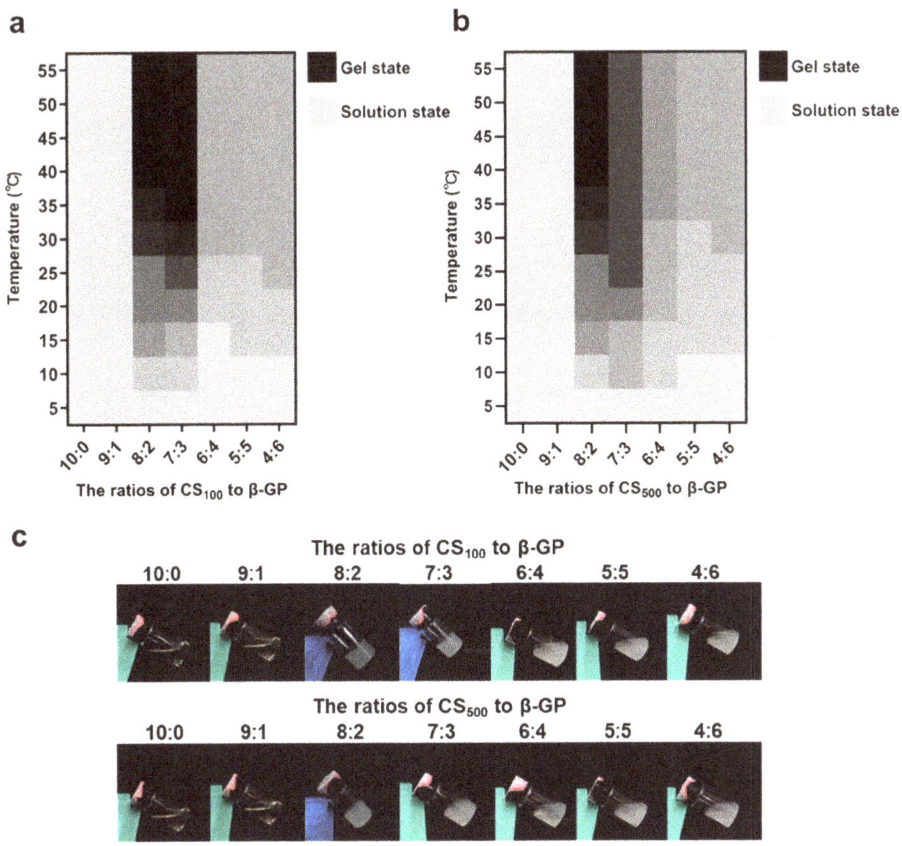

Figure 2. Schematic of optimal formulation selection. (**a**) gel condition of each sample of CS_{100} as a function of temperature, (**b**) gel condition of each sample of CS_{500} as a function of temperature, (**c**) schematic representation of the gel state of each sample.

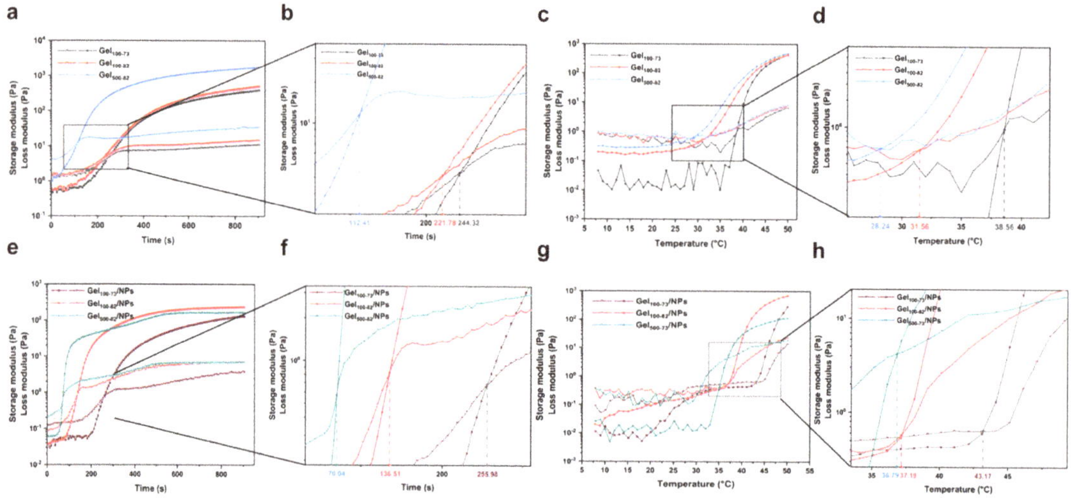

Figure 3. Storage modulus (G′) and loss modulus (G″) of the samples. (a) time scanning of gel samples, (b) detail map of time scanning of gel samples, (c) temperature scanning of gel samples, (d) detail map of temperature scanning of gel samples, (e) time scanning of composite samples, (f) detail map of time scanning of composite samples, (g) temperature scanning of composite samples, (h) detail map of temperature scanning of composite samples.

2.2.3. FTIR Characterization

The gel formation mechanism can be clearly observed in the FTIR characteristic curve. The FTIR results obtained were identical to those obtained in previous hydrogel experiments [44,46]. The stretching of -OH and -NH corresponded to the strong and wide overlapping peak at 3450 cm^{-1} in the infrared spectrum of CS_{100}, and the band at 1657 cm^{-1} was attributed to the bending vibration of -C=O (Figure 4a). The peak of the infrared spectral line of CS_{500} was similar to that of CS_{100} (Figure 4b). For β-GP, the peak of stretching -OH was found at 3273 cm^{-1}. The asymmetric stretching vibration of -PO_4^{3-} appeared at 1095 cm^{-1}, and the symmetrical stretching vibration peak of -PO_4^{3-} was observed at 977 cm^{-1}. The overlapping peaks of stretching -OH and -NH can also be seen on the bands of gel samples. However, the electrostatic interaction between -PO_4^{3-} and -NH caused the peak shape and corresponding wavelength of the peak tip to change slightly. The characteristic peaks of -PO_4^{3-} on the bands of the gel samples coincided with those of β-GP. Moreover, the small magnitude of the shift exhibited by the peak of -C=O may be caused by the formation of hydrogen bonds during gelation.

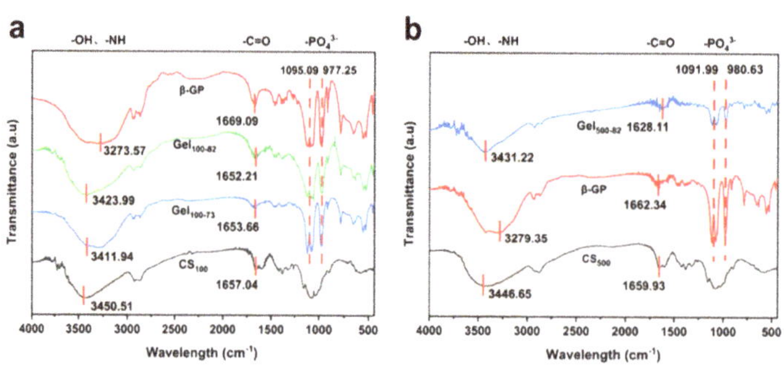

Figure 4. FTIR spectrogram of samples of the CS_{100} progenitor (a) and the CS_{500} progenitor (b).

2.2.4. Microstructure

The morphology and microstructure of each cryogel and freeze-dried composite can be seen in the SEM images. Cryogels and freeze-dried composites were obtained from original samples by freeze-drying, and the structures were related to the expected structure of the hydrogels and composites. As shown in Figure 5, each sample has a network structure distinguished by the size and degree of porosity, which is in agreement with published results [46]. Gel_{100-73} had large and loose pores with an overall elongated appearance (Figure 5a). The network structures of Gel_{100-82} and Gel_{500-82} were relatively dense, and their pores were small and nearly circular (Figure 5b,c). In contrast, the pores of Gel_{500-82} were slightly smaller than those of Gel_{100-82}, which may be related to the larger molecular weight of CS_{500}. The large molecular weight determines the relatively long molecular chain of CS_{500}, leading to enhanced intermolecular interaction forces (hydrogen bonding and hydrophobic interactions). This is the same principle as that in a study in which the introduction of titanium (Ti) assisted in the enhancement of gel strength [47]. The SEM images of each composite (Figure 5d–f) presented similar results to the SEM images of the corresponding hydrogels, with little variation in the grid size of the gel organization; however, the grid structure of Gel_{100-73}/NPs and Gel_{100-82}/NPs was not as pronounced as that of the corresponding hydrogel samples due to the influence of the surface charge of the NPs. The microstructure of the group without gelation showed neatly arranged fibers, without obvious reticular structures or complex winding structures (Figure 6). This indicated that, in the group without gelation, the addition of β-GP formed a hydration layer around the CS molecules; thus, the molecular chains originally in the entangled state were extended and arranged in order, but no cross-linking occurred under the effect of temperature or no network structure was formed after cross-linking.

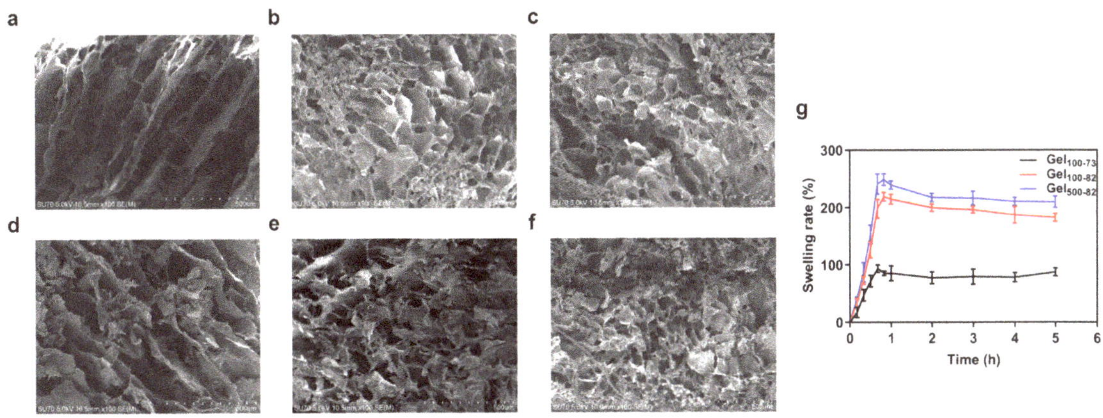

Figure 5. SEM of Gel_{100-73} (**a**), Gel_{100-82} (**b**), and Gel_{500-82} (**c**) SEM of Gel_{100-73}/NPs (**d**) Gel_{100-82}/NPs (**e**) and Gel_{500-82}/NPs (**f**) and swelling rate of gel samples (**g**). Scale bar: 500 μm.

2.2.5. Swelling Ratios

The swelling rate is commonly used to measure the degree of water absorption and swelling of hydrogels which, to a certain extent, determines the application field and specific use of hydrogels [48,49]. Gel_{100-73} and Gel_{100-82} reached the maximum swelling rate at around 50 min. Gel_{500-82} reached the maximum swelling rate at around 40 min. The swelling rate of the gel samples rapidly reached the maximum value within 1 h and then gradually stabilized (Figure 5g). Both Gel_{100-82} and Gel_{500-82} had a maximum swelling of more than twice the original mass, at 220.54 ± 6.20% and 248.55 ± 10.02%, respectively, indicating excellent swelling efficiency. The swelling rate of Gel_{100-73} was relatively small but also close to 100%. An excellent swelling effect indicates that the hydrogel has a strong water-holding capacity (WHC), which enables the hydrogel to provide a moist environment for the surrounding contacts. At the same time, correlating with the SEM images of the samples, it can be inferred that the WHC of the loosely structured reticular structure is not as good as that of the tightly structured reticular structure [45]. Similar to the previous results of Bhuiyan et al., the change in the CS ratio is also one of the factors affecting the swelling rate [43]. The increase in the CS ratio contributes to an increased swelling ratio, likely because the introduction of the polar groups generates a more hydrophilic chemical structure.

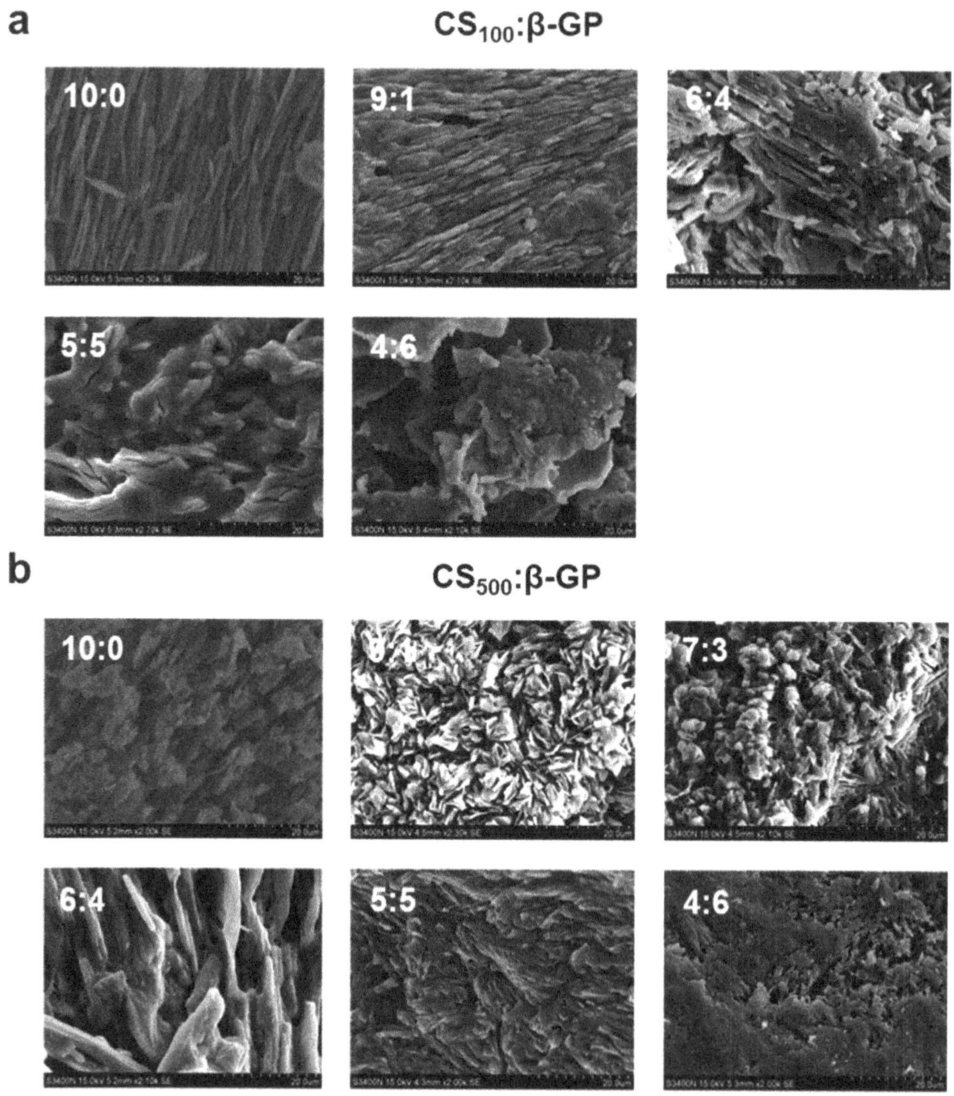

Figure 6. SEM of the non-gel group of CS_{100} (**a**) and CS_{500} (**b**). Scale bar: 20 μm.

2.2.6. In Vitro Release Assays

The composite system released nisin Z into the PBS solution it was in contact with, and the release efficiency was determined using the BCA method. The experimental results show that the nisin Z release rate of the sample reached the maximum within

24 h, after which the nisin Z was released steadily (Figure 7a). It is worth noting that the release rate of $Gel_{100\text{-}82}$ was always higher than that of $Gel_{100\text{-}73}$ and $Gel_{500\text{-}82}$, finally reaching 20.09 ± 1.68%. The release rates of $Gel_{100\text{-}73}$ and $Gel_{500\text{-}82}$ ultimately stabilized at 18.43 ± 0.47% and 19.16 ± 0.57%, respectively. Compared with complexes reported in the literature, the release rates of our samples were low [50,51]. The method used in the release rate calculation was based on the entrapment efficiency of nanoparticles. However, some nisin Z may remain on the surface of the NPs and be bound by intermolecular forces, resulting in a high encapsulation efficiency. This part of nisin Z is also unable to be subsequently released from the gel tissue, thus resulting in a low overall release rate from the sample. The low level and sustained drug release efficiency are more suitable for use in situations where tight control of the drug concentration is required. The ophthalmic drug delivery system developed by Hsu et al. exploited this property [52].

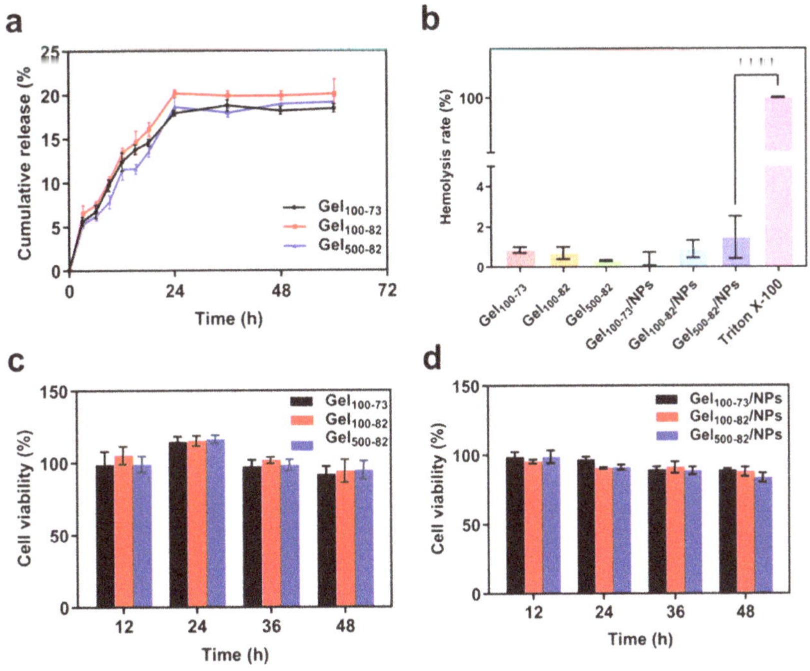

Figure 7. Characterization of nanoparticle–hydrogel composite system. (**a**) efficiency of nisin Z release from $Gel_{100\text{-}73}$, $Gel_{100\text{-}82}$, and $Gel_{500\text{-}82}$, (**b**) hemolysis rate of samples, (**c**) in vitro toxicity of gel samples, (**d**) in vitro toxicity of gel composite system. **** $p < 0.0001$ indicates significant differences between samples and Triton X-100 control group.

2.2.7. In Vitro Toxicity Assays

For the toxicity experiment, the CCK-8 method was used to explore the cytotoxicity of the samples. The cytotoxicity levels of materials are generally classified into four grades: grade 0 (no cytotoxicity, cell viability ≥ 100%), grade 1 (mild cytotoxicity, cell viability ranging from 75% to 99%), grade 2 (moderate cytotoxicity, cell viability ranging from 50% to 74%), and grade 3 (severe cytotoxicity, cell viability ranging from 25% to 49%) [50,51]. The cell viability of the hydrogel samples was more than 100% when they were in contact with the cells for 24 h (Figure 7c). With the extension of time, the cell survival rate decreased, but it was still grade 2 with a level of mild cytotoxicity. The cell viability of the composite systems showed a decreasing trend with the extension of time, which may be related to the introduction of antimicrobial peptides (Figure 7d). However, the overall level was still maintained at grade 2, showing great biocompatibility. Compared to the complex encapsulated with inorganic nanoparticles, our material presents lower cytotoxicity, correlated with the safety of nisin Z [53]. The cytotoxicity of previously developed composites utilized in wound dressings is equivalent to that of this product, which is a decisive factor of the material as a wound dressing [54,55].

2.2.8. Hemolysis Evaluation

The hemolysis rate is an important factor in evaluating the biocompatibility of materials, which represents the degree of damage to erythrocytes. To a certain extent, the lower the hemolysis rate, the safer the material, as hemolysis will not occur when the blood is in contact with the material [56]. It can be seen from Figure 7b that the hemolysis rate of each gel sample and composite system was lower than 2%, and there was no significant difference between the samples. Previous materials typically used 5% as a critical point for the hemolysis rate; from this point of view, our materials have the safety and biocompatibility required for biological dressings [51]. In practical application, the effects of endogenous and exogenous factors on the hemolysis rate should be comprehensively considered. The ability of our material to provide a low hemolysis rate at the point of use enhances the tolerance to hemolysis caused by internal factors.

2.2.9. In Vitro Antibacterial Characterization

The antibacterial activity of the samples was evaluated using the inhibition zone and colony counting method. As shown in Figure 8b, neither the hydrogel nor the complex exhibited antibacterial properties against *E. coli*. According to previous studies, nisin Z is able to inhibit Gram-positive bacteria by forming membrane channels with them, leading to the efflux of cellular contents. However, nisin Z is usually unable to form membrane channels on the surface of Gram-negative bacteria and needs to disrupt the cellular membranes with the help of other active substances before exerting an inhibitory effect through the outer membranes [57–59]. $Gel_{100\text{-}73}$ and each complex exhibited significant inhibitory effects against *S. aureus*, with the diameter of the zone of inhibition for each complex against *S. aureus* provided in Table S2. The results of the colony counting method also showed the same antibacterial effect (Figure 8a). There was no significant difference in the number of *E. coli* colonies on the serial dilution plates, but the inhibitory effect on *S. aureus* was more obvious at a 10^3-fold dilution. In Figure 8c, the antibacterial effect of the nisin Z solution and NPs is also shown, and it can be observed that the antibacterial effect of the above complex mainly comes from nisin Z. The antibacterial properties of $Gel_{100\text{-}73}$ may be related to those of the CS [60–62]. Furthermore, $Gel_{100\text{-}82}$ and $Gel_{500\text{-}82}$ showed similar dense reticulation structures in SEM results. Combined with this, we envisioned that the failure of both $Gel_{100\text{-}82}$ and $Gel_{500\text{-}82}$ to inhibit *S. aureus* could also be due to their overly dense structures, which blocked the inhibitory sites of chitosan in the hydrogels. *S. aureus* is a common foodborne pathogen that also poses a great risk in human wounds and disease infection. The complex shows an excellent inhibitory effect against *S. aureus*, and the loaded antimicrobial peptides can also avoid drug resistance, which has great application prospects in biology [63]. A recently reported nisin nanoparticle pre-formulation, with improved antibacterial effects encapsulated with a special material, has a similar effect to our material [64]. The nano-encapsulation technology significantly enhances the antibacterial efficacy of active substances, thereby ensuring their effectiveness and potentially even improving it. Another experiment utilizing nisin as a gel coating further substantiated its antibacterial efficacy [65]. This investigation not only validated the functionality of NPs but also provided evidence for the encapsulation and subsequent release of NPs via hydrogel.

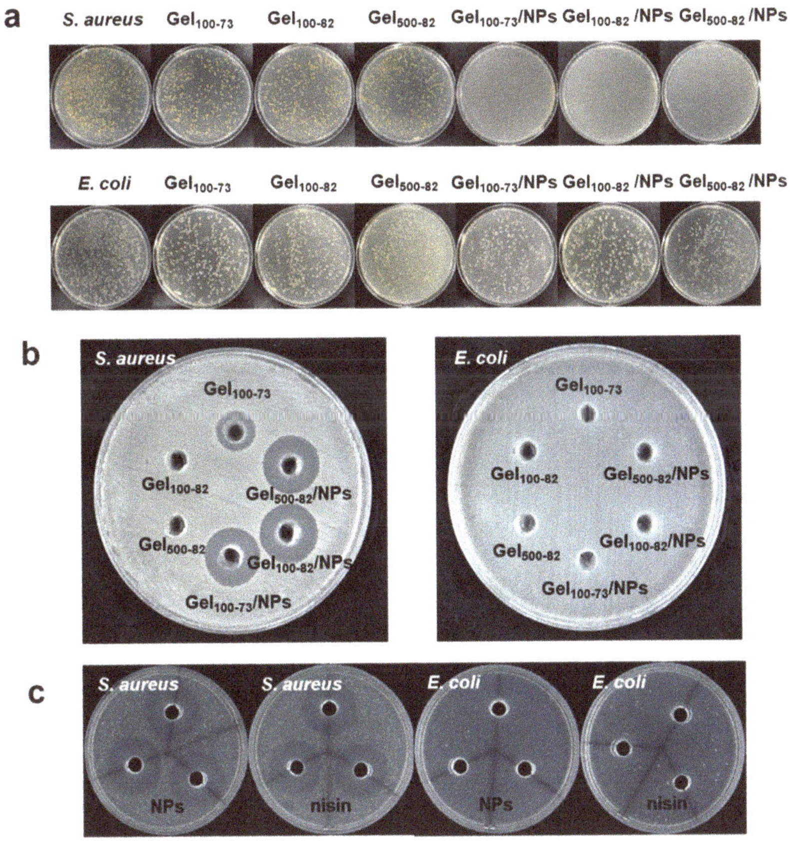

Figure 8. Antibacterial properties of the material. (**a**) results of plate count assay for the inhibition of samples against *S. aureus* and *E. coli*, (**b**) results of inhibition zone test of samples against *S. aureus* and *E. coli*, (**c**) Inhibition zone results of NPs and nisin Z against *S. aureus* and *E. coli*.

3. Materials and Methods

3.1. Materials

For this study, the supplier of nisin (3%) was CHIHONBIO Co., Ltd. (Luoyang, China), and its specific species was nisin Z. γ-Polyglutamic acid (MW > 700 kDa) was provided by Shanghai Yongchuan Biotechnology Co., Ltd. (Shanghai, China). Chitosan and β-Glycerol phosphate disodium salt (β-GP, purity > 95%) were purchased from Shanghai Yongchuan Biotechnology Co., Ltd. (Shanghai, China). The molecular weights of CS were 100 (degree of deacetylation 85%) and 500 (degree of deacetylation 80%) kDa, denoted as CS_{100} and CS_{500}, respectively. The hydrochloric acid (37%) used in the experiment was provided by the Hazardous Chemicals Management Department of Ningbo University, and substances, such as acetic acid glacial (99.5%), PBS buffer, and paraffin oil (99%) were also provided by Shanghai Yongchuan Biotechnology Co., Ltd. (Shanghai, China).

3.2. Synthesis of Nanoparticles

The synthesis of nanoparticles employed a modified approach based on previous methods [66]. After being dialyzed and freeze-dried for 24 h using a SCIENTZ-18N freeze-dryer, purchased from Ningbo Scientz Biotechnology Co., Ltd. (Ningbo, China), the purchased nisin Z samples were stored in a refrigerator at 4 °C for future use. For the synthesis of nanoparticles, the nisin Z aqueous solution was filtered through a disposable

0.22 μm membrane and then progressively added to the equivalent volume of a γ-PGA solution (4 mg/mL) and stirred for 30 min. CS_{100} was added to an acetic acid solution (1%, v/v) to form a 2 mg/mL chitosan acetate solution and agitated at room temperature until completely dissolved. The specific proportions of the various substances are detailed in Table S3. The mixed solution with nisin Z and γ-PGA was added dropwise to the acetic acid solution of CS_{100} and stirred overnight. It was then briefly homogenized using a handheld homogenizer and filtered through a disposable 0.22 μm membrane to obtain the NPs solution.

3.3. Zetasizer Particle Characterization

A Malvern Zetasizer Advance Series Lab (Malvern Panalytical, Malvern, UK) was used to measure the nanoparticle size, zeta potential, and polydispersity index (PDI). The nanoparticle size and polydispersity index were determined at 25 °C and a 173° scattering angle in a disposable cell (DTS0012) [67]. Similarly, potential measurements were also conducted in a specific sample dish (DTS1070). To ensure that the resultant values were within the confidence interval, the nanoparticle solution was diluted 5-fold using ultrapure water before measurement. The measurements were repeated three times for each sample.

3.4. Entrapment Efficiency of Nisin Z

Inspired by the prior approach of testing the encapsulation efficiency of gel beads, in this experiment, we permitted the unencapsulated nisin Z to enter the dialysate from the nanoparticle solution via dialysis and then utilized an Enhanced BCA Protein Assay Kit (Beyotime Biotechnology, Shanghai, China) to quantify the concentration of nisin Z in the dialysate [68]. The encapsulation rate of nisin Z was estimated using the following Equation (1). Dialysate of n times the volume of the nanoparticle solution was added to the dialysis vessel to facilitate calculation of the results. At the same time, we dried and weighed the NPs and calculated the loading capacity of each NPs sample for nisin Z according to the following Formula (2). The experiment was repeated to obtain three readings.

$$Entrapment\ efficiency\ (\%) = \frac{Total\ nisin\ Z\ concentration - Unencapsulated\ nisin\ Z\ concentration}{Total\ nisin\ Z\ concentration} \times 100 \quad (1)$$

$$Loading\ capacity(\%) = \frac{Total\ mass\ of\ nisin\ Z - Unencapsulated\ nisin\ Z\ mass}{Total\ mass\ of\ NPs} \times 100 \quad (2)$$

3.5. Preparation of Hydrogel and Nanoparticle–Hydrogel Composite System

We drew on previous ice bath methods for hydrogel fabrication [69] to identify the optimal ratios for achieving a temperature response hydrogel in subsequent experiments. A quantity of CS was weighed and added to a 0.1 mol/L hydrochloric acid solution with simultaneous magnetic stirring to prepare a hydrochloric acid solution with a CS concentration of 2% (m/v). It was mixed at room temperature until completely dissolved, then sterilized and stored in a refrigerator at 4 °C. An appropriate amount of β-GP was dissolved in sterile water to obtain a 45% (m/v) aqueous solution of β-GP, followed by its filtration through a 0.22 μm disposable sterile needle filter membrane. To create a clear and uniform CS/β-GP combination, the 45% aqueous β-GP solution was added dropwise to the 2% CS solution and stirred continuously in an ice bath for 30 min. The mixture was then incubated in a water bath for a set amount of time at a suitable temperature to form a hydrogel.

In preparing the composites, the CS/β-GP mixture was first stirred in an ice bath for 30 min according to the above steps. Next, the NPs solution was added drop by drop to the CS/β-GP mixture under ice bath conditions. To ensure uniform distribution of NPs, the mixture was again stirred in an ice bath for 30 min. Then, the nanoparticle–hydrogel composite system can be generated following further water bath incubation [70].

3.6. Determination of the Optimal Ratio of Gel Components

Two molecular weights (CS_{100} and CS_{500}) of CS were selected, and the ratio of CS to β-GP was adjusted to investigate the optimal gel composition. Provided that the total volume was the same, the volume ratios chosen for the experiment were 10:0, 9:1, 8:2, 7:3, 6:4, 5:5, and 4:6. Preliminarily, the gel environment temperature was set at 37 °C, with a total testing duration of 15 min [71]. The group with the best gelation was selected visually using the inversion method, which involved inverting the container every 30 s for 15 min to observe the fluidity of each sample, determine whether gelation occurred, and document the time it took for gelation. In addition, the gelation state of the sieved groups at various temperatures was recorded to observe the transition from solution to gel, and temperature gradients were set up between 5 °C and 55 °C. The whole experiment was repeated three times.

3.7. Rheological Characterization

The dynamic oscillation tests were performed using a DHR-2 rheometer (Discovery Hybrid Rheometer, TA Instrument, Newcastle, DE, USA), including time scanning and oscillatory temperature change [72]. For the experiments, we equipped a parallel disk of 40 mm and lowered it to a gap of 1 mm. To prevent moisture loss, a ring of paraffin oil was dripped around the sample to seal it after the disc was lowered. Time scans were performed at 37 °C to record the dynamic modulus versus time over 1000 s, enabling the time required for gelation to be analyzed. Oscillatory temperature changes were performed to record the temperature at which the sol–gel transition behavior occurred, and the temperature was programmed to rise from 5 °C to 55 °C. A constant frequency of 1 Hz was set for all these measurements. The measurements were repeated three times for each sample.

3.8. Fourier Transform Infrared Spectroscopy (FTIR)

The hydrogel was freeze-dried for FTIR spectral analysis using a Nicolet iS50 FTIR spectrometer (Thermo Fisher Scientific Inc., Waltham, MA, USA). The freeze-dried samples were evenly mixed with KBr at a ratio of 1:100 and finely ground using a clean and dry agate mortar. The powder was placed in a tablet press and scanned in the energy range of 400 to 4000 cm^{-1}, provided that the tablet was uniform and crack-free [73]. The measurements were repeated three times for each sample.

3.9. Scanning Electron Microscopy (SEM)

The morphological structures of the samples were observed using a Hitachi S-3400 scanning electron microscope (Hitachi High-Technologies Corporation, Tokyo, Japan). In order to photograph the internal structure of the hydrogel, the freeze-dried sample was cut into cubes of the same size to photograph the cutting section. The sample was loaded on a cylindrical stage with conductive tape and scanned at an accelerating voltage of 15 kV after spraying a layer of gold [74]. The experiment was performed three times in parallel.

3.10. Swelling Ratios

The swelling rate of the sample was determined using the weighing method; that is, the proportion of mass increase in the freeze-dried sample after water absorption over a certain period of time was determined. After weighing, the freeze-dried samples were soaked in distilled water at room temperature and subsequently taken out and weighed every hour (recording measurements every ten minutes for the first hour) after drying the surface moisture with filter paper [75]. The experiment was performed three times in parallel.

3.11. In Vitro Release Assays

The release efficiency of nisin Z in the composite system was monitored for 60 h at 37 °C. The sustained-release test was carried out in a glass bottle using 2 mL gel complex and 18 mL PBS solution. A certain amount of PBS solution was taken out regularly, and the content of nisin Z was determined using the BCA method. The corresponding volume of PBS was supplemented for subsequent measurements [52]. The measurements were repeated three times for each sample.

3.12. In Vitro Toxicity Assays

RAW264.7 macrophages were used for the CCK-8 (2-(2-methoxy-4-nitrophenyl)-3-(4-nitrophenyl)-5-(2,4-disulfophenyl)-2H-tetrazolium monosodium salt) method to evaluate the cytotoxicity of the materials [76]. The RAW264.7 cell line was stored at the Department of Food Science and Engineering, Ningbo University. After activation and passage of RAW cells, a batch of cells with good growth and a differentiation of less than 20% was selected for the experiment. The sample was co-cultured with the cell complete medium (the same volume) in a 6-well plate for 12 h to obtain the hydrogel extract. Cells were seeded into 96-well plates at a density of 5×10^4 cells/well and incubated at 37 °C and 5% CO_2 for 12 h. Subsequently, the medium was discarded, 100 μL of the extract was added to each well, and a control group with complete medium was set. After co-incubation for 12 h, 24 h, 36 h and 48 h, the original culture was discarded and 100 μL of complete culture medium containing 10% CCK-8 was added. Simultaneously, a control group with only CCK-8 and complete culture medium without cells was established. The culture plate was incubated for 1.5 h, and the absorbance at 450 nm was measured using a microplate reader. The experiment involved the conduction of three parallel tests. The cytotoxicity was calculated using the following formula:

$$Cell\ viability\ (\%) = \frac{A_s - A_b}{A_c - A_b} \times 100 \quad (3)$$

where A_s represents the absorbance of the experimental wells (containing cells, medium, CCK-8 and extract), A_b represents the absorbance of the blank wells (containing medium and CCK-8), and A_c refers to the absorbance of the control wells (containing cells, medium and CCK-8). The measurements were repeated three times for each sample.

3.13. Assay for Erythrocyte Hemolysis

The hemolysis assay was determined using the induction method, following a previous protocol with minor modifications [77]. Anticoagulated fresh sheep blood for testing was centrifuged at 2000 rpm and 4 °C for 10 min, and the supernatant was removed. Erythrocytes were resuspended in PBS and washed three times with centrifugation at 2000 rpm and 4 °C for 10 min. The concentration of the washed cells was adjusted to 1×10^8 cells/mL (200 μL of erythrocyte precipitate was added to 10 mL of PBS via suction). Blood cell suspension (100 μL) was mixed with 100 μL of each sample extract in a centrifuge tube; sterile PBS and Triton were used as controls for 0% and 100% hemolysis, respectively. After exposing the mixture for 30 min in an incubator at 37 °C, the mixture was centrifuged at 2000 rpm and 4 °C for 3 min. The absorbance of the supernatant was measured at 540 nm, and the hemolysis rate was calculated using the following formula:

$$Hemolysis\ rate\ (\%) = \frac{A_x - A_{0\%}}{A_{100\%} - A_{0\%}} \times 100 \quad (4)$$

where A_x refers to the absorbance of experimental wells (extract group), $A_{0\%}$ denotes the absorbance of blank wells (PBS group), and $A_{100\%}$ denotes the absorbance of control wells (Triton group). The measurements were repeated three times for each sample.

3.14. In Vitro Antibacterial Assays

The inhibition circle and colony counting methods were used to investigate the antimicrobial properties of materials. Gram-positive bacteria (*S. aureus* ATCC 6538) and Gram-negative bacteria (*E. coli* ATCC 25922) were selected for antibacterial experiments. The specific operation of the inhibition zone method was as follows: first, 100 μL of the bacterial solution (*S. aureus* or *E. coli*, $OD_{600\ nm} = 0.4$) was poured into the agar solid medium and mixed well. The plate was then poured and punched, and 200 µL of the sample was added to each hole, one by one. Finally, it was cultured in an incubator at 37 °C for 24 h, and then the diameter of the inhibition zone was measured [78].

The colony plate counting method is divided into three steps: co-culture, dilution coating, and colony counting [79]. Equal amounts of bacterial solution and samples were co-cultured in an incubator for 18 h, and the group with only bacterial solution was used as a control. After culturing, the bacteria on the surface of the sample were removed using ultrasonic treatment. The bacterial solution from each group was diluted and coated with a gradient dilution. Three dilution factors—10^2, 10^3, and 10^4—were selected. After 24 h of continuous culturing, colony counting was performed. The samples used for bacteriostatic experiments were prepared under sterile conditions, and the sterility of the raw materials and instruments was ensured. The measurements were repeated three times for each sample.

3.13. Statistical Analysis

The statistical analysis was carried out in triplicate. All data were expressed as the mean values and standard deviations (SDs) ($n = 3$). The images were generated using GraphPad Prism 10 (GraphPad Software, San Diego, CA, USA) or Origin 2021 software (OriginLab Corporation, Northampton, MA, USA). Significant differences were defined as $p < 0.05$.

4. Conclusions

We embedded nisin Z in nanoparticles and then used temperature-sensitive hydrogel as a carrier to obtain an effective sustained-release nanoparticle–hydrogel composite system. To achieve efficient encapsulation and release, we optimized the design of both the NPs and hydrogel. Within the experimental range, the NPs with a nisin Z content of 5 mg/mL were ultimately designated for the highest encapsulation effectiveness (86.18%) and optimal particle properties. The thermosensitive hydrogel was fabricated via the electrostatic interaction between CS and β-GP, and the three optimal gel groups ($Gel_{100\text{-}73}$, $Gel_{100\text{-}82}$, and $Gel_{500\text{-}82}$) were determined using the inverted tube method. The thermosensitive gel phenomenon is observed only when the interaction force between CS and β-GP is relatively balanced; otherwise, they will remain in a solution or sol state. The NPs–hydrogel composite system had a strong antibacterial effect on *S. aureus* and great biocompatibility, attributed to the efficient encapsulation and sustained release (20.09 ± 1.68%) of nisin Z. In this study, we attempted to incorporate nisin Z into a nanoparticle–hydrogel composite system, resulting in a novel encapsulation form for nisin Z in the biomedical field, particularly for trauma dressings.

Supplementary Materials: The following supporting information can be downloaded at: https://www.mdpi.com/article/10.3390/md22090428/s1, Table S1: Characterization results of all nanoparticles; Table S2: Circle of inhibition diameters of nisin Z, NPs and composites on *S. aureus*; Table S3: Specific compositions of hydrogels, NPs and composites.

Author Contributions: Conceptualization, Y.L.; methodology, Y.L. and Y.H.; software, Y.H. and S.H.; validation, Y.H., S.H. and J.C.; formal analysis, Y.L., Y.H., S.H., J.C., M.W. and Z.L.; investigation, Y.H., S.H., J.C., M.W. and Z.L.; resources, Y.L.; data curation, Y.H., S.H., J.C. and M.W.; writing—original draft preparation, Y.H. and S.H.; writing—review and editing, Y.L. and Y.H.; project administration, Y.L.; funding acquisition, Y.L.; supervision, Y.L. All authors have read and agreed to the published version of the manuscript.

Funding: This research was funded by the National Training Program of Innovation and Entrepreneurship for Undergraduates, grant number 202211646012.

Institutional Review Board Statement: Not applicable.

Data Availability Statement: The data presented in this study are available on request from the corresponding author.

Acknowledgments: We are grateful for the equipment-sharing service provided by the Ningbo University Large Instrument Sharing Management Platform.

Conflicts of Interest: The authors declare no conflicts of interest. The funders had no role in the design of this study, the collection, analyses, or interpretation of data, in the writing of the manuscript, or in the decision to publish the results.

References

1. Yu, X.; Lu, N.; Wang, J.; Chen, Z.; Chen, C.; Mac Regenstein, J.; Zhou, P. Effect of N-terminal modification on the antimicrobial activity of nisin. *Food Control* **2020**, *114*, 107227. [CrossRef]
2. Chandrasekar, V.; Coupland, J.N.; Anantheswaran, R.C. Characterization of nisin containing chitosan-alginate microparticles. *Food Hydrocoll.* **2017**, *69*, 301–307. [CrossRef]
3. Flynn, J.; Mallen, S.; Durack, E.; O'Connor, P.M.; Hudson, S.P. Mesoporous matrices for the delivery of the broad spectrum bacteriocin, nisin A. *J. Colloid Interface Sci.* **2019**, *537*, 396–406. [CrossRef]
4. Ibarra-Sanchez, L.A.; El-Haddad, N.; Mahmoud, D.; Miller, M.J.; Karam, L. *Invited review*: Advances in nisin use for preservation of dairy products. *J. Dairy Sci.* **2020**, *103*, 2041–2052. [CrossRef]
5. Duchateau, A.L.L.; van Scheppingen, W.B. Stability study of a nisin/natamycin blend by LC-MS. *Food Chem.* **2018**, *266*, 240–244. [CrossRef]
6. Khan, F.; Singh, P.; Joshi, A.S.; Tabassum, N.; Jeong, G.-J.; Bamunuarachchi, N.I.; Mijakovic, I.; Kim, Y.-M. Multiple potential strategies for the application of nisin and derivatives. *Crit. Rev. Microbiol.* **2023**, *49*, 628–657. [CrossRef] [PubMed]
7. LeBel, G.; Piche, F.; Frenette, M.; Gottschalk, M.; Grenier, D. Antimicrobial activity of nisin against the swine pathogen *Streptococcus suis* and its synergistic interaction with antibiotics. *Peptides* **2013**, *50*, 19–23. [CrossRef]
8. Zhao, X.; Kuipers, O.P. Synthesis of silver-nisin nanoparticles with low cytotoxicity as antimicrobials against biofilm-forming pathogens. *Colloids Surf. B-Biointerfaces* **2021**, *206*, 111965. [CrossRef] [PubMed]
9. Zhu, C.; Zou, S.; Rao, Z.; Min, L.; Liu, M.; Liu, L.; Fan, L. Preparation and characterization of hydroxypropyl chitosan modified with nisin. *Int. J. Biol. Macromol.* **2017**, *105*, 1017–1024. [CrossRef]
10. Gedarawatte, S.T.G.; Ravensdale, J.T.; Al-Salami, H.; Dykes, G.A.; Coorey, R. Antimicrobial efficacy of nisin-loaded bacterial cellulose nanocrystals against selected meat spoilage lactic acid bacteria. *Carbohydr. Polym.* **2021**, *251*, 117096. [CrossRef]
11. Luo, L.; Wu, Y.; Liu, C.; Zou, Y.; Huang, L.; Liang, Y.; Ren, J.; Liu, Y.; Lin, Q. Elaboration and characterization of curcumin-loaded soy soluble polysaccharide (SSPS)-based nanocarriers mediated by antimicrobial peptide nisin. *Food Chem.* **2021**, *336*, 127669. [CrossRef]
12. Niaz, T.; Shabbir, S.; Noor, T.; Rahman, A.; Bokhari, H.; Imran, M. Potential of polymer stabilized nano-liposomes to enhance antimicrobial activity of nisin Z against foodborne pathogens. *Lwt-Food Sci. Technol.* **2018**, *96*, 98–110. [CrossRef]
13. Monfared, Y.K.; Mahmoudian, M.; Hoti, G.; Caldera, F.; Lopez Nicolas, J.M.; Zakeri-Milani, P.; Matencio, A.; Trotta, F. Cyclodextrin-Based Nanosponges as Perse Antimicrobial Agents Increase the Activity of Natural Antimicrobial Peptide Nisin. *Pharmaceutics* **2022**, *14*, 685. [CrossRef]
14. Niaz, T.; Shabbir, S.; Noor, T.; Imran, M. Active Composite Packaging Reinforced with Nisin-Loaded Nano-Vesicles for Extended Shelf Life of Chicken Breast Filets and Cheese Slices. *Food Bioprocess Technol.* **2022**, *15*, 1284–1298. [CrossRef]
15. Reczynska-Kolman, K.; Hartman, K.; Kwiecien, K.; Brzychczy-Wloch, M.; Pamula, E. Composites Based on Gellan Gum, Alginate and Nisin-Enriched Lipid Nanoparticles for the Treatment of Infected Wounds. *Int. J. Mol. Sci.* **2022**, *23*, 321. [CrossRef]
16. Zakaria, F.; Ashari, S.E.; Azmi, I.D.M.; Rahman, M.B.A. Recent advances in encapsulation of drug delivery (active substance) in cubosomes for skin diseases. *J. Drug Deliv. Sci. Technol.* **2022**, *68*, 103097. [CrossRef]
17. Li, S.; Zhang, Y.-X. Sensitive delivery systems and novel encapsulation technologies for live biotherapeutic products and probiotics. *Crit. Rev. Microbiol.* **2023**, *50*, 371–384. [CrossRef]
18. Celen, T.; Anumudu, C.; Miri, T.; Onyeaka, H.; Fernandez-Trillo, P. Nisin:Carboxymethylcellulose polyion complex (PIC) nanoparticles. *Prep. Antimicrob. Activity. Carbohydr. Polym.* **2023**, *317*, 121032.
19. Pachaleva, J.; Gruskiene, R.; Bockuviene, A.; Sereikaite, J. The application of pectin and chitooligosaccharides for the preparation of nisin-loaded nanoparticles with different surface charges. *Cellulose* **2023**, *30*, 8985–8996. [CrossRef]
20. Flynn, J.; Culebras, M.; Collins, M.N.; Hudson, S.P. The impact of varying dextran oxidation levels on the inhibitory activity of a bacteriocin loaded injectable hydrogel. *Drug Deliv. Transl. Res.* **2023**, *13*, 308–319. [CrossRef] [PubMed]
21. Soares, R.S.; Santos, R.; Cunha, E.; Tavares, L.; Trindade, A.; Oliveira, M. Influence of Storage on the Antimicrobial and Cytotoxic Activities of a Nisin-biogel with Potential to be Applied to Diabetic Foot Infections Treatment. *Antibiotics* **2020**, *9*, 781. [CrossRef] [PubMed]
22. Cui, G.; Guo, X.; Su, P.; Zhang, T.; Guan, J.; Wang, C. Mussel-inspired nanoparticle composite hydrogels for hemostasis and wound healing. *Front. Chem.* **2023**, *11*, 1154788. [CrossRef] [PubMed]

23. Jing, Y.; Ruan, L.; Jiang, G.; Nie, L.; Shavandi, A.; Sun, Y.; Xu, J.; Shao, X.; Zhu, J. Regenerated silk fibroin and alginate composite hydrogel dressings loaded with curcumin nanoparticles for bacterial-infected wound closure. *Biomater. Adv.* **2023**, *149*, 213405. [CrossRef] [PubMed]
24. Rojas, M.A.; Amalraj, J.; Santos, L.S. Biopolymer-Based Composite Hydrogels Embedding Small Silver Nanoparticles for Advanced Antimicrobial Applications: Experimental and Theoretical Insights. *Polymers* **2023**, *15*, 3370. [CrossRef]
25. Saadh, M.J.; Hsu, C.-Y.; Mustafa, M.A.; Mutee, A.F.; Kaur, I.; Ghildiyal, P.; Ali, A.-J.A.; Adil, M.; Ali, M.S.; Alsaikhan, F.; et al. Advances in chitosan-based blends as potential drug delivery systems: A review. *Int. J. Biol. Macromol.* **2024**, *273*, 132916. [CrossRef]
26. Balogun-Agbaje, O.A.; Odeniyi, O.A.; Odeniyi, M.A. Drug delivery applications of poly-γ-glutamic acid. *Future J. Pharm. Sci.* **2021**, *7*, 125. [CrossRef]
27. Hu, Z.; Liu, D.; Wang, M.; Yu, C.; Han, Z.; Xu, M.; Yue, W.; Nie, G. β-Alanine enhancing the crosslink of chitosan/poly-(γ-glutamic acid) hydrogel for a potential alkaline-adapted wound dressing. *Int. J. Biol. Macromol.* **2023**, *231*, 123157. [CrossRef]
28. Liu, K.; Chen, Y.; Yang, Z.; Jin, J. Preparation and characterization of CS/γ-PGA/PC complex nanoparticles for insulin oral delivery. *Colloid Polym. Sci.* **2023**, *301*, 481–490. [CrossRef]
29. Chen, H.; Zhuang, J.; Wu, X.; Shen, X.; Zhang, Q.; Zhang, W. Preparation of the Chitosan/Poly-γ-Glutamic Acid/Glabrid in Hybrid Nanoparticles and Study on its Releasing Property. *Curr. Drug Deliv.* **2023**, *20*, 1195–1205. [CrossRef]
30. Serrano, I.; Alhinho, B.; Cunha, E.; Tavares, L.; Trindade, A.; Oliveira, M. Bacteriostatic and Antibiofilm Efficacy of a Nisin Z Solution against Co-Cultures of *Staphylococcus aureus* and *Pseudomonas aeruginosa* from Diabetic Foot Infections. *Life* **2023**, *13*, 504. [CrossRef]
31. Shin, J.M.; Gwak, J.W.; Kamarajan, P.; Fenno, J.C.; Rickard, A.H.; Kapila, Y.L. Biomedical applications of nisin. *J. Appl. Microbiol.* **2016**, *120*, 1449–1465. [CrossRef] [PubMed]
32. Oliveira, W.d.S.; Teixeira, C.R.V.; Mantovani, H.C.; Dolabella, S.S.; Jain, S.; Barbosa, A.A.T. Nisin variants: What makes them different and unique? *Peptides* **2024**, *177*, 171220.
33. Murat, D.; Hakki, T.I. Food stabilizing potential of nisin Z produced by wild *Lactococcus lactis* subsp. lactis from raw milk and some fermented products. *Lwt-Food Sci. Technol.* **2021**, *150*, 112065. [CrossRef]
34. Heo, W.-S.; Kim, E.-Y.; Kim, Y.-R.; Hossain, M.T.; Kong, I.-S. Salt effect of nisin Z isolated from a marine fish on the growth inhibition of *Streptococcus iniae*, a pathogen of streptococcosis. *Biotechnol. Lett.* **2012**, *34*, 315–320. [CrossRef]
35. Araujo, C.; Munoz-Atienza, E.; Perez-Sanchez, T.; Poeta, P.; Igrejas, G.; Hernandez, P.E.; Herranz, C.; Ruiz-Zarzuela, I.; Cintas, L.M. Nisin Z Production by *Lactococcus lactis* subsp *cremoris* WA2-67 of Aquatic Origin as a Defense Mechanism to Protect Rainbow Trout (*Oncorhynchus mykiss*, Walbaum) Against *Lactococcus garvieae*. *Mar. Biotechnol.* **2015**, *17*, 820–830. [CrossRef]
36. Araujo, M.K.; Gumiela, A.M.; Bordin, K.; Luciano, F.B.; Freitas de Macedo, R.E. Combination of garlic essential oil, allyl isothiocyanate, and nisin Z as bio-preservatives in fresh sausage. *Meat Sci.* **2018**, *143*, 177–183. [CrossRef]
37. Cohen-Erez, I.; Rapaport, H. Coassemblies of the Anionic Polypeptide γ-PGA and Cationic β-Sheet Peptides for Drug Delivery to Mitochondria. *Biomacromolecules* **2015**, *16*, 3827–3835. [CrossRef]
38. Zhang, L.; Chen, D.; Yu, D.; Regenstein, J.M.; Jiang, Q.; Dong, J.; Chen, W.; Xia, W. Modulating physicochemical, antimicrobial and release properties of chitosan/zein bilayer films with curcumin/nisin-loaded pectin nanoparticles. *Food Hydrocoll.* **2022**, *133*, 107955. [CrossRef]
39. Pandey, P.; Hansmann, U.H.E.; Wang, F. Altering the Solubility of the Antibiotic Candidate Nisin-A Computational Study. *Acs Omega* **2020**, *5*, 24854–24863. [CrossRef]
40. Luo, L.; Wu, Y.; Liu, C.; Huang, L.; Zou, Y.; Shen, Y.; Lin, Q. Designing soluble soybean polysaccharides-based nanoparticles to improve sustained antimicrobial activity of nisin. *Carbohydr. Polym.* **2019**, *225*, 115251. [CrossRef]
41. Dannert, C.; Stokke, B.T.; Dias, R.S. Nanoparticle-Hydrogel Composites: From Molecular Interactions to Macroscopic Behavior. *Polymers* **2019**, *11*, 275. [CrossRef] [PubMed]
42. Lin, L.; Luo, C.; Li, C.; Chen, X.; Cui, H. A Novel Biocompatible Ternary Nanoparticle with High Antibacterial Activity: Synthesis, Characterization, and Its Application in Beef Preservation. *Foods* **2022**, *11*, 438. [CrossRef] [PubMed]
43. Bhuiyan, M.H.; Clarkson, A.N.; Ali, M.A. Optimization of thermoresponsive chitosan/B-glycerophosphate hydrogels for injectable neural tissue engineering application. *Colloids Surf. B-Biointerfaces* **2023**, *224*, 113193. [CrossRef] [PubMed]
44. Tan, W.; Sijie, C.; Xu, Y.; Chen, M.; Liao, H.; Niu, C. Temperature-Sensitive Nanocarbon Hydrogel for Photothermal Therapy of Tumors. *Int. J. Nanomed.* **2023**, *18*, 6137–6151. [CrossRef]
45. Moradi, L.; Witek, L.; Nayak, V.V.; Pereira, A.C.; Kim, E.; Good, J.; Liu, C.-J. Injectable hydrogel for sustained delivery of progranulin derivative Atsttrin in treating diabetic fracture healing. *Biomaterials* **2023**, *301*, 122289. [CrossRef]
46. Dehghan-Baniani, D.; Chen, Y.; Wang, D.; Bagheri, R.; Solouk, A.; Wu, H. Injectable in situ forming kartogenin-loaded chitosan hydrogel with tunable rheological properties for cartilage tissue engineering. *Colloids Surf. B-Biointerfaces* **2020**, *192*, 111059. [CrossRef]
47. Liu, C.; Yang, P.; Li, J.; Cao, S.; Shi, J. NIR/pH-responsive chitosan hydrogels containing Ti_3C_2/AuNRs with NIR-triggered photothermal effect. *Carbohydr. Polym.* **2022**, *295*, 119853. [CrossRef]
48. Uysal, Y.; Dogaroglu, Z.G.; Caylali, Z.; Makas, M.N. Evaluation of swelling properties of different biochar-doped hydrogels. *J. Soils Sediments* **2023**, *23*, 3787–3805. [CrossRef]

49. Wang, H.; Lu, Z.; Wang, F.; Li, Y.; Ou, Z.; Jiang, J. A novel strategy to reinforce double network hydrogels with enhanced mechanical strength and swelling ratio by nano cement hydrates. *Polymer* **2023**, *269*, 125725. [CrossRef]
50. Zang, S.; Mu, R.; Chen, F.; Wei, X.; Zhu, L.; Han, B.; Yu, H.; Bi, B.; Chen, B.; Wang, Q.; et al. Injectable chitosan/β-glycerophosphate hydrogels with sustained release of BMP-7 and ornidazole in periodontal wound healing of class III furcation defects. *Mater. Sci. Eng. C-Mater. Biol. Appl.* **2019**, *99*, 919–928. [CrossRef]
51. Zhang, D.; Ouyang, Q.; Hu, Z.; Lu, S.; Quan, W.; Li, P.; Chen, Y.; Li, S. Catechol functionalized chitosan/active peptide microsphere hydrogel for skin wound healing. *Int. J. Biol. Macromol.* **2021**, *173*, 591–606. [CrossRef] [PubMed]
52. Hsu, X.-L.; Wu, L.-C.; Hsieh, J.-Y.; Huang, Y.-Y. Nanoparticle-Hydrogel Composite Drug Delivery System for Potential Ocular Applications. *Polymers* **2021**, *13*, 642. [CrossRef] [PubMed]
53. Yu, Y.-C.; Hu, M.-H.; Zhuang, H.-Z.; Phan, T.H.M.; Jiang, Y.-S.; Jan, J.-S. Antibacterial Gelatin Composite Hydrogels Comprised of In Situ Formed Zinc Oxide Nanoparticles. *Polymers* **2023**, *15*, 3978. [CrossRef] [PubMed]
54. Hu, J.; Liu, X.; Gao, Q.; Ouyang, C.; Zheng, K.; Shan, X. Thermosensitive PNIPAM-Based Hydrogel Crosslinked by Composite Nanoparticles as Rapid Wound-Healing Dressings. *Biomacromolecules* **2023**, *24*, 1345–1354. [CrossRef]
55. Yuan, Q.; Yin, J.; Li, L.; Bao, B.; Zhang, X.; Li, M.; Tang, Y. Conjugated Polymer Composite Nanoparticles Augmenting Photosynthesis-Based Light-Triggered Hydrogel Promotes Chronic Wound Healing. *Adv. Sci.* **2024**, *11*, 2304048. [CrossRef]
56. Lv, X.; Liu, Y.; Song, S.; Tong, C.; Shi, X.; Zhao, Y.; Zhang, J.; Hou, M. Influence of chitosan oligosaccharide on the gelling and wound healing properties of injectable hydrogels based on carboxymethyl chitosan/alginate polyelectrolyte complexes. *Carbohydr. Polym.* **2019**, *205*, 312–321. [CrossRef]
57. Pattanayaiying, R.; H-Kittikun, A.; Cutter, C.N. Effect of lauric arginate, nisin Z, and a combination against several food-related bacteria. *Int. J. Food Microbiol.* **2014**, *188*, 135–146. [CrossRef]
58. Yoon, J.-H.; Jeong, D.Y.; Lee, S.B.; Kim, S.R. Control of Listeria monocytogenes and *Escherichia coli* O157:H7 in enoki mushrooms (*Flammulina velutipes*) by combined treatments with organic acids, nisin, and ultrasound. *Food Control* **2021**, *129*, 108204. [CrossRef]
59. Costello, K.M.; Velliou, E.; Gutierrez-Merino, J.; Smet, C.; El Kadri, H.; Van Impe, J.F.; Bussemaker, M. The effect of ultrasound treatment in combination with nisin on the inactivation of *Listeria innocua* and *Escherichia coli*. *Ultrason. Sonochem.* **2021**, *79*, 105776. [CrossRef]
60. Liu, X.; Liao, W.; Xia, W. Recent advances in chitosan based bioactive materials for food preservation. *Food Hydrocoll.* **2023**, *140*, 108612. [CrossRef]
61. Huang, P.; Su, W.; Han, R.; Lin, H.; Yang, J.; Xu, L.; Ma, L. Physicochemical, Antibacterial Properties, and Compatibility of ZnO-NP/Chitosan/11-Glycerophosphate Composite Hydrogels. *J. Microbiol. Biotechnol.* **2022**, *32*, 522–530. [CrossRef] [PubMed]
62. Costa, E.M.; Silva, S.; Tavaria, F.K.; Pintado, M.M. Insights into chitosan antibiofilm activity against methicillin-resistant *Staphylococcus aureus*. *J. Appl. Microbiol.* **2017**, *122*, 1547–1557. [CrossRef] [PubMed]
63. Haidari, H.; Bright, R.; Strudwick, X.L.; Garg, S.; Vasilev, K.; Cowin, A.J.; Kopecki, Z. Multifunctional ultrasmall AgNP hydrogel accelerates healing of *S. aureus* infected wounds. *Acta Biomaterialia* **2021**, *128*, 420–434. [CrossRef]
64. Naranjo, M.F.; Kumar, A.; Ratrey, P.; Hudson, S.P. Pre-formulation of an additive combination of two antimicrobial agents, clofazimine and nisin A, to boost antimicrobial activity. *J. Mater. Chem. B* **2024**, *12*, 1558–1568. [CrossRef]
65. Lu, P.; Zhao, H.; Zheng, L.; Duan, Y.; Wu, M.; Yu, X.; Yang, Y. Nanocellulose/Nisin Hydrogel Microparticles as Sustained Antimicrobial Coatings for Paper Packaging. *Acs Appl. Polym. Mater.* **2022**, *4*, 2664–2673. [CrossRef]
66. Sullivan, D.J.; Cruz-Romero, M.; Collins, T.; Cummins, E.; Kerry, J.P.; Morris, M.A. Synthesis of monodisperse chitosan nanoparticles. *Food Hydrocoll.* **2018**, *83*, 355–364. [CrossRef]
67. Ji, S.; Sun, R.; Wang, W.; Xia, Q. Preparation, characterization, and evaluation of tamarind seed polysaccharide-carboxymethylcellulose buccal films loaded with soybean peptides-chitosan nanoparticles. *Food Hydrocoll.* **2023**, *141*, 108684. [CrossRef]
68. Savic Gajic, I.M.; Savic, I.M.; Gajic, D.G.; Dosic, A. Ultrasound-Assisted Extraction of Carotenoids from Orange Peel Using Olive Oil and Its Encapsulation in Ca-Alginate Beads. *Biomolecules* **2021**, *11*, 225. [CrossRef]
69. Tucker, L.J.; Grant, C.S.; Gautreaux, M.A.; Amarasekara, D.L.; Fitzkee, N.C.; Janorkar, A.V.; Varadarajan, A.; Kundu, S.; Priddy, L.B. Physicochemical and Antimicrobial Properties of Thermosensitive Chitosan Hydrogel Loaded with Fosfomycin. *Mar. Drugs* **2021**, *19*, 144. [CrossRef]
70. Li, Y.; Li, G.; Sha, X.; Li, L.; Zhang, K.; Liu, D.; Hao, Y.; Cui, X.; Wang, L.; Wang, H. An intelligent vancomycin release system for preventing surgical site infections of bone tissues. *Biomater. Sci.* **2020**, *8*, 3202–3211. [CrossRef]
71. Liu, Y.; Lang, C.; Ding, Y.; Sun, S.; Sun, G. Chitosan with enhanced deprotonation for accelerated thermosensitive gelation with β-glycerophosphate. *Eur. Polym. J.* **2023**, *196*, 112229. [CrossRef]
72. Khoobbakht, F.; Khorshidi, S.; Bahmanyar, F.; Hosseini, S.M.; Aminikhah, N.; Farhoodi, M.; Mirmoghtadaie, L. Modification of mechanical, rheological and structural properties of agar hydrogel using xanthan and locust bean gum. *Food Hydrocoll.* **2024**, *147*, 109411. [CrossRef]
73. Li, D.; Cui, H.; Hayat, K.; Zhang, X.; Ho, C.-T. Superior environmental stability of gelatin/CMC complex coacervated microcapsules via chitosan electrostatic modification. *Food Hydrocoll.* **2022**, *124*, 107341. [CrossRef]
74. Shu, X.; Liu, J.; Mao, L.; Yuan, F.; Gao, Y. Composite hydrogels filled with rhamnolipid-based nanoemulsion, nanostructured lipid carrier, or solid lipid nanoparticle: A comparative study on gel properties and the delivery of lutein. *Food Hydrocoll.* **2024**, *146*, 109264. [CrossRef]

75. Moradi, S.; Barati, A.; Tonelli, A.E.; Hamedi, H. Chitosan-based hydrogels loading with thyme oil cyclodextrin inclusion compounds: From preparation to characterization. *Eur. Polym. J.* **2020**, *122*, 109303. [CrossRef]
76. Zhang, A.; Fan, X.; Zeng, X.; Xu, J.; Zhou, C.; Xia, Q.; Sun, Y.; Wu, Z.; Pan, D. Enhancing physicochemical, antimicrobial, and release properties of fish skin gelatin films using dual-layer nanoparticles loaded with tea polyphenols/kojic acid for air-dried chicken preservation. *Food Hydrocoll.* **2024**, *149*, 109580. [CrossRef]
77. He, J.-R.; Zhu, J.-J.; Yin, S.-W.; Yang, X.-Q. Bioaccessibility and intracellular antioxidant activity of phloretin embodied by gliadin/sodium carboxymethyl cellulose nanoparticles. *Food Hydrocoll.* **2022**, *122*, 107076. [CrossRef]
78. Tian, B.; Cheng, J.; Zhang, T.; Liu, Y.; Chen, D. Multifunctional chitosan-based film loaded with hops β-acids: Preparation, characterization, controlled release and antibacterial mechanism. *Food Hydrocoll.* **2022**, *124*, 107337. [CrossRef]
79. Dong, Z.; Ma, F.; Wei, X.; Zhang, L.; Ding, Y.; Shi, L.; Chen, C.; Ma, Y.; Ma, Y. Injectable, thermo-sensitive and self-adhesive supramolecular hydrogels built from binary herbal small molecules towards reusable antibacterial coatings. *Rsc Adv.* **2024**, *14*, 2027–2035. [CrossRef]

marine drugs

MDPI

Article

High Degree of Polymerization of Chitin Oligosaccharides Produced from Shrimp Shell Waste by Enrichment Microbiota Using Two-Stage Temperature-Controlled Technique of Inducing Enzyme Production and Metagenomic Analysis of Microbiota Succession

Delong Pan, Peiyao Xiao, Fuyi Li, Jinze Liu, Tengfei Zhang, Xiuling Zhou * and Yang Zhang *

School of Life Science, Liaocheng University, Liaocheng 252059, China; 2210150302@stu.lcu.edu.cn (D.P.); m19861908982@163.com (P.X.); 15653129517@163.com (F.L.); 13176629777@163.com (J.L.); lcubiozy@163.com (T.Z.)
* Correspondence: xiulingzhou@lcu.edu.cn (X.Z.); zhy@lcu.edu.cn (Y.Z.)

Abstract: The direct enzymatic conversion of untreated waste shrimp and crab shells has been a key problem that plagues the large-scale utilization of chitin biological resources. The microorganisms in soil samples were enriched in two stages with powdered chitin (CP) and shrimp shell powder (SSP) as substrates. The enrichment microbiota XHQ10 with SSP degradation ability was obtained. The activities of chitinase and lytic polysaccharide monooxygenase of XHQ10 were 1.46 and 54.62 U/mL. Metagenomic analysis showed that *Chitinolyticbacter meiyuanensis*, *Chitiniphilus shinanonensis*, and *Chitinimonas koreensis*, with excellent chitin degradation performance, were highly enriched in XHQ10. Chitin oligosaccharides (CHOSs) are produced by XHQ10 through enzyme induction and two-stage temperature control technology, which contains CHOSs with a degree of polymerization (DP) more significant than ten and has excellent antioxidant activity. This work is the first study on the direct enzymatic preparation of CHOSs from SSP using enrichment microbiota, which provides a new path for the large-scale utilization of chitin bioresources.

Keywords: directional domestication; enrichment microbiota; metagenomics; high degree of polymerization; chitin oligosaccharides

1. Introduction

Chitin polymerized from N-acetylglucosamine (GlcNAc) is widely found in the exoskeletons of crustaceans (e.g., shrimps and crabs), insects, and the cell walls of fungi. Chitin is the most abundant nitrogen-containing natural biopolymer on earth [1]. Although about 10^{11} tons of chitin are produced globally each year, large accumulation in nature is not a problem, mainly due to the degradation of chitin by microorganisms. The formation and degradation of chitin play an important role in the carbon and nitrogen cycle of the earth. With the booming of the global fishing industry, about 6–8 million tons of crab, shrimp, and lobster shell waste are generated every year [2]. These seafood wastes have not been properly treated, which not only causes a waste of resources but also creates challenges related to environmental protection. Chitin and its deacetylated product, chitosan, extracted from shrimp and crab shell residues, are biocompatible, biodegradable, nontoxic, and antibacterial. Chitin and chitosan have been used in agriculture, medicine, food, cosmetics, and the textile industry [3,4]. However, the water insolubility of chitin and chitosan limits their wide application [5]. Chitin oligosaccharides (CHOSs) are a high value-added degradation product of chitin. They have antibacterial, anti-inflammatory, and immune protective activities, intestinal regulation activity, and other biological activities. In particular, their excellent solubility gives them a broad application prospect [6–8]. As a positively charged basic amino oligosaccharide, the efficient preparation and application of CHOSs can provide a green method for the high value-added utilization of seafood waste. Currently, CHOSs are obtained through two main methods: chitin degradation and biosynthesis [9,10]. Chitin degradation uses chitin as raw material to prepare CHOSs by physical, chemical, and biological degradation methods. Biodegradation can overcome the disadvantages of physical and chemical methods, such as non-environmental protection and poor quality. It is the trend for CHOS production and preparation in the future [11]. However, CHOSs have not yet been fully industrialized due to the poor activity of the single enzyme, low production efficiency, and the limitation of separation methods, thus hindering its biological activity research and application promotion [12].

So far, much experimental data on the enzymatic preparation of CHOSs has been accumulated [13]. The research results on the structure and mechanism of action of chitin-degrading enzymes significantly increased the possibility of controlling the depolymerization process of chitin. Theoretically, various forms of oligosaccharides can be prepared by selecting different substrates and hydrolases [14]. However, most of the existing preparation theories and production methods of CHOSs are based on chitin and its non-crystalline chitin as substrates, and most of the products obtained are low-molecular-weight CHOSs with a degree of polymerization (DP) below 10 [15]. How to realize the direct enzymatic utilization of crystalline chitin and even natural chitin materials, such as shrimp and crab shells, has become a hot issue in the green and efficient preparation of CHOSs.

Shrimp and crab shells can be completely degraded by chitin-degrading enzymes produced by microorganisms in the natural environment, such as chitin, chitosanase, and lytic polysaccharide monooxygenase (LPMO), through synergistic action [1]. Under the action of a specific multi-enzyme system, high-efficiency CHOS preparation from shrimp and crab shells can be realized. Currently, the cognition of the synergistic action mode of chitin-degrading enzymes mainly depends on the summary of the enzymatic properties and mechanism of action of a single enzyme. However, this type of study is relatively scarce. It has become a major bottleneck in resolving the synergistic mechanism of chitin-degrading enzymes due to the complexity of the multi-enzymatic degradation process of chitin and the lack of samples. Obtaining microbial communities that degrade biomass efficiently from nature has become a new idea and scheme to solve the problems of the degradation and biotransformation of complex substances [16,17]. Enrichment, domestication, and other technologies are effective methods to obtain microbial communities, which play an important role in the culture of extreme microorganisms and special microbiota [18–20]. With the rapid development of high-throughput sequencing technology and modern mass spectrometry (MS) technology, metagenome and metaproteome have been applied to the analysis of microbial community functions and mechanisms. Existing metagenomic studies have shown that although the diversity of microbial communities tends to decline during the enrichment process [21,22], it is conducive to the recovery and enrichment of functional strains, especially dormant cells [23]. Therefore, the combined application of enrichment, domestication and metagenomic technologies has great potential in improving the performance of functional microbiota and analyzing potential mechanisms.

In this study, the soil sample (named LNM) was enriched with chitin powder (CP) and shrimp shell powder (SSP) as substrates by using directional enrichment technology to obtain the acclimated bacteria with the ability to degrade SSP to produce CHOSs with a high DP. With the help of metagenomics technology, the community composition and succession rules of the original samples and the enrichment microbiota were systematically compared. The results provide new insights for the enrichment and domestication in improving the performance of functional microbiota and laid the foundation for studying the synergistic mechanism between chitin-degrading enzymes and the large-scale production of CHOSs.

2. Results and Discussions

2.1. Domestication of SSP-Degrading Microbiota

The previously established enrichment and acclimation technology was used to conduct a two-stage directional culture of LNM in soil samples [24]. After 20 experimental cycles of enrichment with CP as the substrate, a microbial community (LNM20) with high CP degradation ability was obtained. During the enrichment process, the activities of chitinase and LPMO gradually increased with the increase in enriched experimental cycles (Figure 1a), and the chitinase and LPMO enzyme activities of LNM20 were 0.76 and 34.98 U/mL, respectively. LNM20 was inoculated onto agar plates with CP as the substrate. After culturing, the CP around the colony dissolved and disappeared, forming a visible dissolution circle (Figure 1b), indicating that LNM20 has good CP degradation ability. LNM20 was further acclimated with SSP as the substrate to obtain a high-efficiency degradation bacteria group with SSP degradation performance. After ten experimental acclimation cycles, an enrichment microbiota (XHQ10) that can directly degrade SSP was obtained. The activities of chitinase and LPMO increased during acclimation, but the upward trend gradually slowed. The highest enzyme activities were 1.46 and 54.62 U/mL, which were 1.92 and 1.56 times that of LNM20. Under the condition of liquid fermentation, XHQ10 could degrade most of the powdery SSP in 3 days (Figure 1c,d). Therefore, XHQ10 has a good SSP degradation ability.

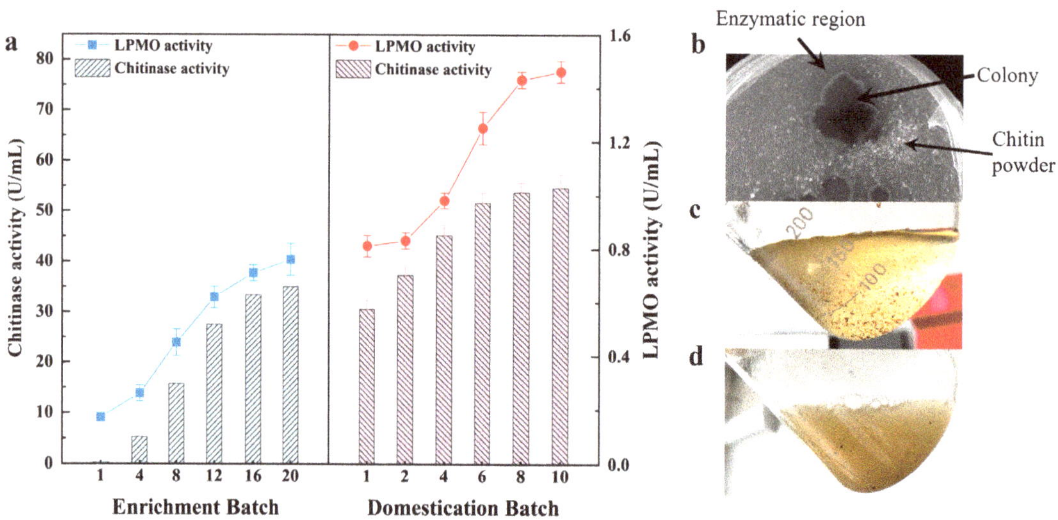

Figure 1. Enrichment effect and degradation performance of different experimental cycles of enrichment microbiota. (**a**) Chitin-degrading enzyme activity of enrichment microbiota in different experimental cycles. (**b**) Enzymatic region degrading CP and producing clear transparent circles on agar plates. (**c**,**d**) State of SSP in the fermentation broth before inoculation with XHQ10 and three days after fermentation.

2.2. Comparison of Chitin Degradation Enzyme Activities of Enrichment Microbiota

Shrimp and crab shells can be completely degraded by microorganisms in the natural environment, and this process is completed by chitin-degrading enzymes produced by microorganisms. The synergistic mechanism of chitin degradation by LPMO and chitinase has opened up a new strategy for the efficient production of CHOSs [25]. The activities of the chitin enzyme and LPMO were determined at different fermentation time points to compare the ability of LNM, LNM20, and XHQ10 to produce the chitin enzyme and LPMO. The results are shown in Figure 2a. The fermentation enzyme production processes of LNM, LNM20, and XHQ10 showed significant differences. The activities of chitinase and LPMO in the whole fermentation process of LNM were close to 0. The chitinase and LPMO activities of LNM20 and XHQ10 were low at the initial stage of fermentation (0–24 h), and then the activity of LPMO showed an obvious upward trend. In contrast, the chitinase activity increased at a greater rate at 60 h. When the fermentation time was 84 h, the two enzyme activities of LNM20 and XHQ10 reached the maximum value, and then the enzyme activities showed a slow downward trend. XHQ10 and LNM20 had the highest chitinase activity of 1.41 and 0.70 U/mL, respectively, and the highest LPMO activity of 55.58 and 34.31 U/mL, respectively. XHQ10 had a higher chitin-degrading enzyme activity than LNM20. The proteins contained in the supernatants of LNM20 and XHQ10 after 84 h of fermentation were analyzed by SDS-PAGE. Figure 2b shows that the proteins secreted by LNM20 and XHQ10 differed greatly in type and concentration, and the number of proteins secreted by XHQ10 that accounted for a higher content was significantly higher. Therefore, enrichment can effectively improve the activities of chitinase and LPMO, and the substrate transition from CP to SSP significantly changes the species and concentration of proteins secreted by the microbiota.

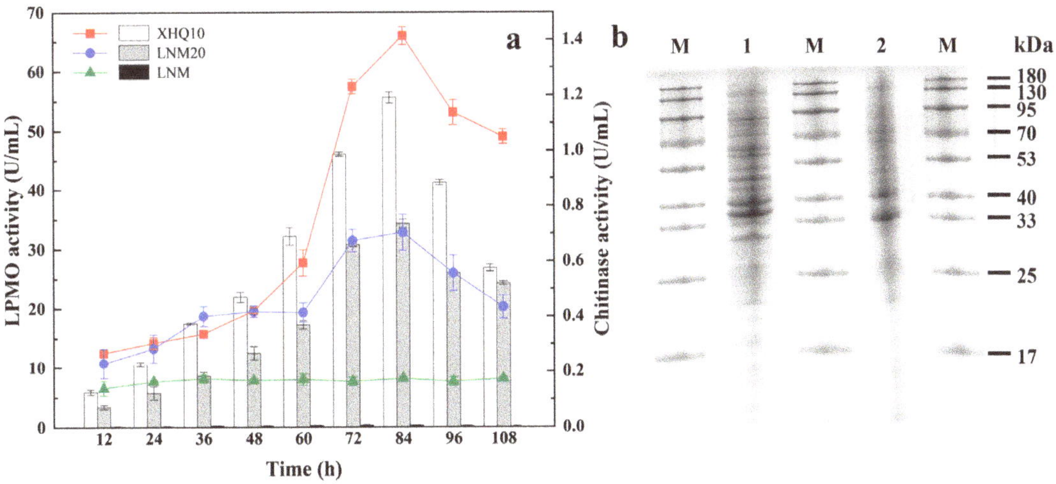

Figure 2. Enzyme production of different bacterial groups during domestication. (**a**) Changes in enzyme activity during different microbial fermentation processes. (**b**) SDS-PAGE analysis of LNM20 and XHQ10 fermentation supernatants. M, molecular mass markers; lane 1, protein contained in XHQ10 supernatant after 84h of fermentation; and lane 2, protein contained in LNM20 supernatant after 84h of fermentation.

2.3. Comparison of Fermentation Products of Enrichment Microbiota

The products obtained by fermentation of LNM20 and XHQ10 after 84 h were analyzed by HPLC, and the results are shown in Figure 3a. With reference to the peak times of DP1-6 in the CHOS standard, the LNM20 and XHQ10 products contained CHOSs at different DPs [26]. The LNM20 product contained DP1-DP6 CHOSs, and the XHQ10 product contained DP3-6 CHOSs. The LNM20 and XHQ10 products still had multiple response peaks after the DP6 standard, and the products were presumed to contain CHOSs with higher DPs. MALDI-TOF MS was used based on HPLC analysis to understand the difference between LNM20 and XHQ10 hydrolysates. As shown in Figure 3b,c, the main peak in the MALDI TOF MS spectrum in positive mode, was attributed to CHOSs with different DPs. The results proved that LNM20 and XHQ10 could degrade SSP effectively and produce CHOSs. The MALDI-TOF MS spectrum of LNM20 showed that CHOSs with a DP of 2–6, determined by HPLC, and CHOSs with a DP of 7–10 were captured. Similarly, CHOSs with DPs of 7–13 were detected in the MALDI-TOF MS spectra of XHQ10. The hydrolysate of XHQ10 contained more CHOSs with high DPs than that of LNM20.

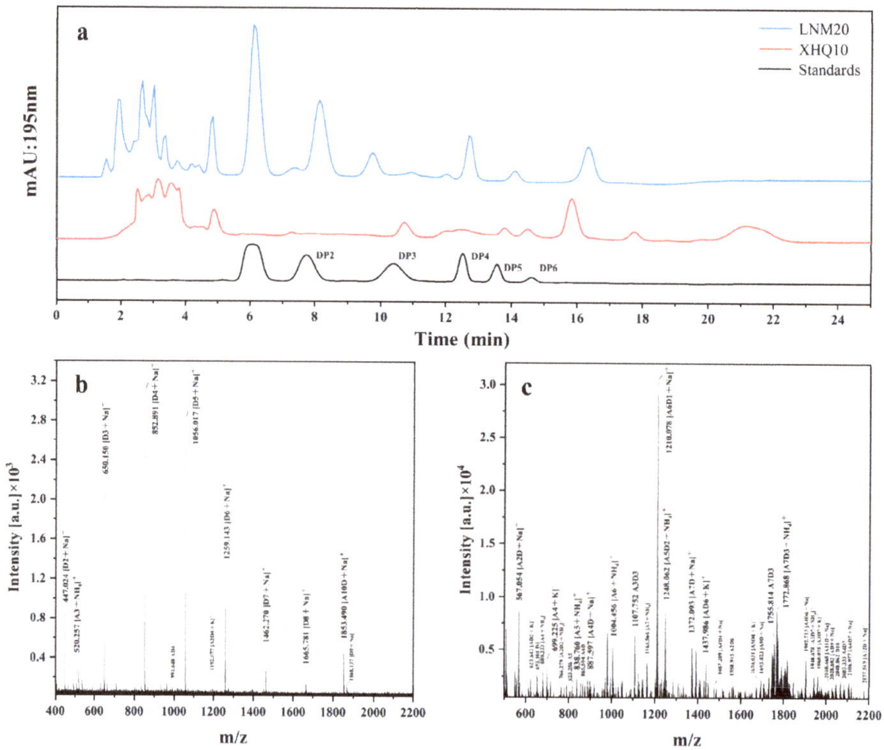

Figure 3. Hydrolysis of CP-, SSP-, and CHOS-generating capacity of LNM20 and XHQ10. (**a**) HPLC profiles of fermentation products of LNM20 and XHQ10. MALDI-TOF MS analysis of LNM20 (**b**) and XHQ10 (**c**) fermentation products.

Chitin degradation by chitinase (endochitinase and exochitinase) is an important way to produce CHOSs [27]. However, the DP of CHOSs obtained by chitin enzymes is mostly below 10 [2]. Studies have shown that the biological activity of CHOSs is closely related to their physical and chemical parameters, such as DP, acetylation degree, and charge distribution [28,29]. CHOSs with slightly higher DPs (DP 5–10) have higher activity in anti-tumor activity, superoxide anion scavenging activity, and immune activation than CHOSs with low DPs (DP 2–4) [13]. High DP CHOSs prepared by XHQ10 are of great significance in promoting CHOS application research.

2.4. *Community Composition and Succession Analysis during Acclimation*

2.4.1. Dataset Overview

Three parallel metagenomic sequences were performed for each of LNM, LNM20, and XHQ10 to reveal the changes in the composition of microbial communities during the enrichment and acclimation of chitin-degrading bacteria. A total of 74,926,897 (22.48 GB) RawReads were obtained for three samples of LNM, with an average of 6.10 GB RawReads per sample; a total of 70,840,368 (21.25 GB) RawReads were obtained for three samples of LNM20, with an average of 7.08 GB RawReads per sample; and XHQ10 received a total of 61,019,577 (18.31 GB) RawReads for three samples, with an average of 7.49 GB RawReads per sample. After low-quality data were removed and host contamination was filtered out, 71,610,263, 67,902,878, and 58,569,366 clean reads were obtained for samples of LNM, LNM20, and XHQ10, respectively, (Table 1).

Table 1. Overview of the quality of metagenomic sequencing data.

Sample ID	Grouping	RawReads	Raw Base (GB)	Clean Reads	Cleaned (%)
LNM	SS11	19784882	5.94	18925514	95.66
	SS12	26947750	8.08	25736535	95.51
	SS13	28194265	8.46	26948214	95.58
LNM20	FS21	22424308	6.73	21512100	95.93
	FS22	26312463	7.89	25221987	95.86
	FS23	22103597	6.63	21168791	95.77
XHQ10	FS11	20613776	6.18	19727621	95.7
	FS12	20716845	6.22	19935093	96.23
	FS13	19688956	5.91	18906652	96.03

2.4.2. Community Composition and Taxonomic Changes

The similarity and uniqueness of species composition among sample groups can be intuitively expressed by comparing the number of common and unique species among LNM, LNM20, and XHQ10. As shown in Figure 4a, 68 species were common in the domestication process. In addition, 279 of the 555 species obtained from the 2811 species included in LNM after 20 enrichments were absent from the species composition before domestication. Among the 208 species obtained by XHQ10, 45 species reappeared after the disappearance of LNM20, and 63 species were specific to XHQ10 (Supplementary Table S1). The results showed that enrichment and domestication significantly changed the number and species composition of OTUs. The microbial community structure of three sample phylum and genus levels were analyzed. At the phylum level, Pseudomonadota and Actinomycetota were the most abundant taxonomic groups in LNM, accounting for 33.78% and 26.34%, respectively, whereas LNM20 and XHQ10 were dominated by Pseudomonadota, accounting for 98.51% and 99.95%, respectively (Figure 4b,c, Table S2). At the genus level, *Chitinolyticbacter*, *Laribacter*, *Sinorhizobium*, *Chitiniphilus*, *Streptomyces*, *Herbaspirillum*, *Chitinimonas*, *Microvirga*, and *Stenotrophomonas* were the dominant types in the three samples (Figure 4b,d, Table S2). Significant differences between the samples at all levels are shown in Figure 4b–d. These results suggest significant changes in the community during enrichment and acclimation.

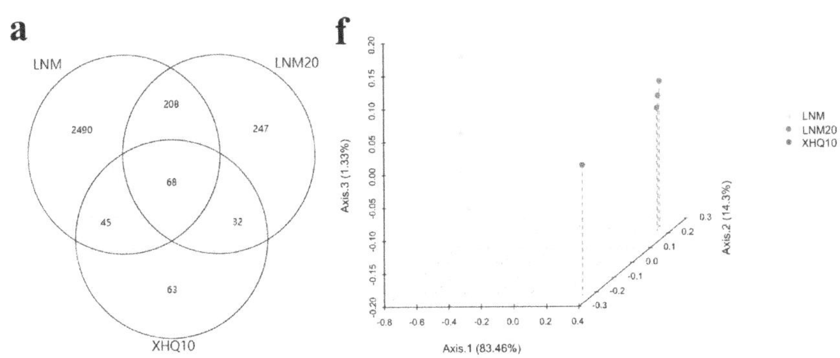

Figure 4. *Cont.*

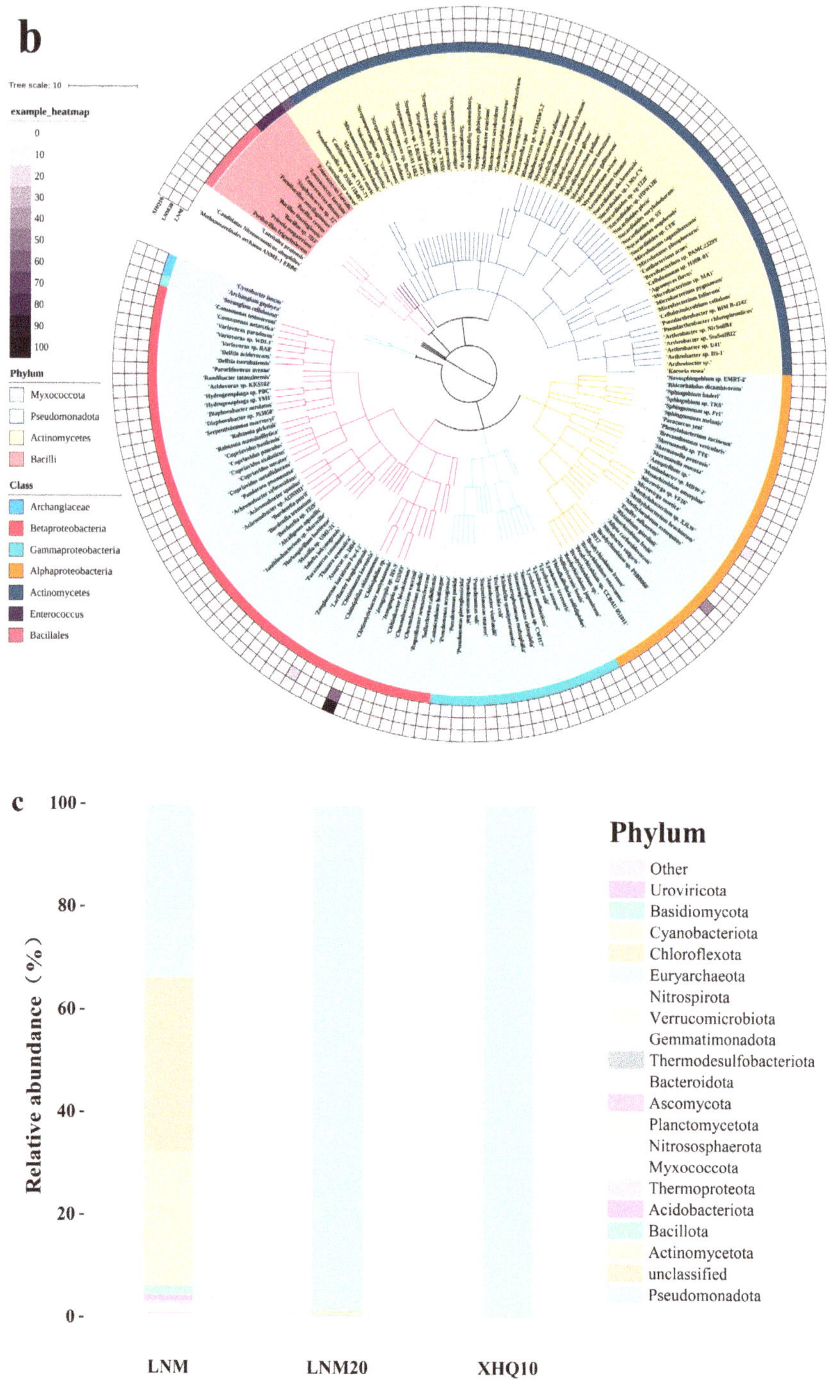

Figure 4. *Cont.*

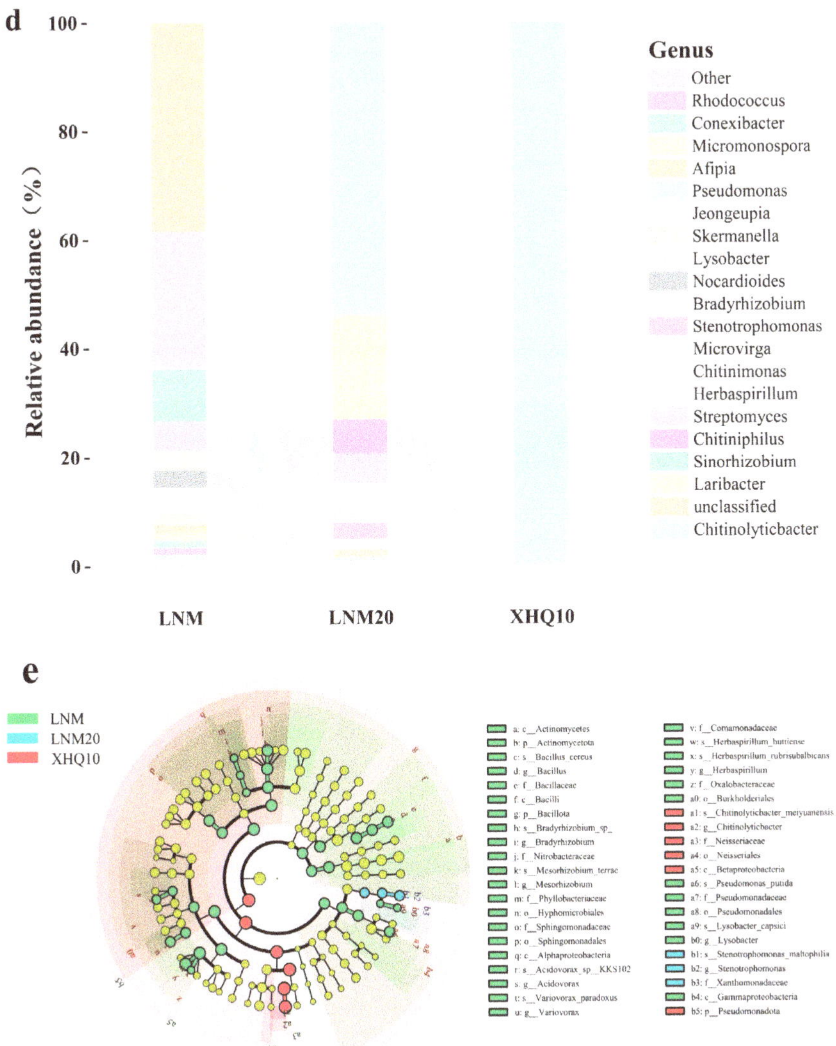

Figure 4. Metagenomic analysis of the microbiota during domestication. (**a**) Venn diagram of OTUs for three samples. (**b**) Three sample species-level phylogenetic trees. The color of the heatmap represents the number of strains contained by each species in the three samples. (**c**) Relative abundance at the phylum level. (**d**) Relative abundance at the genus level. (**e**) Dynamic branching diagram of microbial phylogeny of three samples. Nodes with different colors represent microorganisms that play a key role in the grouping indicated by the color, and yellow nodes indicate that they are not significant. Microbial markers with LDA score ≥ 4 in the three samples are listed. Volcano plot of the differences between LNM and LNM20. (**f**) Between-group analysis of PCoA based on the Bray-Curtis distance. Different colors represent different groups.

2.4.3. Excavation of Key Microbiota

The species level and abundance of each macrogenomic sample were analyzed using the linear discriminant analysis effect size method [30] to identify further the microbiota in which the domestication process plays a role. The branching diagram contains six levels from phylum to species, revealing iconic microbiota, with three bacterial taxa enriched in LNM20, six bacterial taxa enriched in XHQ10, and thirty-three bacterial taxa enriched in LNM (Figure 4e). The specifications *Stenotropionas maltohilia*, genus *Stenotropionas*, and family Xanthomonadaceae may play important roles in LNM20, whereas the specifications *Chitolyticbacter meiyuanensis*, genus chitolytic bacteria, family Neisseriaceae, order Neisseriales, class Betaprotecteria, and phylum Pseudomonas may play important roles in XHQ10. A species-level volcano plot (Figures S1 and S2) showed that compared with LNM, *Laribacter hongkongensis*, *Chitinolyticbacter meiyuanensis*, *Chitiniphilus shinanonensis*, *Chitinimonas koreensis*, and *Jeongeupia* sp. USM3 were significantly upregulated in LNM20. The top three bacterial groups significantly upregulated in XHQ10 were *Chitinolyticbacter meiyuanensis*, *Chitiniphilus shinanonensis*, and *Chitinimonas koreensis*. *Stenotrophomonas maltophilia* contains nine genes related to chitin turnover [31]. The *Jeongeupia* species has a complete chitin degradation system and shows good chitin degradation performance [32,33]. Although no detailed report on chitin degradation by *Jeongeupia* sp. USM3 is available, 11 potential chitinase genomes were found in its genes [34]. *Laribacter hongkongensis* has chitinase and chitin deacetylase activities, but its chitin degradation ability is relatively weak [35]. *Chitinolyticbacterium meiyuanensis*, *Chitiniphilus shinanonensis*, and *Chitinimonas koreensis* have good chitin degradation ability [36–38]. In summary, the proportion of genera related to chitin degradation gradually increased during the enrichment process, with an increase in the enrichment of chitin-degrading bacteria in LNM20 and ultimately, the species in XHQ10 were dominated by chitin-degrading bacteria.

2.4.4. Analysis of Microbiota Diversity

PCoA analysis was performed based on Bray–Curtis distance to determine the diversity and community structure differences of the whole dataset [39]. Figure 4f shows that three parallel datasets from three samples were clustered, with some overlap and dispersion within each cluster. LNM showed considerable dispersion, LNM20 was relatively concentrated, and XHQ10 completely overlapped. The differences in microbial diversity among LNM, LNM20, and XHQ10 samples were revealed. The LNM20 had low abundance and diversity, and the XHQ10 had even lower abundance and diversity.

2.4.5. Functional Annotation Analysis

During acclimation, the microbiota can encode various complex Carbohydrate-Active enZYmes (CAZymes), and these CAZymes contain many enzymes involved in chitin degradation. The relative abundance changes of Glycoside Hydrolase (GHS), Glycosyl-Transferases (GTs), Polysaccharide Lyases (PLs), Carbohydrate Esterases (CEs), Auxiliary Activities (AAs), and Carbohydrate-Binding Modules (CBMs) in the three samples are summarized in Figure 5a to understand the differences at the gene level in the domestication process. In previous studies, GH, CBM, and AA have been identified as the main CAZymes involved in chitin degradation [40,41]. The GH family is the main component of CAZyme involved in chitin degradation. They can directly hydrolyze the glycosidic bonds of carbohydrate CBMs, which are not directly involved in catalysis, but they help relax the crystal structure of the substrate, improving the overall catalytic efficiency of chitin degradation [40]. The AA family is a newly introduced CAZyme class in recent years, and the multi-enzyme synergism with GHs has an important potential for application in biorefineries. It has become a new direction for developing chitin resources [25]. The relative abundance of GHs was maintained at a high level in the three samples, and no obvious change was found in the relative abundance during domestication (Figure 5a and Supplementary Table S3). The relative abundance of CBMs significantly increased at LNM20 and further increased at XHQ10, indicating that the domestication process increased the relative abundance of CBMs (Figure 5a and Table S3). A notable detail is that the relative abundance of AAs in XHQ10 increased to some extent compared with that of LNM and LNM20 (Figure 5a and Table S3).

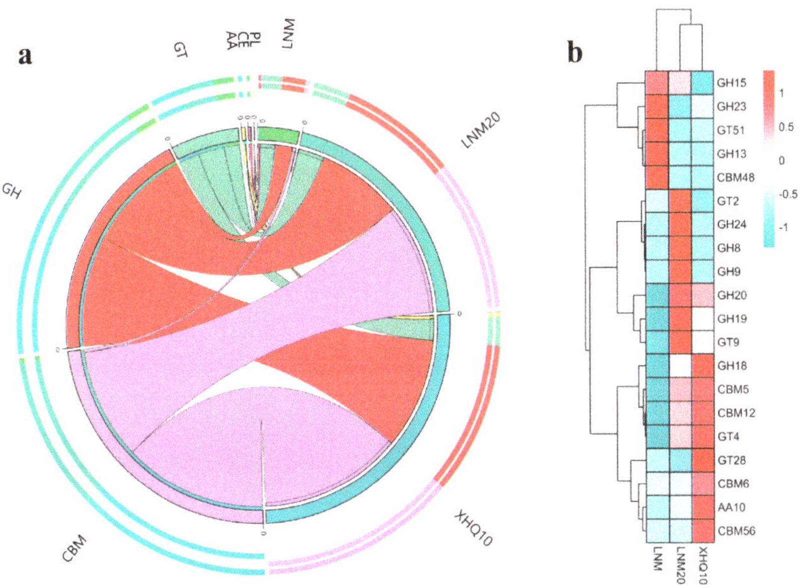

Figure 5. Analysis of CAZyme composition. (**a**) Dynamic changes in CAZyme class-level composition of three samples. (**b**) Heatmap of the top 20 CAZymes' relative abundance at the family-level of three samples.

Figure 5b shows the CAZyme heatmap of the top 20 relative abundances. Significant differences were observed in the CAZyme family with the change in the enrichment process. Among these CAZyme families, the GH18 family has chitinase, exo-β-1,4-n-acetylglucosaminidase, and β-n-acetylhexosaminidase activities; the GH19 family has chitinase and exochitinase activities; and the GH20 family has β-1,6-n-acetylglucosaminidase and β-n-acetylhexosaminidase activities. The chitin-degrading enzymes of the three GH families, especially the GH18 family of chitinases, play an important role in degrading chitin [42]. Its relative abundance increased significantly during acclimation (Figure 5b and Table S4). CBM5 and CBM12 are distantly related, and both have chitin-binding functions that promote the hydrolysis of chitin, whose relative abundance increased significantly during acclimation (Figure 5b and Table S4). The AA10 family has lytic chitin monooxygenase activity, which plays an important role in synergizing other chitin-degrading enzymes to crystallize chitin [25]. AA10s were significantly enriched in LNM20 and more abundant in XHQ10.

In general, the enrichment of the CAZymes family related to chitin degradation indicates that the hydrolysis ability is improved, which also shows that the domestication enrichment effect is significant. *Chitinolyticbacter meiyuanensis*, *Chitiniphilus shinanonensis*, and *Chitinimonas koreensis* are the main contributors of CAZymes in chitin degradation. Therefore, changes in microbial communities during domestication cause changes in the enzyme family, further improving the chitin degradation ability of enrichment microbiota.

2.5. Optimization of Enzyme Production and Degradation Conditions of SSP by XHQ10 and Characterization of Products

2.5.1. Induction of Enzyme Production and Construction of Two-Stage Temperature Control Technology

Except for a few strains, chitinase expression is constitutive and inducible, and most microbial chitinases belong to inducible enzymes. The expression of the microbial chitinase gene significantly increased when it was induced by chitin and its degradation products (such as CHOSs, Nacetylglucosamine, and D-glucosamine) [43,44]. XHQ10 fermentation broth contains chitin and its degradation products, so XHQ10 (XFP) fermentation products were tried as inducers. After XHQ10 was cultured for 12 h, CP, CC, SSP, and XFP (calculated as CHOSs) were added at a final concentration of 1g/L to evaluate their effect on the induction of the enzyme production of XHQ10. By evaluating the induction results, CC and XFP were selected as inducers of XHQ10. The two-stage temperature-controlled fermentation process curve of XHQ10 with XFP and CC as inducers is shown in Figure 6a. The chitinase activity was lower in the early stage of XFP induction than in the control, but the chitinase activity increased significantly after 24 h. The chitinase activity of XFP as an inducer reached a maximum value of 1.61 U/mL at 60 h, which was 1.34 and 2.78 times higher than that of CC as an inducer and control, respectively. After the temperature was adjusted to 50 °C, the chitinase activity of XFP as an inducer showed a slight downward trend. Although the chitinase activity induced by control and CC as inducer still increased, it was consistently lower than that of XFP as an inducer.

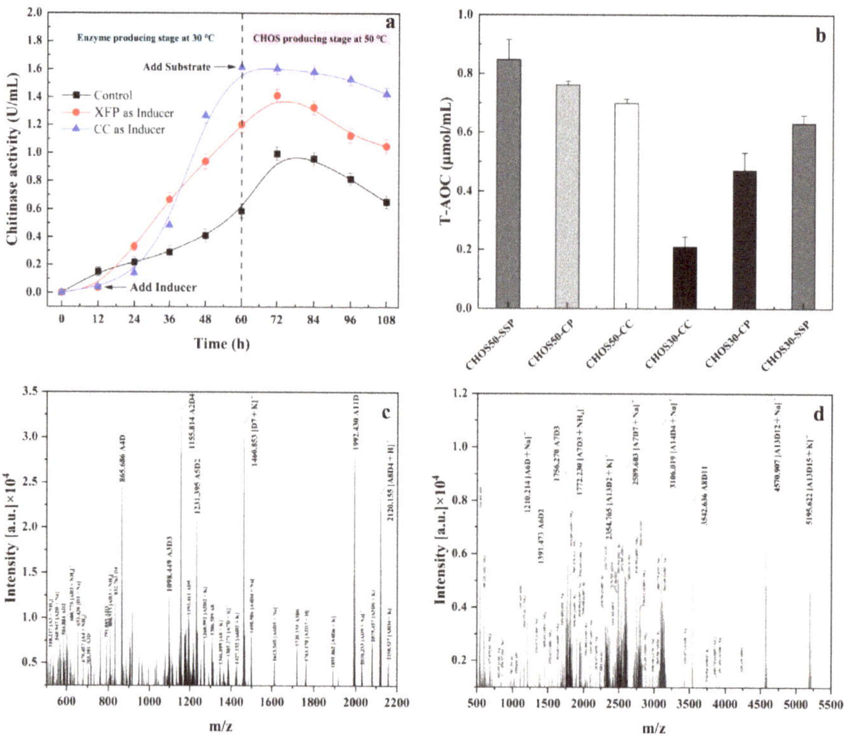

Figure 6. Changes in chitinase enzyme activity under enzyme-producing induction and two-stage temperature control techniques. (**a**) Changes in enzyme activity of different substrates under a two-step mechanism. (**b**) Antioxidant capacity. (**c**) CHOS30-SSP MALDI-TOF MS analysis. (**d**) CHOS50-SSP MALDI-TOF MS analysis.

2.5.2. Antioxidant Activity of XHQ10 Hydrolysates

CP, CC, and SSP were selected as substrates for enzymatic hydrolysis at 30 °C and 50 °C to investigate the degradation effect of XFP-induced XHQ10 on different substrates at different temperatures. The lyophilized hydrolysates of the three substrates were named CHOS30-CP, CHOS30-CC, and CHOS30-SSP at 30 °C, and the lyophilized hydrolysates at 50 °C were named CHOS50-CP, CHOS50-CC, and CHOS50-SSP. Reactive oxygen species (ROS) are oxygen-active molecules formed during metabolism. Excess ROS and its oxidative stress can cause damage to the body, inducing cancer, cardiovascular disease, and inflammatory diseases [45]. Antioxidants can delay or block the oxidation process to reduce the damage caused by oxidative stress. Existing studies have shown that CHOSs have a certain antioxidant activity [8,46,47]. Therefore, antioxidant capacity is an important indicator of CHOSs. A total antioxidant capacity (T-AOC) assay was performed on lyophilized enzymatic hydrolysates obtained at 30 °C and 50 °C. As shown in Figure 6b, all enzymatic hydrolysates had a certain antioxidant capacity. The order of T-AOC was CHOS50-SSP > CHOS50-CP > CHOS50-CC. The changing trend of T-AOC of the hydrolysates obtained by the enzymatic hydrolysis of the three substrates at 30 °C was the same as that obtained at 50 °C, but the T-AOC of the three hydrolysates was lower than that of CHOS50-CC.

2.5.3. MALDI-TOF/TOF MS Analysis of XHQ10 Enzymatic Hydrolysates

The CHOS50-SSP obtained after 8 h of enzymatic hydrolysis at 50 °C was analyzed by MALDI-TOF/TOF MS and compared with CHOS30-SSP. Figure 6b,c show that CHOS30-SSP and CHOS50-SSP were composed of multiple types of CHOSs. The CHOS30-SSP had a higher percentage of CHOSs, with DPs of 5, 6, 7, and 12, whereas the CHOSs comprising CHOS50-SSP had a higher degree of aggregation, and most of these were CHOSs with DPs > 10. The highest degree of aggregation that could be detected was 28, which is a rarity in studies of enzymatically dissociated gibberellic acid materials.

XFP was selected as the inducer of XHQ10, and a two-stage process technology was constructed, which induced enzyme production culture at 30 °C and the enzymatic hydrolysis of SSP at 50 °C. This technology can effectively improve the enzyme production level of XHQ10 and effectively degrade SSP to obtain CHOSs with good antioxidant activity and a high DP, which has significant potential in the high-value utilization of discarded shrimp and crab shells.

3. Materials and Methods

3.1. Chemical Reagents and Enrichment Microbiota

CP was purchased from Shanghai Yuanye Biotechnology Co., Ltd. (Shanghai, China). Shrimp (*Litopenaeus vannamei*) was purchased from local seafood markets. Soil samples were collected in Liaocheng, Shandong, China (36°43′36″ N, 116°01′67″ E). The shells were obtained, thoroughly washed with running water to remove residual shrimp meat and dried in a 60 °C vacuum oven for 24 h. The dried material was crushed in a pulverizer, and then SSP was prepared through a 40-mesh sieve. Colloidal chitin (CC) was prepared using the previous method [48]. Briefly, dissolve 10 g chitin powder in 200 mL pre-cooled concentrated HCl, stir at 4 °C for 2 h, add 2.5 L pre-cooled deionized water, and continue stirring for 10 min. The suspension was centrifuged at 10,000 rpm at 4 °C for 10 min, and the precipitate was repeatedly washed with deionized water to neutral pH and stored at 4 °C for future use. The remaining reagents were analytical grade reagents, and they can be used without additional purification. In accordance with previous research methods [24], 20 experimental cycles of microbiota (LNM20) were enriched and cultured on CP as the initial enrichment microbiota, and LNM20 was acclimated by SSP. The acclimation medium consisted of 2 g/L peptone, 1 g/L glucose, 0.7 g/L K_2HPO_4, 0.3 g/L KH_2PO_4, 0.5 g/L $MgSO_4$, and 4 g/L SSP. After ten experimental cycles of enrichment, the XHQ10 with the ability to degrade SSP was obtained. Soil samples (LNM), LNM20, and XHQ10 were selected for comparison with the fermentation experiment and metagenomic sequencing analysis.

3.2. Comparison of Fermentation Performance of LNM, LNM20, and XHQ10

LNM, LNM20, and XHQ10 were individually fermented for 108 h. The fermentation medium consisted of 1 g/L glucose, 2 g/L peptone, 2 g/L yeast extract, 0.7 g/L K_2HPO_4, 0.3 g/L KH_2PO_4, 0.5 g/L $MgSO_4$, and 4 g/L SSP. The culture conditions were as follows: a 250 mL Erlenmeyer flask was filled with 50 mL of fermentation medium, the inoculation volume was 4% (*v*/*v*), the temperature was 30 °C, the pH was 7.0, and the rotating speed was 200 rpm. The fermentation experiment was set up three times in parallel. During the culture process, 1 mL of fermentation broth was sampled every 12 h to determine chitinase and LPMO activities, and the products were analyzed.

3.3. Chitin Enzyme and LPMO Activity Analysis

Fixed-point fermentation broth (1 mL) was centrifuged at 4 °C for 10 min at 12,000× *g*, and the supernatant was used for chitin enzyme and LPMO activity assays. Using the method described by Rojas-Avelizapa [49], the chitinase activity was determined by the DNS method under the condition of the chitinase enzymatic hydrolysis of colloidal chitin to release N-acetyl-D-glucosamine (GlcNAc). The absorbance was recorded at 540 nm. The readings were compared with the GlcNAc standard curve. The linear regression equation for the GlcNAc standard curve was $Y = 1.2075X - 0.0887$ ($R^2 = 0.9994$). The unit chitin enzyme activity (U) was defined as the amount of enzyme required to produce 1 mmol of GlcNAc during 1 min at 37 °C. The activity of LPMO was determined in accordance with a previous method [24]. The secreted proteins of LNM20 and XHQ10 after 84 h of fermentation were analyzed by SDS-PAGE with 5% stacked gel and 12.5% resolved gel in Tris-glycine buffer (pH 8.3).

3.4. Analysis of Fermentation and Enzymatic Hydrolysis Products

When sampling at a specific time or after reaching the enzyme reaction time, an equal amount of acetonitrile was added to the fermentation broth or enzyme reaction system to terminate the fermentation or enzyme reaction and centrifuged at 4 °C at 12,000× *g* for 10 min through membrane treatment. Samples were separated on a Prevail Carbohydrate ES column (4.6 × 250 mm) at a column temperature of 30 °C, a flow rate of 1.0 mL/min, and a detection wavelength of 195 nm. The mobile phase was acetonitrile and water, and the detection method was as follows: 0 min, acetonitrile: water (75:25, *v*/*v*); 7 min, acetonitrile: water (75:25, *v*/*v*); 8 min, acetonitrile: water (65:35, *v*/*v*); 15 min, acetonitrile: water (65:35, *v*/*v*); 16 min, acetonitrile: water (75:25, *v*/*v*); and 22 min, acetonitrile: water (75:25, *v*/*v*). The detection time was 30 min.

An appropriate amount of freeze-dried enzymatic hydrolysate was obtained, dissolved in ultrapure water with 2,5-DHB as a substrate, and analyzed using UltrafleXtreme MALDI-TOF/TOF (Bruker Daltonics GmbH, Bremen, Germany) forward collection mode. The collection adopted the reflective type, the acceleration voltage was 20 kV, the reflective layer voltage was 21.1 kV, and the collection range was m/z 400–6000.

3.5. Metagenomic Sequencing

The total DNA of LNM, LNM20, and XHQ10 was extracted using the E.Z.N.A. Soil DNA Kit (Omega Bio-Tek, Norcross, GA, USA). The metagenomic library was prepared and sequenced by Shenzhen Microecology Technology Co., Ltd. (Shenzhen, China) using the HiSeq2000 platform (Illumina PE150, San Diego, CA, USA).

3.6. Bioinformatics and Data Analysis

Kraken2 was used to compare with the self-built microbial nucleic acid database (screening the sequences belonging to bacteria, fungi, archaea, and viruses in the NCBI NT nucleic acid database and RefSeq whole genome database) to calculate the sequence number of species contained in the sample. Bracken was then used to estimate the actual abundance of species in the sample. Read-based metagenomic species annotation is more comprehensive and accurate than assembly-based species annotation. Starting from quality control and removing the reads of host genes, the reads of each sample were compared to the database (UniRef90) by using HUMAnN 3 3.6 software (based on DIAMOND, and the annotation information and relative abundance table of the CAZY database were obtained in accordance with the corresponding relationship between UniRef90 ID and the CAZY database.

3.7. Two-Stage Temperature-Controlled Degradation of SSP by XHQ10 to Prepare CHOSs

XHQ10 was cultured in a culture medium without SSP at 30 °C and 200 rpm. After 12 h, CC and XFP were added and continued to be induced to culture for 60 h. After the temperature was increased to 50 °C, SSP with a final concentration of 8 g/L was added. After eight hours of incubation, it was centrifuged at 4 °C at 12,000× g for 10 min. The supernatant was lyophilized, and the CHOS product was obtained.

3.8. Analysis of the Antioxidant Activity of CHOSs

The modified FRAP method [50] was used to detect the total antioxidant capacity (T-AOC) of different CHOS samples. The specific operation and analysis steps were carried out in accordance with the specification of the T-AOC Assay Kit (BIOBOX, AKAO012C, Beijing, China).

4. Conclusions

In this study, LNM from soil samples was enrichment cultured with CP as a substrate to obtain chitin-degrading microbiota LNM20, and LNM20 was acclimated with SSP to obtain the enrichment microbiota XHQ10 with high-efficiency SSP degradation. The community composition and succession rules of the original samples and enrichment microbiota were systematically compared using metagenomics technology. The high antioxidant activity of CHOSs was obtained by optimizing the conditions of XHQ10 enzymatic hydrolysis and constructing the enzyme induction and two-stage temperature control technology. The results provide new insights for enrichment and acclimation in improving the performance of functional microbiota. They lay the foundation for studying the synergistic mechanism between degrading enzymes in chitin and the large-scale production of CHOSs.

Supplementary Materials: The following supporting information can be downloaded at: https://www.mdpi.com/article/10.3390/md22080346/s1. Volcano plot of the differences between LNM and LNM20 (Figure S1) and LNM and XHQ10 (Figure S2) at the species level. The X coordinate was |log2 (fold change)| and the Y coordinate was −log 10 (P adj), $p < 0.05$, log2 (fold change) >2. Each point represented a species. Table S1: All OTU classification information table; Table S2: Table of relative abundance of each sample at each taxonomic level; Table S3: Cazy database class level abundance table; Table S4: Cazy database abundance table.

Author Contributions: Conceptualization, X.Z. and Y.Z.; methodology, D.P. and Y.Z.; software, D.P. and Y.Z.; validation, P.X. and F.L.; formal analysis, D.P. and P.X.; investigation, D.P., P.X., F.L., J.L., T.Z. and X.Z.; resources, X.Z. and Y.Z.; data curation, D.P.; writing—original draft preparation, D.P., P.X. and Y.Z.; writing—review and editing, X.Z. and Y.Z.; visualization, D.P. and Y.Z.; supervision, X.Z.; project administration, Y.Z.; funding acquisition, Y.Z. and X.Z. All authors have read and agreed to the published version of the manuscript.

Funding: This research was funded by the National Natural Science Foundation of China (32100065 and 32300031), Shandong Province Youth Entrepreneurship Technology Support Program for Higher Education Institutions (2023KJ207), and the Natural Science Foundation of Shandong Province of China (ZR2023MB095).

Institutional Review Board Statement: Not applicable.

Data Availability Statement: The data from the present study are available in the article; further inquiries can be directed to the corresponding author.

Conflicts of Interest: The authors declare no conflicts of interest.

References

1. Jiang, W.-X.; Li, P.-Y.; Chen, X.-L.; Zhang, Y.-S.; Wang, J.-P.; Wang, Y.-J.; Sheng, Q.; Sun, Z.-Z.; Qin, Q.-L.; Ren, X.-B.; et al. A Pathway for Chitin Oxidation in Marine Bacteria. *Nat. Commun.* **2022**, *13*, 5899. [CrossRef]
2. Zhang, R.; Zhao, Q.; Yi, Z.; Zhang, K.; Shi, J.; Zhu, L.; Chen, Y.; Jin, J.; Zhao, L. Chitin Oligosaccharides for the Food Industry: Production and Applications. *Syst. Microbiol. Biomanuf.* **2022**, *3*, 49–74. [CrossRef]

3. Roy, V.C.; Islam, M.R.; Sadia, S.; Yeasmin, M.; Park, J.-S.; Lee, H.-J.; Chun, B.-S. Trash to Treasure: An Up-to-Date Understanding of the Valorization of Seafood By-Products, Targeting the Major Bioactive Compounds. *Mar. Drugs* **2023**, *21*, 485. [CrossRef]
4. Yadav, M.; Goswami, P.; Paritosh, K.; Kumar, M.; Pareek, N.; Vivekanand, V. Seafood Waste: A Source for Preparation of Commercially Employable Chitin/Chitosan Materials. *Bioresour. Bioprocess.* **2019**, *6*, 8. [CrossRef]
5. Aranaz, I.; Alcántara, A.R.; Civera, M.C.; Arias, C.; Elorza, B.; Heras Caballero, A.; Acosta, N. Chitosan: An Overview of Its Properties and Applications. *Polymers* **2021**, *13*, 3256. [CrossRef]
6. Sun, Y.; Cui, J.; Tian, L.; Mi, Y.; Guo, Z. Phenolic Acid Functional Quaternized Chitooligosaccharide Derivatives: Preparation, Characterization, Antioxidant, Antibacterial, and Antifungal Activity. *Mar. Drugs* **2023**, *21*, 535. [CrossRef]
7. Benchamas, G.; Huang, G.; Huang, S.; Huang, H. Preparation and Biological Activities of Chitosan Oligosaccharides. *Trends Food Sci. Technol.* **2021**, *107*, 38–44. [CrossRef]
8. Li, B.; Cui, J.; Xu, T.; Xu, Y.; Long, M.; Li, J.; Liu, M.; Yang, T.; Du, Y.; Xu, Q. Advances in the Preparation, Characterization, and Biological Functions of Chitosan Oligosaccharide Derivatives: A Review. *Carbohydr. Polym.* **2024**, *332*, 121914. [CrossRef]
9. Shi, J.; Deng, C.; Zhang, C.; Quan, S.; Fan, L.; Zhao, L. Combinatorial Metabolic Engineering of Escherichia Coli for de Novo Production of Structurally Defined and Homogeneous Amino Oligosaccharides. *Synth. Syst. Biotechnol.* **2024**, *9*, 713–722. [CrossRef]
10. Yamabhai, M.; Khamphio, M.; Min, T.T.; Soem, C.N.; Cuong, N.C.; Aprilia, W.R.; Luesukprasert, K.; Teeranitayatarn, K.; Maneedaeng, A.; Tuveng, T.R.; et al. Valorization of Shrimp Processing Waste-Derived Chitosan into Anti-Inflammatory Chitosan-Oligosaccharides (CHOS). *Carbohydr. Polym.* **2024**, *324*, 121546. [CrossRef]
11. Subramanian, K.; Balaraman, D.; Panangal, M.; Nageswara Rao, T.; Perumal, E.; Kumarappan, A.; Sampath Renuga, P.; Arumugam, S.; Thirunavukkarasu, R.; Aruni, W.; et al. Bioconversion of Chitin Waste through *Stenotrophomonas maltophilia* for Production of Chitin Derivatives as a Seabass Enrichment Diet. *Sci. Rep.* **2022**, *12*, 4792. [CrossRef] [PubMed]
12. Kumar, M.; Madhuprakash, J.; Balan, V.; Kumar Singh, A.; Vivekanand, V.; Pareek, N. Chemoenzymatic Production of Chitooligosaccharides Employing Ionic Liquids and *Thermomyces lanuginosus* Chitinase. *Bioresour. Technol.* **2021**, *337*, 125399. [CrossRef]
13. Zhou, J.; Wen, B.; Xie, H.; Zhang, C.; Bai, Y.; Cao, H.; Che, Q.; Guo, J.; Su, Z. Advances in the Preparation and Assessment of the Biological Activities of Chitosan Oligosaccharides with Different Characteristics. *Food Funct.* **2021**, *12*, 926–951. [CrossRef] [PubMed]
14. Aktuganov, G.E.; Melent'ev, A.I. Specific Features of Chitosan Depolymerization by Chitinases, Chitosanases, and Nonspecific Enzymes in the Production of Bioactive Chitooligosaccharides (Review). *Appl. Biochem. Microbiol.* **2017**, *53*, 611–627. [CrossRef]
15. Liu, Y.; Qin, Z.; Wang, C.; Jiang, Z. N-Acetyl-d-Glucosamine-Based Oligosaccharides from Chitin: Enzymatic Production, Characterization and Biological Activities. *Carbohydr. Polym.* **2023**, *315*, 121019. [CrossRef]
16. Dai, Y.; Yang, F.; Liu, X.; Wang, H. The Discovery and Characterization of a Novel Chitinase with Dual Catalytic Domains from a Qinghai-Tibetan Plateau Wetland Soil Metagenome. *Int. J. Biol. Macromol.* **2021**, *188*, 482–490. [CrossRef]
17. Sethupathy, S.; Morales, G.M.; Li, Y.; Wang, Y.; Jiang, J.; Sun, J.; Zhu, D. Harnessing Microbial Wealth for Lignocellulose Biomass Valorization through Secretomics: A Review. *Biotechnol. Biofuels* **2021**, *14*, 154. [CrossRef] [PubMed]
18. Wang, J.; Cao, L.; Liu, Y.; Zhang, Q.; Ruan, R.; Luo, X. Effect of Acclimatized Paddy Soil Microorganisms Using Swine Wastewater on Degradation of Rice Straw. *Bioresour. Technol.* **2021**, *332*, 125039. [CrossRef]
19. Barbosa, R.G.; van Veelen, H.P.J.; Pinheiro, V.; Sleutels, T.; Verstraete, W.; Boon, N. Enrichment of Hydrogen-Oxidizing Bacteria from High-Temperature and High-Salinity Environments. *Appl. Environ. Microbiol.* **2021**, *87*, e02439-20–20. [CrossRef]
20. Ibáñez, A.; Barreiro, C.; Diez-Galán, A.; Cobos, R.; Calvo-Peña, C.; Coque, J.J.R. Molecular Identification and Acid Stress Response of an *Acidithiobacillus thiooxidans* Strain Isolated from Rio Tinto (Spain). *IJMS* **2023**, *24*, 13391. [CrossRef]
21. Niu, L.; Chen, Y.; Li, Y.; Wang, Y.; Shen, J.; Wang, L.; Zhang, W.; Zhang, H.; Zhao, B. Diversity, Abundance and Distribution Characteristics of Potential Polyethylene and Polypropylene Microplastic Degradation Bacterial Communities in the Urban River. *Water Res.* **2023**, *232*, 119704. [CrossRef] [PubMed]
22. Mendes, I.V.; Garcia, M.B.; Bitencourt, A.C.A.; Santana, R.H.; de Castro Lins, P.; Silveira, R.; Simmons, B.A.; Gladden, J.M.; Kruger, R.H.; Quirino, B.F. Bacterial Diversity Dynamics in Microbial Consortia Selected for Lignin Utilization. *PLoS ONE* **2021**, *16*, e0255083. [CrossRef] [PubMed]
23. Mu, D.-S.; Liang, Q.-Y.; Wang, X.-M.; Lu, D.-C.; Shi, M.-J.; Chen, G.-J.; Du, Z.-J. Metatranscriptomic and Comparative Genomic Insights into Resuscitation Mechanisms during Enrichment Culturing. *Microbiome* **2018**, *6*, 230. [CrossRef] [PubMed]
24. Zhang, Y.; Pan, D.; Xiao, P.; Xu, Q.; Geng, F.; Zhang, X.; Zhou, X.; Xu, H. A Novel Lytic Polysaccharide Monooxygenase from Enrichment Microbiota and Its Application for Shrimp Shell Powder Biodegradation. *Front. Microbiol.* **2023**, *14*, 1097492. [CrossRef] [PubMed]
25. Pan, D.; Liu, J.; Xiao, P.; Xie, Y.; Zhou, X.; Zhang, Y. Research Progress of Lytic Chitin Monooxygenase and Its Utilization in Chitin Resource Fermentation Transformation. *Fermentation* **2023**, *9*, 754. [CrossRef]
26. Chen, Y.; Zhou, N.; Chen, X.; Wei, G.; Zhang, A.; Chen, K.; Ouyang, P. Characterization of a New Multifunctional GH20 β-N-Acetylglucosaminidase From *Chitinibacter* sp. GC72 and Its Application in Converting Chitin Into N-Acetyl Glucosamine. *Front. Microbiol.* **2022**, *13*, 874908. [CrossRef] [PubMed]
27. Wang, J.; Zhu, M.; Wang, P.; Chen, W. Biochemical Properties of a Cold-Active Chitinase from Marine Trichoderma Gamsii R1 and Its Application to Preparation of Chitin Oligosaccharides. *Mar. Drugs* **2023**, *21*, 332. [CrossRef] [PubMed]

28. Aam, B.B.; Heggset, E.B.; Norberg, A.L.; Sørlie, M.; Vårum, K.M.; Eijsink, V.G.H. Production of Chitooligosaccharides and Their Potential Applications in Medicine. *Mar. Drugs* **2010**, *8*, 1482–1517. [CrossRef] [PubMed]
29. Kumar, M.; Brar, A.; Vivekanand, V.; Pareek, N. Bioconversion of Chitin to Bioactive Chitooligosaccharides: Amelioration and Coastal Pollution Reduction by Microbial Resources. *Mar. Biotechnol.* **2018**, *20*, 269–281. [CrossRef]
30. Segata, N.; Izard, J.; Waldron, L.; Gevers, D.; Miropolsky, L.; Garrett, W.S.; Huttenhower, C. Metagenomic Biomarker Discovery and Explanation. *Genome Biol.* **2011**, *12*, R60. [CrossRef]
31. Suma, K.; Podile, A.R. Chitinase A from Stenotrophomonas Maltophilia Shows Transglycosylation and Antifungal Activities. *Bioresour. Technol.* **2013**, *133*, 213–220. [CrossRef] [PubMed]
32. Arnold, N.D.; Garbe, D.; Brück, T.B. Isolation, Biochemical Characterization, and Genome Sequencing of Two High-quality Genomes of a Novel Chitinolytic *Jeongeupia* Species. *Microbiologyopen* **2023**, *12*, e1372. [CrossRef] [PubMed]
33. Arnold, N.D.; Garbe, D.; Brück, T.B. Proteomic and Transcriptomic Analyses to Decipher the Chitinolytic Response of *Jeongeupia* spp. *Mar. Drugs* **2023**, *21*, 448. [CrossRef] [PubMed]
34. Zain, N.-A.A.; Ng, L.-M.; Foong, C.P.; Tai, Y.T.; Nanthini, J.; Sudesh, K. Complete Genome Sequence of a Novel Polyhydroxyalkanoate (PHA) Producer, *Jeongeupia* sp. USM3 (JCM 19920) and Characterization of Its PHA Synthases. *Curr. Microbiol.* **2020**, *77*, 500–508. [CrossRef] [PubMed]
35. Yuan, P.-B.; Zhan, Y.; Zhu, J.-H.; Ling, J.-H.; Chen, E.-Z.; Liu, W.-T.; Wang, L.-J.; Zhong, Y.-X.; Chen, D.-Q. Pan-Genome Analysis of *Laribacter hongkongensis*: Virulence Gene Profiles, Carbohydrate-Active Enzyme Prediction, and Antimicrobial Resistance Characterization. *Front. Microbiol.* **2022**, *13*, 862776. [CrossRef] [PubMed]
36. Kim, B.-Y.; Weon, H.-Y.; Yoo, S.-H.; Chen, W.-M.; Kwon, S.-W.; Go, S.-J.; Stackebrandt, E. *Chitinimonas Koreensis* sp. Nov., Isolated from Greenhouse Soil in Korea. *Int. J. Syst. Evol. Microbiol.* **2006**, *56*, 1761–1764. [CrossRef] [PubMed]
37. Zhang, A.; Mo, X.; Zhou, N.; Wang, Y.; Wei, G.; Hao, Z.; Chen, K. Identification of Chitinolytic Enzymes in *Chitinolyticbacter meiyuanensis* and Mechanism of Efficiently Hydrolyzing Chitin to N-Acetyl Glucosamine. *Front. Microbiol.* **2020**, *11*, 572053. [CrossRef]
38. Rani, T.S.; Madhuprakash, J.; Podile, A.R. Chitinase-E from *Chitiniphilus shinanonensis* Generates Chitobiose from Chitin Flakes. *Int. J. Biol. Macromol.* **2020**, *163*, 1037–1043. [CrossRef] [PubMed]
39. Roesch, L.F.W.; Fulthorpe, R.R.; Riva, A.; Casella, G.; Hadwin, A.K.M.; Kent, A.D.; Daroub, S.H.; Camargo, F.A.O.; Farmerie, W.G.; Triplett, E.W. Pyrosequencing Enumerates and Contrasts Soil Microbial Diversity. *ISME J.* **2007**, *1*, 283–290. [CrossRef]
40. Nguyen, S.T.C.; Freund, H.L.; Kasanjian, J.; Berlemont, R. Function, Distribution, and Annotation of Characterized Cellulases, Xylanases, and Chitinases from CAZy. *Appl. Microbiol. Biotechnol.* **2018**, *102*, 1629–1637. [CrossRef]
41. Moon, M.; Lee, J.-P.; Park, G.W.; Lee, J.-S.; Park, H.J.; Min, K. Lytic Polysaccharide Monooxygenase (LPMO)-Derived Saccharification of Lignocellulosic Biomass. *Bioresour. Technol.* **2022**, *359*, 127501. [CrossRef] [PubMed]
42. Ren, X.-B.; Dang, Y.-R.; Liu, S.-S.; Huang, K.-X.; Qin, Q.-L.; Chen, X.-L.; Zhang, Y.-Z.; Wang, Y.-J.; Li, P.-Y. Identification and Characterization of Three Chitinases with Potential in Direct Conversion of Crystalline Chitin into N, N′-Diacetylchitobiose. *Mar. Drug* **2022**, *20*, 165. [CrossRef] [PubMed]
43. Colson, S.; van Wezel, G.P.; Craig, M.; Noens, E.E.E.; Nothaft, H.; Mommaas, A.M.; Titgemeyer, F.; Joris, B.; Rigali, S. The Chitobiose-Binding Protein, DasA, Acts as a Link between Chitin Utilization and Morphogenesis in Streptomyces Coelicolor. *Microbiology* **2008**, *154*, 373–382. [CrossRef] [PubMed]
44. López-García, C.L.; Guerra-Sánchez, G.; Santoyo-Tepole, F.; Olicón-Hernández, D.R. Chitinase Induction in *Trichoderma harzianum*: A Solid-State Fermentation Approach Using Shrimp Waste and Wheat Bran/Commercial Chitin for Chitooligosaccharides Synthesis. *Prep. Biochem. Biotechnol.* **2024**, *12*, 1–11. [CrossRef] [PubMed]
45. Zhang, J.; Wang, L.; Tan, W.; Li, Q.; Dong, F.; Guo, Z. Preparation of Chitosan-Rosmarinic Acid Derivatives with Enhanced Antioxidant and Anti-Inflammatory Activities. *Carbohydr. Polym.* **2022**, *296*, 119943. [CrossRef] [PubMed]
46. Wang, K.; Yu, D.; Bai, Y.; Cao, H.; Guo, J.; Su, Z. Isolation and Purification of Chitosan Oligosaccharides (Mw $\leq$ 1000) and Their Protective Effect on Acute Liver Injury Caused by CCl4. *Mar. Drugs* **2024**, *22*, 128. [CrossRef] [PubMed]
47. Xia, W.; Wei, X.Y.; Xie, Y.Y.; Zhou, T. A Novel Chitosan Oligosaccharide Derivative: Synthesis, Antioxidant and Antibacterial Properties. *Carbohydr. Polym.* **2022**, *291*, 119608. [CrossRef]
48. Zhang, Y.; Zhou, X.; Ji, L.; Du, X.; Sang, Q.; Chen, F. Enzymatic Single-Step Preparation and Antioxidant Activity of Hetero-Chitooligosaccharides Using Non-Pretreated Housefly Larvae Powder. *Carbohydr. Polym.* **2017**, *172*, 113–119. [CrossRef]
49. Rojas-Avelizapa, L.I.; Cruz-Camarillo, R.; Guerrero, M.I.; Rodríguez-Vázquez, R.; Ibarra, J. Selection and Characterization of a Proteo-Chitinolytic Strain of Bacillus Thuringiensis, Able to Grow in Shrimp Waste Media. *World J. Microbiol. Biotechnol.* **1999**, *15*, 299–308. [CrossRef]
50. Benzie, I.F.F.; Strain, J.J. The Ferric Reducing Ability of Plasma (FRAP) as a Measure of "Antioxidant Power": The FRAP Assay. *Anal. Biochem.* **1996**, *239*, 70–76. [CrossRef]

marine drugs

MDPI

Article

Tunable Oxidized-Chitin Hydrogels with Customizable Mechanical Properties by Metal or Hydrogen Ion Exposure

Angelica Mucaria, Demetra Giuri, Claudia Tomasini, Giuseppe Falini and Devis Montroni *

Department of Chemistry "Giacomo Ciamician", University of Bologna, Via F. Selmi 2, 40126 Bologna, Italy; angelica.mucaria2@unibo.it (A.M.); demetra.giuri2@unibo.it (D.G.); claudia.tomasini@unibo.it (C.T.); giuseppe.falini@unibo.it (G.F.)
* Correspondence: devis.montroni@unibo.it

Abstract: This study focuses on the optimization of chitin oxidation in C6 to carboxylic acid and its use to obtain a hydrogel with tunable resistance. After the optimization, water-soluble crystalline β-chitin fibrils (β-chitOx) with a degree of functionalization of 10% were obtained. Diverse reaction conditions were also tested for α-chitin, which showed a lower reactivity and a slower reaction kinetic. After that, a set of hydrogels was synthesized from β-chitOx 1 wt.% at pH 9, inducing the gelation by sonication. These hydrogels were exposed to different environments, such as different amounts of Ca^{2+}, Na^{+} or Mg^{2+} solutions, buffered environments such as pH 9, PBS, pH 5, and pH 1, and pure water. These hydrogels were characterized using rheology, XRPD, SEM, and FT-IR. The notable feature of these hydrogels is their ability to be strengthened through cation chelation, being metal cations or hydrogen ions, with a five- to tenfold increase in their storage modulus (G′). The ions were theorized to alter the hydrogen-bonding network of the polymer and intercalate in chitin's crystal structure along the a-axis. On the other hand, the hydrogel dissolved at pH 9 and pure water. These bio-based tunable hydrogels represent an intriguing material suitable for biomedical applications.

Keywords: chitin; gel; oxidation; calcium; pH

1. Introduction

Hydrogels are materials characterized by a crosslinked 3D network of polymeric molecules, fibers, or particles, in which the aqueous phase acts as a dispersing medium. Due to their nature, these materials have the ability to absorb and retain large quantities of water without dissolving or undergoing significant changes in shape [1,2].

Hydrogels can be classified based on their framework molecules, method of synthesis, and the type of interactions (i.e., chemical or physical) that define them. However, a relevant distinction is made between synthetic and natural hydrogels, depending on the source of the constituent molecule or molecules [3]. The latter are frequently preferred over their synthetic counterparts due to their reproducibility, biodegradability, non-toxicity, and biocompatibility. Currently, a variety of natural biopolymers are used to prepare hydrogels including cellulose, hemicellulose, chitin, chitosan, gelatin, agarose, starch, hyaluronic acid, alginate, collagen, and DNA [2,4]. Among these biopolymers, chitin has attracted great interest due to its huge reserves in nature, being the second most abundant biopolymer on Earth [5,6], high hydration, and strong network of hydrogen bonds [7]. All these properties make it a promising candidate for the preparation of highly swelling hydrogel systems [8,9]. Chitin exists in nature as three polymorphs, α-chitin, β-chitin, and γ-chitin, which differ in their arrangements of polymeric chains: anti-parallel, parallel, and alternating, respectively [10]. The α-chitin is the most abundant polymorph and can be found in arthropods cuticles, fungi cell walls, or the cysts of *Entamoeba* [11,12]. The β-chitin is still abundant in nature and can be found in mollusks, or other biomineralizing organisms, as *Foraminifera* [13,14]. Finally, the γ-chitin is extremely rare and its actual existence is still debatable. To date, this last polymorph has been documented in the cocoon fibers of the *Orgya dubia*, the stomach of *Loligo*, and some insects' peritrophic membranes [5,15].

The α-chitin is the most crystalline polymorph, with a crystallinity generally greater than 80% [16]. It is characterized by a strong network of hydrogen bonds that lead to a high mechanical modulus. On the other hand, the β-chitin has a lower crystallinity and fewer intermolecular forces. This has led to a special interest in this polymorph since it is more prone to functionalization and solubilization compared to α-chitin [15].

Chitin-based hydrogels have shown promising application prospects in the field of biomedicine [2]. The usual method for gelling native chitin involves cross-linking chitin molecules in solutions using physical or chemical methods [17]. However, the main difficulty in making chitin hydrogels is the lack of appropriate solvents due to chitin's insolubility. Several methods have been proposed to address this issue, including the use of specialized solvents to dissolve native chitin [18–23], hydrophilic modification [24,25], and the defibrillation of the material [1].

In Fan et al. (2008), a series of LiOH/KOH/urea aqueous solutions with different weight ratios were used to dissolve chitin structures with degrees of acetylation ranging from 5% to 94%. Then, coagulants were applied to these chitin/chitosan solutions to form robust hydrogels [16]. In Dang et al. (2020), the solubilization of chitin was carried out in a $CaCl_2$ methanol solution. Chitin hydrogels were then formed by adding excess water to the chitin solution and removing the methanol and calcium ions through dialysis or filtration [19]. In all these studies, organic solvents or other potentially toxic chemicals were used to achieve chitin solubilization.

Chitosan, an acid-soluble deacetylated derivative of chitin, allows hydrogels to be obtained by exploiting the interactions of the cationic amino groups of chitosan, with negatively charged molecules and anions. Chitosan can form an ionic complexation with small anionic molecules, such as sulfates, citrates, and phosphates [26,27], or metal anions such as ${MoO_4}^{2-}$ [28,29].

In contrast, in Ma, Qinyan, et al. (2019), a β-chitin fiber solution was obtained using ammonium persulfate as an oxidizing agent to introduce a carboxylic group at the C6 position of chitin [30].

Other oxidation methods are also reported in the literature. 2,2,6,6-Tetramethylpiperidine 1-oxyl (TEMPO) is often used as a primary oxidizing agent in combination with other co-oxidants, as in the system TEMPO/NaClO/NaBr [31,32]. However, the use of TEMPO to obtain soluble fibrils is a combined process often involving other steps such as mechanical disintegration, sonication, or the enzymatic, or acid hydrolysis method [33].

Oxidation with periodate is also reported. The reaction involves pre-treating chitin with KOH for 24 h and using periodic acid to ensure the correct amount of periodate ions in the reaction solution. The reaction requires long reaction times, such as 14 days, as reported by Liu et al. (2021) [34].

The carboxylic acid introduced with these methods facilitates the defibrillation of the system by creating negative charges that repel each other in a basic environment.

We chose to follow the oxidation method with ammonium persulfate as reported in Ma, Qinyan, et al. (2019), since our work represents a continuation of theirs. Additionally, the use of ammonium persulfate allows for chitin fibers to be obtained with fewer reaction steps and less time required, compared to other oxidants used.

The oxidized chitin fibers obtained by Ma, Qinyan, et al. (2019) were able to form hydrogels upon sonication. Following this method, the hydrogels could also potentially be formed by physical interactions (anion–cation and/or hydrogen bonding), avoiding the use of potentially toxic cross-linkers [35]. This potential was not explored in the 2019 paper.

Among biopolymers, oxidized chitin shows a similarity with alginate, as both structures have the C6 oxidized to a carboxylic group. However, in oxidized chitin, the degree of C6 oxidation is controlled and not complete as in alginate, which means the contribution of hydrogen bonding and apolar interactions is prevalent during gelation [36]. Moreover, alginate is solubilized as single molecules, while oxidized chitin dissolves/disperses as crystalline nano-fibrils. This difference affects the gelling capability, even if its mechanism is, in part, based on analogue interactions. Moreover, the presence of crystalline fibrils is expected to positively affect the mechanical resistance of the gel.

In this study, we aim to explore the potential preparation of hydrogels using both β-chitin and α-chitin as the starting material, aiming to explore their different reactivity. In Ma, Qinyan, et al. (2019), only β-chitin was studied. In addition to their work, our aim is to optimize their synthesis by controlling the starting material grain size and the oxidation conditions. By doing so, we expect to oxidize preferentially the chitin's low crystalline regions, leaving the core of the crystalline motifs almost unaltered. This would allow us to obtain hydrogel building blocks with a higher value. Finally, we want to explore the tunability of the hydrogel's mechanical properties by exposing it to diverse environments, as in the presence of metal ions or variable pHs.

These optimized hydrogels could lead to potential applications in diverse fields, including the biomedical one where a material's adaptation to a changing environment could be crucial.

2. Results and Discussion

2.1. Chitin Grain Size Optimization

The first step of the work was to optimize the material synthesis, to obtain soluble crystalline fibrils of oxidized chitin. To maximize the yield of production and fasten the kinetic of oxidation, we decided to work with chitin powder, rather than flakes, to have a higher initial exposed surface. While α-chitin was already purchased as a powder, the purchased β-chitin was milled to obtain a fine powder. Both powders were sieved to obtain diverse grain size: (i) from 600 μm to 150 μm, (ii) from 150 μm to 45 μm, and (iii) below 45 μm.

A study on the crystallinity and structure of α-chitin and β-chitin powders with diverse grain sizes was carried out. The aim was to reduce the grain size of the powder while preserving the crystallinity as much as possible. Table 1 reports the diffraction angles and the crystallite size (d) of the main reflections obtained from the analyses of X-ray powder diffraction (XRPD) patterns of α-chitin and β-chitin powders sieved at different grain sizes. The diffractograms are reported in Figure S1. Miller indices were attributed according to literature data [37,38]. Chitin's diffraction peaks are a convolution of different signals, and the Miller indices reported refers only to the most intense one, which contributes to most of the signal intensity.

Table 1. Diffraction angles and crystallite sizes (d) of the main reflections for the various grain sizes of α-chitin and β-chitin. The standard deviation of the measurements is reported in parentheses. Chitin's diffraction peaks are a convolution of different signals, the Miller indices reported refer only to the most intense signals which account for most of the signal intensity.

		600–150 μm			150–45 μm			<45 μm		
Sample		2θ (°)	d (nm)	CI (%)	2θ (°)	d (nm)	CI (%)	2θ (°)	d (nm)	CI (%)
α-chitin	(020)	9.26 (0.01)	7.05 (0.07)	79.7 (0.4)	9.16 (0.06)	7.2 (0.2)	78.0 (0.2)	9.09 (0.03)	7.0 (0.2)	70.7 (0.7)
	(110)	19.22 (0.01)	4.82 (0.03)	83.9 (0.2)	19.13 (0.06)	4.8 (0.2)	81 (1)	19.07 (0.04)	4.48 (0.03)	75.1 (0.7)
β-chitin	(010)	8.2 (0.1)	4.04 (0.09)	69 (1)	8.38 (0.06)	3.78 (0.08)	59 (3)	8.07 (0.01)	3.52 (0.07)	46 (1)
	(100)	19.7 (0.1)	2.34 (0.08)	68.7 (0.4)	19.92 (0.03)	2.17 (0.01)	61 (1)	19.62 (0.01)	1.78 (0.01)	49.5 (0.5)

In the α-chitin powders, a shift of the (020) and (110) reflections towards lower diffraction angles was observed when reducing the grain size dimension from 600 μm down to below 45 μm. This indicates an increase in the unit cell parameters, suggesting a weakening of interchain interactions. This process was also associated with a mild decrease in crystallinity in the sample with a grain size < 45 μm, as indicated by the decrease in the crystal size along the zone axis associated to the (110) plane and the decrease in the CI calculated on both planes.

The β-chitin powder samples showed lower crystallinity than α-chitin, when compared in the same grain size ranges. Similarly, to that observed for α-chitin, a progressive reduction in the crystallinity occurred also for β-chitin. In this case, the decrease was observed both along the [010] and [100] zone axes. Moreover, for this polymorph, an increase in the lattice cell parameters was revealed by the shift of the diffraction peaks to lower angles which, as for α-chitin, can be associated with a weakening of the interplane and interchain interactions.

From the XRD analysis, for the β polymorph, a decrease in crystallite size corresponding to the (100) reflection is observed as the particle size decreases. An overall decrease in the material crystallinity was observed from the CI of both peaks decreasing the grain size.

A decrease in crystallinity decreasing particle size was also reported in Jorge A. et al. (2018) [39]. Both α-chitin and β-chitin were ground in a knife mill and sieved according to the particle sizes.

However, for the α polymorph, less significant variations are observed. This is consistent with the different crystalline packing of the two polymorphs, more compact for α-chitin [39].

Based on the obtained results, samples of α-chitin and β-chitin powders having grain sizes between 600 μm and 150 μm were used for further experiments, being the ones retaining the highest crystallinity.

2.2. Optimization of the Oxidation Time for β-Chitin

The selected chitin samples, with a powder grain size of 600–150 μm, were oxidized by reacting with ammonium persulfate at a temperature of 40 °C following the procedure described in Section 3.4. This synthetic step was optimized in the reaction time.

An initial evaluation of the reaction progress was monitored through turbidity measurements, as the heterogeneous reaction proceeds with the dissolution of chitin particles. After just 3 h of reaction, the initial grains in the β-chitin powder reduced their dimensions, as evidenced by the rapid decrease in transmittance of the reaction batch (Figure S2). After 24 h, the reaction mixture showed a complete dissolution of chitin, while, after 45 h, an increment in turbidity was observed due to aggregate formation. In parallel, the yield of the reaction decreased progressively while increasing the reaction time; see Figure 1. This can be attributed to the formation of chitin fibers of nanometric sizes that were lost during the purification of the product or the degradation of chitin due to excessive oxidation. At 45 h, it was no longer possible to collect material, even by dialysis, suggesting a complete degradation of the polysaccharide chain. Considering the turbidity profile obtained and the yield associated with that, we decided to focus our attention on the following reaction times, 3 h, 6 h, 9 h, and 24 h, to sample along different positions of the turbidity profile. No time over 24 h was examined since the yield was considered too low.

The obtained oxidized chitin powders were examined using solid-state nuclear magnetic resonance (ss-NMR), Fourier-transform infrared spectroscopy (FT-IR), and XRPD.

The FT-IR spectra (see Figure S3b) showed all the absorption bands of chitin, plus a new weak absorption band at 1730 cm^{-1}. This band has been documented as a stretching of the C=O of carboxylic acids [30], demonstrating a positive functionalization of chitin. A ratio between this band and the band at 1073 cm^{-1} (a C–O stretching associated to the sugar ring) was initially used to monitor the oxidation along the time; the results are reported in Figure S3c.

The ss-NMR spectra (see Figure S4) were used to calculate the degree of functionalization (%DF) of chitin. The %DF are reported in Figure 2 and show an incremental trend increasing the reaction time, showing an oxidation of about 10% of the C6 of the chitin monomers at 24 h. The trend observed is in accordance with the one extrapolated from the FT-IR spectra.

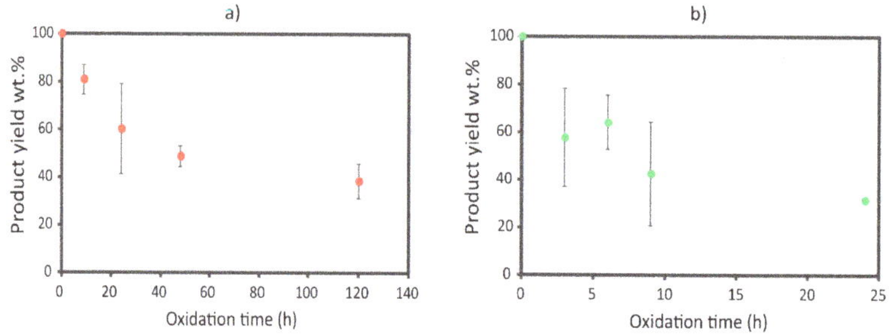

Figure 1. Product yield expressed as weight percentage obtained at the various oxidation times of (**a**) α-chitin and (**b**) β-chitin.

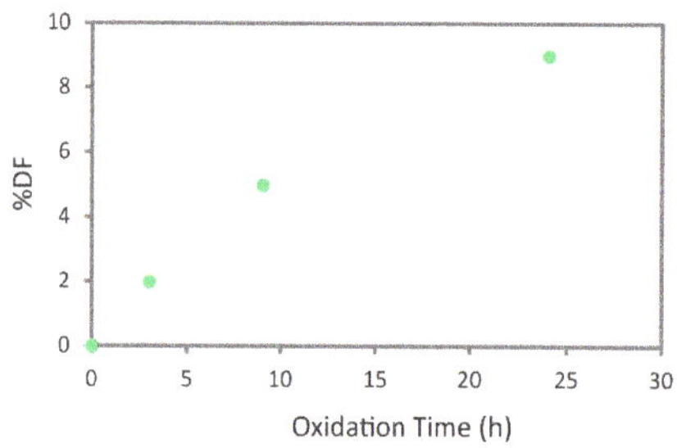

Figure 2. Degree of functionalization percentage (%DF) of C6 of β-chitin at different reaction times.

The XRPD analyses are reported in Table 2 and were used to identify the influence of oxidation on chitin's crystallographic organization. The diffraction patterns show that, along the [010] direction, a reduction in the cell parameter is observed, increasing the reaction time. This reduction is also associated with a decrease in crystallinity, revelated by a decrease in the crystal size. Conversely, along the [100] direction, the crystallinity and the dimension of the lattice parameter increased. This structural reorganization is likely due to the oxidation of C6 to a carboxylate group and a successive rearrangement of the hydrogen bond network in the crystal during the oxidation reaction.

Table 2. Diffraction angles and crystallite sizes (d) of the main reflections for the various oxidation time of β-chitin. The standard deviation of the measurements is reported in parentheses. Chitin's diffraction peaks are a convolution of different signals; the Miller indices reported refer only to the most intense signals which account for most of the signal intensity.

	(010)			(100)		
Sample	**2θ (°)**	**d (nm)**	**CI (%)**	**2θ (°)**	**d (nm)**	**CI (%)**
Control	8.21 (0.04)	4.0 (0.1)	69 (1)	19.70 (0.01)	2.4 (0.1)	68.7 (0.4)
3 h	8.32 (0.02)	3.74 (0.08)	74 (2)	19.50 (0.05)	2.60 (0.09)	73 (3)
6 h	8.24 (0.01)	3.65 (0.08)	76 (2)	19.44 (0.01)	2.70 (0.04)	73 (3)
9 h	8.62 (0.09)	3.5 (0.3)	65 (5)	19.35 (0.07)	2.73 (0.02)	71 (1)
24 h	8.63 (0.07)	3.3 (0.1)	66 (3)	19.26 (0.04)	2.78 (0.07)	75.4 (0.4)

An increase in the average distance among the (100) plane agrees with what was expected with the C6 oxidation. In β-chitin, the C6 hydroxyl group is pointed along the a-axis and is involved in hydrogen bonding with the carbonyl group of the amide of the adjacent polymeric chain. Increasing the degree of functionalization is expected to increase the average distance along this direction since the carboxylate will require more space compared to a primary alcohol. Despite that, the carboxylate may act as a hydrogen bond receiver from the N-H of the amide. The hydrogen bond associated with a carboxylate group is expected to be stronger and more stable compared to that of an alcohol. Moreover, carboxylate has lower mobility due to its carbon hybridization and its higher steric hinderance compared to a primary alcohol. The combination of stabilization, new potential interactions, and a decrease in the mobility of this group may lead to the increase in crystallinity along this crystallographic axis.

On the other hand, a shrinkage of the cell parameters was observed along the [010] direction. This direction is perpendicular to the polymeric chains and does not present any direct hydrogen bonding between adjacent chains. The oxidation of the C6 may lead to the formation of direct hydrogen bonding in between polymeric chains, such as new hydrogen bonds involving the C=O of the carboxylate group, leading to a shortening of their average interplanar distance. Simultaneously, the relatively low degree of functionalization achieved will create small variations or kinks of the chains along this distance where the carboxylate group is, leading to an increase in the overall disorder of the crystal.

Based on the XRPD results, none of the rearrangement observed induced a significant loss in crystallinity.

In conclusion, combining the degree of functionalization calculated from the ss-NMR spectra (Figure 2) and the relative intensity of the FT-IR bands (Figure S3c), the 24 h time point was selected as the optimal reaction time. In fact, at 24 h, the reaction still shows a relevant product yield (32 wt.%), while the product has the highest degree of functionalization, approximately 10% of the C6 of β-chitin, with no significant loss in crystallinity.

2.3. Optimization of the Oxidation Time for α-Chitin

In contrast to β-chitin, α-chitin shows a markedly different reactivity, coherent with what has previously been documented in the literature [5,15]. According to the turbidity measurements, the transmittance of the reaction batch decreases much more slowly compared to the β-chitin; see Figure S2. This suggests that α-chitin fibers are less accessible to ammonium persulfate, and defibrillation occurs over longer periods. At about 72 h, a complete dissolution of the chitin was observed. Basing on the turbidity measurement results, we decided to focus our attention on the following reaction times, 9 h, 24 h, 48 h, and 120 h, to sample along different positions of the turbidity profile. No time longer than 120 h were tested (even if it still showed a yield of about 40 wt.%.) since the reaction time would be too long and not compatible with practical applications.

A lower reactivity of α-chitin can also be deduced by FT-IR spectra analyses (Figure S3). As for β-chitin, an absorption band at 1730 cm^{-1} is observed in the oxidized samples. A ratio between the intensity of this signal and the one at 1073 cm^{-1} was calculated, as was carried out for the other polymorph. After 120 h, the relative intensity of the carboxylic acid band increases by only 27%, a value inferior to that obtained for β-chitin after 24 h of reaction (30%).

Compared to β-chitin, the FT-IR analyses on the oxidized α-chitin samples showed a smaller increment in oxidation in the first 24 h. No differences in the FT-IR relative intensity and yield (Figure 1a) were observed between 48 h and 120 h, suggesting the reaction reached a plateau, and the decrease in turbidity was probably due to slow solubilization processes.

The XRPD pattern was acquired for each of the α-chitin powder synthetized and the data are reported in Table 3 and Figure S5. Comparing the diffraction pattern, no significant variation was observed along the (110) or the (020) directions in the peak position. A mild increase in crystallinity was observed along the (020) plane, while no relevant trend in the crystallinity was identified along the (110) plane. The same result was also obtained by Jiang J. et al. [40], in which α-chitin is oxidized using the O_2/Laccase/TEMPO system, testing various concentrations of TEMPO. From the XRD analysis, it is observed that the structure of chitin remains mostly unchanged. Here, too, a slight increase in crystallinity is observed, probably due to the removal of the water-soluble amorphous zones during oxidation. These results suggest that oxidation occurs on the chitin fiber surface [40].

Table 3. Diffraction angles and crystallite sizes (d) of the main reflections for the various oxidation time of α-chitin. The standard deviation of the measurements is reported in parentheses. Chitin's diffraction peaks are a convolution of different signals; the Miller indices reported refer only to the most intense signals which account for most of the signal intensity.

	(020)			(110)		
Sample	**2θ (°)**	**d (nm)**	**CI (%)**	**2θ (°)**	**d (nm)**	**CI (%)**
Control	9.26 (0.04)	7.2 (0.1)	79.7 (0.4)	19.25 (0.01)	5.23 (0.04)	83.9 (0.2)
9 h	9.18 (0.07)	7.6 (0.1)	83.6 (0.4)	19.13 (0.07)	5.1 (0.1)	89.2 (0.8)
24 h	9.21 (0.04)	8.1 (0.4)	83 (2)	19.13 (0.01)	5.3 (0.1)	89.8 (0.8)
48 h	9.21 (0.04)	8.1 (0.4)	83.0 (0.1)	19.2 (0.1)	5.37 (0.06)	91.6 (0.9)
5 gg	9.25 (0.09)	8.0 (0.6)	79.9 (0.2)	19.22 (0.05)	5.0 (0.3)	91.0 (0.2)

We chose to test only the β-chitin to produce oxidized chitin hydrogels, as α-chitin requires a long oxidation time to achieve a good degree of functionalization.

2.4. Rheological Analysis

Following the procedure described in Section 3.9, a hydrogel was obtained from the β-chitin grains in the range of 600–150 μm, oxidized for 24 h. The hydrogel prepared (1 wt.% concentration) appeared as a homogeneous material with a soft consistency, as is visible in the picture in Figure 3.

Figure 3. A β-chitin hydrogel camera picture. The hydrogel cylinder has a diameter of about 1 cm.

With the aim to modulate its mechanical resistance and test its chemical stability, we exposed the hydrogel to diverse environments. The conditions studied were mostly the following two: in the presence of calcium ions, and under different pH values.

The samples were prepared as described in Sections 3.9.1 and 3.9.2.

For the first study, various molar ratios between carboxylate groups and calcium ions (Ca^{2+}) were tested. Different addition methods of Ca^{2+} were tested: (i) the surface addition of the calcium solution on the hydrogel, (ii) the mixing of the solution with the hydrogel, and (iii) the addition of Ca^{2+} before sonication. Among them, the surface addition method was selected, as it produced samples with a higher storage modulus (G′) and better reproducibility compared with (ii) (as reported in Table S1), while method (iii) showed the formation of aggregates and hydrogel clumps.

2.4.1. Amplitude Sweep Tests

The viscoelastic behavior of the hydrogels in terms of storage (G′) and loss (G″) moduli, representing the elastic and the viscous portion of the viscoelastic behavior, respectively, was analyzed using rheological experiments. For all the tested samples, the G′ modulus is higher than the loss modulus G″, thus indicating that all the tested samples have a solid structure and can be classified as gels [41,42]. The linear viscoelastic region (LVE) indicates the range of strain where the sample is elastic and can recover its original state when the strain is removed. All the tested samples are very elastic, presenting a large LVE range and no crossover point (G′ = G″): none of the samples reaches the breaking point, where the material converts from solid to liquid.

Ca^{2+} were added in different molar ratios compared to the carboxylate groups, quantified by ss-NMR (See Section 2.2). The molar ratios studied were Ca^{2+}: COO^- equal to 1:3; 1:1; 3:1, and 5:1. These concentrations were chosen to explore different binding ratios between the carboxylate and the Ca^{2+}, assuming they would bind in a 1:1 or 1:2 ratio. Higher concentrations were studied to explore the presence of other binding sites.

The amplitude sweep curves of all tested samples are shown in Figure 4, while the numerical values derived from the analyses of the Ca^{2+}-treated samples are reported in Table 4. The data reveal that the incorporation of Ca^{2+} results in a notable enhancement of the stiffness of the gels, which may be the consequence of a probable physical crosslinking between the Ca^{2+} and the carboxylates present in the chain of the oxidized chitin [43]. Compared to the control, where the same volume of water was added, all the samples with Ca^{2+} have a storage modulus one order of magnitude higher and show a better reproducibility, with reduced standard deviation values compared to the control.

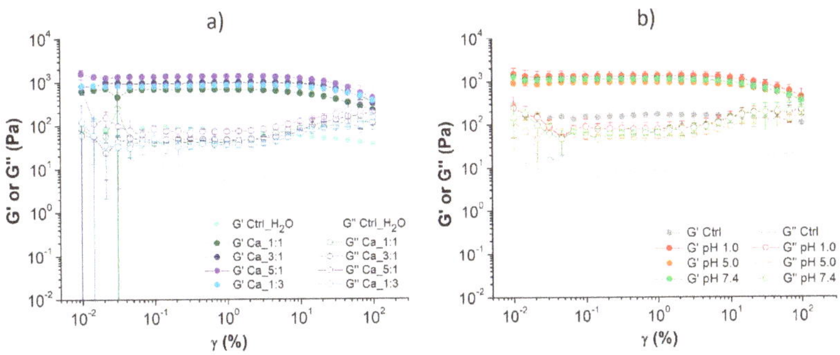

Figure 4. Amplitude sweep tests of the samples tested with (**a**) different molar ratios of Ca^{2+} and (**b**) different pH conditions. The standard deviation of the Ctrl and Ctrl_H_2O sample was omitted for clarity and its full graph can be seen in Figure S6.

Table 4. Table shows the values of G′ and G″ calculated at a shear strain value equal to 0.046%.

Sample	Storage Modulus G′; (ɣ = 0.046%)	Loss Modulus G″; (ɣ = 0.046%)
	[Pa]	[Pa]
Ctrl_H_2O	80 ± 50	20 ± 10
Ca_1:3	820 ± 30	30 ± 10
Ca_1:1	690 ± 30	40 ± 20
Ca_3:1	940 ± 50	50 ± 20
Ca_5:1	1100 ± 300	60 ± 40

The increase in storage modulus rises with the increasing amount of added Ca^{2+}, resulting in a mechanically stronger material [44].

A similar increase in storage modulus was observed by modifying the type of bivalent metal ion, using magnesium ions in a 3:1 ratio (800 Pa). Furthermore, slightly lower G′ values (690 Pa) was observed when a NaCl solution with the same ionic strength was added (Table S2).

Comparing with the results found in the literature, the hydrogels obtained here show a superior performance compared to other hydrogels from natural polymers. In the work by Syverud, Kristin, et al. (2015), a cellulose nanofibril hydrogel obtained with TEMPO achieves a storage modulus G′ of 275 ± 62 Pa when using a 0.98 wt.% dispersion [45]. Additionally, in the work by Cuomo et al. (2019), an increase in the strength of the alginate hydrogel is reported, correlated to the quantity of added Ca^{2+}. However, using a 1 wt.% alginate solution, the maximum storage modulus reached is only 100 Pa [46]. Lower G′ values (about 10^2 Pa) were also reported for gelatin gels (1 wt.%) blended with increasing amounts of chitosan (0.1–0.8 wt.%) with pH values between 3 and 4 [47]. Few examples of chitin gels were also reported, obtained by heating the polymer in ionic liquids. In these cases, the stiffness is highly dependent on the solvent used. Weak gels (G′ values in the range 10^1–10^2 Pa) [48] were obtained from IL (1-allyl-3-methylimidazolium bromide) for relatively high chitin concentrations (5–7 wt.%). G′ values slightly higher than ours (about 10^4 Pa) were instead reported by Deng et al. (2020) for 1 wt.% chitin gels obtained in [BMIM]Ac (1-butyl-3-metlimidazolium acetate) [19].

Successively the stability of the hydrogel at different pHs, specifically 1.0, 5.0, 7.4, and 9.0, was tested. These pHs were chosen due to their biological relevance, as pH 1.0 is the pH associated with digestion (specifically in the stomach), 7.4 is generally considered a healthy physiological pH, and pH 5.0 is usually associated with inflammation. pH 9.0 was tested as an alkaline counterpart. In order to reproduce the hydrogel stability in a physiological environment, PBS was used as a buffer solvent at pH 7.4.

For this test, each hydrogel was immersed in a buffered solution for 24 h, while a hydrogel stored in milliQ water was measured as a control. At the end of this exposure, the hydrogels immersed in milliQ water and carbonate buffer at pH 9.0 were not tested because the gel lost its structure and dissolved completely. Due to the absence of a control experiment in the same condition, the initial hydrogel was used as the control experiment.

The remaining samples were measured for their rheological properties and the results of the amplitude test are shown in the Figure 4b, while the numerical data are reported in Table 5. The hydrogels immersed in pH 1.0, pH 5.0, and pH 7.4 show a storage modulus (G′) five times greater than the starting hydrogel used as a control, as seen in Figure 4b, but no significant difference in the rheological properties was observed among these conditions. A mild increase in the loss modulus was also observed in these samples compared to the control. Even in this case, no crossover points (G′ = G″) are detected: none of the tested samples reaches the breaking point in the tested shear strain range.

Table 5. Table shows the values of G′ and G″ calculated at a shear strain value equal to 0.046%.

Sample	Storage Modulus G′; (ɣ = 0.046%)	Loss Modulus G″; (ɣ = 0.046%)
	[Pa]	[Pa]
Ctrl	200 ± 100	20 ± 20
pH 1	1300 ± 400	50 ± 10
pH 5	950 ± 60	70 ± 10
pH 7.4	1150 ± 30	50 ± 30

These results suggest that, within the oxidized chitin hydrogel, favorable interactions are formed at pH 1.0, 5.0, and 7.4. However, at very basic pH levels (pH 9.0), the interactions are so weak that they cause the collapse of the entire hydrogel structure. Indeed, at acidic pH levels, the deacetylated amines (pK_a 6.3) [5] of chitin protonate and can interact with the carboxylate ions (with a pK_a likely similar to that of alginate, 3.2 [49]) present in the polymer chains. Alternatively, since we do not know the actual pK_a of this compound, at basic pH levels, the carboxylic acids may deprotonate, and the electrostatic repulsion between polymeric chains, combined with the higher polymer solubility, may lead to the collapse of the hydrogel structure.

The collapse of the hydrogel in water, on the other hand, suggests that it is probably unstable at pHs much lower than pH 9. Acid pHs may stabilize the material, inducing positive interactions by protonation, while PBS may stabilize the structure by cation ionic bridges, as already observed by adding NaCl, $MgCl_2$, or $CaCl_2$. Considering the composition of PBS, the ion bridges formed are likely due to Na^+ or K^+. Specifically, in the previous section, it was observed that Na^+ induced a toughening of the hydrogel, but its efficiency is lower compared to Ca^{2+}. The higher G' observed at pH 7.4 (PBS) is likely due to the abundance of Na^+ in the PBS, which allowed us to saturate all the binding sites, maximizing the interactions in the hydrogel.

2.4.2. Frequency Sweep Tests

Frequency sweep tests were carried out on the different hydrogel samples to describe their time-dependent behavior in the non-destructive deformation range (LVE). The shear strain value (γ= 0.1%) to perform the test was chosen within the LVE region, obtained from the amplitude sweep (see Section 3.10.1).

The results of the frequency sweep experiments (Figures S7 and S8) confirm the solid-like behavior of all the hydrogels ($G' > G''$). Some of the samples show a crossover point at high frequencies (always $\omega > 10$ rad/s) where $G' < G''$ and a probable breaking of the gel network occurs. The only samples where G' and G'' are completely independent from the frequency and exhibit purely elastic properties [50] are the gel with Ca 5:1 and the one at pH 5.0, apart from the Ctrl, which has a lower value of G'. From this, it can also be deduced that the hydrogel right after the synthesis, Ctrl, is elastic, while the control sample with the addition of water, Ctrl_H_2O, shows a viscoelastic behavior. This is coherent with the solubilization observed when fully immersed in water. In all samples, G' exhibits a constant trend for the majority of the tested frequencies. This indicates the presence of a stable cross-linked network [51].

2.5. *Structural Characterization*

2.5.1. Infrared Spectroscopy

The attenuated total reflectance (ATR) FT-IR spectra were performed on the freeze-dried hydrogels and used to investigate variations in the interactions in the hydrogel network in the different environments. In all samples, the same pattern of absorption bands was observed and only shifts and changes in relative intensity were observed.

In Figure 5, the spectra of the samples with an increased amount of Ca^{2+} are reported. The addition of Ca^{2+} was observed to induce changes in the O-H stretching (from 3435 to 3372 cm^{-1}) and in the shape of the amide I band, which exhibits two peaks (at 1630 and 1655 cm^{-1}), in the control that merge into a single peak as more Ca^{2+} were added. Additionally, a shift in the peak of C-O stretching (from 1029 to 1025 cm^{-1} and from 1067 to 1064 cm^{-1}) and N-H stretching is observed (from 3099 to 3108 cm^{-1}; from 3280 to 3255 cm^{-1}). No shift was observed in the absorption bands associated with the carbon backbone of the polymer (i.e., sugar ring vibration).

The shifts observed are mostly associated with functional groups involved in hydrogen bonding as a donor or receiver. According to that, Ca^{2+} ions seem to interact in between polymeric chains and mediate hydrogen bonding between the chitin polymeric chains involving hydroxyl groups, and the amide N-H and carbonyl group. This intercalation does not appear to interfere with the polymer chain structure which appears unaltered, meaning no intramolecular difference in the degree of freedom of the polymeric chains is observed.

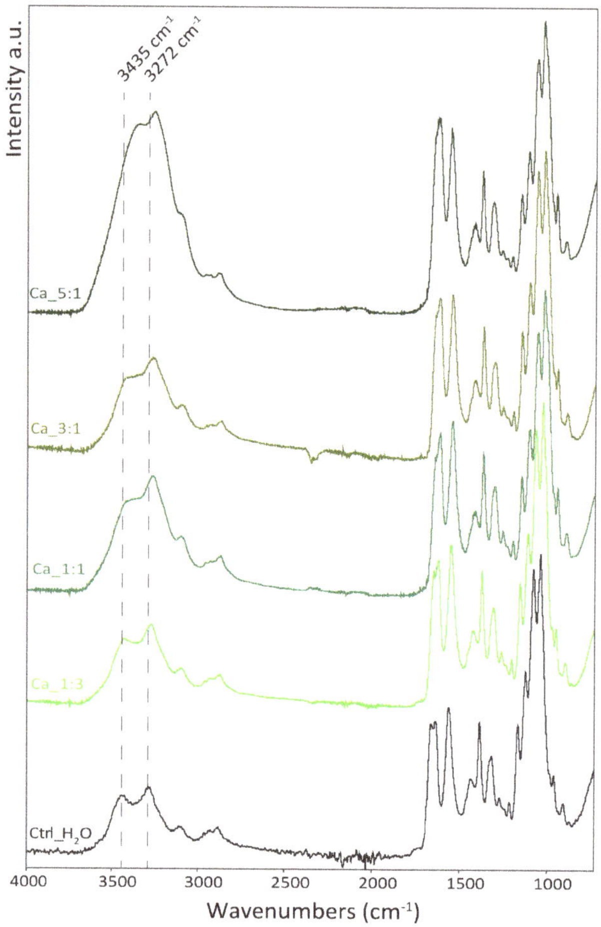

Figure 5. ATR-FTIR spectra of the samples with different Ca^{2+}/carboxylate groups molar ratios. The dashed lines indicate the stretching motion of O-H (3435 cm^{-1}) and N-H stretching (3281 cm^{-1}) referred to the control sample (Ctrl_H_2O).

In the hydrogels stored at different pH conditions, the O-H stretching band shifted to lower wavenumbers (from 3280 to 3255 cm^{-1}) along with the N-H stretching (from 3094 to 3090 cm^{-1}) as the pH decreases. Along with this shift, the two bands tend to merge into one; see Figure 6.

The amide I band exhibits the same trend observed in the hydrogels treated with calcium where the control shows a band split into two peaks that merge into one as the pH decreases.

The pH variation also affects the ring vibration, as evidenced by the shift of the ring stretching band from 894 to 900 cm^{-1}, increasing the pH. Other bands associated with the polymer backbone do not appear to show significant shifts (i.e., asymmetric bridge oxygen stretching, or asymmetric in phase ring stretching).

Contrary to the Ca^{2+}-treated hydrogels, a shift in the CH_2 bending and CH_3 deformation was observed from 1429 cm^{-1} in the control to about 1419 cm^{-1} for all the pH-treated samples. No other significant shift was observed in other apolar moieties.

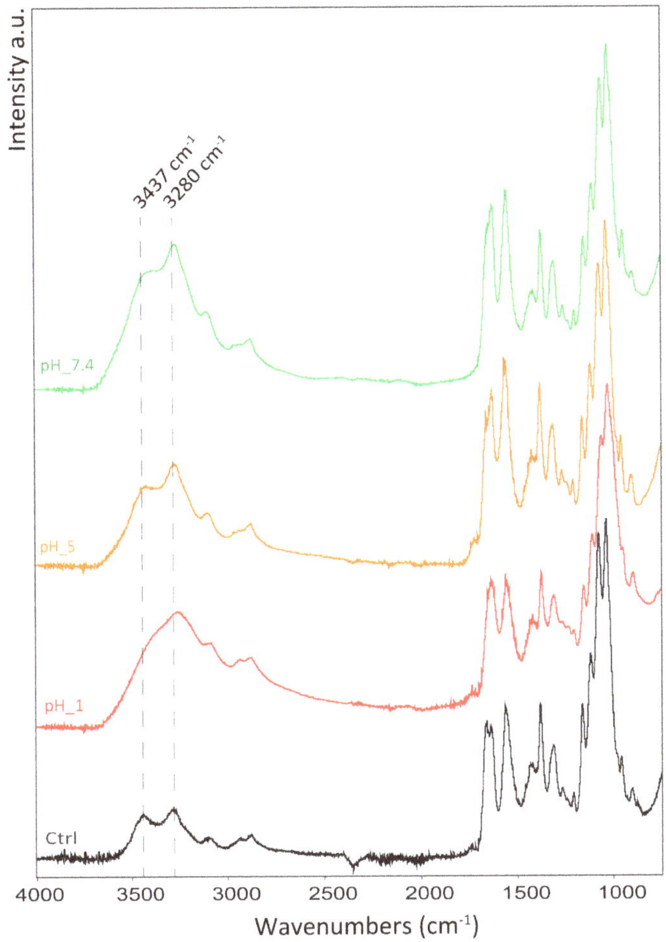

Figure 6. ATR-FTIR spectra of samples in different pH conditions. The dashed lines indicate the stretching motion of O-H (3437 cm^{-1}) and N-H stretching (3280 cm^{-1}) referred to the control sample (Ctrl).

As observed with the hydrogel treated with Ca^{2+}, an increase in the hydrogel rheological properties is associated with shifts to lower wavenumbers in the functional groups involved in hydrogen bonding (O-H, N-H, and C=O). This suggests a shortening of the hydrogen-bonding interactions, leading to more polarized functional groups and higher interactions between polymeric chains. Along with that, pH appears to also force additional interactions between apolar groups that affects the mobility of the chitin ring. This interaction is likely positively driven by the increment in the polarity and ionic strength of the solution.

All peak assignments are provided in the supporting information Table S3 [52].

2.5.2. X-ray Powder Diffraction Analysis

X-ray powder diffraction (XRPD) analysis was performed on the freeze-dried hydrogels to investigate their structural organization.

In the XRPD analyses of the hydrogels obtained with different Ca^{2+} additions, reported in Figure 7a, an overall decrease in the crystallinity was observed, increasing the Ca amount.

This was observed as a decrease in the overall signal intensity; the XRPD were acquired on a similar amount of the sample.

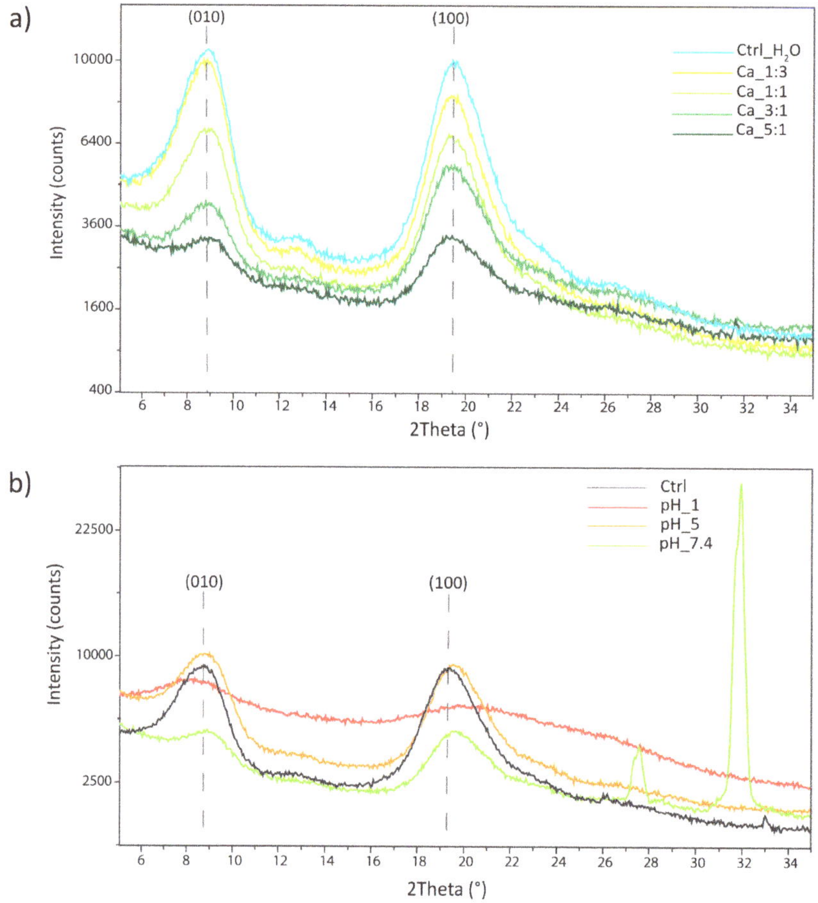

Figure 7. XRPD analyses on freeze-dried hydrogels treated (**a**) with different carboxylate group/calcium ion molar ratios (**b**) in different pH conditions. Chitin's diffraction peaks are a convolution of different signals; the Miller indices reported refer only to the most intense signals which account for most of the signal intensity. In the sample treated at pH 7.4, two additional peaks are visible; those peaks have been attributed to inorganic salts derived from traces of PBS in the hydrogel during freeze-drying.

A reduction in the unit cell parameter along the [010] direction was observed, increasing the Ca^{2+}. Indeed, the peak corresponding to the (010) reflection shifts from a value of 8.82° in the control sample treated with water to a value of 9.06° in the sample treated with Ca 5:1. This corresponds to a decrease in interplanar distance of 0.26 Å. The shift was associated with an increase in the crystallinity of this diffraction peak, visible as an increase in the crystal size. The results are reported in Table 6.

An increment in the a-axis unit cell parameter (0.02 Å), from 19.48° to 19.39°, was observed along the [100] direction, increasing the Ca^{2+}. In this case, a reduction in the crystallinity was observed along this direction. In this direction, an initial increase in crystallinity was observed, moving from the control to the sample treated with Ca 1:3 or 1:1.

Table 6. Diffraction angles and crystallite sizes (d) of the main reflections in the diffractograms obtained for the hydrogels containing different Ca^{2+}/carboxylate group molar ratios. Chitin's diffraction peaks are a convolution of different signals; the Miller indices reported refer only to the most intense signals which account for most of the signal intensity. * The values reported for the Ca_5:1 should be considered indicative, given the low quality of the diffraction peaks.

		(010)			(100)	
Sample	**2θ (°)**	**d (nm)**	**CI (%)**	**2θ (°)**	**d (nm)**	**CI (%)**
Ctrl_H_2O	8.82	3.59	73	19.48	3.08	72
Ca_1:3	8.78	3.59	74	19.41	3.21	70
Ca_1:1	8.87	3.69	69	19.41	3.17	68
Ca_3:1	8.88	3.94	51	19.43	2.89	62
Ca_5:1 *	9.06	4.38	42	19.39	2.77	43

The results suggest that Ca^{2+} ions intercalate in the crystal structure, disrupting its order and decreasing the overall polymer crystallinity. This intercalation is likely initially due to the chelation of the Ca^{2+} by the carboxylate groups, and then probably affects different non-specific binding sites. A similar effect has also been reported in the preparation of a hydrogel starting from carboxymethylcellulose (CMC) crosslinked with calcium ions. In their work, the XRD analysis showed a crystallinity index (CI) of 32.77% for the CMC which becomes equal to only 4.24% when calcium is added [53].

In our case, this interaction is probably occurring along the a-axis (between the two planes of the sugar rings of adjacent polymeric chains) where we theorized the carboxylic group to be mostly located and involved as a hydrogen-bonding acceptor with the N-H of the amide in the adjacent chain. Ca^{2+} is probably inserting in between the two parts of this hydrogen bond, inducing an enlargement of the cell parameter and introducing big defects in the crystal structure. The insertion of calcium ions along the a-axis in between polymeric chains could induce important modifications in the chain-to-chain interaction along the b-axis (where polymeric chains are positioned side-to-side). The presence of Ca^{2+} could seize the carboxylate, eliminating the kinks and defects accumulated along this axis after the functionalization. On the other hand, the presence of a positive charge is probably acting as a bridge between the chains, inducing a contraction of the unit cell along this axis. Due to a lower steric hindrance along this plane and a higher interaction, the cell shrinks and becomes more ordered along this direction.

The XRPD acquired on the hydrogels exposed to different pHs are reported in Figure 7b. This set of samples showed a less clear trend compared to the Ca^{2+}-treated ones.

The overall intensity appeared to be comparable between the control and the sample at pH 5, while the sample at pH 7.4 showed an intensity intermediate to that of the samples treated with Ca 3:1 and 5:1. Finally, the sample treated at pH 1 showed the lowest overall crystallinity among all the hydrogel tested.

The same trend is also observed along the [100] direction where a reduction in the unit cell parameter is observed, moving from the control to pH 5, pH 7.4, and pH 1. The shift was also associated with a reduction in the crystal size. The results are reported in Table 7.

Along the [010] direction, a contraction of the cell parameter is observed at pH 5 and 7.4 along with an increase in crystal size. At pH 1, instead, a strong expansion of the unit cell and a decrease in crystallinity is observed along this direction.

In the sample at pH 7.4, two additional signals were observed at 27.51° and 31.87°. These signals were attributed to inorganic salts derived from the PBS solution entrapped in the hydrogel.

Table 7. Diffraction angles and crystallite sizes (d) of the main reflections in the diffractograms obtained for the hydrogels treated in different pH conditions. Chitin's diffraction peaks are a convolution of different signals; the Miller indices reported refer only to the most intense signals which account for most of the signal intensity. * The values reported for the pH_1 sample should be considered indicative, given the low quality of the diffraction peaks.

	(010)			(100)		
Sample	**2θ (°)**	**d (nm)**	**CI (%)**	**2θ (°)**	**d (nm)**	**CI (%)**
Ctrl	8.71	3.59	73	19.39	3.14	73
pH 1 *	8.30	2.48	52	20.67	0.97	52
pH 5	8.81	3.65	67	19.58	3.07	64
pH 7.4	8.94	4.15	31	19.64	2.91	13

The XRPD results suggest a different interaction is governing the aggregation at pH 7.4 and 5 since a contraction is observed along the [100] direction, suggesting a positive contribution by the presence of ions to the interchain interactions. The carboxylate protonation in acid environment could, in fact, induce a change in the hydrogen-bonding pattern along the a-axis. It would likely interact with the C=O of the amide of the adjacent polymeric chain, restoring the native hydrogen bond pattern of β-chitin. Most likely, both ions' interactions, coherent with the observations in the (010) plane, and protonation are occurring, leading to the increase in disorder observed.

Contrary to all the previous samples, the sample at pH 1 appears strongly altered, up to almost complete amorphization. Due to that, the results concerning the peak position or the crystal size cannot be considered reliable. This strong alteration may be the result of a complete protonation of the carboxylic acids and/or the co-ordination of positive charges in the crystal in non-specific binding sites. In fact, β-chitin has already been reported as strongly swellable in acid pHs [7,54] and this amorphization may have been guided to the same interactions.

2.5.3. Scanning Electron Microscopy Analyses

Scanning electron microscopy (SEM) was used to analyze the morphology and aggregation of the freeze-dried hydrogels. After freeze-drying, all the samples retained their shape as hydrogels and showed a soft consistency; see Figure S9. The hydrogel treated at pH 1, instead, showed a strong shrinkage and a brittle consistency.

Observing the samples with SEM, all samples showed a porous structure with smooth sheet-like walls and filaments of fibers embedded in the matrix. The increment in the Ca^{2+} addition was observed to be associated with a thickening of the pore walls and a reduction in the free fiber presence (Figure 8).

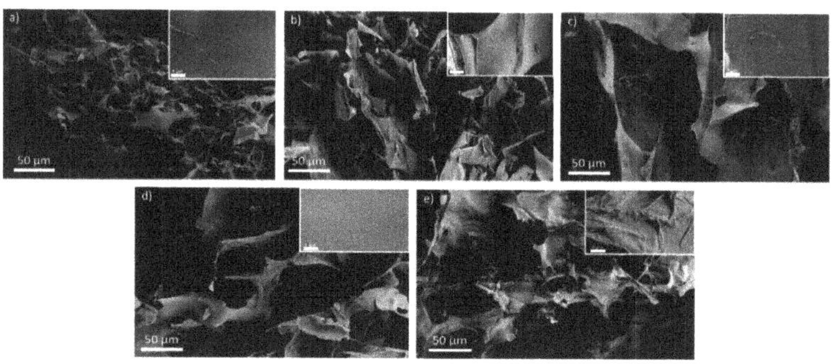

Figure 8. SEM images of the freeze-dried hydrogels: (**a**) Ctrl_H_2O; (**b**) Ca_1:3; (**c**) Ca_1:1; (**d**) Ca 3:1; and (**e**) Ca_5:1. For each condition, an insight with a higher magnification is reported.

A similar wall thickening and fiber reduction was observed also in the samples exposed to different pHs, as is visible in Figure 9. This effect followed the same trend observed in the XRPD, control sample, pH 5, pH 7.4, and, finally, pH 1. The latter showed the thickest walls among all samples and exhibited sharper and more delineated fractures.

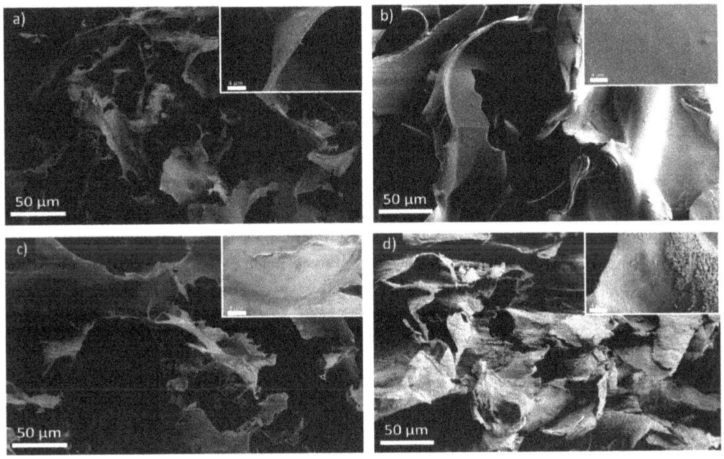

Figure 9. SEM images of the freeze-dried hydrogels exposed to different pHs: (**a**) Ctrl; (**b**) pH_1; (**c**) pH_5; and (**d**) pH_7.4. For each condition, an insight with a higher magnification is reported.

The sheet-like morphology observed could be an artifact of the sample freezing and may not be representative of the hydrogel organization in solution. Despite that, the SEM analyses suggests that an increase in the interactions of the hydrogel occurs, increasing the cation concentration in solution—we see a thickening of the laminar structures. This thickening is probably due to the aggregation of loosely bound fibers, which are then forced to interact with the hydrogel network.

2.6. Discussion on Hydrogel Tunability

The characterizations performed on the hydrogels exposed to different environments showed how the rheological properties and the structure of the material can be rearranged based on diverse triggers. Such a rearrangement can increase the storage modulus of the gel from 5 to 10 times. These triggers are generally related to positively charged chemical species, such as metal cations or hydrogen ions.

These cations interact with the polymeric chains and even intercalate in the chitin's crystal structure. The cations seem to interact with groups generally involved in hydrogen bonding, namely, alcohols, amines, and amides (both N-H and C=O). A strong interaction is also likely occurring with the carboxylic group. These interactions probably see the cation involved in the dipolar interaction between these groups.

When the ions intercalate in the crystalline regions, this seems to happen along the a-axis, in between the sugar ring plane of the polymeric chains, causing a decrease in crystallinity. When Ca^{2+} ions are intercalating, an expansion of the crystallographic unit cell in this direction is observed. On the other hand, a contraction is observed when monovalent cations, like Na^{+}, K^{+}, or H^{+}, interact with the hydrogel crystal units. This contraction is probably the reason why the CH_2 bending and the sugar ring stretching are affected.

Simultaneously, a contraction over the b-axis, where the polymeric chains are arranged side-to-side, is observed in all the conditions tested, along with an increase in crystallinity. This suggests an increment of the interactions over this direction, generally presenting very little direct interactions in β-chitin. Overall, the intercalation of cations in the crystal is associated with a decrease in crystallinity due to a disruption of the crystal structure order; this loss in crystallinity increases, increasing the ion concentration in solution.

The morphological examination of the freeze-dried hydrogels suggests an increase in the hydrogel network interaction, increasing the cation presence. It also shows how the fibrils in the hydrogel are only partially interacting in the control, and then increase their interaction when exposed to cations. This observation suggests that the cations induce the formation of additional linkage in the matrix, or favor the formation of the one already possible, forcing new fibrils to aggregate in the hydrogel network.

It is worth mentioning that the frequency sweep tests identified four samples with viscoelastic behavior instead of an elastic one. Except for one of the controls, the samples exhibiting a viscoelastic behavior are the ones exposed to the lowest and highest ion amounts—respectively, the Ca 1:3, the sample in PBS (pH 7.4), and the one at pH 1. This suggests an optimal ionic exposure that maximize the properties of the hydrogel, giving it an elastic behavior. This condition varies when different ions are used. Such a change in the rheological properties is probably due to an excessive aggregation of the fibrils, as visible in the SEM of the sample at pH 1, which confer a more brittle behavior.

3. Materials and Methods

3.1. Materials

The β-chitin flakes from the squid pens of *Loligo vulgaris* were purchased from BioLog Heppe® GmbH (Landsberg, Germany). The α-chitin powder from shrimp shells was purchased from Glentham Life Sciences (Planegg, Germany). All other chemical reagents were purchased from Merck and used without any additional purification.

3.2. Chitin Ball Milling

The initial β-chitin powder was ground with a ball mill 8000 Mixer Mill™ SPEX CertiPrep™ (Metuchen, NJ, USA). The jar used has a diameter of 3.85 cm and height 5.70 cm. The grinding was carried out with three small spheres with a diameter of 0.62 cm and three spheres with a diameter of 1.27 cm. Both the jar and the spheres are made of hardened steel. Then, 2 g of powder were ground at a time for 10 min. Subsequently, both β-chitin and α-chitin were sieved through 600, 150, and 45 μm sieves (GIULIANI Tecnologie srl, Turin, Italy) to separate different powder meshes.

3.3. X-ray Powder Diffraction Analysis

X-ray diffraction patterns were collected using a PanAnalytical X'Pert Pro diffractometer equipped with a multiarray X'Celerator detector using Cu Kα radiation generated at 40 kV and 40 mA (λ = 1.54056 Å).

The diffraction patterns were collected in the 2θ range between 5° and 35° with a step size (Δ2θ) of 0.05° and time per step of 123 s.

The peak maximum and full width at half maximum (FWHM) measurements were calculated using the PanAnalytical X'Pert Data Viewer software (version 1.2d). The value was then used to determine the crystallite size (d) using the Sherrer equation.

The crystal index (CI) was determined as a percentage using the following equation [55,56]:

$$CI = [(I_{cryst} - I_{am})/I_{cryst}] \times 100$$

where I_{cryst} is the intensity in counts at the maximum of the reflex and I_{am} is the intensity of amorphous diffraction at $2\theta \cong 14°$.

3.4. Chitin Oxidation

For the oxidation reaction, 0.5 g of chitin powder (600–150 μm) were added in 50 mL of a 45 wt.% ammonium persulfate solution (APS) at 40 °C under stirring for diverse reaction times in a round bottom flask. For gel production, β-chitin was stirred for 24 h.

After that, the solution appears as an opalescent dispersion. The material in suspension was collected and washed with deionized water by consecutive centrifugation with an Eppendorf Centrifuge 5810 R at 8000 rpm for 5 min until it reached neutrality.

The powder was then resuspended in 30 mL of milliQ water and left under magnetic stirring overnight, to eliminate possible reaction residues. The next day, the powder was centrifuged, as was carried out previously, and resuspended in about 15 mL of milliQ water, frozen in the freezer, and desiccated by freeze-drying using a FreeZone 1 (Labconco Corp., Kansas City, MO, USA). The product was conserved dry at room temperature.

3.5. ss-NMR Measurements

The ^{13}C solid-state nuclear magnetic resonance (ss-NMR) experiments were performed on a Bruker 700 MHz Wide Bore spectrometer equipped with a 4 mm CP MAS X/Y/H DVT probe in a double-resonance H-X configuration. The analyses were performed operating at 260 K and MAS speed of 12 kHz. The CP contact time was 2 ms, and the recycle delay was 10 s. A typical number of 512 or 2048 scans were acquired for each spectrum.

The degree of functionalization (%*DF*) was calculated, setting the total integral for C3, C4, and C5 to 3 and measuring the C6 integral (I_t). The %*DF* was then calculated using the following formula:

$$\%DF = \left(1 - \frac{I_t}{I_{t0}}\right) \times 100$$

where I_t is the integral of the C6 signal attributed to the various oxidized chitin samples and it is the one of the initial chitins. The NMR signal attribution is reported in Figure S4 [15,57].

3.6. UV–Visible Spectroscopy

Spectrophotometric measurements were carried out with a UV–vis spectrophotometer (Varian Cary 300 Bio UV–Visible Spectrophotometer, Santa Clara, CA, USA) using a spectral range of 300–900 nm, scan rate of 600 nm·min^{-1}, with a resolution of 1 nm. Turbidity measurements were measured at 500 nm.

3.7. FTIR-Spectroscopy

The infrared spectra were collected using a Fourier-transform infrared spectroscopy (FTI) (Thermo Scientific Nicolet iS10, Waltham, MA, USA) and processed with Omnic software 9.8.286 (Thermo Electron Corp., Woburn, MA, USA).

The samples were prepared in KBr pellets (1 wt.% of sample) and the spectra were collected with 2 cm^{-1} resolution and 70 scans.

The intensity of each peak was obtained by correcting the intensity on the baseline [30]. The oxidation degree was then evaluated by calculating the ratio between the carboxylic acid peak (1730 cm^{-1}) and the most intense peak of the glucosidic ring which falls at 1074 cm^{-1} [58].

3.8. Product Yeld

The product yield was calculated based on the initial mass (mg_i) of chitin powder and the final mass (mg_f) obtained from the oxidation reaction and calculated using the following formula:

$$product\ yield = \frac{mg_f}{mg_i} \times 100$$

3.9. Hydrogel Preparation

Oxidized β-chitin powder (βchitOx) was dispersed in milliQ water to obtain a 1 wt.% dispersion. This was vigorously stirred using a magnetic stirrer for about 2 h at room temperature.

Once no visible aggregates were observed by the naked eye, the pH of the dispersion was between 3 and 4. A required volume of a 1 M NaOH solution was added to the dispersion to bring the pH between 8 and 9.

The obtained dispersion was divided into 2 mL aliquots inside sterilin tubes (Thermo Scientific™ Sterilin™ 7 mL Polystyrene Bijou Containers, Waltham, MA, USA) and sonicated at 10% amplitude with a tip sonicator (Branson Ultrasonics™ Brookfield, WI, USA; Sonifier 250; 20 kHz) equipped with a semimicro tip, with the container immersed in an ice bath to prevent overheating. The sonicated β-chitOx dispersion was then left to rest to obtain the βchitOx-hydrogel. Since only one type of unaltered (with Ca^{2+} or pH) hydrogel was prepared in this study, we will refer to this sample as hydrogel or control.

3.9.1. Metal Ion Addition to the Hydrogel

To investigate the effect of Ca^{2+} on the hydrogel, different $CaCl_2$ solutions were prepared at concentrations of 16.6, 50, 150, and 250 mM. A volume equal to 196 μL of these solutions was used to obtain molar ratios of 1:3, 1:1, 3:1, and 5:1 with the carboxylate groups formed after the reaction of oxidation (quantified by ss-NMR measurement). The tests on Mg^{2+} and Na^{+} were performed using $MgCl_2$ and NaCl solutions.

Different addition methods were tested:

- Surface addition: Allowing the solution to drip from the walls of the sterilin tubes containing the hydrogel.
- Mixed addition: Inserting the solution into the hydrogel, and then mixing with the pipette tip. This process disrupts the hydrogel to better blend the solution.
- Calcium solution addition before the sonication step.

The hydrogel is then left to rest for 24 h to allow the Ca^{2+} to diffuse homogenously.

3.9.2. Study of pH Influence on the Hydrogel

The sterilins containing the hydrogels were immersed in beakers containing 90 mL of solutions at different pH values: 1 (0.1 M HCl); 5 (10 mM citrate buffer $Na_3C_6H_5O_7/C_6H_8O_7$); 7.4 (phosphate buffered saline, PBS); 9 (10 mM carbonate buffer $NaHCO_3$; Na_2CO_3); and MilliQ water used as control. The sterilins were left in solution with a mild stirring on a rocking table for 24 h; along this time, the buffer solution was changed twice to maintain a constant environment.

3.10. Hydrogel Characterization

3.10.1. Rheological Properties

The rheological properties of the different hydrogels were assessed using an Anton Paar (Graz, Austria) MCR102 rheometer. The tested hydrogels (2 mL) were directly prepared in 7 mL Sterilin™ polystyrene containers (Thermo Scientific™, Waltham, MA, USA) which fit in the rheometer. A cup and vane geometry was used, setting a gap of 2.1 cm. Oscillatory amplitude sweep experiments (ɣ: 0.01–100%) were performed using a constant angular frequency of 10 rad·s^{-1}. The frequency sweep experiments were performed at a constant shear strain of ɣ= 0.1% (within the LVE region) with an increasing angular frequency (ω) from 0.1 to 100 rad·s^{-1}. All tests were performed at a fixed temperature of 23 °C, controlled by an integrated Peltier system.

3.10.2. FTIR-Spectroscopy

The hydrogels were analyzed using the FTIR equipment described in Section 3.7. The hydrogels were frozen with liquid nitrogen, and then dried by freeze-drying using a FreeZone 1 (Labconco Corp., Kansas City, MO, USA). The dry samples were measured in attenuated total reflectance (ATR) mode using 1 cm^{-1} resolution and 200 scans. The data analysis was then performed using the OMNIC software 9.8.286.

3.10.3. X-ray Powder Diffraction Analysis

For the X-ray powder analysis of the freeze-dried hydrogels (see Section 3.10.2), the same instrumentation described in Section 3.3 was used. The diffraction patterns were collected in the 2θ range between 5° and 35° with a step size (Δ2θ) of 0.05° and a time per step of 360 s.

3.10.4. Scanning Electron Microscopy Images

For the scanning electron microscopy (SEM) imaging, the samples were dried, as reported; see Section 3.10.2 (Figure S8) A piece of the dry samples was then cut with a scalpel, glued on a carbon tape, and coated with 10–20 nm of gold prior analyses. The images were collected with a SEM Zeiss LEOc1530 FEG using a voltage of 5 kV and an aperture of 20 μm.

4. Conclusions

In this study, we prepared tunable chitin hydrogels, exploring in detail, optimizing, and customizing a known synthetic process. The grain size of the starting material (between 600 μm and 150 μm) and the reaction time (24 h) were optimized to achieve highly crystalline oxidized β-chitin fibrils (having a %DF of 10%) with a yield of a 32 wt.%. Different reaction conditions were also tested on α-chitin, which exhibited a lower reactivity and a much longer reaction kinetic. Structural characterization (FT-IR and XRPD) was performed on all the reaction products to identify how the functionalization affects the structure of the biopolymer.

Successively, a hydrogel was obtained from the oxidized β-chitin. Such a hydrogel exhibited tunable rheological properties (up to a 10-times increment in G') when exposed to cations (Ca^{2+}, Mg^{2+}, Na^{+}, or H^{+}), which induce structural rearrangements. The initial gelation is induced by a few minutes of sonication and occurs along the time, but it can be fastened if cations are present. Such a process could be compatible with an application as injectable gels for wound sealing or surgery since the physiological ion concentration would be enough to induce the formation of a tough hydrogel (as observed in PBS). Moreover, the hydrogel is stable in physiological environments, both healthy (pH 7.4), inflamed or unhealthy [59,60] (pH 5), and those connected to digestion (pH 1). This allows its use as a biomedical material for application with almost every human tissue. This also makes them a material of potential interest for the development of wound dressings. On the other hand, the instability of these hydrogels in alkaline environments and pure water make their disposal extremely easy and eco-friendly. Potentially, the hydrogel could even be dissolved at pH 9, and the fibrils could be recovered and recycled to produce a new hydrogel. Such a solubility could also be exploited to deliver biologically active compounds into the stomach of animals having an alkaline stomach pH, such as pests, like insects, or economically relevant animals, like sea urchins [61,62].

In conclusion, these negatively charged β-chitin fibrils could act as complementary materials to positively charged chitosan fibers. Additionally, their potential to obtain acid-stable tough hydrogels, that, alternatively, could be reinforced using metal cations, could represent an alternative to the structurally similar alginate hydrogels. This study sheds light on an alternative material with peculiar properties that could represent an important resource for the development of future technological materials, especially in the biomedical field.

Supplementary Materials: The following supporting information can be downloaded at: https://www.mdpi.com/article/10.3390/md22040164/s1, the full characterization of the oxidized chitins, missing rheological data or characterization of the hydrogel not reported in the main text (including frequency sweep tests), and the interpretation of the FT-IR bands.

Author Contributions: Conceptualization, G.F.; methodology, A.M., D.G., C.T., G.F. and D.M.; validation, A.M., D.G., C.T., G.F. and D.M.; formal analysis, A.M., D.G. and D.M.; investigation, A.M. and D.M.; resources, C.T. and G.F.; data curation, A.M., D.G. and D.M.; writing—original draft preparation, A.M.; writing—review and editing, A.M., D.G., C.T., G.F. and D.M.; visualization, G.F. and D.M.; supervision, D.G., C.T., G.F. and D.M.; project administration, G.F. and D.M.; funding acquisition, G.F. All authors have read and agreed to the published version of the manuscript.

Funding: G.F. thanks the National Recovery and Resilience Plan (NRRP), Mission 4 Component 2 Investment 1.4—Call for tender No. 3138 of 16 December 2021, rectified by Decree n.3175 of 18 December 2021 of Italian Ministry of University and Research funded by the European

Union—NextGenerationEU, project code CN_00000033, Concession Decree No. 1034 of 17 June 2022 adopted by the Italian Ministry of University and Research, Project title "National Biodiversity Future Center—NBFC".

Data Availability Statement: The data can be made available upon request to the corresponding author.

Acknowledgments: This work benefited from the support and the use of the resources of Instruct-ERIC, and, specifically, of the CERM/CIRMMP Italy Centre (VID:35740).

Conflicts of Interest: The authors declare no conflicts of interest.

References

1. Liu, L.; Wang, R.; Yu, J.; Jiang, J.; Zheng, K.; Hu, L.; Wang, Z.; Fan, Y. Robust Self-Standing Chitin Nanofiber/Nanowhisker Hydrogels with Designed Surface Charges and Ultralow Mass Content via Gas Phase Coagulation. *Biomacromolecules* **2016**, *17*, 3773–3781. [CrossRef] [PubMed]
2. Liao, J.; Hou, B.; Huang, H. Preparation, Properties and Drug Controlled Release of Chitin-Based Hydrogels: An Updated Review. *Carbohydr. Polym.* **2022**, *283*, 119177. [CrossRef]
3. Ullah, F.; Othman, M.B.H.; Javed, F.; Ahmad, Z.; Akil, H.M. Classification, Processing and Application of Hydrogels: A Review. *Mater. Sci. Eng. C* **2015**, *57*, 414–433. [CrossRef]
4. Catoira, M.C.; Fusaro, L.; Di Francesco, D.; Ramella, M.; Boccafoschi, F. Overview of Natural Hydrogels for Regenerative Medicine Applications. *J. Mater. Sci. Mater. Med.* **2019**, *30*, 115. [CrossRef]
5. Montroni, D.; Marzec, B.; Valle, F.; Nudelman, F.; Falini, G. β-Chitin Nanofibril Self-Assembly in Aqueous Environments. *Biomacromolecules* **2019**, *20*, 2421–2429. [CrossRef]
6. Montroni, D. Hierarchically Organized Chitin-Based Matrices. Ph.D. Thesis, University of Bologna, Bologna, Italy, 2020.
7. Barbalinardo, M.; Biagetti, M.; Valle, F.; Cavallini, M.; Falini, G.; Montroni, D. Green Biocompatible Method for the Synthesis of Collagen/Chitin Composites to Study Their Composition and Assembly Influence on Fibroblasts Growth. *Biomacromolecules* **2021**, *22*, 3357–3365. [CrossRef] [PubMed]
8. Tamura, H.; Furuike, T.; Nair, S.V.; Jayakumar, R. Biomedical Applications of Chitin Hydrogel Membranes and Scaffolds. *Carbohydr. Polym.* **2011**, *84*, 820–824. [CrossRef]
9. Peralta Ramos, M.L.; González, J.A.; Albornoz, S.G.; Pérez, C.J.; Villanueva, M.E.; Giorgieri, S.A.; Copello, G.J. Chitin Hydrogel Reinforced with TiO2 Nanoparticles as an Arsenic Sorbent. *Chem. Eng. J.* **2016**, *285*, 581–587. [CrossRef]
10. Ahmad, S.I.; Ahmad, R.; Khan, M.S.; Kant, R.; Shahid, S.; Gautam, L.; Hasan, G.M.; Hassan, M.I. Chitin and Its Derivatives: Structural Properties and Biomedical Applications. *Int. J. Biol. Macromol.* **2020**, *164*, 526–539. [CrossRef]
11. Huang, W.; Montroni, D.; Wang, T.; Murata, S.; Arakaki, A.; Nemoto, M.; Kisailus, D. Nanoarchitected Tough Biological Composites from Assembled Chitinous Scaffolds. *Acc. Chem. Res.* **2022**, *55*, 1360–1371. [CrossRef]
12. Lee, J.E.; Connolloy, J.; Yang, W.; Freychet, G.; Wang, T.; Herrera, S.A.; Murata, S.; Dasika, P.S.; Montroni, D.; Pohl, A.; et al. Fibrous Anisotropy and Mineral Gradients Within the Radula Stylus of Chiton: Controlled Stiffness and Damage Tolerance in a Flexible Biological Composite. *J. Compos. Mater.* **2023**, *57*, 565–574. [CrossRef]
13. Montroni, D.; Zhang, X.; Leonard, J.; Kaya, M.; Amemiya, C.; Falini, G.; Rolandi, M. Structural Characterization of the Buccal Mass of Ariolimax Californicus (Gastropoda; Stylommatophora). *PLoS ONE* **2019**, *14*, e212249. [CrossRef] [PubMed]
14. Montroni, D.; Leonard, J.; Rolandi, M.; Falini, G. Morphology and Organization of the Internal Shell of Ariolimax Californicus (Gastropoda; Stylommatophora), an Asymmetric Two-Face Biomineralized Matrix. *J. Struct. Biol.* **2021**, *2013*, 107764. [CrossRef] [PubMed]
15. Jang, M.K.; Kong, B.G.; Jeong, Y., II; Lee, C.H.; Nah, J.W. Physicochemical Characterization of α-Chitin, β-Chitin, and γ-Chitin Separated from Natural Resources. *J. Polym. Sci. A Polym. Chem.* **2004**, *42*, 3423–3432. [CrossRef]
16. Fan, Y.; Saito, T.; Isogai, A. Chitin Nanocrystals Prepared by TEMPO-Mediated Oxidation of α-Chitin. *Biomacromolecules* **2008**, *9*, 192–198. [CrossRef]
17. Shen, X.; Shamshina, J.L.; Berton, P.; Gurau, G.; Rogers, R.D. Hydrogels Based on Cellulose and Chitin: Fabrication, Properties, and Applications. *Green Chem.* **2015**, *18*, 53–75. [CrossRef]
18. Fang, Y.; Zhang, R.; Duan, B.; Liu, M.; Lu, A.; Zhang, L. Recyclable Universal Solvents for Chitin to Chitosan with Various Degrees of Acetylation and Construction of Robust Hydrogels. *ACS Sustain. Chem. Eng.* **2017**, *5*, 2725–2733. [CrossRef]
19. Deng, L.; Zhang, L.M. Rheological Characteristics of Chitin/Ionic Liquid Gels and Electrochemical Properties of Regenerated Chitin Hydrogels. *Colloids Surf. A Physicochem. Eng. Asp.* **2020**, *586*, 124220. [CrossRef]
20. Tamura, H.; Nagahama, H.; Tokura, S. Preparation of Chitin Hydrogel under Mild Conditions. *Cellulose* **2006**, *13*, 357–364. [CrossRef]
21. Yilmaz, E.; Bengisu, M. Preparation and Characterization of Physical Gels and Beads from Chitin Solutions. *Carbohydr. Polym.* **2003**, *54*, 479–488. [CrossRef]
22. Mukesh, C.; Mondal, D.; Sharma, M.; Prasad, K. Choline Chloride-Thiourea, a Deep Eutectic Solvent for the Production of Chitin Nanofibers. *Carbohydr. Polym.* **2014**, *103*, 466–471. [CrossRef]

23. Montroni, D.; Kobayashi, T.; Hao, T.; Lublin, D.; Yoshino, T.; Kisailus, D. Direct Ink Write Printing of Chitin-Based Gel Fibers with Customizable Fibril Alignment, Porosity, and Mechanical Properties for Biomedical Applications. *J. Funct. Biomater.* **2022**, *13*, 83. [CrossRef] [PubMed]
24. Liang, K.; Chang, B.; Tsai, G.; Lee, J.; Fu, W.-R. Heterogeneous N-Deacetylation of Chitin in Alkaline Solution. *Carbohydr. Res.* **1997**, *303*, 327–332.
25. Liu, H.; Yang, Q.; Zhang, L.; Zhuo, R.; Jiang, X. Synthesis of Carboxymethyl Chitin in Aqueous Solution and Its Thermo- and PH-Sensitive Behaviors. *Carbohydr. Polym.* **2016**, *137*, 600–607. [CrossRef] [PubMed]
26. Shu, X.Z.; Zhu, K.J. Controlled Drug Release Properties of Ionically Cross-Linked Chitosan Beads: The Influence of Anion Structure. *Int. J. Pharm.* **2002**, *233*, 217–225. [CrossRef] [PubMed]
27. Shen, E.C.; Wang, C.; Fu, E.; Chiang, C.Y.; Chen, T.T.; Nieh, S. Tetracycline Release from Tripolyphosphate-Chitosan Cross-Linked Sponge: A Preliminary in Vitro Study. *J. Periodontal. Res.* **2008**, *43*, 642–648. [CrossRef] [PubMed]
28. Brack, H.P.; Tirmizi, S.A.; Risen, W.M. A Spectroscopic and Viscometric Study of the Metal Ion-Induced Gelation of the Biopolymer Chitosan. *Polymer* **1997**, *38*, 2351–2362. [CrossRef]
29. Dambies, L.; Vincent, T.; Domard, A.; Guibal, E. Preparation of Chitosan Gel Beads by Ionotropic Molybdate Gelation. *Biomacromolecules* **2001**, *2*, 1198–1205. [CrossRef] [PubMed]
30. Ma, Q.; Pang, K.; Wang, K.; Huang, S.; Ding, B.; Duan, Y.; Zhang, J. Ultrafine and Carboxylated β-Chitin Nanofibers Prepared from Squid Pen and Its Transparent Hydrogels. *Carbohydr. Polym.* **2019**, *211*, 118–123. [CrossRef]
31. Fan, Y.; Saito, T.; Isogai, A. TEMPO-Mediated Oxidation of β-Chitin to Prepare Individual Nanofibrils. *Carbohydr. Polym.* **2009**, *77*, 832–838. [CrossRef]
32. Kato, Y.; Kaminaga, J.; Matsuo, R.; Isogai, A. TEMPO-Mediated Oxidation of Chitin, Regenerated Chitin and N-Acetylated Chitosan. *Carbohydr. Polym.* **2004**, *58*, 421–426. [CrossRef]
33. Oun, A.A.; Rhim, J.W. Effect of Oxidized Chitin Nanocrystals Isolated by Ammonium Persulfate Method on the Properties of Carboxymethyl Cellulose-Based Films. *Carbohydr. Polym.* **2017**, *175*, 712–720. [CrossRef] [PubMed]
34. Liu, P.; Liu, H.; Schäfer, T.; Gutmann, T.; Gibhardt, H.; Qi, H.; Tian, L.; Zhang, X.C.; Buntkowsky, G.; Zhang, K. Unexpected Selective Alkaline Periodate Oxidation of Chitin for the Isolation of Chitin Nanocrystals. *Green Chem.* **2021**, *23*, 745–751. [CrossRef]
35. Montembault, A.; Tahiri, K.; Korwin-Zmijowska, C.; Chevalier, X.; Corvol, M.T.; Domard, A. A Material Decoy of Biological Media Based on Chitosan Physical Hydrogels: Application to Cartilage Tissue Engineering. *Biochimie* **2006**, *88*, 551–564. [CrossRef] [PubMed]
36. Ching, S.H.; Bansal, N.; Bhandari, B. Alginate Gel Particles–A Review of Production Techniques and Physical Properties. *Crit. Rev. Food Sci. Nutr.* **2017**, *57*, 1133–1152. [CrossRef] [PubMed]
37. Moussian, B. Chitin: Structure, Chemistry and Biology. In *Advances in Experimental Medicine and Biology*; Springer New York LLC.: New York, NY, USA, 2019; Volume 1142, pp. 5–18.
38. Nagahama, H.; Higuchi, T.; Jayakumar, R.; Furuike, T.; Tamura, H. XRD Studies of β-Chitin from Squid Pen with Calcium Solvent. *Int. J. Biol. Macromol.* **2008**, *42*, 309–313. [CrossRef] [PubMed]
39. De, J.A.; Delezuk, M.; Pavinatto, A.; Campana-Filho, S.P. Influence of the Process Parameters on β-Chitin and α-Chitin Extraction: Probing about the Grinding and Particles Size. *Mater. Today Proc.* **2019**, *14*, 722–732.
40. Jiang, J.; Ye, W.; Yu, J.; Fan, Y.; Ono, Y.; Saito, T.; Isogai, A. Chitin Nanocrystals Prepared by Oxidation of α-Chitin Using the O2/Laccase/TEMPO System. *Carbohydr. Polym.* **2018**, *189*, 178–183. [CrossRef] [PubMed]
41. Giuri, D.; Jurković, L.; Fermani, S.; Kralj, D.; Falini, G.; Tomasini, C. Supramolecular Hydrogels with Properties Tunable by Calcium Ions: A Bio-Inspired Chemical System. *ACS Appl. Bio. Mater.* **2019**, *2*, 5819–5828. [CrossRef]
42. Mezger, T.G. *Applied Rheology-with Joe Flow on Rheology Road*; Anton Paar: Graz, Austria, 2014.
43. Uraki, Y.; Fujii, T.; Matsuoka, T.; Miura, Y.; Tokura, S. Site Specific Binding of Calcium Ions to Anionic Chitin Derivatives. *Carbohydr. Polym.* **1993**, *20*, 139–143. [CrossRef]
44. Cofelice, M.; Messia, M.C.; Marconi, E.; Cuomo, F.; Lopez, F. Effect of the Xanthan Gum on the Rheological Properties of Alginate Hydrogels. *Food Hydrocoll.* **2023**, *142*, 108768. [CrossRef]
45. Syverud, K.; Pettersen, S.R.; Draget, K.; Chinga-Carrasco, G. Controlling the Elastic Modulus of Cellulose Nanofibril Hydrogels—Scaffolds with Potential in Tissue Engineering. *Cellulose* **2015**, *22*, 473–481. [CrossRef]
46. Cuomo, F.; Cofelice, M.; Lopez, F. Rheological Characterization of Hydrogels from Alginate-Based Nanodispersion. *Polymers* **2019**, *11*, 259. [CrossRef] [PubMed]
47. Derkach, S.R.; Voron'ko, N.G.; Sokolan, N.I. The Rheology of Hydrogels Based on Chitosan–Gelatin (Bio)Polyelectrolyte Complexes. *J. Dispers. Sci. Technol.* **2017**, *38*, 1427–1434. [CrossRef]
48. Prasad, K.; Murakami, M.A.; Kaneko, Y.; Takada, A.; Nakamura, Y.; Kadokawa, J.I. Weak Gel of Chitin with Ionic Liquid, 1-Allyl-3-Methylimidazolium Bromide. *Int. J. Biol. Macromol.* **2009**, *45*, 221–225. [CrossRef]
49. Sondermeijer, H.P.; Witkowski, P.; Woodland, D.; Seki, T.; Aangenendt, F.J.; Van Der Laarse, A.; Itescu, S.; Hardy, M.A. Optimization of Alginate Purification Using Polyvinylidene Difluoride Membrane Filtration: Effects on Immunogenicity and Biocompatibility of Three-Dimensional Alginate Scaffolds. *J. Biomater. Appl.* **2016**, *31*, 510–520. [CrossRef] [PubMed]
50. Tunn, I.; Harrington, M.J.; Blank, K.G. Bioinspired Histidine-Zn2+ Coordination for Tuning the Mechanical Properties of Self-Healing Coiled Coil Cross-Linked Hydrogels. *Biomimetics* **2019**, *4*, 25. [CrossRef] [PubMed]

51. Moura, M.J.; Figueiredo, M.M.; Gil, M.H. Rheological Study of Genipin Cross-Linked Chitosan Hydrogels. *Biomacromolecules* **2007**, *8*, 3823–3829. [CrossRef] [PubMed]
52. Montroni, D.; Fermani, S.; Morellato, K.; Torri, G.; Naggi, A.; Cristofolini, L.; Falini, G. β-Chitin Samples with Similar Microfibril Arrangement Change Mechanical Properties Varying the Degree of Acetylation. *Carbohydr. Polym.* **2019**, *207*, 26–33. [CrossRef]
53. Tuan Mohamood, N.F.A.Z.; Abdul Halim, A.H.; Zainuddin, N. Carboxymethyl Cellulose Hydrogel from Biomass Waste of Oil Palm Empty Fruit Bunch Using Calcium Chloride as Crosslinking Agent. *Polymers* **2021**, *13*, 4056. [CrossRef]
54. Montroni, D.; Sparla, F.; Fermani, S.; Falini, G. Influence of Proteins on Mechanical Properties of a Natural Chitin-Protein Composite. *Acta Biomater.* **2021**, *120*, 81–90. [CrossRef] [PubMed]
55. Pacheco, N.; Garnica-Gonzalez, M.; Gimeno, M.; Bárzana, E.; Trombotto, S.; David, L.; Shirai, K. Structural Characterization of Chitin and Chitosan Obtained by Biological and Chemical Methods. *Biomacromolecules* **2011**, *12*, 3285–3290. [CrossRef] [PubMed]
56. Kaya, M.; Baran, T.; Mentes, A.; Asaroglu, M.; Sezen, G.; Tozak, K.O. Extraction and Characterization of α-Chitin and Chitosan from Six Different Aquatic Invertebrates. *Food Biophys.* **2014**, *9*, 145–157. [CrossRef]
57. Wang, Q.; Yan, X.; Chang, Y.; Ren, L.; Zhou, J. Fabrication and Characterization of Chitin Nanofibers through Esterification and Ultrasound Treatment. *Carbohydr. Polym.* **2018**, *180*, 81–87. [CrossRef] [PubMed]
58. Duarte, M.L.; Ferreira, M.C.; Marvão, M.R.; Rocha, J. An Optimised Method to Determine the Degree of Acetylation of Chitin and Chitosan by FTIR Spectroscopy. *Int. J. Biol. Macromol.* **2002**, *31*, 1–8. [CrossRef] [PubMed]
59. Steen, K.H.; Steen, A.E.; Reeh', P.W. A Dominant Role of Acid PH in Inflammatory Excitation and Sensitization of Nociceptors in Rat Skin, In Vitro. *J. Neurosci.* **1995**, *15*, 3982–3989. [CrossRef] [PubMed]
60. Magnabosco, G.; Ianiro, A.; Stefani, D.; Soldà, A.; Rapino, S.; Falini, G.; Calvaresi, M. Doxorubicin-Loaded Squid Pen Plaster: A Natural Drug Delivery System for Cancer Cells. *ACS Appl. Bio. Mater.* **2020**, *3*, 1514–1519. [CrossRef] [PubMed]
61. Boudko, D.Y.; Moroz, L.L.; Linser, P.J.; Trimarchi, J.R.; Smith, P.J.S.; Harvey, W.R. In Situ Analysis of PH Gradients in Mosquito Larvae Using Non-Invasive, Self-Referencing, PH-Sensitive Microelectrodes. *J. Exp. Biol.* **2001**, *204*, 691–699. [CrossRef]
62. Stumpp, M.; Hu, M.Y.; Tseng, Y.C.; Guh, Y.J.; Chen, Y.C.; Yu, J.K.; Su, Y.H.; Hwang, P.P. Evolution of Extreme Stomach PH in Bilateria Inferred from Gastric Alkalization Mechanisms in Basal Deuterostomes. *Sci. Rep.* **2015**, *5*, 10421. [CrossRef]

marine drugs

MDPI

Article

Porous Nano-Fiber Structure of Modified Electrospun Chitosan GBR Membranes Improve Osteoblast Calcium Phosphate Deposition in Osteoblast-Fibroblast Co-Cultures

Hengjie Su [1,2], Tomoko Fujiwara [3], Omar Skalli [4], Gretchen Schreyack Selders [2], Ting Li [1], Linna Mao [1] and Joel D. Bumgardner [2,*]

1 Institute of Biomedical Engineering, Chinese Academy of Medical Sciences & Peking Union Medical College, Tianjin 300192, China
2 Department of Biomedical Engineering, University of Tennessee Health Science Center-Memphis Joint Graduate Biomedical Engineering Program, The University of Memphis, Memphis, TN 38152, USA
3 Department of Chemistry, The University of Memphis, Memphis, TN 38152, USA; tfjiwara@memphis.edu
4 Integrated Microscopy Center, The University of Memphis, Memphis, TN 38152, USA
* Correspondence: jbmgrdnr@memphis.edu

Abstract: Desirable characteristics of electrospun chitosan membranes (ESCM) for guided bone regeneration are their nanofiber structure that mimics the extracellular fiber matrix and porosity for the exchange of signals between bone and soft tissue compartments. However, ESCM are susceptible to swelling and loss of nanofiber and porous structure in physiological environments. A novel post-electrospinning method using di-tert-butyl dicarbonate (tBOC) prevents swelling and loss of nanofibrous structure better than sodium carbonate treatments. This study aimed to evaluate the hypothesis that retention of nanofiber morphology and high porosity of tBOC-modified ESCM (tBOC-ESCM) would support more bone mineralization in osteoblast-fibroblast co-cultures compared to Na_2CO_3 treated membranes (Na_2CO_3-ESCM) and solution-cast chitosan solid films (CM-film). The results showed that only the tBOC-ESCM retained the nanofibrous structure and had approximately 14 times more pore volume than Na_2CO_3-ESCM and thousands of times more pore volume than CM-films, respectively. In co-cultures, the tBOC-ESCM resulted in a significantly greater calcium-phosphate deposition by osteoblasts than either the Na_2CO_3-ESCM or CM-film ($p < 0.05$). This work supports the study hypothesis that tBOC-ESCM with nanofiber structure and high porosity promotes the exchange of signals between osteoblasts and fibroblasts, leading to improved mineralization in vitro and thus potentially improved bone healing and regeneration in guided bone regeneration applications

Keywords: chitosan; electrospinning; guided bone regeneration; calcium phosphate deposition; co-culture

1. Introduction

Guided bone regeneration (GBR) membranes are used in dental applications to prevent the invasion of gingival tissues into the bone-regenerating compartments. In addition to the shielding/barrier function of the membranes, the interconnected porosity and structure of the barrier membranes are reported as also playing important roles in healing by enabling an exchange of nutrients and growth factors between the gingival and bone compartments and mimicking extracellular matrix structure of newly forming tissues [1–5]. Electrospinning is one method to make GBR membranes that mimic the native nanofibrous extracellular matrix (ECM) structure and provide an interconnected porosity to allow the diffusion of nutrition, metabolites, and soluble factors between compartments that support healing [6].

Chitosan is a natural polysaccharide that has been widely researched as a GBR material due to its biocompatibility, homology to native ECM component hyaluronan, degradability, and ability to be electrospun [7]. However, electrospun chitosan membranes can be highly susceptible to swelling and loss of nanofibers and the interconnected porous structure [8,9]. Our group has successfully developed a novel post-electrospinning method that retains the nanofibrous structure of electrospun chitosan membranes (ESCM) better than other methods based on alkali treatments [7]. The post-spinning modification involves first removing acetate ions left over from the electrospinning process in the fibers using triethylamine (TEA) and then reacting di-tert-butyl dicarbonate (tBOC) with the chitosan amino groups to create a hydrophobic wrap on the outside of the electrospun chitosan fibers that prevents fiber swelling [7]. We demonstrated that the TEA/tBOC modified ESCM (TEA/tBOC-ESCM) exhibited little swelling and retained their nanofiber structure in physiological solutions, whereas a typical Na_2CO_3 treatment used to remove acetate ion salts resulted in extensive swelling and loss of nanofiber structure. The TEA/tBOC-ESCM were shown to be compatible with cultured cells, to degrade in vitro, to possess adequate mechanical properties for GBR applications, to provide effective barrier function, and to support bone healing in a rodent model.

This study aimed to further examine the hypothesis that the retained nanofiber morphology and porosity of the TEA/tBOC-ESCM improves in vitro cell performance as compared to Na_2CO_3 treated ESCMs (Na_2CO_3-ESCM) and solution-cast chitosan solid film membranes (CM-film). The membranes were characterized for surface morphology, porosity, water contact angle and ash and endotoxin content. Fibroblast and osteoblast cells in individual and co-cultures were used to evaluate the effect of nanofiber structure and porosity on cell growth as well as osteoblastic mineralization. The co-culture model was introduced to simulate the clinical application of GBR membrane treatments that aimed to separate and prevent the ingrowth of fibroblasts into sites of osteoblast growth and mineralization.

2. Results and Discussion

2.1. Membrane Characterizations

Results for the characterizations of the test membranes for fiber structure and diameter, porosity, hydrophobicity endotoxins and ash content are shown in Figure 1 and Table 1. The fiber diameter of the non-treated ESCM was 172 ± 48 nm, while that of the TEA/tBOC-ESCM was 219 ± 93 nm. There were no obvious fibers in the Na_2CO_3-ESCM, and no discernible structure in the CM-film. Consistent with previous reports, SEM microphotographs and analysis revealed that only the TEA/tBOC-ESCM were composed of well-defined nanometer diameter fibers, whereas the Na_2CO_3-ESCM exhibited significant fiber swelling, making the determination of individual fiber diameters difficult [7,10]. Because of the retained nanofiber structure, the TEA/tBOC-ESCM exhibited approximately 14 times greater porosity than the Na_2CO_3-ESCM (Table 1). As the CM-films were produced using a solution casting method, they lacked both fiber structure and significant porosity. The porosity of the GBR membranes is reported to be important for allowing nutrient and cell signaling exchange between osteogenic and soft/epithelial tissue compartments [11–14]. The significantly greater porosity of the TEA/tBOC-ESCM due to the retention of the nanofiber structure may be an advantage for improved nutrient and signal exchange as compared to the lower porosity Na_2CO_3-ESCM or essentially non-porous CM-film.

There was no significant difference in the median pore size of the two ESCM membranes, and even the solution-cast CM-film exhibited a range of pore sizes (Table 1). Since the CM-film becomes extremely brittle and stiff after neutralization, the small pores in the CM-film were attributed to minor defects that may arise in the casting, drying and neutralization of the films and or in manipulating films using forceps. The similarity in pore size between the TEA/tBOC-ESCM and Na_2CO_3-ESCM is more indicative of spaces between the randomly oriented fibers. Nevertheless, the range of pore sizes (~5 to ~44 μm) of both ESCMs is similar to other studies of electrospun membranes for GBR applications [11–14]. These studies report that these small pore sizes and the circuitous connections of the pores due to non-woven random fiber networks limit cell migration through the membranes, which is important to their barrier function while also allowing for nutrient and signal exchange.

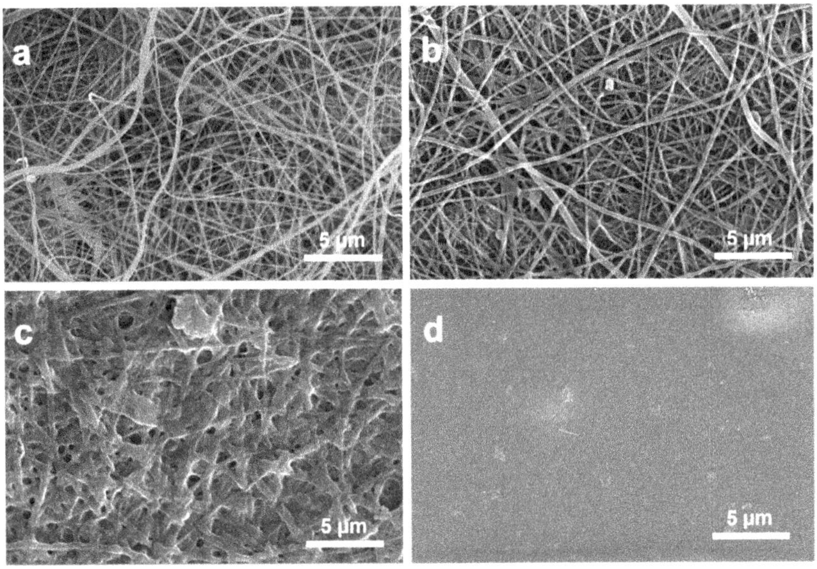

Figure 1. SEM images of the (**a**) non-treated chitosan electrospun membrane, (**b**) TEA/tBOC-ESCM, (**c**) Na_2CO_3-ESCM, and (**d**) the CM-film. The TEA/tBOC-ESCM exhibited similar nanofiber morphology and diameters as the as-spun membranes, whereas the Na_2CO_3-ESCM showed significant fiber swelling and loss of nanofiber morphology. The CM-film exhibited a smooth surface and no fibrous structure.

Table 1. Characteristics of TEA/tBOC-modified and Na_2CO_3-treated ESCM and solution-cast chitosan membrane films (CM-film). Data are expressed as mean (n = 3) ± standard deviation except for pore size, which is expressed as a median and percentiles due to a skewed pore size distribution.

	CM-Film	Na_2CO_3-ESCM	TEA/tBOC-ESCM
Median pore size (25th percentile, 75th percentile) (μm)	11.4 (4.7, 41.7)	12.2 (4.9, 42.5)	15.1 (6.6, 44.1)
Total pore volume (mm^3/g)	0.1 ± 0.2 [a]	32.4 ± 47.1 [b]	461.0 ± 96.6 [c]
Water contact angle (degree)	96.2 ± 2.4 [a]	76.0 ± 10.8 [b]	87.4 ± 9.0 [b]
Endotoxin (EU/mL)	<0.1	<0.1	<0.1
Ash (wt%) [&]	<0.5% [a]	1.9 ± 0.5% [b]	<0.5% [a]

Letter superscripts indicate significant differences at α = 0.05 level of significance; [&] original chitosan powder ash content < 0.5%.

The CM-film had the statistically highest water contact angle, followed by the TEA/tBOC-ESCM and then the Na_2CO_3-ESCM, which had the statistically lowest contact angle (Table 1). The high hydrophobic character of the CM-film is attributed in part to the strong alkali treatment, which deprotonates the amino groups in the chitosan polymer in addition to neutralizing and removing acetate ions from the solution casting process. The increased hydrophobicity of TEA/tBOC-ESCM stems from the methyl groups within the tBOC group, contrasting with Na_2CO_3-ESCM, which retains residual carbonate salts contributing to its previously reported higher hydrophilicity [7]. Although hydrophobic biomaterial surfaces are often linked to reduced cellular attachment and spreading, this may not pose a significant concern for GBR applications, where the primary role is to serve as a barrier against cellular migration. However, due to the gradual hydrolysis of tBOC from the ESCM over a period of 7–10 days, and given that the membranes feature pores within the range of 40μm in diameter, which is sufficient for tissue ingrowth, the integration of the membrane into local tissues does occur [7,15,16]. The integration of the membrane within tissues is advantageous for stabilizing its placement and guiding tissue growth during healing.

Membranes were tested for endotoxin using a Limulus Amebocyte Lysate (LAL) based assay (Table 1). Endotoxin is a component of the outer cell wall of Gram-negative bacteria and may persist as a contaminant in chitosan powders after processing from the original biological source material and impact cellular and host tissue responses such as inflammation and or sepsis [17]. The results indicated that the endotoxin concentrations of the membranes were very low, remaining below the 0.1 EU/mL test kit limit. This value is also beneath the 0.5 EU/mL limit recommended by the US FDA for devices that come into contact with the cardiovascular and lymphatic systems, whether directly or indirectly [18].

The original source material for the chitosan used in this research is shrimp shells, which contain calcium carbonate as the main mineral component of the exoskeleton matrix. Since ash content represents the amount of residual minerals in chitosan after processing from the original source material, the ash content is largely due to residual calcium carbonate minerals but may also contain other elements from the environment like lead and mercury that can have adverse effects on host cells and tissues [17]. The ash content of the starting chitosan material as well as for both the CM-films and TEA/tBOC-ESCMs was determined to be <0.5%. This low ash content is indicative of chitosan with a low amount of contaminating minerals. On the other hand, the Na_2CO_3-ESCM exhibited an ash content of 2.0 ± 0.6%, which was a significant four-fold increase over the other groups. This increase in ash content is attributed to residual Na_2CO_3 salts in or on the membrane after soaking in the highly concentrated Na_2CO_3 solution. Compared with other studies, the ash contents of the chitosan powders were generally less than 2% [19–23]. Hence, the ash contents of this study were within the range of other chitosan materials.

2.2. *Fibroblast and Osteoblast Co-Culture*

The dual cell culture setup was an initial attempt to mimic the cell growth environment of periodontal treatment, aiming to investigate whether a chitosan membrane with a porous structure is superior to one without pores. The insert of the dual cell culture provided a relatively individual fibroblast growth environment, with the only communication pathway being the bottom membrane, facilitating the nutrient exchange between fibroblasts and osteoblasts. When the chitosan membrane fully covered the insert bottom, the membrane's porosity determined the level of exchange. Despite the potential presence of chitosan extraction residues in all chitosan groups, the only variable across these groups was the presence of a porous structure in the membrane. A more porous membrane structure indicated a higher level of nutrient exchange between the two cell growth environments.

Focused on the aim of this study, which was to investigate whether a chitosan membrane with a porous structure was superior to one without pores, the selection of groups was aimed at minimizing variables. One control group consisted of samples without a chitosan membrane. This control was established to assess whether chitosan membranes effectively facilitated nutrient exchange, considering that the insert provides a clear pathway. The CM-film group was selected as another control to eliminate the potential influence of chitosan extraction residues and to focus on the variable of chitosan membrane porosity.

In the dual cell culture results, both the NIH 3T3 and MC3T3 E1 cells exhibited significant cell proliferation after 14 days, indicating that all the membranes were compatible with cell growth (Figure 2). In the previous study, Saos-2 osteosarcoma cells demonstrated significant proliferation after 5 days on both the TEA/tBOC-ESCM and the Na_2CO_3-ESCM [7]. Therefore, it can be inferred that the electrospun chitosan membranes and the cast film were compatible with both fibroblasts and osteoblasts.

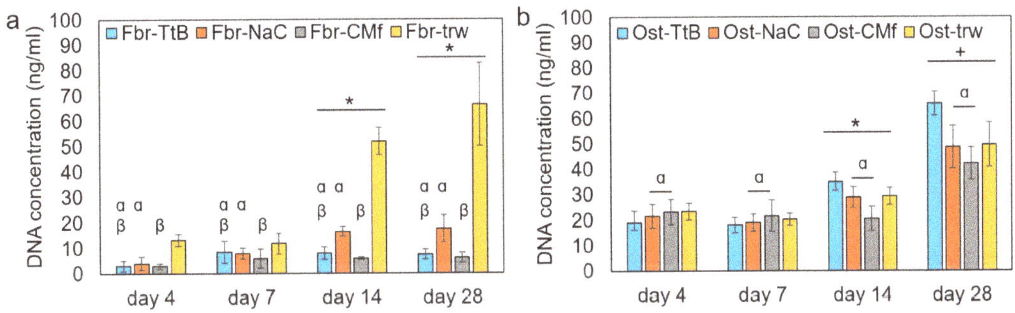

Figure 2. The dual cell culture results of (**a**) the top NIH 3T3 cell proliferation in the cell culture inserts and (**b**) the bottom MC3T3 cell proliferation on the well bottoms. * and + denote the significant difference in time points. α and β denote the significant differences among the cell culture groups. For abbreviations used in the graph, "Fbr-" indicates information regarding fibroblasts, while "Ost-" indicates information regarding osteoblasts. Regarding membrane types, "TtB" indicates the group with NIH 3T3 cells on the TEA/tBOC-ESCM, "NaC" indicates the group with NIH 3T3 cells on the Na_2CO_3-ESCM, "CMf" indicates the group with NIH 3T3 cells on the CM-Film, and "trw" indicates the group with NIH 3T3 cells on the insert with no membrane.

The NIH 3T3 fibroblasts exhibited significantly more growth in the control inserts (without chitosan membrane/film) compared to the cells in inserts with chitosan membranes (Figure 2a). This difference is primarily attributed to the transwell polycarbonate membranes, which are specifically treated to promote cell attachment and growth. Significantly greater fibroblast growth was observed on the Na_2CO_3-ESCM membranes (Fbr-NaC) compared to the CM-film. However, no difference in growth was detected between the fibroblasts on the TEA/tBOC-ESCM (Fbr-TtB) and the other two membranes, suggesting that the difference was small. Regarding porosity, the TEA/tBOC-ESCM exhibited approximately 10 times greater porosity, while the CM-films had approximately 100 times less porosity than the Na_2CO_3-ESCM. This is consistent with previous research, indicating that higher porosity contributes to increased cell growth [24]. Moreover, the hydrophobic nature of chitosan membranes may have influenced fibroblast growth. The Na_2CO_3-ESCM group, being more hydrophilic, showed significantly higher growth compared to the CM-film group, which is more hydrophobic. This aligns with the theory that a more hydrophilic surface promotes increased cell growth [25].

The MC3T3 E1 cells in co-culture with the TEA/tBOC-ESCM exhibited significantly different cell growth as compared to other chitosan membrane groups (Figure 2b). It showed that the cell proliferation below the TEA/tBOC-ESCM was lower than the other membrane groups at day 4 and day 7. However, it showed higher cell proliferation than the other membrane groups after 14 and 28 days. At the same time, there was no significant difference between the Ost-TtB group and Ost-trw group, which were prior to the non-porous structure membrane groups. It suggested that the TEA/tBOC-ESCM did not have a negative effect on bone cell proliferation. Its porous structure, similar to the ECM structure, facilitated a nutrient exchange between the NIH 3T3 and MC3T3 E1 cell growth environments, potentially promoting long-term MC3T3 E1 cell proliferation.

The ALP concentration normalized to the DNA amount exhibited significant increases in all groups after 14 and 28 days (Figure 3a). Osteoblasts in co-culture with the fibroblasts on the CM-film showed the highest ALP activity of the test groups on days 14 and 28, while osteoblasts in co-culture with fibroblasts on the ESCMs tended to have ALP activities that were either comparable or lower than osteoblasts cultured alone or in control co-culture.

However, the differences between these groups were not large since the osteoblasts cultured alone were not significantly different between all three membrane groups. It is further noted that ALP expression varies in a temporal fashion, and peaks in ALP expression may have been missed due to gaps between time points used. In a study by Ghuman et al., the ALP normalized to the control of the MC3T3 cells in the co-culture with gingival fibroblasts was significantly lower than the pure MC3T3 cell growth group [26], which implies that cross-talk between the fibroblasts and the MC3T3 cells might not stimulate ALP expression. In contrast, Zhu et al. showed that the relative ALP expression in MC3T3 E1 co-cultured with ephrinB2 transgenic periodontal ligament cells was significantly higher than in the pure MC3T3 E1 cell growth group [27]. Differences in these studies may be related to time points (day 3 vs. day 14) used to evaluate ALP in addition to differences in types of fibroblast cells [26,27]. Further research is still needed to explore the communication between fibroblasts and osteoblasts on osteoblastic differentiation and mineralization. Future studies may also be needed to consider the communication between gingival epithelial cells and osteoblasts.

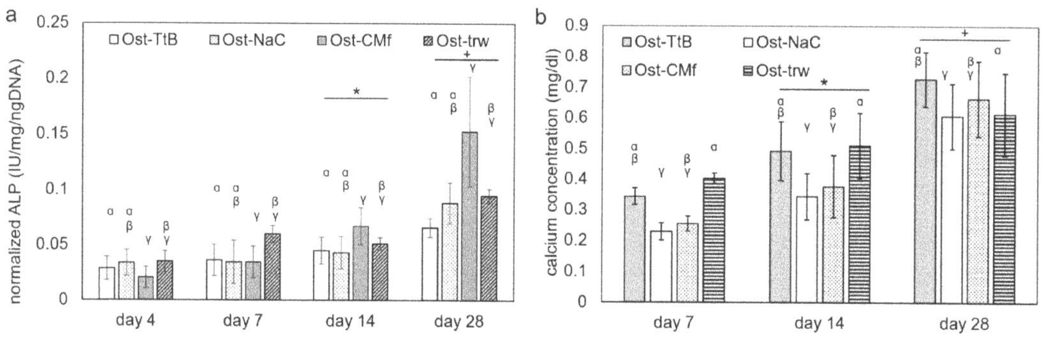

Figure 3. The dual cell culture results of (**a**) the ALP expression of the MC3T3 E1 cells and (**b**) the calcium deposit of the MC3T3 E1 cells at each time point. * and + denote the significant difference between day 4 and day 7. α, β and γ denote the significant difference among the cell culture groups. For abbreviations used in the graph, while "Ost-" indicates information regarding osteoblasts. Regarding membrane types, "TtB" indicates the group with NIH 3T3 cells on the TEA/tBOC-ESCM, "NaC" indicates the group with NIH 3T3 cells on the Na_2CO_3-ESCM, "CMf" indicates the group with NIH 3T3 cells on the CM-Film, and "trw" indicates the group with NIH 3T3 cells on the insert with no membrane.

The amount of calcium, serving as an indicator of calcium-phosphate deposition, exhibited a significant increase in all groups after 14 and 28 days (Figure 3b), indicating cells were able to differentiate and elaborate a mineralized matrix. Among the groups, the Ost-TtB and the Ost-trw groups displayed a significantly higher calcium concentration than the Ost-$NaCO_3$ group, while the Ost-CMf group showed a significantly lower calcium concentration than the Ost-trw group. The fluorescent stain graphs of osteocalcin (Figure 4) revealed increased production of osteocalcin after 14 days, serving as another indicator of heightened osteogenic activity similar to the calcium-phosphate deposition. The Ost-TtB group exhibited a significantly higher calcium concentration than the Ost-NaC group, suggesting that the TEA/tBOC-ESCM promotes cell mineralization better than the Na_2CO_3-ESCM. Although not statistically significant, the Ost-TtB group showed a trend of higher calcium concentration compared to the Ost-CMf group. Therefore, the TEA/tBOC-ESCM demonstrated similar or superior performance in cell mineralization compared to the CM-film. Since the TEA/tBOC-ESCM was the only type that preserved the fibrous structure, it could be inferred that the group with the nanostructure membrane showed improved support for osteogenic differentiation and elaboration of a mineralized matrix, which are important for regenerating bone. This supports the hypothesis that the porous electrospun membranes allow for better nutrient, small molecule and cell signal communication between the NIH 3T3 cells and the MC3T3 E1 cells. The higher calcium deposition observed in the Ost-trw group compared to the Ost-NaC and Ost-CMf groups further demonstrated the contribution of the mesh structure of the transwell, which provided optimal communication between the fibroblast and osteoblast growth environments. This suggests that the existing nutrient exchange between NIH 3T3 cells and MC3T3 E1 cells may have contributed to promoting calcium deposition. In Yang's study, MC3T3 E1 cells showed more mineralization ability on the fibrous membrane with a higher chitosan content compared to the fibrous membrane with more PCL [28]. This suggested that chitosan itself may promote cell mineralization beyond the porous structure. In this study, the MC3T3 E1 cells were not directly in contact with the chitosan membranes, which could explain why chitosan did not significantly promote cell mineralization.

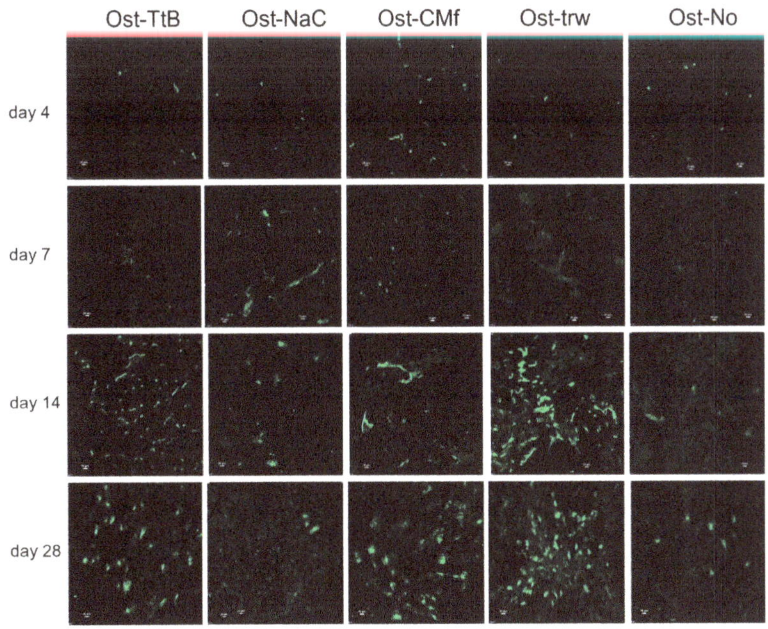

Figure 4. The fluorescent graphs of the anti-osteocalcin stained MC3T3 E1 cells at day 4, 7, 14 and 28. The osteocalcin increased from day 4 to day 28. For abbreviations used in the graph, while "Ost-" indicates information regarding osteoblasts. Regarding membrane types, "TtB" indicates the group with NIH 3T3 cells on the TEA/tBOC-ESCM, "NaC" indicates the group with NIH 3T3 cells on the Na_2CO_3-ESCM, "CMf" indicates the group with NIH 3T3 cells on the CM-Film, and "trw" indicates the group with NIH 3T3 cells on the insert with no membrane.

3. Materials and Methods

3.1. Chitosan Membrane and Film Preparation

Three types of chitosan-based membranes were prepared for this study: (a) TEA/tBOC-ESCM, (b) Na_2CO_3-ESCM and (c) a control CM-film treated with NaOH. The electrospun membranes were spun at 26 kV as described in previous studies using a 71% DDA chitosan (molecular weight = 311.5 kDa, Primex, Siglufjörður, Iceland) dissolved at 5.5 (w/v)% in 7:3 (v/v) trifluoroacetic acid (TFA, Thermo Fisher Scientific, Waltham, MA, USA) and dichloromethane (DCM, Sigma Aldrich, St. Louis, MI, USA). To make the TEA/tBOC-ESCM, membranes were first immersed in 10% (v/v) triethylamine/acetone solution for 24 h. After washing with acetone 3 times, membranes were immersed in the 0.1 g/mL tBOC solution for 48 h and then again washed with acetone 3 times [7]. Membranes were dried between two pieces of nylon mesh and under heavy weight to exclude the moisture in the air. To make the Na_2CO_3-ESCM, electrospun membranes were immersed in 5 M Na_2CO_3 solution for 3 h at room temperature and then dried between two pieces of nylon mesh [10].

To prepare CM-films, a 2 % (*w*/*v*) chitosan (71% DDA) in 2% acetic acid solution was prepared by stirring overnight. Twenty-eight millimeters of chitosan solution was pipetted into a petri dish (diameter = 8.5 cm) and allowed to dry for a few days at ambient conditions. For the neutralization, 2 M NaOH/DI H_2O solution was pipetted into the petri dish to immerse films for 1 h. CM-film was rinsed with DI H_2O for 30 min 3 to 5 times to remove excess NaOH and then dried in air at room temperature.

3.2. Membrane Characterizations

ESCM and solution-cast membranes were characterized for fiber structure and diameter, porosity, hydrophobicity, ash and endotoxin content. For fiber structure and diameter, specimens were examined using a scanning electron microscope (SEM, model EVO HD15, Carl Zeiss AG, Jena, Germany) from 2500× to 6000× after coating with 8 nm gold-palladium [7]. Fiber diameters were measured using image analysis software of the SEM. The porosity of test specimens was evaluated via porosimeter (Pascal 140 Mercury Porosimeter, Thermo Fisher Scientific, Waltham, MA, USA). Hydrophilic and hydrophobic character was measured by water contact angle using a VCA Optima measurement system (AST products, Inc, Billerica, MA, USA). Ash content of the test membranes was measured via combustion at 550 ± 20 °C to determine residual minerals from the original chitosan powder or as a result of the manufacture of ESCMs or the CM-films [29]. For endotoxin testing, gas-sterilized 1.5 cm diameter discs of test membrane samples were incubated in pyrogen-free water at a ratio of 1:100 (μg sample/μL water) for 24 h at 50 °C with constant shaking [30]. Endotoxin levels in pyrogen-free water extracts were determined using the Pierce LAL chromogenic endotoxin quantitation kit (Thermo Fisher Scientific, Waltham, MA, USA). For each material characterization method, three specimens of each test membrane type (n = 3/membrane) were evaluated.

3.3. Fibroblast and Osteoblast Co-Culture

Before the co-culture, the membranes underwent gas sterilization with ethylene oxide. The culture media was formulated by mixing α-MEM-media (Corning, Cellgro) with 5 mM β-glycerophosphate, 50 μg/mL ascorbic acid, 10 nM dexamethasone, 10% fetal bovine serum (FBS), 500 I.U./mL penicillin, 500 μg/mL streptomycin, and 25 μg/mL amphotericin-B.

Standard 24-well plates (Costar^{TM} flat bottom cell culture microplates, Corning, Glendale, AZ, USA) were used to seed osteoblasts and support inserts. First, MC3T3 E1 cells were seeded at 2×10^4 cells/well of a 24-well plate in the medium and allowed to attach overnight. Medium was removed, and a Falcon^{TM} Transwell cell culture insert (pore size = 0.4 μm, Corning, Glendale, AZ, USA) was placed into each well. Next, a disc-shaped test or control chitosan membrane (diameter = 6 mm) was placed in the insert, and then fresh culture medium was added to the wells to cover the membranes. After about 24 h, the medium was removed, and 1×10^4 NIH 3T3 cells/insert were seeded on either the test or control membranes and fresh medium was added to the wells to cover the cells in the wells and in the transwell inserts (Figure 5).

The co-culture groups are described in Table 2. Cells were co-cultured for 28 days, with the medium being changed every 2–3 days. At days 4, 7, 14, and 28, osteoblast and fibroblast cell growth were evaluated using the Quant-iT^{TM} PicoGreen^{TM} dsDNA Assay kit (Thermo Fisher Scientific, Waltham, MA, USA) (n = 3/membrane or film/time point). The osteoblast cells were evaluated for alkaline phosphatase (ALP) enzyme activity as an early marker of bone cell differentiation using the QuantiChrome^{TM} Alkaline Phosphatase Assay Kit (BioAssay Systems) normalized to dsDNA, and mineral deposition on days 7, 14 and 28 via the Calcium Assay (Pointe Scientific, Inc.,

Canton, MI, USA) as a terminal indicator of differentiation. (n = 3/membrane or film/time point). For qualitative assessment of a late marker for bone cell differentiation (n = 1/membrane or film/time point), immunostaining of the bone cell matrix for osteocalcin was performed using a primary anti-osteocalcin antibody (BGLAP Picoband™, Boster Biological Technology, Pleasanton, CA, USA), followed by a Donkey anti-rabbit IgG ReadyProbes™ secondary antibody conjugated with Alexa Fluor 488 (Thermo Fisher Scientific, Waltham, MA, USA). Wells were then treated with ProLong™ Diamond Antifade Mountant with DAPI (Thermo Fisher Scientific, Waltham, MA, USA) prior to viewing on a Nikon Eclipse microscope (Nikon, Melville, NY, USA). Images were collected under the same brightness and contrast conditions to facilitate qualitative comparisons.

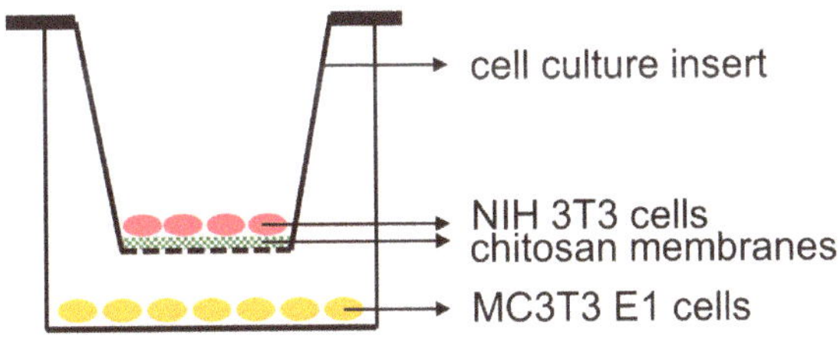

Figure 5. Diagram of co-culture arrangement of bone and fibroblast cells separated by a chitosan membrane. The MC3T3 E1 osteoblasts were seeded on the bottom of wells. After over-night incubation for osteoblast cell attachment, culture inserts containing a test chitosan membrane or no chitosan membrane (insert only) as a control were placed into the wells. Then, the NIH3T3 fibroblasts were seeded on the membrane/film samples.

Table 2. Co-culture group descriptions.

Group	TEA/tBOC-ESCM	Na_2CO_3-ESCM	CM-Film	Control
Cell culture insert	NIH 3T3 cells on the TEA/tBOC-ESCM (Fbr-TtB)	NIH 3T3 cells on the Na_2CO_3-ESCM (Fbr-NaC)	NIH 3T3 cells on the CM-film (Fbr-CMf)	No membrane, NIH 3T3 cells only (Fbr-trw)
Well plate bottom	MC 3T3 (Ost-TtB)	MC 3T3 (Ost-NaC)	MC 3T3 (Ost-CMf)	MC 3T3 (Ost-trw)

3.4. Statistical Analysis

The data for fiber diameter, water contact angle and ash content were analyzed by one-way analysis of variance (ANOVA) at the 0.05 level of significance. The cell culture data were analyzed by the two-way ANOVA ($\alpha = 0.05$). As appropriate, Tukey's post-hoc tests were used to distinguish significantly different groups at $\alpha = 0.05$ level of significance.

4. Conclusions

In conclusion, the TEA/tBOC-ESCM preserved the nanofibrous and highly porous structure, whereas the Na_2CO_3-ESCM did not, and the CM-film did not have any fibrous structure and lacked porosity. Hence, the TEA/tBOC-ESCM had more pore volume than the other two. The CM-film was more hydrophobic than the other two electrospun membranes, and the TEA/tBOC-ESCM was more hydrophobic than the Na_2CO_3-ESCM. The ash contents of all the membranes were under 0.5% except for the Na_2CO_3-ESCM, which was under 2%. All the membranes/films had extremely low endotoxin concentrations, which were considered suitable for the FDA requirements. The in vitro evaluation showed that all the membranes were osteoblast- and fibroblast-compatible. The TEA/tBOC-ESCM showed more MC3T3 E1 cell proliferation and more or similar deposited calcium amount than the other two membranes/films. The higher amount of deposited calcium indicated faster mineralization with the TEA/tBOC-ESCM membrane. In clinical treatment, faster mineralization indicates faster bone formation. Hence, these results indicate that the highly porous nanofiber structure of the TEA/tBOC membranes may have an advantage in improving the bone healing/regeneration time in GBR clinical treatments.

Author Contributions: Conceptualization, methodology, paper revision, and funding acquisition, J.D.B.; conceptualization, methodology, project administration, data analysis, paper writing, and revision, H.S.; porosity test, G.S.S.; chemical analysis, T.F.; osteocalcin staining test, O.S.; resources, L.M. and T.L. All authors have read and agreed to the published version of the manuscript.

Funding: This research was supported by the Biomaterials Applications of Memphis (BAM) Laboratories (Project #2013_0124) in the Biomedical Engineering Department at the University of Memphis and a grant (Project# 09 2016-2017) from the FedEx Institute of Technology, Memphis, TN, USA, Chinese Academy of Medical Science health innovation project (grant nos. 2021-I2M-1-058 and 2021-I2M-1-042), CAMS Innovation Fund for medical sciences (2021-I2M-C&T-B-095), Key Project of Construction of Drug Regulatory Science System (RS2024X007), STI 2030-Major Projects (2021ZD0200406) and Fundamental Research Funds for the Central Universities (3332023070).

Data Availability Statement: The data presented in this study are available on request from the corresponding author.

Acknowledgments: This work was performed as part of the requirements to complete a Ph.D. in Biomedical Engineering at the University of Memphis and in the UofM-UTHSC Joint BME Graduate Program.

Conflicts of Interest: Authors Hengjie Su, Tomoko Fujiwara and Joel D Bumgardner have a patent on the TEA/tBOC-ESCMs that were used in this study (U.S. Patent No. 11,878,088; Chitosan Nanofiber Compositions, Compositions Comprising Modified Chitosan, and Methods of Use).

References

1. Bee, S.-L.; Hamid, Z.A.A. Asymmetric Resorbable-Based Dental Barrier Membrane for Periodontal Guided Tissue Regeneration and Guided Bone Regeneration: A Review. *J. Biomed. Mater. Res. Part B Appl. Biomater.* **2022**, *110*, 2157–2182. [CrossRef]
2. Sanz, M.; Dahlin, C.; Apatzidou, D.; Artzi, Z.; Bozic, D.; Calciolari, E.; De Bruyn, H.; Dommisch, H.; Donos, N.; Eickholz, P.; et al. Biomaterials and Regenerative Technologies Used in Bone Regeneration in the Craniomaxillofacial Region: Consensus Report of Group 2 of the 15th European Workshop on Periodontology on Bone Regeneration. *J. Clin. Periodontol.* **2019**, *46*, 82–91. [CrossRef]
3. Kim, K.; Su, Y.; Kucine, A.J.; Cheng, K.; Zhu, D. Guided Bone Regeneration Using Barrier Membrane in Dental Applications. *ACS Biomater. Sci. Eng.* **2023**, *9*, 5457–5478. [CrossRef]
4. Caballé-Serrano, J.; Munar-Frau, A.; Ortiz-Puigpelat, O.; Soto-Penaloza, D.; Peñarrocha, M.; Hernández-Alfaro, F. On the Search of the Ideal Barrier Membrane for Guided Bone Regeneration. *J. Clin. Exp. Dent.* **2018**, *10*, e477–e483. [CrossRef]
5. Liang, C.; Wang, G.; Liang, C.; Li, M.; Sun, Y.; Tian, W.; Liao, L. Hierarchically Patterned Triple-Layered Gelatin-Based Electrospun Membrane Functionalized by Cell-Specific Extracellular Matrix for Periodontal Regeneration. *Dent. Mater.* **2024**, *40*, 90–101. [CrossRef]
6. Qian, M.; Li, S.; Xi, K.; Tang, J.; Shen, X.; Liu, Y.; Guo, R.; Zhang, N.; Gu, Y.; Xu, Y.; et al. ECM-Engineered Electrospun Fibers with an Immune Cascade Effect for Inhibiting Tissue Fibrosis. *Acta Biomater.* **2023**, *171*, 308–326. [CrossRef]
7. Su, H.; Liu, K.-Y.; Karydis, A.; Abebe, D.G.; Wu, C.; Anderson, K.M.; Ghadri, N.; Adatrow, P.; Fujiwara, T.; Bumgardner, J.D. In Vitro and in Vivo Evaluations of a Novel Post-Electrospinning Treatment to Improve the Fibrous Structure of Chitosan Membranes for Guided Bone Regeneration. *Biomed. Mater.* **2016**, *12*, 015003. [CrossRef]
8. Cooper, A.; Bhattarai, N.; Kievit, F.M.; Rossol, M.; Zhang, M. Electrospinning of Chitosan Derivative Nanofibers with Structural Stability in an Aqueous Environment. *Phys. Chem. Chem. Phys.* **2011**, *13*, 9969–9972. [CrossRef]
9. Sangsanoh, P.; Supaphol, P. Stability Improvement of Electrospun Chitosan Nanofibrous Membranes in Neutral or Weak Basic Aqueous Solutions. *Biomacromolecules* **2006**, *7*, 2710–2714. [CrossRef]
10. Norowski, P.A.; Fujiwara, T.; Clem, W.C.; Adatrow, P.C.; Eckstein, E.C.; Haggard, W.O.; Bumgardner, J.D. Novel Naturally Crosslinked Electrospun Nanofibrous Chitosan Mats for Guided Bone Regeneration Membranes: Material Characterization and Cytocompatibility. *J. Tissue Eng. Regen. Med.* **2015**, *9*, 577–583. [CrossRef]
11. Masoudi Rad, M.; Nouri Khorasani, S.; Ghasemi-Mobarakeh, L.; Prabhakaran, M.P.; Foroughi, M.R.; Kharaziha, M.; Saadatkish, N.; Ramakrishna, S. Fabrication and Characterization of Two-Layered Nanofibrous Membrane for Guided Bone and Tissue Regeneration Application. *Mater. Sci. Eng. C Mater. Biol. Appl.* **2017**, *80*, 75–87. [CrossRef]
12. Yang, F.; Both, S.K.; Yang, X.; Walboomers, X.F.; Jansen, J.A. Development of an Electrospun Nano-Apatite/PCL Composite Membrane for GTR/GBR Application. *Acta Biomater.* **2009**, *5*, 3295–3304. [CrossRef]
13. Dos Santos, V.I.; Merlini, C.; Aragones, Á.; Cesca, K.; Fredel, M.C. In Vitro Evaluation of Bilayer Membranes of PLGA/Hydroxyapatite/β-Tricalcium Phosphate for Guided Bone Regeneration. *Mater. Sci. Eng. C Mater. Biol. Appl.* **2020**, *112*, 110849. [CrossRef]

14. Ebrahimi, L.; Farzin, A.; Ghasemi, Y.; Alizadeh, A.; Goodarzi, A.; Basiri, A.; Zahiri, M.; Monabati, A.; Ai, J. Metformin-Loaded PCL/PVA Fibrous Scaffold Preseeded with Human Endometrial Stem Cells for Effective Guided Bone Regeneration Membranes. *ACS Biomater. Sci. Eng.* **2021**, *7*, 222–231. [CrossRef]
15. Su, H.; Fujiwara, T.; Anderson, K.M.; Karydis, A.; Ghadri, M.N.; Bumgardner, J.D. A Comparison of Two Types of Electrospun Chitosan Membranes and a Collagen Membrane In Vivo. *Dent. Mater.* **2021**, *37*, 60–70. [CrossRef]
16. Su, H. In Vitro and In Vivo Evaluation and Mechanical Improvement of the Electrospun Chitosan Membrane. Ph.D. Thesis, The University of Memphis, Memphis, TN, USA, 2020.
17. Jennings, J.A.; Bumgardner, J.D. (Eds.) *Chitosan Based Biomaterials*; Woodhead Publishing Series in Biomaterials; Elsevier/Woodhead Publishing: Amsterdam, The Netherlands, 2017; ISBN 978-0-08-100230-8.
18. Pang, E.-K.; Paik, J.-W.; Kim, S.-K.; Jung, U.-W.; Kim, C.-S.; Cho, K.-S.; Kim, C.-K.; Choi, S.-H. Effects of Chitosan on Human Periodontal Ligament Fibroblasts In Vitro and on Bone Formation in Rat Calvarial Defects. *J. Periodontol.* **2005**, *76*, 1526–1533. [CrossRef]
19. Islam, M.d.S.; Yeum, J.H. Electrospun Pullulan/Poly(Vinyl Alcohol)/Silver Hybrid Nanofibers: Preparation and Property Characterization for Antibacterial Activity. *Colloids Surf. A Physicochem. Eng. Asp.* **2013**, *436*, 279–286. [CrossRef]
20. Pipattanawarothai, A.; Suksai, C.; Srisook, K.; Trakulsujaritchok, T. Non-Cytotoxic Hybrid Bioscaffolds of Chitosan-Silica: Sol-Gel Synthesis, Characterization and Proposed Application. *Carbohydr. Polym.* **2017**, *178*, 190–199. [CrossRef]
21. Sibaja, B.; Culbertson, E.; Marshall, P.; Boy, R.; Broughton, R.M.; Solano, A.A.; Esquivel, M.; Parker, J.; Fuente, L.D.L.; Auad, M.L. Preparation of Alginate–Chitosan Fibers with Potential Biomedical Applications. *Carbohydr. Polym.* **2015**, *134*, 598–608. [CrossRef]
22. Haider, S.; Park, S.-Y. Preparation of the Electrospun Chitosan Nanofibers and Their Applications to the Adsorption of Cu(II) and Pb(II) Ions from an Aqueous Solution. *J. Membr. Sci.* **2009**, *328*, 90–96. [CrossRef]
23. Phongying, S.; Aiba, S.; Chirachanchai, S. Direct Chitosan Nanoscaffold Formation via Chitin Whiskers. *Polymer* **2007**, *48*, 393–400. [CrossRef]
24. Wang, Y.; Xu, R.; Luo, G.; Lei, Q.; Shu, Q.; Yao, Z.; Li, H.; Zhou, J.; Tan, J.; Yang, S.; et al. Biomimetic Fibroblast-Loaded Artificial Dermis with "Sandwich" Structure and Designed Gradient Pore Sizes Promotes Wound Healing by Favoring Granulation Tissue Formation and Wound Re-Epithelialization. *Acta Biomater.* **2016**, *30*, 246–257. [CrossRef]
25. Boyan, B.D.; Lotz, E.M.; Schwartz, Z. Roughness and Hydrophilicity as Osteogenic Biomimetic Surface Properties. *Tissue Eng. Part A* **2017**, *23*, 1479–1489. [CrossRef]
26. Ghuman, M.S.; Al-Masri, M.; Xavier, G.; Cobourne, M.T.; McKay, I.J.; Hughes, F.J. Gingival Fibroblasts Prevent BMP-Mediated Osteoblastic Differentiation. *J. Periodontal Res.* **2019**, *54*, 300–309. [CrossRef]
27. Zhu, S.Y.; Wang, P.L.; Liao, C.S.; Yang, Y.Q.; Yuan, C.Y.; Wang, S.; Dissanayaka, W.L.; Heng, B.C.; Zhang, C.F. Transgenic Expression of ephrinB2 in Periodontal Ligament Stem Cells (PDLSCs) Modulates Osteogenic Differentiation via Signaling Crosstalk between ephrinB2 and EphB4 in PDLSCs and between PDLSCs and Pre-Osteoblasts within Co-Culture. *J. Periodontal. Res.* **2017**, *52*, 562–573. [CrossRef]
28. Yang, X.; Chen, X.; Wang, H. Acceleration of Osteogenic Differentiation of Preosteoblastic Cells by Chitosan Containing Nanofibrous Scaffolds. *Biomacromolecules* **2009**, *10*, 2772–2778. [CrossRef]
29. Hamilton, V.; Yuan, Y.; Rigney, D.A.; Puckett, A.D.; Ong, J.L.; Yang, Y.; Elder, S.H.; Bumgardner, J.D. Characterization of Chitosan Films and Effects on Fibroblast Cell Attachment and Proliferation. *J. Mater. Sci. Mater. Med.* **2006**, *17*, 1373–1381. [CrossRef]
30. Nakagawa, Y.; Murai, T.; Hasegawa, C.; Hirata, M.; Tsuchiya, T.; Yagami, T.; Haishima, Y. Endotoxin Contamination in Wound Dressings Made of Natural Biomaterials. *J. Biomed. Mater. Res. Part B Appl. Biomater.* **2003**, *66*, 347–355. [CrossRef]

marine drugs

MDPI

Article

Effects of Shrimp Shell-Derived Chitosan on Growth, Immunity, Intestinal Morphology, and Gene Expression of Nile Tilapia (*Oreochromis niloticus*) Reared in a Biofloc System

Nguyen Vu Linh [1,2], Anisa Rilla Lubis [1], Nguyen Dinh-Hung [3], Supreya Wannavijit [1], Napatsorn Montha [1], Camilla Maria Fontana [1], Phattawin Lengkidworraphiphat [4], Orranee Srinual [1,2], Won-Kyo Jung [5], Marina Paolucci [6] and Hien Van Doan [1,2,*]

[1] Department of Animal and Aquatic Sciences, Faculty of Agriculture, Chiang Mai University, Chiang Mai 50200, Thailand; linhvu.n@cmu.ac.th (N.V.L.); anisarillalubis@gmail.com (A.R.L.); phoooooooooo10@gmail.com (S.W.); napatsorn.m@cmu.ac.th (N.M.); camillamaria.fontana@phd.unipd.it (C.M.F.); orranee.s@cmu.ac.th (O.S.)

[2] Functional Feed Innovation Center (FuncFeed), Faculty of Agriculture, Chiang Mai University, Chiang Mai 50200, Thailand

[3] Aquaculture Pathology Laboratory, School of Animal & Comparative Biochemical Sciences, The University of Arizona, Tucson, AZ 85721, USA; dinhhung@arizona.edu

[4] Multidisciplinary Research Institute, Chiang Mai University, 239 Huay Keaw Rd., Suthep, Muang, Chiang Mai 50200, Thailand; junjira_l@cmu.ac.th

[5] Marine Integrated Biomedical Technology Center, The National Key Research Institutes in Universities, Pukyong National University, Busan 48513, Republic of Korea; wkjung@pknu.ac.kr

[6] Department of Science and Technologies, University of Sannio, 82100 Benevento, Italy; paolucci@unisannio.it

[*] Correspondence: hien.d@cmu.ac.th; Tel.: +66-992-971888

Abstract: Chitosan (CH) shows great potential as an immunostimulatory feed additive in aquaculture. This study evaluates the effects of varying dietary CH levels on the growth, immunity, intestinal morphology, and antioxidant status of Nile tilapia (*Oreochromis niloticus*) reared in a biofloc system. Tilapia fingerlings (mean weight 13.54 ± 0.05 g) were fed diets supplemented with 0 (CH0), 5 (CH5), 10 (CH10), 20 (CH20), and 40 (CH40) mL·kg^{-1} of CH for 8 weeks. Parameters were assessed after 4 and 8 weeks. Their final weight was not affected by CH supplementation, but CH at 10 mL·kg^{-1} significantly improved weight gain (WG) and specific growth rate (SGR) compared to the control ($p < 0.05$) at 8 weeks. Skin mucus lysozyme and peroxidase activities were lower in the chitosan-treated groups at weeks 4 and 8. Intestinal villi length and width were enhanced by 10 and 20 mL·kg^{-1} CH compared to the control. However, 40 mL·kg^{-1} CH caused detrimental impacts on the villi and muscular layer. CH supplementation, especially 5–10 mL·kg^{-1}, increased liver and intestinal expressions of interleukin 1 (*IL-1*), interleukin 8 (*IL-8*), LPS-binding protein (*LBP*), glutathione reductase (*GSR*), glutathione peroxidase (*GPX*), and glutathione S-transferase (*GST-α*) compared to the control group. Overall, dietary CH at 10 mL·kg^{-1} can effectively promote growth, intestinal morphology, innate immunity, and antioxidant capacity in Nile tilapia fingerlings reared in biofloc systems.

Keywords: by-product; feed additive; growth performance; immune response; mRNA expression

1. Introduction

The rapid growth of aquaculture has led to an increased demand for improved diets and feed supplements for farmed fish [1]. Feed represents one of the largest costs for aquaculture producers. Determining the specific nutritional requirements, optimal feeding strategies, and nutrient utilization of each fish species is, therefore, critical to enable sustainable and scalable production [2]. Nile tilapia (*Oreochromis niloticus*) has become one of the most widely farmed aquaculture species because of its rapid growth, ability to adapt to different environments, disease resistance, and high protein content in the flesh [3,4]. Recently, tilapia farming practices have shifted from extensive to intensive commercial production systems [5,6]. However, disease outbreaks have emerged as the main obstacle to sustainable intensive tilapia farming globally. Finding ways to prevent diseases will be crucial to address this challenge as tilapia production continues to intensify [6]. Antibiotics have traditionally been used in commercial fish farms to prevent disease transmission. However, concerns over antibiotic usage in aquaculture have led to investigating replacements to reduce reliance on these treatments [7,8]. This has sparked great interest in finding new, innovative feed additives for tilapia, such as probiotics [9,10], prebiotics [11–13], and synbiotics [14,15]. These additives have been reported to enhance growth performance, health status, immune function, antioxidant defenses, and immune-related gene expression in tilapia. Such improvements could positively impact overall production. Replacing antibiotics with alternative feed supplements, particularly bioactive compounds derived from seafood products, may provide health benefits for tilapia while addressing issues surrounding antibiotic use in aquaculture.

Shrimp, one of the many varieties of seafood, is a popular and healthful dietary choice globally. Its output reached 8.25 million metric tons in 2015 and reached 9.66 million metric tons in 2019, with an annual growth rate of 2–3% [16], resulting in 6–8 million tons of waste [17,18]. The majority of shrimp waste is discarded publicly in landfills [19], burned, or dumped into the oceans [20,21]. Only a small portion is utilized as food and feed for animals and aquaculture [22,23]. The shrimp disposal sites could be major sources of offensive odors, as well as dust, gases, and fumes [18,24]. The rapid breakdown of shrimp waste can result in the appearance and spread of infections by flies, mosquitoes, and rats, endangering human health [25,26]. Nonetheless, this waste stream also includes beneficial natural substances, chief among them being chitin, which is an essential component in the production of chitosan [27–29]. Chitosan (CH) has been shown to have anti-cancer [30,31], anti-inflammatory [32,33], and neuroprotective [34] activities, in addition to having antioxidant, anti-diabetic, anti-hypertensive, and wound-healing [35] properties. Additionally, chitosan has antibacterial properties against the majority of bacteria, molds, and yeasts [36]. Furthermore, chitosan is a nontoxic, biodegradable, and biocompatible biopolymer. These characteristics make chitosan and its derivatives suitable for usage in a wide range of sectors, including the food, pharmaceutical, and agricultural industries [37–39].

Biofloc technology (BFT) has emerged as a sustainable aquaculture practice that enables fishponds to self-nitrify without water exchange [40,41]. In BFT systems, flocs formed from organic particulate matter and diverse microorganisms serve as an in situ food source. Fish can directly consume these protein-rich flocculants, reducing the need for fishmeal and soybean meal in feeds [42–45]. By substituting commercial diets with biofloc, the risks of mycotoxin and antinutrient exposure are also decreased, lowering feed costs [46,47]. Tilapia is especially well-suited for biofloc farming, as the species can effectively utilize biofloc for nutrition [48,49]. Given the benefits of BFT for tilapia production, this study aimed to evaluate how chitosan feed supplementation influences the growth, immune function, intestinal histology, and expression of key immune-antioxidant genes in Nile tilapia reared in a biofloc system. The overarching goal was to assess the potential of CH as a feed additive for enhancing tilapia health and productivity under sustainable BFT conditions.

2. Results

2.1. Growth Performance

The growth performance of Nile tilapia fingerlings fed the chitosan supplemented diets is shown in Table 1. After 4 weeks, FW was significantly higher ($p < 0.05$) in the CH10 group compared to the control group, while no significant differences were detected among groups at 8 weeks ($p > 0.05$). Weight gain did not differ significantly between the control and treatments at 4 weeks ($p > 0.05$). However, at 8 weeks, fish fed the CH10 diet showed significantly increased weight gain compared to the control group ($p < 0.05$) (Table 1). The

CH10 group also exhibited the highest SGR at both 4 weeks (2.83 ± 0.07) and 8 weeks (2.29 ± 0.03) (Table 1). No significant differences in FCR were observed between groups at any time (Table 1). The survival rate exceeded 95% in all treatments after the 8-week feeding trial.

Table 1. Growth performances and feed efficiency in Nile tilapia (*Oreochromis niloticus*) fingerlings fed diets with increasing chitosan (CH) levels for 4 and 8 weeks. Data is shown as mean ± SEM. Different letters (a–b) indicate significant differences between dietary groups. Diets: CH0, 0 mL·kg^{-1} COS (control); CH5, 5 mL·kg^{-1}; CH10, 10 mL·kg^{-1}; CH20, 20 mL·kg^{-1}; CH40, 40 mL·kg^{-1} of chitosan.

	CH0	CH5	CH10	CH20	CH40	*p*-Value
IW (g)	13.48 ± 0.03	13.66 ± 0.05	13.55 ± 0.03	13.56 ± 0.08	13.47 ± 0.05	0.499
FW (g)						
4 weeks	29.45 ± 1.33	28.97 ± 0.98	31.72 ± 0.28	29.06 ± 0.38	28.87 ± 1.27	0.470
8 weeks	50.44 ± 0.67 ^{b}	50.23 ± 0.62 ^{b}	50.69 ± 0.54 ^{a}	50.60 ± 1.20 ^{b}	50.29 ± 1.95 ^{b}	0.049
WG (g)						
4 weeks	15.97 ± 1.26	15.30 ± 0.93	18.17 ± 0.44	15.49 ± 0.45	15.40 ± 1.20	0.228
8 weeks	36.96 ± 1.30 ^{b}	36.56 ± 0.04 ^{b}	40.14 ± 0.83 ^{a}	37.03 ± 0.85 ^{b}	36.81 ± 0.47 ^{b}	0.050
SGR (%/day)						
4 weeks	2.60 ± 0.13 ^{ab}	2.50 ± 0.10 ^{b}	2.83 ± 0.07 ^{a}	2.54 ± 0.06 ^{ab}	2.54 ± 0.13 ^{ab}	0.032
8 weeks	2.20 ± 0.04 ^{b}	2.17 ± 0.00 ^{b}	2.29 ± 0.03 ^{a}	2.19 ± 0.04 ^{b}	2.19 ± 0.01 ^{b}	0.047
FCR						
4 weeks	0.76 ± 0.03 ^{a}	0.76 ± 0.01 ^{a}	0.75 ± 0.03 ^{a}	0.75 ± 0.02 ^{a}	0.75 ± 0.02 ^{a}	0.479
8 weeks	1.02 ± 0.01 ^{a}	1.07 ± 0.04 ^{a}	1.08 ± 0.04 ^{a}	1.02 ± 0.03 ^{a}	1.07 ± 0.08 ^{a}	0.855
SR (%)						
4 weeks	96.67 ± 1.67 ^{a}	95.00 ± 2.89 ^{a}	98.33 ± 3.33 ^{a}	98.33 ± 1.67 ^{a}	98.33 ± 1.67 ^{a}	0.046
8 weeks	96.67 ± 1.67 ^{a}	95.00 ± 2.89 ^{a}	98.33 ± 3.33 ^{a}	98.33 ± 1.67 ^{a}	98.33 ± 1.67 ^{a}	0.046

IW (g) = initial weight; FW (g) = final weight; WG (g) = weight gain; SGR (%) = specific growth rate; FCR = feed conversion ratio; SR (%) = survival rate.

2.2. Immunological Response

Lysozyme and peroxidase activities of skin mucus in Nile tilapia after 4 and 8 weeks of feeding are shown in Table 2. At both time points, skin mucus lysozyme activity (SMLA) and skin mucus peroxidase activity (SMPA) were significantly higher in the control group compared to all dietary CH treatments ($p < 0.05$). No significant differences were detected between the various CH-supplemented diets for either enzyme activity ($p > 0.05$).

Table 2. Skin mucus lysozyme and peroxidase activities in Nile tilapia (*Oreochromis niloticus*) fingerlings fed diets with increasing chitosan (CH) levels for 4 and 8 weeks. Data is shown as mean ± SEM. Different letters (a–b) indicate significant differences between dietary groups. Diets: CH0, 0 mL·kg^{-1} (control); CH5, 5 mL·kg^{-1}; CH10, 10 mL·kg^{-1}; CH20, 20 mL·kg^{-1}; CH40, 40 mL·kg^{-1} of chitosan.

		CH0	CH5	CH10	CH20	CH40	*p*-Value
4 weeks	SMLA	0.217 ^{a} ± 0.01	0.205 ^{b} ± 0.01	0.211 ^{ab} ± 0.01	0.195 ^{b} ± 0.01	0.201 ^{b} ± 0.01	0.048
	SMPA	0.313 ^{a} ± 0.02	0.287 ^{b} ± 0.01	0.304 ^{ab} ± 0.01	0.261 ^{b} ± 0.02	247 ^{b} ± 0.03	0.050
8 weeks	SMLA	0.249 ^{a} ± 0.01	0.215 ^{b} ± 0.01	0.220 ^{ab} ± 0.01	0.201 ^{b} ± 0.01	0.215 ^{b} ± 0.01	0.001
	SMPA	0.213 ^{a} ± 0.04	0.193 ^{b} ± 0.06	0.204 ^{ab} ± 0.06	0.181 ^{b} ± 0.02	147 ^{b} ± 0.02	0.036

SMLA (μg mL^{-1}) = skin mucus lysozyme activity; SMPA (μg mL^{-1}) = skin mucus peroxidase activity.

Serum lysozyme and peroxidase activities are illustrated in Table 3. Serum lysozyme activity (SL) was higher in the control group at 4 weeks. Serum peroxidase activity (SP) was significantly higher in CH 10 at week 4 ($p < 0.05$). However, this difference was not significant at 8 weeks. In contrast, the CH20 and CH40 groups exhibited notably reduced SP at both sampling times.

Table 3. Serum lysozyme and peroxidase activities in Nile tilapia (*Oreochromis niloticus*) fingerlings fed diets with increasing chitosan (CH) levels for 4 and 8 weeks. Data is shown as mean ± SEM. Different letters (a–c) indicate significant differences between dietary groups by one-way ANOVA and Duncan's test ($p < 0.05$). Diets: CH0, 0 mL·kg^{-1} COS (control); CH5, 5 mL·kg^{-1}; CH10, 10 mL·kg^{-1}; CH20, 20 mL·kg^{-1}; CH40, 40 mL·kg^{-1} of chitosan.

		CH0	CH5	CH10	CH20	CH40	*p*-Value
4 weeks	SL	0.297 ^{a} ± 0.01	0.275 ^{bc} ± 0.01	0.289 ^{ab} ± 0.01	0.258 ^{c} ± 0.01	0.286 ^{b} ± 0.01	0.038
	SP	0.443 ^{b} ± 0.02	0.449 ^{b} ± 0.02	0.501 ^{a} ± 0.01	0.388 ^{c} + 0.01	0.392 ^{c} ± 0.01	0.015
8 weeks	SL	0.248 ^{a} ± 0.01	0.225 ^{bc} ± 0.01	0.240 ^{ab} ± 0.01	0.212 ^{c} ± 0.01	0.231 ^{b} ± 0.01	0.049
	SP	0.477 ^{b} ± 0.05	0.438 ^{abc} ± 0.04	0.522 ^{a} ± 0.03	0.377 ^{c} ± 0.02	0.392 ^{bc} ± 0.03	0.041

SL: serum lysozyme activity (µg mL^{-1}); SP: serum peroxidase activity (µg mL^{-1}).

2.3. Histological Analysis

Intestinal morphology and related parameters of Nile tilapia fingerlings are presented in Figures 1 and 2. Villus length and width were significantly increased in the CH10 treatment group compared to the control group and the other treatment groups ($p < 0.05$). Fish fed the CH20 diet also exhibited greater villus length and width compared to the control. Additionally, the muscularis layer was the thickest in the CH10 group among all diets ($p < 0.05$). In contrast, the CH40 diet resulted in noticeable morphological alterations, including decreased villus length, villus width, and reduced muscularis thickness ($p < 0.05$).

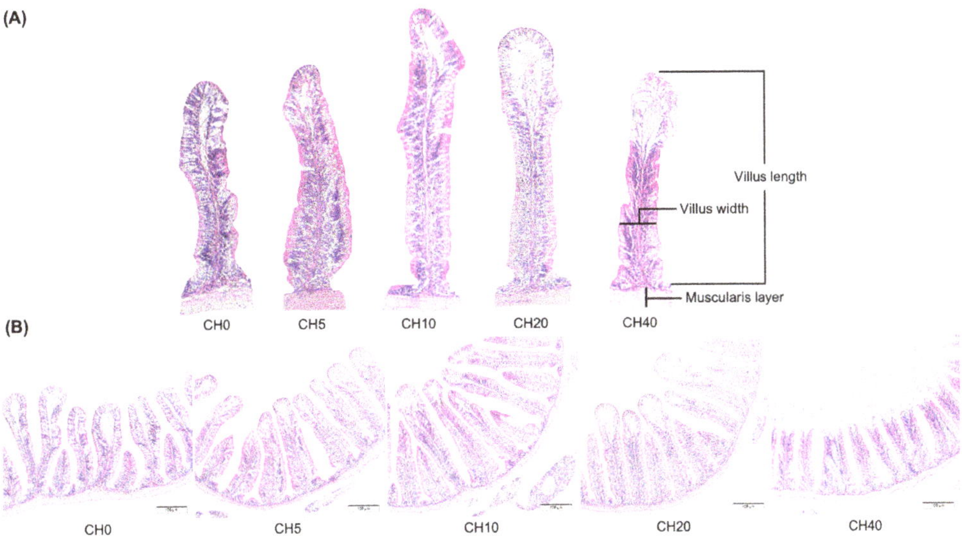

Figure 1. Intestinal morphology in Nile tilapia (*Oreochromis niloticus*) fingerlings fed diets with increasing chitosan (CH) levels after 8 weeks compared to the non-supplemented control diet. (**A**) A comparison of the length and width of the villus and thickness of the muscularis layer. (**B**) The cross-section through the microanatomy of the anterior intestine. Diets: CH0, 0 mL·kg^{-1} COS (control); CH5, 5 mL·kg^{-1}; CH10, 10 mL·kg^{-1}; CH20, 20 mL·kg^{-1}; CH40, 40 mL·kg^{-1} of chitosan. The tissue was stained with hematoxylin and eosin (H&E). The bars in the pictures are 100 µm.

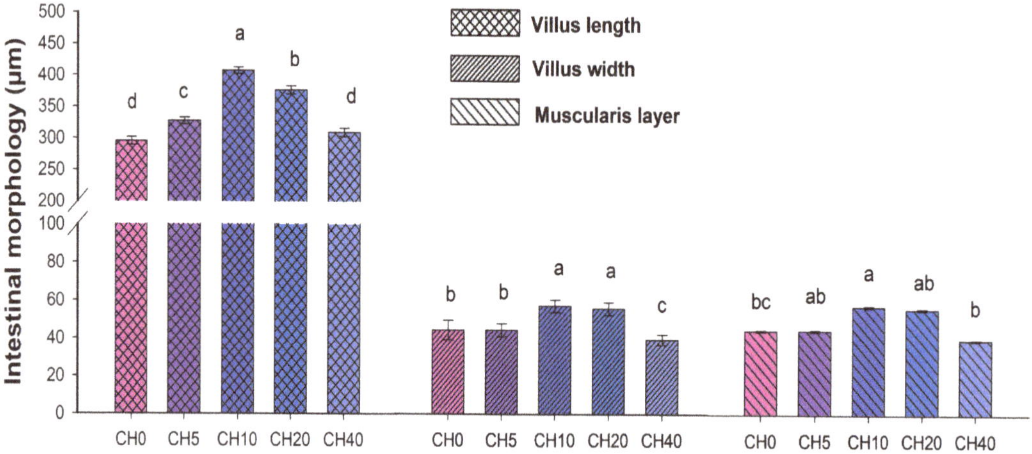

Figure 2. The measurements of the intestinal morphology of Nile tilapia (*Oreochromis niloticus*) fingerlings fed diets with increasing chitosan (CH) levels. Diets: CH0, 0 mL·kg^{-1} COS (control); CH5, 5 mL·kg^{-1}; CH10, 10 mL·kg^{-1}; CH20, 20 mL·kg^{-1}; CH40, 40 mL·kg^{-1} of chitosan. Five sections were randomly selected for measurement for each fish, with three fish per treatment. Values are means ± SEM (n = 15, microfields). Values in the same row with different superscripts indicate a significant difference between the CH-containing groups ($p < 0.05$).

2.4. Immune and Antioxidant-Related Gene Expressions

The effects of dietary CH-supplemented on the expression of immune-related (*IL-1*, *IL-8*, *LBP*) and antioxidant-related (*GSR*, *GPX*, *GST-α*) genes in the liver and intestine of tilapia are shown in Figures 3 and 4. In both tissues, all supplemented diets appeared to upregulate these genes compared to the control. In the liver, the CH5 diet induced the greatest increase in most of the genes, with IL-8 expression being significantly higher than the control and the other diets ($p < 0.05$). For *IL-1*, *GSR*, and *GST-α*, there were no significant differences among supplemented groups ($p > 0.05$). *GPX* expression was significantly higher in the CH5 compared to the CH40 group ($p < 0.05$). Additionally, the CH5 and CH10 diet groups elicited increased *LBP* expression compared to the control and CH20 and CH40 groups ($p < 0.05$).

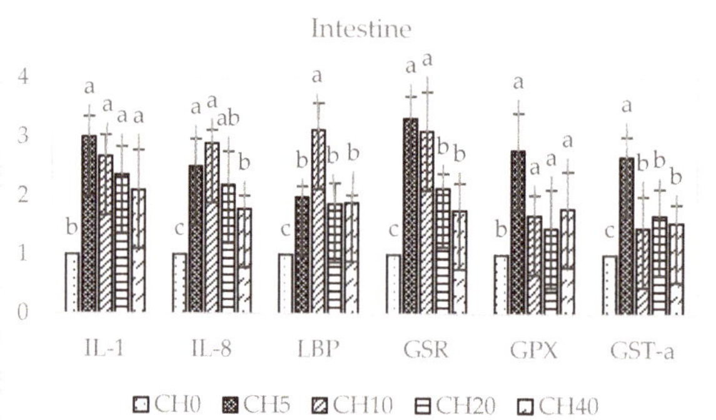

Figure 3. Expression transcript levels of interleukin-1 (*IL-1*), interleukin-8 (*IL-8*), lipopolysaccharide-binding protein (*LBP*), glutathione reductase (*GSR*), glutathione peroxidase (*GPX*), and glutathione

S-transferase-α (*GST-α*) in the liver of Nile tilapia (*Oreochromis niloticus*) fingerlings fed diets with increasing chitosan (CH) levels for 4 and 8 weeks ($n = 6$). Data is shown as mean ± SEM. Different letters (a–c) indicate significant differences between dietary groups by one-way ANOVA and Duncan's test ($p < 0.05$). Diets: CH0, 0 mL·kg^{-1} COS (control); CH5, 5 mL·kg^{-1}; CH10, 10 mL·kg^{-1}; CH20, 20 mL·kg^{-1}; CH40, 40 mL·kg^{-1} of chitosan.

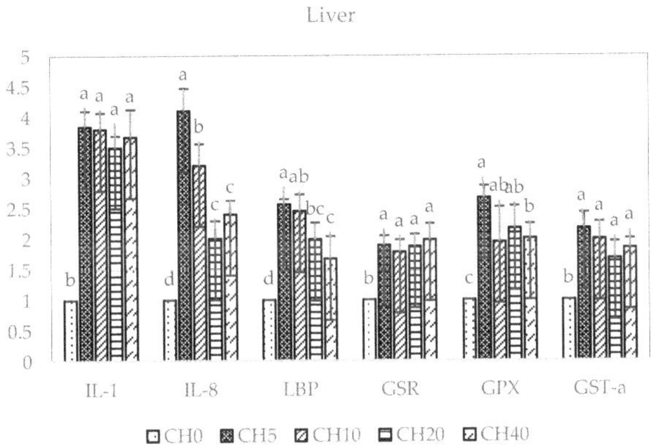

Figure 4. Expression transcript levels of interleukin-1 (*IL-1*), interleukin-8 (*IL-8*), lipopolysaccharide-binding protein (*LBP*), glutathione reductase (*GSR*), glutathione peroxidase (*GPX*), and glutathione S-transferase-α (*GST-α*) in intestine of Nile tilapia (*Oreochromis niloticus*) fingerlings fed diets with increasing chitosan (CH) levels for 4 and 8 weeks ($n = 6$). Data is shown as mean ± SEM. Different letters (a–c) indicate significant differences between dietary groups by one-way ANOVA and Duncan's test ($p < 0.05$). Diets: CH0, 0 mL·kg^{-1} COS (control); CH5, 5 mL·kg^{-1}; CH10, 10 mL·kg^{-1}; CH20, 20 mL·kg^{-1}; CH40, 40 mL·kg^{-1} of chitosan.

In the intestine, expression of *IL-1* and *GPX* was significantly higher in all dietary CH-treated groups compared to the control group ($p < 0.05$), with no statistically significant differences among supplemented diets ($p > 0.05$). Interestingly, the highest *LBP* expression occurred in the CH10 group, which was significantly different from the control and other treated groups ($p < 0.05$). *GSR* expression was significantly higher in the CH5 and CH10 groups compared to the control and CH20 and CH40 groups ($p < 0.05$). Additionally, CH5 and CH10 diets elicited clear increases in *IL-8* and CH-5 increased *GST-α* expression compared to the control and the other CH-supplemented groups.

3. Discussion

The current study demonstrated the beneficial effects of dietary-supplemented CH on the growth of Nile tilapia (*Oreochromis niloticus*) fingerlings cultured in a biofloc system. Influences of dietary chitosan on growth have been evaluated in various aquatic species with variable results [10,50–52]. Specifically in Nile tilapia, previous findings on chitosan's effects as a feed additive have been heterogeneous. Shiau and Yu [53] found that 2–10% of dietary chitosan inhibited tilapia growth, while Romana-Eguia et al. [54] showed no impact on growth. However, other studies [55,56] reported improved growth and meat quality with chitosan supplementation in Nile tilapia. Interestingly, this study indicated that only the 10 mL·kg^{-1} CH diet significantly increased FW, WG, and SGR of Nile tilapia fingerlings, suggesting that the benefits are dose-dependent, with excessive amounts conferring no added growth effects. Indeed, multiple studies have evidenced the detrimental impacts of immunostimulant over-supplementation on aquaculture species, including immune exhaustion and slowed growth [57,58]. Shiau and Yu [53] reported decreased weight gain in Nile tilapia with chitosan, potentially due to reduced nutrient digestibility and absorption. Chitosan particle size may also influence Nile tilapia growth [59]. Several lines of evidence suggest that chito-oligosaccharides can improve growth performance in tilapia through various interrelated mechanisms. As prebiotics, CH can modify intestinal microbial communities in a beneficial manner, supporting gut health and likely enhancing nutrient digestion and absorption [60]. The immunostimulatory effects of chito-oligosaccharides are also thought to play a key role by reducing the susceptibility to infectious diseases, allowing tilapia to allocate more energy towards growth rather than mounting inflammatory responses [61]. Chito-oligosaccharides have additionally been shown to increase the activities of digestive enzymes like protease, lipase, and amylase in tilapia, which could lead to greater utilization of feed for growth [62]. Finally, mitigation of oxidative stress by the antioxidant properties of CH enables available energy to be used for anabolism rather than neutralizing reactive oxygen species, supporting growth [63,64]. The growth-promoting effects of immunostimulants like CH are influenced by numerous factors, including dosage, molecular weight, feeding duration, temperature, administration route, and species difference [65]. Our results demonstrated that the benefits of CH on growth were dose-dependent in Nile tilapia, with 10 $mL \cdot kg^{-1}$ being the optimal supplementation level for improving performance. While the precise modes of action have yet to be elucidated, this study provides valuable insights into appropriate CH dosing strategies for maximizing growth in tilapia aquaculture. Further research is still needed to fully understand the biological pathways and key factors mediating the effects of CH on increasing fish growth rate.

Innate immunity serves as the first line of defense against pathogens in fish [66]. The body's surface mucosa provides a physical and immunological barrier, playing crucial roles in protection, sensory function, and ion regulation [67,68]. Mucosal responses are key in early infection control, as many pathogens initially adhere to mucosal surfaces during invasion [69]. Lysozyme, found in mucus, fluids, and tissues, is an important component of teleost innate immunity due to its bactericidal and opsonizing effects [70,71]. Serum lysozyme can indicate the innate status of the host by initiating the complement cascade [69,70]. CH's free radical scavenging amino groups can boost unstable lysozyme [72], and oral chitosan has been shown to increase lysozyme activity in various fish species [73,74]. Peroxidase, another key innate immune enzyme, helps to maintain redox homeostasis in immune cells and acts as a microbicidal agent by destroying H_2O_2 [75,76]. In this study, our results demonstrated that dietary CH supplementation at 5–40 $mL \cdot kg^{-1}$ significantly decreased peroxidase and lysozyme activities in the skin mucus after 4 weeks, declining further by 8 weeks. Interestingly, 10 $mL \cdot kg^{-1}$ CH increased blood serum peroxidase activity at 4 and 8 weeks of the feeding trial. These results align with Yu et al. [58], who showed 10 g/kg CH reduced lysozyme in golden pompano (*Trachinotus ovatus*). The contrasting blood and mucus enzyme responses reveal the complex immunomodulatory effects of CH in fish. Achieving optimal benefits likely requires careful dosage optimization. Our findings highlight the need for further research into appropriate CH supplementation strategies to support mucosal and systemic innate defenses in aquaculture species.

Oligosaccharide supplementation in fish diets promotes feed conversion and enhances intestinal microanatomy, improves mucosal epithelium health, and defends against opportunistic bacterial infections [77]. This optimization of intestinal morphology may increase the absorption area of the intestine, facilitating efficient nutrient absorption [78]. In the present study, it was observed that the treatment with CH resulted in an increase in villus height and width, along with an increase in the thickness of the muscularis layer, which was particularly pronounced in the CH10 and CH20 groups. Since the proportion of villi is related to the ability to absorb nutrients through the available surface area, the surface increase could potentially improve nutrient utilization and storage [79]. Our results are consistent with previous studies, such as those showing an increase in villus length following supplementation of hybrid grouper diets with CH oligosaccharides [80] or after supplementation of hybrid catfish diets with mannan oligosaccharides [81]. It has also been reported that various other oligosaccharides can significantly increase villus length in numerous fish species, as shown by 1% galacto-oligosaccharide in the diet for red drum [82] or 2% fructo-oligosaccharide in the diet for bluntnose seabream [83]. However, it is worth noting that the effects of oligosaccharides on the villi structure may vary depending on the fish species, oligosaccharide type and concentration, and the fish species' own microbiota. In this study, the CH40-containing treatments resulted in some undesirable changes in the intestinal morphology of the fish, including degeneration of the villi and morphological disorders. This outcome may be attributed to an imbalance of amino acids in the diet, possibly triggered by an incorrect ratio when replacing fish meal with CH, which is consistent with previous studies [80,84]. Since dietary amino acids are primarily used to meet growth requirements and build fish tissue, an imbalance of these amino acids may lead to dysplasia in fish intestinal morphology [85,86]. Overall, our results indicate that CH supplementation at 10–20 mL·kg^{-1} may beneficially enhance the intestinal morphology and the absorptive capacity in tilapia, yet higher doses could negatively impact the intestine structure.

Pro-inflammatory cytokine *IL-1* is essential for innate immunity, stimulating lymphocytes, phagocytes, and infection resistance in fish [87]. *IL-8*, released during inflammation, activates inflammatory cells as a neutrophil chemoattractant and mediator [88]. *IL-1* and *IL-8* coordinate innate inflammatory defenses and pathogen clearance by phagocytes [89,90]. The acute-phase protein lipopolysaccharide-binding protein (*LBP*) also has key innate immune functions, binding lipopolysaccharides and eliciting responses to Gram-negative bacteria [91]. In our study, all CH doses markedly increased hepatic and intestinal expression of the immune genes *IL-1*, *IL-8*, and *LBP* compared to the control, indicating activation of innate immune responses in tilapia. These results align with previous observations reporting *IL-1* and *IL-8* upregulation following immunostimulant feeding in tilapia [92]. However, in golden pompano, CH reduced *IL-8* expression [57], highlighting species-specific differences. Antioxidant supplements can improve fish health by reducing oxidative stress. Glutathione peroxidase (*GPX*) and glutathione reductase (*GSR*) remove hydrogen peroxide using glutathione [13,93]. Glutathione S-transferase (*GST*) detoxifies electrophiles, enhancing their elimination [94]. In this study, we found dietary CH-supplemented significantly increased antioxidant gene (*GSR*, *GPX*, *GST-α*) expression in tilapia liver and intestine, similar to previous tilapia studies [13] and golden pompano [57]. This suggests that CH may mitigate oxidative damage. Overall, our gene expression analyses indicate that CH can stimulate innate immune and antioxidant responses in tilapia.

The growth benefits of CH in this study may have been enhanced by using a biofloc production system [95], which consists of suspended microbial biomass that acts as a natural food source [46,96–98]. This seems to stem from the fact that the prebiotic effects of CH selectively enriched beneficial biofloc species, maximizing natural productivity. Their immunostimulatory properties likely complemented immune activation by biofloc microbes. The combination of bioavailable nutrients from biofloc consumption and improved digestibility and gut health from CH may have synergistically augmented tilapia growth.

The limitations of this study include the absence of a priori power analysis and the use of non-standard reporting of chitosan concentration in volume units (mL/kg diet) instead of mass units (g/kg diet). The lack of a priori power analysis may serve to hide the true effects of chitosan supplementation, potentially leading to underpowered circumstances that are unable to identify statistically significant results. To address these issues, we propose the use of power analysis in the study design phase as a mean to precisely ascertain the necessary sample sizes. Furthermore, standardizing the reporting of chitosan concentrations in mass units will greatly enhance the reproducibility of research findings and streamline the process of comparing them across different investigations.

4. Materials and Methods

4.1. Nile Tilapia Husbandry

Healthy Nile tilapia fingerlings were acquired from a tilapia farm in Chiang Mai Province, Thailand. The fish were first acclimated for two weeks under standard aquaculture conditions and fed on commercial diets twice daily. The tilapia was then moved into fifteen 150 L fiberglass tanks for the feeding trials. Water quality parameters, including temperature (°C), pH, dissolved oxygen (mg·L^{-1}), and ammonium, were maintained within optimal ranges for Nile tilapia [99] throughout the experiment as follows: T° = 28.5 ± 0.07; pH = 7.81 ± 0.03; dissolved oxygen = 5.76 ± 0.02 mg·L^{-1}, and ammonium = 0.12 ± 0.002 mg·L^{-1}.

4.2. Diet Preparation and Experimental Design

4.2.1. Preparation of Chitosan (CH)

The CH supplement used in this study was obtained from Olizac Technologies Co., Ltd., Khlong Nueng, Khlong Luang District, Pathum Thani, Thailand. It was extracted from shrimp shell via enzymatic hydrolysis as described previously [100]. Briefly, shrimp shells underwent deproteinization, demineralization, and depigmentation before being deacetylated with 50% sodium hydroxide to achieve over 90% degree of deacetylation. The CH was then precipitated and lyophilized. A mixture of chitinase and chitosanase enzymes was applied to produce CH with a molecular weight of approximately 10 kDa, as determined by gel permeation chromatography.

4.2.2. Experimental Design

Five experimental diets containing different levels of CH were prepared. The diet formulations are shown in Table 4 The dry ingredients were thoroughly mixed and then pelletized with the addition of oil and water into 2 mm pellets. Feeds were stored at 4 °C until use. After a two-week acclimation, 300 healthy tilapia fingerlings (13.54 ± 0.05 g) were randomly distributed into the following treatment groups (n = 20 fish per tank, 3 replicate tanks per treatment): CH0, 0 mL·kg^{-1} CH as control; CH5, 5 mL·kg^{-1}; CH10, 10 mL·kg^{-1}; CH20, 20 mL·kg^{-1}; and CH40, 40 mL·kg^{-1}. Fish were fed the experimental diets twice daily for 8 weeks while water quality parameters were monitored daily as described previously [99].

Table 4. The formulation and proximate composition of the experimental diets (g/kg of the basal diets).

	CH0	CH5	CH10	CH20	CH40
Fish meal	200	200	200	200	200
Corn meal	150	150	150	150	150
Soybean meal	390	390	390	390	390
Wheat flour	70	70	70	70	70
Rice bran	150	150	150	150	150
Soybean oil	2	2	2	2	2
Chitosan solution (mL)	0	5	10	20	40
Binder	20	20	20	20	20
Premix [1]	10	10	10	10	10
Vitamin C 98%	8	8	8	8	8
Total (g)	1000	1000	1000	1000	1000
	Proximate composition of the experimental diets (%)				
Crude protein	32.80	32.00	32.60	32.40	32.50
Crude lipid	2.85	2.75	2.63	2.78	2.88
Fiber	3.68	3.74	3.44	3.72	3.55
Ash	7.59	7.86	7.75	7.35	7.91
Dry matter	99.16	98.40	98.35	97.77	97.54
Gross Energy (cal/g)	4273.00	4261.50	4253.90	4262.00	4245.00

[1] Vitamin and trace mineral mix supplemented as follows (IU kg^{-1} or g kg^{-1} diet): retinyl acetate 1,085,000 IU; cholecalciferol 217,000 IU; D, L-a-tocopherol acetate 0.5 g; thiamin nitrate 0.5 g; pyridoxine hydrochloride 0.5 g; niacin 3 g; folic 0.05 g; cyanocobalamin 10 g; Ca pantothenate 1 g kg^{-1}; inositol 0.5 g; zinc 1 g; copper 0.25 g; manganese 1.32 g; iodine 0.05 g; sodium 7.85 g.

4.3. Biofloc Water Preparation

Biofloc was established in the experimental tanks 3 weeks prior to starting the feeding trial. Coarse salt (400 g), molasses (5 g), dolomite (5 g), and control feed (2 g) were added to each tank to initiate floc formation. The carbon-to-nitrogen (C:N) ratio was maintained at 15:1 by supplementing with molasses (40% carbon) 2 h after each feeding [40]. The C/N ratio was monitored by measuring residual nitrogen levels in the tanks and determining the carbon and nitrogen content of the feed.

4.4. Growth Performance

After 4 and 8 weeks of feeding the experimental diets, all fish were weighed to assess growth performance. Parameters were calculated as follows:

Weight gain (WG, g) = final weight (FW) − initial weight (IW);

Specific growth rate (SGR, %) = 100 × (ln FW − ln IW)/number of experimental days;

Feed conversion ratio (FCR) = amount of feed given (dry weight)/WG (wet weight);

Survival rate (SR, %) = (final number of fish/initial number of fish) × 100.

4.5. Immunological Analysis

4.5.1. Sample Collection

Skin mucus and serum samples were collected to analyze immunological parameters. For skin mucus, 3 fish were randomly selected from each tank and anesthetized with clove oil (200 ppm) to minimize stress and discomfort. Following anesthesia, the fish were humanely euthanized in accordance with ethical guidelines for the collection of skin mucus and serum samples. Individuals were placed in plastic bags containing 10 mL of 50 mM NaCl. The fish were gently rubbed for 1 min to collect skin mucus. The mucus–salt mixture was centrifuged at 1500× g for 10 min at 4 °C. The supernatant was stored at −80 °C until analysis.

Blood samples were collected as previously described [99]. Briefly, 1 mL of blood was drawn from the caudal vein of each fish using a 1 mL syringe and immediately transferred into new sterilized tubes (without anticoagulants). Blood samples were kept at room temperature for 1 h and then incubated for 4 h at 4 °C. Serum samples were collected after centrifugation (15 min, 4 °C at 10,000× g) and stored at −80 °C until analysis.

4.5.2. Immunological Parameter Analysis

Lysozyme and peroxidase activities in undiluted serum and skin mucus samples were performed according to the previously described method [101]. Briefly, 25 µL of serum or 100 µL of skin mucus from each fish was added in triplicate to 96-well plates, followed by 175 µL of a 0.3 mg.mL^{-1} Micrococcus lysodeikticus suspension (in 0.1 M citrate phosphate buffer, pH 5.8). Plates were rapidly mixed, and the decrease in turbidity was measured every 30 s for 10 min at 540 nm using a microplate reader (Synergy H1, BioTek, Santa Clara, CA 95051, USA). A standard curve was generated using known concentrations of hen egg white lysozyme (0–20 µg.mL^{-1}, Sigma-Aldrich Inc., St. Louis, MO 68178, USA).

4.6. Histopathology Analysis

To examine intestinal morphology, the anterior intestine from 3 fish per treatment was sampled at the end of the trial. Tissues were fixed in 10% neutral buffered formalin for 24 h, then transferred to 70% ethanol. Samples were processed using an automated tissue processor, involving dehydration in graded ethanol, clearing with xylene, and embedding in paraffin wax. The tissues were sectioned at 4–5 µm thickness using a microtome (Leica Biosystems, Deer Park, IL 60010, USA) and stained with hematoxylin and eosin (H and E). Slides were viewed and photographed using a light microscope (BX51 Olympus, Tokyo, Japan). Morphometric analysis was performed by measuring villus length, villus width, and muscularis thickness on 5 randomly selected microfields per fish.

4.7. Quantitative Real-Time PCR (qPCR)

4.7.1. Tissue Sampling, Total RNA Isolation, and cDNA Synthesis

Expressions of immune-related (*IL-1*, *IL-8*, and *LBP*) and antioxidant-related (*GST-α*, *GPX*, and *GSR*) genes were analyzed in the liver and intestine after 8 weeks. Examined organs (20–40 mg) were collected from two fish in each tank (n = 6) and stored in sterilized tubes supplemented with 500 µL Trizol (Invitrogen, Waltham, MA, USA) at −80 °C for further analysis. Total RNA was isolated using the PureLink™ RNA Mini Kit (Invitrogen, Thermo Fisher Scientific, Waltham, MA, USA) following the manufacturer's protocol. RNA quantity and quality were assessed by spectrophotometry (NanoDrop™ 2000, Thermo Scientific, Waltham, MA, USA). One µg of total RNA was used for cDNA synthesis with the iScript™ cDNA kit (BIO-RAD, Hercules, CA, USA).

4.7.2. Quantitative Real-Time PCR

Gene expression was quantified by qPCR using the primer sequences listed in Table 5. Reactions contained 1 µL cDNA (100 ng), 0.4 µL each primer (10 µM), 10 µL 2× SYBR Green Mastermix (BIO-RAD, USA) and nuclease-free water to 20 µL total volume. qPCR was performed on a CFX ConnectTM system (BIO-RAD, USA) as described previously [13]. Relative mRNA levels were calculated using the $2^{-\Delta\Delta Ct}$ method [102] with *18S rRNA* as the internal reference gene.

Table 5. Primer sequences used for quantitative real-time PCR.

Genes	Primer Sequence (5′-3′)	Tm (°C)	Product Size (bp)	Reference
18S-rRNA	GTGCATGGCCGTTCTTAGTT CTCAATCTCGTGTGGCTGAA	60	150	XR_003216134
IL-1	GTCTGTCAAGGATAAGCGCTG ACTCTGGAGCTGGATGTTGA	59	200	XM_019365844
IL-8	CTGTGAAGGCATGGGTGTG GATCACTTTCTTCACCCAGGG	59	196	NM_001279704
LBP	ACCAGAAACTGCGAGAAGGA GATTGGTGGTCGGAGGTTTG	59	200	XM_013271147
GST-α	ACTGCACACTCATGGGAACA TTAAAAGCCAGCGGATTGAC	60	190	NM_001279635
GPX	GGTGGATGTGAATGGAAAGG CTTGTAAGGTTCCCCGTCAG	60	190	NM_001279711
GSR	CTGCACCAAAGAACTGCAAA CCAGAGAAGGCAGTCCACTC	60	172	XM_005467348

4.8. Statistical Analyses

The Shapiro–Wilk test was used to assess the normality of the data. One-way analysis of variance (ANOVA) was performed to determine statistically significant differences among the dietary treatment groups. The distribution of the sample variables was considered normal ($p > 0.05$) and was evaluated using a one-way ANOVA. Statistical significance among groups ($p < 0.05$) was compared using post hoc LSD analysis and non-normal distribution ($p < 0.05$). All data were analyzed using Statistix (Analytical Software, v10.0 Tallahassee, FL 32312, USA) statistical software.

5. Conclusions

In summary, this study demonstrates that dietary supplementation with 10 mL·kg^{-1} CH can effectively improve the growth, health, and productivity of Nile tilapia fingerlings reared in biofloc systems. CH also stimulated innate immunity, as shown by increased serum peroxidase activity at 4 weeks. Most notably, CH feeding markedly upregulated the expression of immune and antioxidant genes in the liver and intestine. This indicates that CH can beneficially modulate the immune status and oxidative stress resistance in Nile tilapia. Our findings highlight the potential of CH as a feed additive to improve Nile tilapia fingerlings' health and productivity in sustainable biofloc aquaculture.

Author Contributions: Methodology, Writing, review, and editing, formal analysis, data curation, software analysis, validation, N.V.L.; writing—original draft, A.R.L.; writing, review, and editing, N.D.-H.; investigation, methodology, S.W.; formal analysis, methodology, N.M.; writing, review, and editing, C.M.F.; methodology, P.L.; methodology, O.S.; supervision, validation, W.-K.J.; review and editing, M.P.; conceptualization, data curation, funding acquisition, supervision, validation, and project administration, H.V.D. All authors have read and agreed to the published version of the manuscript.

Funding: This research was funded by the National Research Council of Thailand (PMUC), grant number 640000060128, and partially supported by Chiang Mai University, grant number CoE2566.

Institutional Review Board Statement: All animal care and experimental procedures followed AAALAC (Association for Assessment and Accreditation of Laboratory Animal Care) guidelines and were approved by the Chiang Mai University Research Ethics Committee (Approved No. AG03002-2565).

Data Availability Statement: The data presented in this study are available on request from the corresponding author.

Acknowledgments: The authors would like to thank the National Research Council of Thailand (PMUC) for their support and assistance with this study.

Conflicts of Interest: The authors declare no conflicts of interest.

References

1. Sampathkumar, K.; Yu, H.; Loo, S.C.J. Valorisation of Industrial Food Waste into Sustainable Aquaculture Feeds. *Future Foods* **2023**, *7*, 100240. [CrossRef]
2. Gule, T.T.; Geremew, A. Dietary Strategies for Better Utilization of Aquafeeds in Tilapia Farming. *Aquac. Nutr.* **2022**, *2022*, 9463307. [CrossRef]
3. Abdel-Aziz, M.F.A.; Hassan, H.U.; Yones, A.-M.; Abdel-Tawwab, Y.A.; Metwalli, A.A.A.-T. Assessing the Effect of Different Feeding Frequencies Combined with Stocking Density, Initial Weight, and Dietary Protein Ratio on the Growth Performance of Tilapia, Catfish and Carp. *Sci. Afr.* **2021**, *12*, e00806. [CrossRef]
4. Syed, R.; Masood, Z.; Hassan, H.U.; Khan, W.; Mushtaq, S.; Ali, A.; Gul, Y.; Jafari, H.; Habib, A.; Shah, M.I.A. Growth Performance, Haematological Assessment and Chemical Composition of Nile Tilapia, *Oreochromis niloticus* (Linnaeus, 1758) Fed Different Levels of Aloe Vera Extract as Feed Additives in a Closed Aquaculture System. *Saudi J. Biol. Sci.* **2022**, *29*, 296–303. [CrossRef] [PubMed]
5. El-Sayed, A.-F.M.; Fitzsimmons, K. From Africa to the World—The Journey of Nile Tilapia. *Rev. Aquac.* **2023**, *15*, 6–21. [CrossRef]
6. Delphino, M.; Joshi, R.; Alvarez, A.T. Economic Appraisal of Using Genetics to Control *Streptococcus agalactiae* in Nile Tilapia under Cage and Pond Farming System in Malaysia. *Sci. Rep.* **2022**, *12*, 8754. [CrossRef] [PubMed]
7. Suyamud, B.; Chen, Y.; Quyen, D.T.T.; Dong, Z.; Zhao, C.; Hu, J. Antimicrobial Resistance in Aquaculture: Occurrence and Strategies in Southeast Asia. *Sci. Total Environ.* **2024**, *907*, 167942. [CrossRef]
8. Henriksson, P.J.; Rico, A.; Troell, M.; Klinger, D.H.; Buschmann, A.H.; Saksida, S.; Chadag, M.V.; Zhang, W. Unpacking Factors Influencing Antimicrobial Use in Global Aquaculture and Their Implication for Management: A Review from a Systems Perspective. *Sustain. Sci.* **2018**, *13*, 1105–1120. [CrossRef]
9. Guimarães, M.C.; Cerezo, I.M.; Fernandez-Alarcon, M.F.; Natori, M.M.; Sato, L.Y.; Kato, C.A.; Moriñigo, M.A.; Tapia-Paniagua, S.; Dias, D.d.C.; Ishikawa, C.M. Oral Administration of Probiotics (*Bacillus subtilis* and *Lactobacillus plantarum*) in Nile Tilapia (*Oreochromis niloticus*) Vaccinated and Challenged with *Streptococcus agalactiae*. *Fishes* **2022**, *7*, 211. [CrossRef]
10. Nurmalasari; Liu, C.-H.; Maftuch, I.M.; Hu, S.-Y. Dietary Supplementation with Prebiotic Chitooligosaccharides Enhances the Growth Performance, Innate Immunity and Disease Resistance of Nile Tilapia (*Oreochromis niloticus*). *Fishes* **2022**, *7*, 313. [CrossRef]
11. Wang, T.; Wu, H.-X.; Li, W.; Xu, R.; Qiao, F.; Du, Z.-Y.; Zhang, M.-L. Effects of Dietary Mannan Oligosaccharides (MOS) Supplementation on Metabolism, Inflammatory Response and Gut Microbiota of Juvenile Nile Tilapia (*Oreochromis niloticus*) Fed with High Carbohydrate Diet. *Fish Shellfish Immunol.* **2022**, *130*, 550–559. [CrossRef] [PubMed]
12. Kakavand, F.; Iri, A.; Rezaei Shadegan, M.; Bigdeli, M.; Heydarzadeh Barzegar, D.; Zamani, V.; Barkhordar, M.; Houshmand, P.; Zare Mehrabadi, E.; Hedayati, A. Toxicity Effect of Silver Nitrate on Some Hematological Indices of Nil Tilapia (*Oreochromis niloticus*) Fed with Different Levels of Prebiotic Oyster Mushroom (*Pleurotus ostreatus*). *J. Aquat. Physiol. Biotechnol.* **2022**, *9*, 141–160.
13. Outama, P.; Le Xuan, C.; Wannavijit, S.; Lumsangkul, C.; Linh, N.V.; Montha, N.; Tongsiri, S.; Chitmanat, C.; Van Doan, H. Modulation of Growth, Immune Response, and Immune-Antioxidant Related Gene Expression of Nile Tilapia (*Oreochromis niloticus*) Reared under Biofloc System Using Mango Peel Powder. *Fish Shellfish Immunol.* **2022**, *131*, 1136–1143. [CrossRef] [PubMed]

14. del Valle, J.C.; Bonadero, M.C.; Gimenez, A.V.F. Saccharomyces Cerevisiae as Probiotic, Prebiotic, Synbiotic, Postbiotics and Parabiotics in Aquaculture: An Overview. *Aquaculture* **2023**, *569*, 739342. [CrossRef]
15. Khanjani, M.H.; da Silva, L.O.B.; Fóes, G.K.; Vieira, F.d.N.; Poli, M.A.; Santos, M.; Emerenciano, M.G.C. Synbiotics and Aquamimicry as Alternative Microbial-Based Approaches in Intensive Shrimp Farming and Biofloc: Novel Disruptive Techniques or Complementary Management Tools? A Scientific-Based Overview. *Aquaculture* **2023**, *567*, 739273. [CrossRef]
16. Amiri, H.; Aghbashlo, M.; Sharma, M.; Gaffey, J.; Manning, L.; Moosavi Basri, S.M.; Kennedy, J.F.; Gupta, V.K.; Tabatabaei, M. Chitin and Chitosan Derived from Crustacean Waste Valorization Streams Can Support Food Systems and the UN Sustainable Development Goals. *Nat. Food* **2022**, *3*, 822–828. [CrossRef] [PubMed]
17. FAO. *The State of World Fisheries and Aquaculture 2022*; FAO: Rome, Italy, 2022.
18. Wani, A.K.; Akhtar, N.; Mir, T.u.G.; Rahayu, F.; Suhara, C.; Anjli, A.; Chopra, C.; Singh, R.; Prakash, A.; El Messaoudi, N.; et al. Eco-Friendly and Safe Alternatives for the Valorization of Shrimp Farming Waste. *Environ. Sci. Pollut. Res.* **2023**. [CrossRef] [PubMed]
19. Ravanipour, M.; Bagherzadeh, R.; Mahvi, A.H. Fish and Shrimp Waste Management at Household and Market in Bushehr, Iran. *J. Mater. Cycles Waste Manag.* **2021**, *23*, 1394–1403. [CrossRef]
20. Mao, X.; Guo, N.; Sun, J.; Xue, C. Comprehensive Utilization of Shrimp Waste Based on Biotechnological Methods: A Review. *J. Clean. Prod.* **2017**, *143*, 814–823. [CrossRef]
21. Yadav, M.; Goswami, P.; Paritosh, K.; Kumar, M.; Pareek, N.; Vivekanand, V. Seafood Waste: A Source for Preparation of Commercially Employable Chitin/Chitosan Materials. *Bioresour. Bioprocess.* **2019**, *6*, 8. [CrossRef]
22. Mansyur, N.; Hanudin, E.; Purwanto, B.; Utami, S. The Nutritional Value of Shrimp Waste and Its Response to Growth and N Uptake Efficiency by Corn. *IOP Conf. Ser. Earth Environ. Sci.* **2021**, *748*, 012013. [CrossRef]
23. Liu, Z.; Liu, Q.; Zhang, D.; Wei, S.; Sun, Q.; Xia, Q.; Shi, W.; Ji, H.; Liu, S. Comparison of the Proximate Composition and Nutritional Profile of Byproducts and Edible Parts of Five Species of Shrimp. *Foods* **2021**, *10*, 2603. [CrossRef] [PubMed]
24. Susetyaningsih, R.; Suntoro, S.; Gunawan, T.; Budiastuti, M.T.S. Impact of Shrimp Pond Waste on Water Quality (Case Study of Trisik Lagoon in Yogyakarta). *AIP Conf. Proc.* **2020**, *2296*, 020050.
25. Dauda, A.B.; Ajadi, A.; Tola-Fabunmi, A.S.; Akinwole, A.O. Waste Production in Aquaculture: Sources, Components and Managements in Different Culture Systems. *Aquac. Fish.* **2019**, *4*, 81–88. [CrossRef]
26. El Amri, H.; Boukharta, M.; Zakham, F.; Ennaji, M.M. Emergence and Reemergence of Viral Zoonotic Diseases: Concepts and Factors of Emerging and Reemerging Globalization of Health Threats. In *Emerging and Reemerging Viral Pathogens*; Elsevier: Amsterdam, The Netherlands, 2020; pp. 619–634.
27. Yan, N.; Chen, X. Sustainability: Don't Waste Seafood Waste. *Nature* **2015**, *524*, 155–157. [CrossRef]
28. Hosney, A.; Ullah, S.; Barčauskaitė, K. A Review of the Chemical Extraction of Chitosan from Shrimp Wastes and Prediction of Factors Affecting Chitosan Yield by Using an Artificial Neural Network. *Mar. Drugs* **2022**, *20*, 675. [CrossRef] [PubMed]
29. de Queiroz Antonino, R.S.C.M.; Lia Fook, B.R.P.; de Oliveira Lima, V.A.; de Farias Rached, R.Í.; Lima, E.P.N.; da Silva Lima, R.J.; Peniche Covas, C.A.; Lia Fook, M.V. Preparation and Characterization of Chitosan Obtained from Shells of Shrimp (*Litopenaeus vannamei* Boone). *Mar. Drugs* **2017**, *15*, 141. [CrossRef] [PubMed]
30. Xu, Q.; Wang, W.; Qu, C.; Gu, J.; Yin, H.; Jia, Z.; Song, L.; Du, Y. Chitosan Oligosaccharides Inhibit Epithelial Cell Migration through Blockade of N-Acetylglucosaminyltransferase V and Branched GlcNAc Structure. *Carbohydr. Polym.* **2017**, *170*, 241–246. [CrossRef] [PubMed]
31. Zou, P.; Yuan, S.; Yang, X.; Zhai, X.; Wang, J. Chitosan Oligosaccharides with Degree of Polymerization 2–6 Induces Apoptosis in Human Colon Carcinoma HCT116 Cells. *Chem.-Biol. Interact.* **2018**, *279*, 129–135. [CrossRef]
32. Jitprasertwong, P.; Khamphio, M.; Petsrichuang, P.; Eijsink, V.G.H.; Poolsri, W.; Muanprasat, C.; Rangnoi, K.; Yamabhai, M. Anti-Inflammatory Activity of Soluble Chitooligosaccharides (CHOS) on VitD3-Induced Human THP-1 Monocytes. *PLoS ONE* **2021**, *16*, e0246381. [CrossRef]
33. Muanprasat, C.; Wongkrasant, P.; Satitsri, S.; Moonwiriyakit, A.; Pongkorpsakol, P.; Mattaveewong, T.; Pichyangkura, R.; Chatsudthipong, V. Activation of AMPK by Chitosan Oligosaccharide in Intestinal Epithelial Cells: Mechanism of Action and Potential Applications in Intestinal Disorders. *Biochem. Pharmacol.* **2015**, *96*, 225–236. [CrossRef] [PubMed]
34. Dai, X.; Chang, P.; Li, X.; Gao, Z.; Sun, Y. The Inhibitory Effect of Chitosan Oligosaccharides on β-Site Amyloid Precursor Protein Cleaving Enzyme 1 (BACE1) in HEK293 APPswe Cells. *Neurosci. Lett.* **2018**, *665*, 80–85. [CrossRef] [PubMed]
35. Muanprasat, C.; Chatsudthipong, V. Chitosan Oligosaccharide: Biological Activities and Potential Therapeutic Applications. *Pharmacol. Ther.* **2017**, *170*, 80–97. [CrossRef] [PubMed]
36. Bezrodnykh, E.A.; Antonov, Y.A.; Berezin, B.B.; Kulikov, S.N.; Tikhonov, V.E. Molecular Features of the Interaction and Antimicrobial Activity of Chitosan in a Solution Containing Sodium Dodecyl Sulfate. *Carbohydr. Polym.* **2021**, *270*, 118352. [CrossRef] [PubMed]
37. Tabatabaei, M.; Ebrahimi, B.; Rajaei, A.; Movahednejad, M.H.; Rastegari, H.; Taghavi, E.; Aghbashlo, M.; Gupta, V.K.; Lam, S.S. Producing Submicron Chitosan-Stabilized Oil Pickering Emulsion Powder by an Electrostatic Collector-Equipped Spray Dryer. *Carbohydr. Polym.* **2022**, *294*, 119791. [CrossRef] [PubMed]
38. Liang, F.; Li, C.; Hou, T.; Wen, C.; Kong, S.; Ma, D.; Sun, C.; Li, S. Effects of Chitosan–Gentamicin Conjugate Supplement on Non-Specific Immunity, Aquaculture Water, Intestinal Histology and Microbiota of Pacific White Shrimp (*Litopenaeus vannamei*). *Mar. Drugs* **2020**, *18*, 419. [CrossRef] [PubMed]

39. Vázquez, J.A.; Rodríguez-Amado, I.; Montemayor, M.I.; Fraguas, J.; del Pilar González, M.; Murado, M.A. Chondroitin Sulfate, Hyaluronic Acid and Chitin/Chitosan Production Using Marine Waste Sources: Characteristics, Applications and Eco-Friendly Processes: A Review. *Mar. Drugs* **2013**, *11*, 747–774. [CrossRef] [PubMed]
40. Avnimelech, Y. *Biofloc Technology: A Practical Guide Book*; World Aquaculture Society: Sorrento, LA, USA, 2009; ISBN 1-888807-16-4.
41. McCusker, S.; Warberg, M.B.; Davies, S.J.; Valente, C.d.S.; Johnson, M.P.; Cooney, R.; Wan, A.H. Biofloc Technology as Part of a Sustainable Aquaculture System: A Review on the Status and Innovations for Its Expansion. *Aquac. Fish Fish.* **2023**, *33*, 331–352. [CrossRef]
42. Kuhn, D.D.; Boardman, G.D.; Craig, S.R.; Flick, G.J., Jr.; McLean, E. Use of Microbial Flocs Generated from Tilapia Effluent as a Nutritional Supplement for Shrimp, *Litopenaeus vannamei*, in Recirculating Aquaculture Systems. *J. World Aquac. Soc.* **2008**, *39*, 72–82. [CrossRef]
43. Crab, R.; Defoirdt, T.; Bossier, P.; Verstraete, W. Biofloc Technology in Aquaculture: Beneficial Effects and Future Challenges. *Aquaculture* **2012**, *356*, 351–356. [CrossRef]
44. Khanjani, M.H.; Mohammadi, A.; Emerenciano, M.G.C. Microorganisms in Biofloc Aquaculture System. *Aquac. Rep.* **2022**, *26*, 101300. [CrossRef]
45. Ogello, E.O.; Outa, N.O.; Obiero, K.O.; Kyule, D.N.; Munguti, J.M. The Prospects of Biofloc Technology (BFT) for Sustainable Aquaculture Development. *Sci. Afr.* **2021**, *14*, e01053. [CrossRef]
46. Luo, G.; Gao, Q.; Wang, C.; Liu, W.; Sun, D.; Li, L.; Tan, H. Growth, Digestive Activity, Welfare, and Partial Cost-Effectiveness of Genetically Improved Farmed Tilapia (*Oreochromis niloticus*) Cultured in a Recirculating Aquaculture System and an Indoor Biofloc System. *Aquaculture* **2014**, *422*, 1–7. [CrossRef]
47. Xu, W.-J.; Morris, T.C.; Samocha, T.M. Effects of C/N Ratio on Biofloc Development, Water Quality, and Performance of *Litopenaeus vannamei* Juveniles in a Biofloc-Based, High-Density, Zero-Exchange, Outdoor Tank System. *Aquaculture* **2016**, *453*, 169–175. [CrossRef]
48. Yu, Y.-B.; Lee, J.-H.; Choi, J.-H.; Choi, Y.J.; Jo, A.-H.; Choi, C.Y.; Kang, J.-C.; Kim, J.-H. The Application and Future of Biofloc Technology (BFT) in Aquaculture Industry: A Review. *J. Environ. Manag.* **2023**, *342*, 118237. [CrossRef]
49. Chutia, A.; Xavier, K.M.; Shamna, N.; Rani, A.B. Application of Bioflocculating Agent in Inoculum Enhances Quality of Biofloc and Influences Growth, Feed Utilization and Stress Responses of GIFT Tilapia Reared *in-situ*. *Aquaculture* **2022**, *553*, 738050. [CrossRef]
50. Qin, C.; Zhang, Y.; Liu, W.; Xu, L.; Yang, Y.; Zhou, Z. Effects of Chito-Oligosaccharides Supplementation on Growth Performance, Intestinal Cytokine Expression, Autochthonous Gut Bacteria and Disease Resistance in Hybrid Tilapia *Oreochromis niloticus*♀ × *Oreochromis aureus*♂. *Fish Shellfish Immunol.* **2014**, *40*, 267–274. [CrossRef]
51. Rahimnejad, S.; Yuan, X.; Wang, L.; Lu, K.; Song, K.; Zhang, C. Chitooligosaccharide Supplementation in Low-Fish Meal Diets for Pacific White Shrimp (*Litopenaeus vannamei*): Effects on Growth, Innate Immunity, Gut Histology, and Immune-Related Genes Expression. *Fish Shellfish Immunol.* **2018**, *80*, 405–415. [CrossRef]
52. Abdel-Ghany, H.M.; Salem, M.E.-S. Effects of Dietary Chitosan Supplementation on Farmed Fish; a Review. *Rev. Aquac.* **2020**, *12*, 438–452. [CrossRef]
53. Shiau, S.-Y.; Yu, Y.-P. Dietary Supplementation of Chitin and Chitosan Depresses Growth in Tilapia, *Oreochromis niloticus* × *O. aureus*. *Aquaculture* **1999**, *179*, 439–446. [CrossRef]
54. Romana-Eguia, M.R.R.; Parado-Estepa, F.D.; Salayo, N.D.; Lebata-Ramos, M.J.H. *Resource Enhancement and Sustainable Aquaculture Practices in Southeast Asia: Challenges in Responsible Production of Aquatic Species: Proceedings of the International Workshop on Resource Enhancement and Sustainable Aquaculture Practices in Southeast Asia 2014 (RESA)*; Aquaculture Department, Southeast Asian Fisheries Development Center: Bangkok, Thailand, 2015; ISBN 971-9931-04-3.
55. Kamali Najafabad, M.; Imanpoor, M.R.; Taghizadeh, V.; Alishahi, A. Effect of Dietary Chitosan on Growth Performance, Hematological Parameters, Intestinal Histology and Stress Resistance of Caspian Kutum (*Rutilus frisii Kutum* Kamenskii, 1901) Fingerlings. *Fish Physiol. Biochem.* **2016**, *42*, 1063–1071. [CrossRef]
56. Wang, Y.; Li, J. Effects of Chitosan Nanoparticles on Survival, Growth and Meat Quality of Tilapia, *Oreochromis nilotica*. *Nanotoxicology* **2011**, *5*, 425–431. [CrossRef]
57. Yu, W.; Yang, Y.; Chen, H.; Zhou, Q.; Zhang, Y.; Huang, X.; Huang, Z.; Li, T.; Zhou, C.; Ma, Z. Effects of Dietary Chitosan on the Growth, Health Status and Disease Resistance of Golden Pompano (*Trachinotus ovatus*). *Carbohydr. Polym.* **2023**, *300*, 120237. [CrossRef]
58. Yu, W.; Yang, Y.; Zhou, Q.; Huang, X.; Huang, Z.; Li, T.; Wu, Q.; Zhou, C.; Ma, Z.; Lin, H. Effects of Dietary Astragalus Polysaccharides on Growth, Health and Resistance to *Vibrio harveyi* of *Lates calcarifer*. *Int. J. Biol. Macromol.* **2022**, *207*, 850–858. [CrossRef]
59. Abd El-Naby, F.S.; Naiel, M.A.; Al-Sagheer, A.A.; Negm, S.S. Dietary Chitosan Nanoparticles Enhance the Growth, Production Performance, and Immunity in *Oreochromis niloticus*. *Aquaculture* **2019**, *501*, 82–89. [CrossRef]
60. Denji, K.A.; Mansour, M.R.; Akrami, R.; Ghobadi, S.; Jafarpour, S.; Mirbeygi, S. Effect of Dietary Prebiotic Mannan Oligosaccharide (MOS) on Growth Performance, Intestinal Microflora, Body Composition, Haematological and Blood Serum Biochemical Parameters of Rainbow Trout (*Oncorhynchus mykiss*) Juveniles. *J. Fish. Aquat. Sci.* **2015**, *10*, 255.
61. Wang, Y.; Zhao, K.; Li, L.; Song, X.; He, Y.; Ding, N.; Li, L.; Wang, S.; Liu, Z. A Review of the Immune Activity of Chitooligosaccharides. *Food Sci. Technol.* **2023**, *43*, e97822. [CrossRef]

62. Abd El-Naby, A.S.; Al-Sagheer, A.A.; Negm, S.S.; Naiel, M.A. Dietary Combination of Chitosan Nanoparticle and Thymol Affects Feed Utilization, Digestive Enzymes, Antioxidant Status, and Intestinal Morphology of *Oreochromis niloticus*. *Aquaculture* **2020**, *515*, 734577. [CrossRef]
63. Fernandes, J.C.; Eaton, P.; Nascimento, H.; Gião, M.S.; Ramos, Ó.S.; Belo, L.; Santos-Silva, A.; Pintado, M.E.; Malcata, F.X. Antioxidant Activity of Chitooligosaccharides upon Two Biological Systems: Erythrocytes and Bacteriophages. *Carbohydr. Polym.* **2010**, *79*, 1101–1106. [CrossRef]
64. Anraku, M.; Gebicki, J.M.; Iohara, D.; Tomida, H.; Uekama, K.; Maruyama, T.; Hirayama, F.; Otagiri, M. Antioxidant Activities of Chitosans and Its Derivatives in in Vitro and in Vivo Studies. *Carbohydr. Polym.* **2018**, *199*, 141–149. [CrossRef]
65. Dalmo, R.A.; Bøgwald, J. SS-Glucans as Conductors of Immune Symphonies. *Fish Shellfish Immunol.* **2008**, *25*, 384–396. [CrossRef]
66. Alves, A.P.d.C.; Paulino, R.R.; Pereira, R.T.; da Costa, D.V.; e Rosa, P.V. Nile Tilapia Fed Insect Meal: Growth and Innate Immune Response in Different Times under Lipopolysaccharide Challenge. *Aquac. Res.* **2021**, *52*, 529–540. [CrossRef]
67. Conforto, E.; Vílchez-Gómez, L.; Parrinello, D.; Parisi, M.G.; Esteban, M.Á.; Cammarata, M.; Guardiola, F.A. Role of Mucosal Immune Response and Histopathological Study in European Eel (*Anguilla anguilla* L.) Intraperitoneal Challenged by *Vibrio anguillarum* or *Tenacibaculum soleae*. *Fish Shellfish Immunol.* **2021**, *114*, 330–339. [CrossRef]
68. Marshall, W.S.; Bellamy, D. The 50 Year Evolution of in Vitro Systems to Reveal Salt Transport Functions of Teleost Fish Gills. *Comp. Biochem. Physiol. Part A Mol. Integr. Physiol.* **2010**, *155*, 275–280. [CrossRef]
69. McNeilly, T.N.; Naylor, S.W.; Mahajan, A.; Mitchell, M.C.; McAteer, S.; Deane, D.; Smith, D.G.; Low, J.C.; Gally, D.L.; Huntley, J.F. Escherichia Coli O157: H7 Colonization in Cattle Following Systemic and Mucosal Immunization with Purified H7 Flagellin. *Infect. Immun.* **2008**, *76*, 2594–2602. [CrossRef]
70. Magnadóttir, B. Innate Immunity of Fish (Overview). *Fish Shellfish Immunol.* **2006**, *20*, 137–151. [CrossRef]
71. Galagarza, O.A.; Smith, S.A.; Drahos, D.J.; Eifert, J.D.; Williams, R.C.; Kuhn, D.D. Modulation of Innate Immunity in Nile Tilapia (*Oreochromis niloticus*) by Dietary Supplementation of *Bacillus subtilis* Endospores. *Fish Shellfish Immunol.* **2018**, *83*, 171–179. [CrossRef]
72. Park, Y.-K.; Kim, M.-H.; Park, S.-C.; Cheong, H.-S.; Jang, M.-K.; Nah, J.-W.; Hahm, K.-S. Investigation of the Antifungal Activity and Mechanism of Action of LMWS-Chitosan. *J. Microbiol. Biotechnol.* **2008**, *18*, 1729–1734.
73. Chang, Q.; Liang, M.; Wang, J.; Sun, J. Influence of Chitosan on the Growth and Non-Specific Immunity of Japanese Sea Bass (*Lateolabrax japonicus*). *Mar. Fish. Res.* **2006**, *27*, 17–22.
74. Gopalakannan, A.; Arul, V. Immunomodulatory Effects of Dietary Intake of Chitin, Chitosan and Levamisole on the Immune System of *Cyprinus carpio* and Control of *Aeromonas hydrophila* Infection in Ponds. *Aquaculture* **2006**, *255*, 179–187. [CrossRef]
75. Guardiola, F.A.; Cuesta, A.; Arizcun, M.; Meseguer, J.; Esteban, M.A. Comparative Skin Mucus and Serum Humoral Defence Mechanisms in the Teleost Gilthead Seabream (*Sparus aurata*). *Fish Shellfish Immunol.* **2014**, *36*, 545–551. [CrossRef]
76. Srichaiyo, N.; Tongsiri, S.; Hoseinifar, S.H.; Dawood, M.A.; Jaturasitha, S.; Esteban, M.Á.; Ringø, E.; Van Doan, H. The Effects Gotu Kola (*Centella asiatica*) Powder on Growth Performance, Skin Mucus, and Serum Immunity of Nile Tilapia (*Oreochromis niloticus*) Fingerlings. *Aquac. Rep.* **2020**, *16*, 100239. [CrossRef]
77. Dimitroglou, A.; Merrifield, D.; Moate, R.; Davies, S.; Spring, P.; Sweetman, J.; Bradley, G. Dietary Mannan Oligosaccharide Supplementation Modulates Intestinal Microbial Ecology and Improves Gut Morphology of Rainbow Trout, *Oncorhynchus mykiss* (Walbaum). *J. Anim. Sci.* **2009**, *87*, 3226–3234. [CrossRef]
78. Geda, F.; Rekecki, A.; Decostere, A.; Bossier, P.; Wuyts, B.; Kalmar, I.; Janssens, G. Changes in Intestinal Morphology and Amino Acid Catabolism in Common Carp at Mildly Elevated Temperature as Affected by Dietary Mannanoligosaccharides. *Anim. Feed Sci. Technol.* **2012**, *178*, 95–102. [CrossRef]
79. Daniels, C.L.; Merrifield, D.L.; Boothroyd, D.P.; Davies, S.J.; Factor, J.R.; Arnold, K.E. Effect of Dietary *Bacillus* spp. and Mannan Oligosaccharides (MOS) on European Lobster (*Homarus gammarus* L.) Larvae Growth Performance, Gut Morphology and Gut Microbiota. *Aquaculture* **2010**, *304*, 49–57. [CrossRef]
80. Chen, G.; Yin, B.; Liu, H.; Tan, B.; Dong, X.; Yang, Q.; Chi, S.; Zhang, S. Supplementing Chitosan Oligosaccharide Positively Affects Hybrid Grouper (*Epinephelus fuscoguttatus*♀ × *E. lanceolatus*♂) Fed Dietary Fish Meal Replacement with Cottonseed Protein Concentrate: Effects on Growth, Gut Microbiota, Antioxidant Function and Immune Response. *Front. Mar. Sci.* **2021**, *8*, 707627.
81. Hahor, W.; Thongprajukaew, K.; Suanyuk, N. Effects of Dietary Supplementation of Oligosaccharides on Growth Performance, Gut Health and Immune Response of Hybrid Catfish (*Pangasianodon gigas* × *Pangasianodon hypophthalmus*). *Aquaculture* **2019**, *507*, 97–107. [CrossRef]
82. Zhou, Q.-C.; Buentello, J.A.; Gatlin III, D.M. Effects of Dietary Prebiotics on Growth Performance, Immune Response and Intestinal Morphology of Red Drum (*Sciaenops ocellatus*). *Aquaculture* **2010**, *309*, 253–257. [CrossRef]
83. Wu, Y.; Liu, W.; Li, H.; Xu, W.; He, J.; Li, X.; Jiang, G. Effects of Dietary Supplementation of Fructooligosaccharide on Growth Performance, Body Composition, Intestinal Enzymes Activities and Histology of Blunt Snout Bream (*Megalobrama amblycephala*) Fingerlings. *Aquac. Nutr.* **2013**, *19*, 886–894. [CrossRef]
84. Wang, K.; Jiang, W.; Wu, P.; Liu, Y.; Jiang, J.; Kuang, S.; Tang, L.; Zhang, Y.; Zhou, X.; Feng, L. Gossypol Reduced the Intestinal Amino Acid Absorption Capacity of Young Grass Carp (*Ctenopharyngodon idella*). *Aquaculture* **2018**, *492*, 46–58. [CrossRef]
85. Yamamoto, T.; Matsunari, H.; Sugita, T.; Furuita, H.; Masumoto, T.; Iwashita, Y.; Amano, S.; Suzuki, N. Optimization of the Supplemental Essential Amino Acids to a Fish Meal-Free Diet Based on Fermented Soybean Meal for Rainbow Trout *Oncorhynchus mykiss*. *Fish. Sci.* **2012**, *78*, 359–366. [CrossRef]

86. Morales, A.; Gómez, T.; Villalobos, Y.D.; Bernal, H.; Htoo, J.K.; González-Vega, J.C.; Espinoza, S.; Yáñez, J.; Cervantes, M. Dietary Protein-Bound or Free Amino Acids Differently Affect Intestinal Morphology, Gene Expression of Amino Acid Transporters, and Serum Amino Acids of Pigs Exposed to Heat Stress. *J. Anim. Sci.* **2020**, *98*, skaa056. [CrossRef]
87. Sakai, M.; Hikima, J.; Kono, T. Fish Cytokines: Current Research and Applications. *Fish. Sci.* **2021**, *87*, 1–9. [CrossRef]
88. Zlotnik, A.; Yoshie, O. The Chemokine Superfamily Revisited. *Immunity* **2012**, *36*, 705–716. [CrossRef]
89. Dinarello, C.A. Overview of the IL-1 Family in Innate Inflammation and Acquired Immunity. *Immunol. Rev.* **2018**, *281*, 8–27. [CrossRef]
90. Hirayama, D.; Iida, T.; Nakase, H. The Phagocytic Function of Macrophage-Enforcing Innate Immunity and Tissue Homeostasis. *Int. J. Mol. Sci.* **2017**, *19*, 92. [CrossRef]
91. Fu, G.H.; Liu, F.; Xia, J.H.; Yue, G.H. The LBP Gene and Its Association with Resistance to *Aeromonas hydrophila* in Tilapia. *Int. J. Mol. Sci.* **2014**, *15*, 22028–22041. [CrossRef]
92. Fath El-Bab, A.F.; Majrashi, K.A.; Sheikh, H.M.; Shafi, M.E.; El-Ratel, I.T.; Neamat-Allah, A.N.; El-Raghi, A.A.; Elazem, A.Y.A.; Abd-Elghany, M.F.; Abdelnour, S.A. Dietary Supplementation of Nile Tilapia (*Oreochromis niloticus*) with β-Glucan and/or *Bacillus coagulans*: Synergistic Impacts on Performance, Immune Responses, Redox Status and Expression of Some Related Genes. *Front. Vet. Sci.* **2022**, *9*, 1011715. [CrossRef]
93. Imai, H.; Nakagawa, Y. Biological Significance of Phospholipid Hydroperoxide Glutathione Peroxidase (PHGPx, GPx4) in Mammalian Cells. *Free Radic. Biol. Med.* **2003**, *34*, 145–169. [CrossRef]
94. Ahlf, W.; Heise, S. Sediment Toxicity Assessment: Rationale for Effect Classes (5 Pp). *J. Soils Sediments* **2005**, *5*, 16–20. [CrossRef]
95. De Schryver, P.; Crab, R.; Defoirdt, T.; Boon, N.; Verstraete, W. The Basics of Bio-Flocs Technology: The Added Value for Aquaculture. *Aquaculture* **2008**, *277*, 125–137. [CrossRef]
96. Azim, M.E.; Little, D.C. The Biofloc Technology (BFT) in Indoor Tanks: Water Quality, Biofloc Composition, and Growth and Welfare of Nile Tilapia (*Oreochromis niloticus*). *Aquaculture* **2008**, *283*, 29–35. [CrossRef]
97. Mirzakhani, N.; Ebrahimi, E.; Jalali, S.A.H.; Ekasari, J. Growth Performance, Intestinal Morphology and Nonspecific Immunity Response of Nile Tilapia (*Oreochromis niloticus*) Fry Cultured in Biofloc Systems with Different Carbon Sources and Input C: N Ratios. *Aquaculture* **2019**, *512*, 734235. [CrossRef]
98. Khanjani, M.H.; Sharifinia, M.; Hajirezaee, S. Recent Progress towards the Application of Biofloc Technology for Tilapia Farming. *Aquaculture* **2022**, *552*, 738021. [CrossRef]
99. Linh, N.V.; Van Nguyen, D.; Khongdee, N.; Wannavijit, S.; Outama, P.; Le Xuan, C.; Mahatheeranont, S.; Sookwong, P.; Le, T.D.; Hoseinifar, S.H. Influence of Black Rice (*Oryza sativa* L.) Bran Derived Anthocyanin-Extract on Growth Rate, Immunological Response, and Immune-Antioxidant Gene Expression in Nile Tilapia (*Oreochromis niloticus*) Cultivated in a Biofloc System. *Fish Shellfish Immunol.* **2022**, *128*, 604–611. [CrossRef] [PubMed]
100. Suthongsa, S.; Pichyangkura, R.; Kalandakanond-Thongsong, S.; Thongsong, B. Effects of Dietary Levels of Chito-Oligosaccharide on Ileal Digestibility of Nutrients, Small Intestinal Morphology and Crypt Cell Proliferation in Weaned Pigs. *Livest. Sci.* **2017**, *198*, 37–44. [CrossRef]
101. Keereelang, J.; Mangumphan, K.; Chitmanat, C.; Tongsiri, S.; Linh, N.V.; Van Doan, H. Dietary Effect of *Lactobacillus plantarum* (TISTR 912) on Digestive Enzyme Activity, Growth Performance, Immune Response, and Disease Resistance of Black Sharkminnow (*Labeo chrysophekadion*) against *Aeromonas hydrophila* Infection. *Aquac. Rep.* **2022**, *27*, 101409. [CrossRef]
102. Livak, K.J.; Schmittgen, T.D. Analysis of Relative Gene Expression Data Using Real-Time Quantitative PCR and the $2^{-\Delta\Delta Ct}$ Method. *Methods* **2001**, *25*, 402–408. [CrossRef]

marine drugs

Article

Isolation and Purification of Chitosan Oligosaccharides ($Mw \leq 1000$) and Their Protective Effect on Acute Liver Injury Caused by CCl_4

Kai Wang [1,2,†], Dawei Yu [1,2,†], Yan Bai [3], Hua Cao [4], Jiao Guo [2,*] and Zhengquan Su [1,*]

[1] Guangdong Provincial University Engineering Technology Research Center of Natural Products and Drugs, Guangdong Pharmaceutical University, Guangzhou 510006, China; wk972023826@163.com (K.W.); ydw65038@163.com (D.Y.)
[2] Guangdong Metabolic Disease Research Center of Integrated Chinese and Western Medicine, Guangdong Pharmaceutical University, Guangzhou 510006, China
[3] School of Public Health, Guangdong Pharmaceutical University, Guangzhou 510310, China; angell_bai@163.com
[4] School of Chemistry and Chemical Engineering, Guangdong Pharmaceutical University, Zhongshan 528458, China; caohua@gdpu.edu.cn
* Correspondence: gyguoyz@163.com (J.G.); suzhq@scnu.edu.cn (Z.S.); Tel.: +86-20-3935-2345 (Z.S.); Fax: +86-20-3935-2067 (Z.S.)
† These authors contributed equally to this work.

Abstract: Chitosan oligosaccharides are the degradation products of chitin obtained from the shell extracts of shrimps and crabs. Compared with chitosan, chitosan oligosaccharides have better solubility and a wider application range. In this study, high-molecular-weight chitosan oligosaccharides (COST, chitosan oligosaccharides, $MW \leq 1000$) were isolated and purified by a GPC gel column, and the molecular weight range was further reduced to obtain high-purity and low-molecular-weight chitosan (COS46). Compared with COST, COS46 is better at inhibiting CCl_4-induced cell death, improving cell morphology, reducing ALT content, and improving cell antioxidant capacity. The effects of COST and COS46 on CCl_4-induced acute liver injury were further verified in mice. Both COS46 and COST improved the appearance of the liver induced by CCl_4, decreased the levels of ALT and AST in serum, and decreased the oxidation/antioxidant index in the liver. From the liver pathological section, the effect of COS46 was better. In addition, some indicators of COS46 showed a dose-dependent effect. In conclusion, compared with COST, low-molecular-weight COS46 has better antioxidant capacity and a better therapeutic effect on CCl_4-induced acute liver injury.

Keywords: acute liver injury; CCl_4; chitosan oligosaccharide; oxidation resistance

1. Introduction

At present, the shells of crabs and shrimps in marine arthropods are often ignored and underutilized, which are important sources of chitin. After a series of compound chemical treatments, chitosan and chitosan oligosaccharides are obtained, which have important application value in medical care and food [1]. Chitosan oligosaccharide itself has a variety of physiological activities, such as antibacterial properties [2,3], anti-tumor properties [4,5], antioxidant [3,6] and anti-inflammatory effects [7], etc. There are also studies using chitosan oligosaccharides as a drug carrier [8,9] or as a prebiotic to explore their influence on intestinal flora in different diseases [10,11]. It has also been revealed to have anti-diabetes [12], anti-obesity [13], anti-HIV-1 [14], and anti-Alzheimer's disease [15] effects and promote calcium absorption [16].

The excellent antioxidant properties of chitosan oligosaccharides have been the focus of related research, and it has been proven that it can reduce oxidative stress injury by scavenging free radicals and reactive oxygen species and maintaining the activity of antioxidant enzymes [17,18]. Therefore, the antioxidant properties of chitosan oligosaccharides can be identified from two aspects: the ability to scavenge free radicals and the ability to improve the level of antioxidant substances in vivo. However, the exact mechanism has not been explained. According to existing studies and the chemical structure of chitosan oligosaccharides, it is speculated that its amino group can combine with unstable free radicals to form a stable structure, which shows antioxidant properties.

Recent studies have indicated that the biological activity of chitosan oligosaccharides is primarily influenced by DP (degree of polymerization) and Mw (molecular weight). Therefore, it is difficult to determine which parts of chitosan oligosaccharides play a major role in biological activity [19]. Although there is no exact evidence to directly show the relationship between chitosan oligosaccharides' antioxidant properties and molecular weight, it is speculated from the existing relevant studies that a low molecular weight and a high degree of deacetylation can equip chitosan oligosaccharides with a better antioxidant effect [20], and the relationship between the two needs to be further explored.

Currently, chitosan oligosaccharides available in the market are typically synthesized from monosaccharides or extracted from hydrolyzed products. Researchers selectively synthesize the necessary oligosaccharides through specific chemical or enzymatic reactions. However, the synthesis process involves numerous steps and is costly. Therefore, it is more convenient to extract chitosan oligosaccharides [21–23]. Combined with the relationship between the properties of chitosan oligosaccharides and molecular weight, it is of great significance to narrow the molecular weight range through specific technical means, and even obtain chitosan oligosaccharides with a specific degree of polymerization. It is important not only to clarify the pharmacodynamic mechanism of chitosan oligosaccharides with different polymerization degrees but also to develop and utilize chitosan oligosaccharides more effectively. Due to the small difference in charge density and molecular weight between chitosan oligosaccharide monomers, the separation and purification of chitosan oligosaccharides are very challenging, especially when the degree of polymerization is greater than 4. This makes it a hotspot in the field of sugar science.

In the existing research, many methods for recovering and purifying chitosan oligosaccharides, such as gel filtration or size exclusion chromatography, have been found to purify and obtain specific chitosan oligosaccharides, and the molecular weight difference between different polymerization degrees of chitosan oligosaccharides is the basis for separation. It has been reported that chitosan oligosaccharides with a polymerization degree of 2–6 were separated from the hydrolysate of chitosan by gel filtration chromatography and ion exchange chromatography [24]. In an experiment of enzymatic hydrolysis of chitosan, the separation ability of size-exclusion chromatography (SEC) was used to help researchers fully understand the composition of the enzymatic hydrolysate [25]. Ultrafiltration [26] and nanofiltration [27] are also applied to the purification of chitosan oligosaccharides. Ion exchange chromatography [28] and metal affinity chromatography [29] are also widely used to detect and separate chitosan oligosaccharides.

High-performance liquid chromatography (HPLC) is often used for chitosan oligosaccharide analysis and purification [30]. HPLC is often combined with an ultraviolet detector, and the acetyl amino group of chitosan oligosaccharides can form a conjugated structure with the sugar ring, and the absorbance of chitosan oligosaccharides can be detected at a wavelength of approximately 210 nm. A differential refraction detector is preferred for the HPLC analysis of N-acetylated and N-deacetylated chitosan oligosaccharides, but its sensitivity is low and it is not suitable for gradient washing [31]. It is difficult to separate chitosan oligosaccharides by conventional HPLC, mainly because of its strong hydrophilicity and poor retention on HPLC columns [32]. Hydrophilic-interaction chromatography (HILIC) is a technique that can effectively separate polar and hydrophilic compounds. Some studies have successfully separated and purified chitosan oligosaccharides with a polymerization degree of 2–6 from a chitosan oligosaccharides mixture by using HILIC technology and used it to judge the influence of each component on RSC96 cells [32].

N- acetyl chitotriose and N, N′- diacetyl chitotriose can be separated from acetylated chitotriose by a cation exchange gel column [21]. For partially acetylated chitosan oligosaccharides, HPLC-ESI-MS can be used for separation and detection [33].

Capillary Electrophoresis (CE) has the advantage of high separation ability. Compared with HPLC, it has the characteristics of less sample loading, but it is precisely because of the small sample loading that a large number of chitosan oligosaccharides cannot be obtained by CE technology. When used to separate and detect chitosan oligosaccharides, electrophoretic mobility depends on the number of unit structures in an acidic aqueous solution, which is similar to other oligomeric electrolytes [28,34]. In many studies, chitosan oligosaccharides have been successfully separated using this technique [34,35].

CCl_4 is a typical substance with hepatotoxicity, which will remove halogen atoms in liver microsomes through a redox reaction to generate the trichloromethyl radical and its derivative trichloromethyl peroxyradical active intermediate. These free radicals bind to phospholipid molecules on the endoplasmic reticulum (ER) through covalent bonds. This can inhibit protein synthesis, destroy the function and morphology of the cell membrane, and cause hepatotoxicity, fibrosis, lipid peroxidation, and ERS (Endoplasmic reticulum stress) reaction [36,37].

Acute liver injury induced by CCl_4 in animals is very similar to acute chemical liver injury in humans, so it is widely used in the study of potential hepatoprotective drug activity [38]. CCl_4 not only exerts strong toxicity on the liver but also causes damage to other parts of the body [39], including the kidneys [40], nervous system [41], and testes [42].

Currently, there is no research on the role of DP4-6 chitosan oligosaccharides in liver metabolic diseases. This paper proposes a hypothesis that the antioxidant properties of chitosan oligosaccharides are related to their molecular weight. It is speculated that low-molecular-weight chitosan oligosaccharides have a better protective effect on the liver. The paper also describes the process of obtaining DP4-6 chito-oligosaccharides by separating and purifying COST and named the product COS46. The hypothesis was confirmed through cell and animal experiments. The study investigated the difference in antioxidant properties and efficacy of COST in protecting against liver injury before and after separation. These findings are expected to aid in the further development and utilization of chitosan oligosaccharides.

2. Results

2.1. Isolation and Purification of COST

2.1.1. COST Component and DP4-6 COS Content

The results of the TLC analysis of COST are shown in Figure 1A. The selected developer system has a good separation effect on COST, with obvious separation at each point and no obvious tailing phenomenon. According to TLC, compared with the mixed standard (1–7), COST contains chitosan oligosaccharides with a degree of polymerization of 3–7, but there are still a large number of substances piled up at the starting line of the thin layer, which cannot be separated. It is speculated that chitosan oligosaccharides with a higher degree of polymerization than chito-seven sugars exist in COST.

As shown in Table S1, the retention times of COS4, COS5, and COS6 are approximately 6 min, 8 min, and 12 min, respectively, and the standard curve is drawn according to the peak area of the concentration of each chitosan oligosaccharide standard series. The target principal component of COST is integral, and the peak area was 6166, 2267, and 3521, which was substituted into the standard curve for calculation. See the Supplementary Materials for the LC-MS spectrum and data on each standard product and COST (Figure S7, Table S2).

The component concentration of DP = 4(COS4) in COST was 0.99 mg/mL, and the content was 28.78%. The concentration of DP = 5(COS5) was 0.37 mg/mL and the content was 10.76%. The component concentration of DP = 6(COS6) was 0.61 mg/mL and the content was 17.73%. Therefore, in the COST of raw materials, the component content of DP4-6(COS46) is 57.27%.

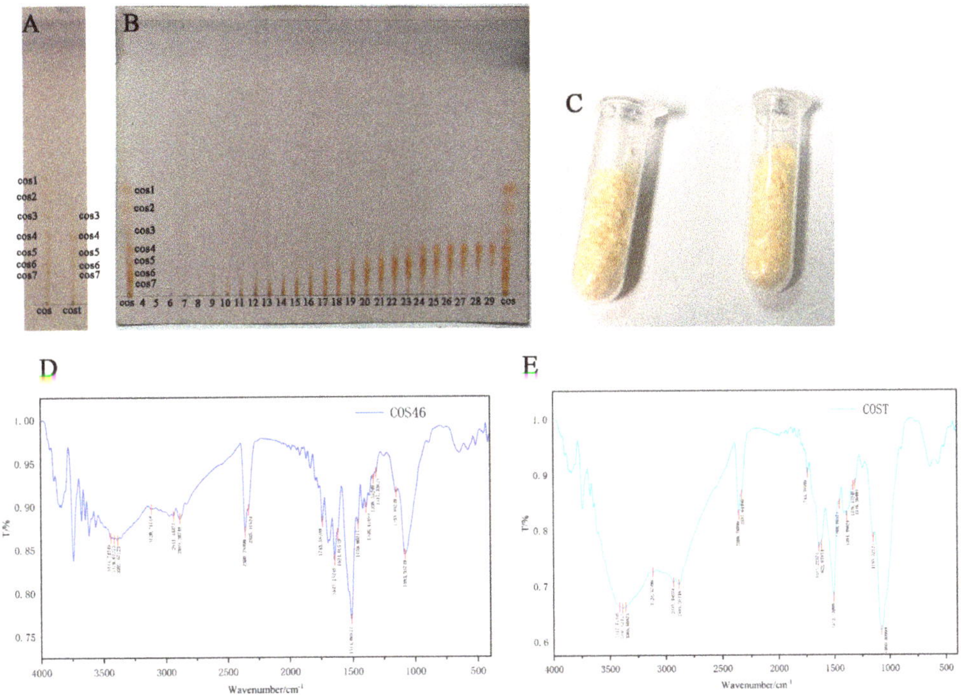

Figure 1. Thin layer diagram of COST and COS (Chitosan oligosaccharides mixed standard DP 1–7) (**A**); TLC thin-layer chromatography at different time periods. Serial number of collected eluate (4–29) (**B**); isolated COS46 freeze-dried powder (**C**); COS46 infrared spectrum (**D**); COST infrared spectrum (**E**).

2.1.2. Gel Chromatographic Separation Yield

TLC chromatography of each tube is shown in Figure 1B. Chitosan oligosaccharides of the target polymerization degree appeared from 19#. The lyophilized product was light yellow, as shown in Figure 1C. As can be seen from Table 1, when the sample size at COST is approximately 0.3 g, approximately 0.1 g of COS46 can be harvested. The COS46 obtained by five times of separation is calculated, and the yield is 30.39 ± 1.57%.

Table 1. Yield of COS46 obtained from GPC separation and purification.

$Mass_{COST}$ (g)	$Mass_{COS46}$ (g)	Yield (%)	Average Yield (%)
0.3059	0.0891	29.13	
0.3343	0.1002	29.97	
0.3290	0.1055	28.72	30.39 ± 1.57
0.3289	0.1055	32.08	
0.3228	0.1035	32.06	

2.1.3. COS46 Purity and Infrared Spectrum

The target component of COS46 was also integrated, and the peak area was 4524, 2340, and 2783, which was substituted into the standard curve for calculation. See the Supplementary Materials for the LC-MS atlas and data on COS46 (Figure S6, Table S2).

In COS46, the component concentration of COS4 was 0.73 mg/mL and the content was 40.11%. The concentration of COS5 was 0.39 mg/mL and the content was 21.43%. The component concentration of COS6 was 0.49 mg/mL and the content was 26.92%. Therefore, the total proportion of the target component in COS46 was 88.46%.

The infrared spectra of COS46 and COST are shown in Figure 1D,E. Compared with the COST of raw materials, COS46 shows an obvious peak shape change at 3400 cm^{-1}, where infrared absorption is mainly generated by hydrogen bonds. It is speculated that the reason is that the polymerization degree of COS46 decreases, and the chain length becomes shorter after purification by gel chromatography. The decrease in intramolecular and intermolecular hydrogen bonds in the molecular chains of chitosan oligosaccharides resulted in increased infrared absorption at this wave number. The C-O-C stretching vibration in the range of 1200–1050 cm^{-1} is the ether bond stretching vibration band in the pyran ring of the chitosan oligosaccharide unit structure, and the band intensity of COST is slightly higher in the figure. This may be due to the separation process affecting the ether bond, resulting in a decrease in the intensity of stretching vibrations in the separated product. Finally, the existence of such a wide peak in the figure may be caused by some water remaining in the sample.

2.1.4. Determination of Deacetylation Degree

The results are shown in Figure 2A. Among the three, only GlcNAc had an obvious maximum absorption peak, GlcN had no obvious absorption, and COS46 showed a weak absorption peak. Then the three were scanned at a 200–215 nm wavelength and the first derivative was plotted. The results are shown in Figure 2B. The maximum absorption peak of GlcNAc and COS46 appeared near 202 nm, and the absorption of GlcNAc was significantly greater than that of COS46.

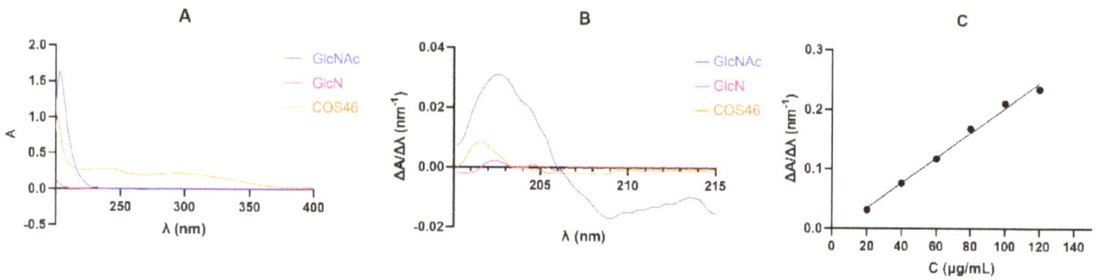

Figure 2. (**A**) Full sweep spectra of GlcNAc, GlcN, and COS46 at 200−400 nm wavelength; (**B**) GlcN and COS46 were mapped with the first derivative at 200–215 nm wavelength scanning; (**C**) the first derivative of GlcNAc (20.0−120.0 μg/mL) in 0.3 mol/L HCl ($y = 0.002097x - 0.007467$, $R^2 = 0.9924$).

The content of GlcNAc In COS46 was calculated using the standard curve. According to the formula, the degree of deacetylation of COS46 is 97.71%, as shown in Table 2.

Table 2. Degrees of deacetylation of COS46.

Sample	Concentration cos_{46} (μg/mL)	Deacetylation Degree (DD)%	Average DD (%)
COS46-1	705.3	97.23	
COS46-2	705.3	97.33	97.71
COS46-3	705.3	98.58	

2.1.5. Molecular Weight of COS46 and COST

The established photometric standard curve is shown in Figure S20. The linear regression equation of the standard curve is $y = 0.4555x + 0.05216$, and $R^2 = 0.9789$.

The measurement results of COS46 and COST are shown in Tables 3 and 4. Both OD1 and OD2 are within the linear range of the standard curve. By substituting the formula, the average molecular weight of COS46 is 628.50 and that of COST is 906.12.

The average molecular weight decreased, which proved that chitosan oligosaccharides with a narrower molecular weight range were successfully obtained in the previous gel chromatography separation.

Table 3. Average molecular weight of COS46.

OD1-COS46		OD2-COS46		DP	Average Molecular Weight
Absorbance	Average	Absorbance	Average		
0.2706 0.2676 0.2752	0.2711	0.1021 0.1040 0.1023	0.1028	3.79	628.50

Table 4. Average molecular weight of COST.

OD1-COST		OD2-COST		DP	Average Molecular Weight
Absorbance	Average	Absorbance	Average		
0.2342 0.2536 0.1445	0.2108	0.1131 0.1175 0.1098	0.1135	5.38	906.12

2.2. *Oxidation Resistance of COS46 and COST*

The scavenging rates of hydroxyl radicals and DPPH radicals by COS46 and COST are shown in Figure 3. For hydroxyl radical scavenging efficiency, within the experimental concentration range, except for the initial concentration of 0.2 mg/mL, the scavenging capacity of COS46 was higher than that of COST. The DPPH radical scavenging efficiency was better on COST and gradually improved with the increase in concentration. As one of the reactive oxygen species in the human body, the harm caused by the hydroxyl radical in the human body has long been recognized by the public. COS46 is more efficient than COST in reducing the formation of the hydroxyl radical, and it is more suitable as an exogenous antioxidant.

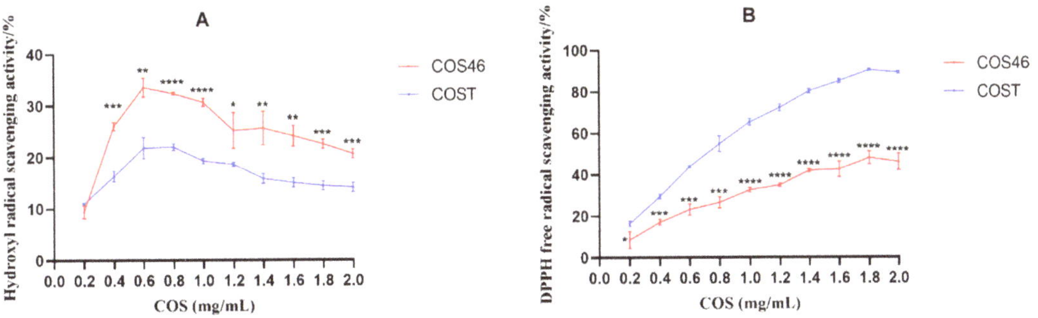

Figure 3. The hydroxyl radical scavenging activity (**A**); the DPPH free radical scavenging ability (**B**); * $p < 0.05$, ** $p < 0.01$, *** $p < 0.001$, **** $p < 0.0001$.

2.3. *Effects of COS46 and COST on Hepatocyte Injury of L02*

2.3.1. Modeling Conditions of Hepatocyte Injury and Dose Concentration

As shown in Figure 4A, with the increase in the DMSO concentration, the cell viability decreased gradually. When it reached 0.12%, the cell survival rate showed a significant difference compared with no DMSO administration. Therefore, in the follow-up experiment, the proportion of DMSO selected in the modeling solution system was 0.10%.

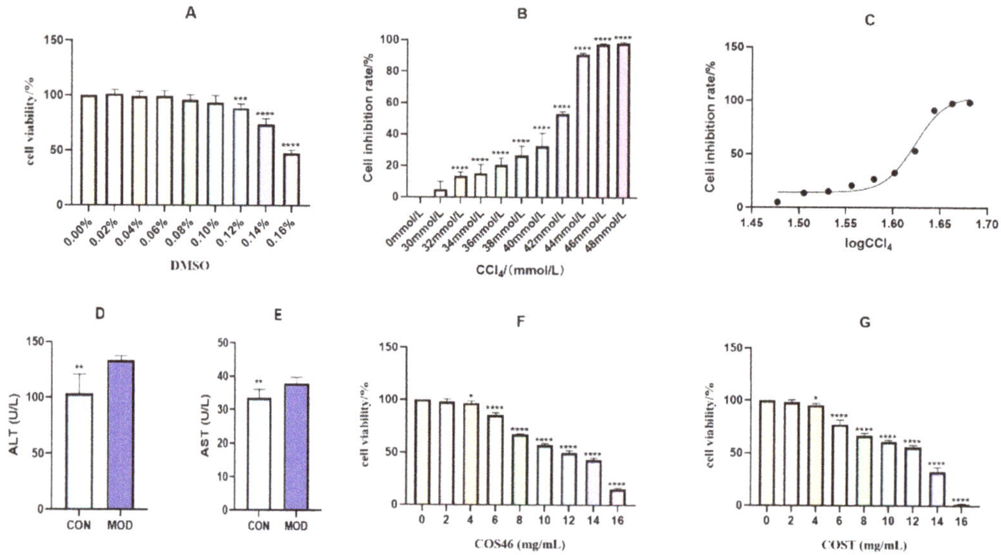

Figure 4. Effect of different percentage concentrations of DMSO on cell viability (**A**); inhibition rate of CCl_4 in different concentrations (**B**); the IC_{50} curve of CCl_4 (**C**); contents of ALT (**D**) and AST (**E**) in cell medium after 24 h treatment with CCl_4; effect of different concentrations of COS46 (**F**) and COST (**G**) on viability of L02 cells (n = 6, mean ± SD); * $p < 0.05$, ** $p < 0.01$, *** $p < 0.001$, **** $p < 0.0001$.

As shown in Figure 4B,C, the cell inhibition rate increased with the increase in the CCl_4 concentration. The modeling concentration is selected according to the principle of IC_{50}, which is 42 mmol/L, and this concentration is selected as the modeling concentration.

After the modeling concentration was determined according to the IC50 curve, CCl_4 cells attached to the wall for 24 h were further incubated for 24 h, and ALT and AST were measured in the culture medium. As shown in Figure 4D,F, ALT and AST in the MOD group were significantly increased, L02 hepatocytes were damaged, and a large number of transaminases were transferred to the culture medium. The results showed that the model of hepatocyte injury was successfully established.

As shown in Figure 4F,G. The cell survival rate decreased with the increase in the drug concentration. When the concentration of COS46 and COST reached 4 mg/mL, the cell survival rate was reduced by approximately 5% compared with the cells without drug administration, and the difference was statistically significant. When the dose was 2 mg/mL, the cell survival rate was more than 98%. Taking this as the selection standard, the high-, medium-, and low-dose concentrations of COS46 and COST were determined to be 2 mg/mL, 1 mg/mL, and 0.5 mg/mL, respectively.

2.3.2. Cell Viability and Biochemical Indices

As shown in Figure 5A, the cell survival rate in the groups given COS46 and COST was significantly improved compared with the MOD group. Among them, the COS46-H,M,L had the same improvement effect on cell survival rate, and the effect was better than that of COST-H,M.

As shown in Figure 5B,C, compared with the MOD group, AST activity in each group decreased, but there was no significant difference. COS46-L and COST-M had the most obvious effect on the reduction in AST, and the effect was similar. COS46-H,M,L could reduce ALT activity, and the three doses were significantly different from the MOD group, and COS46-L was the best. Although COST-H,M,L can also affect ALT, only COST-M showed a significant difference.

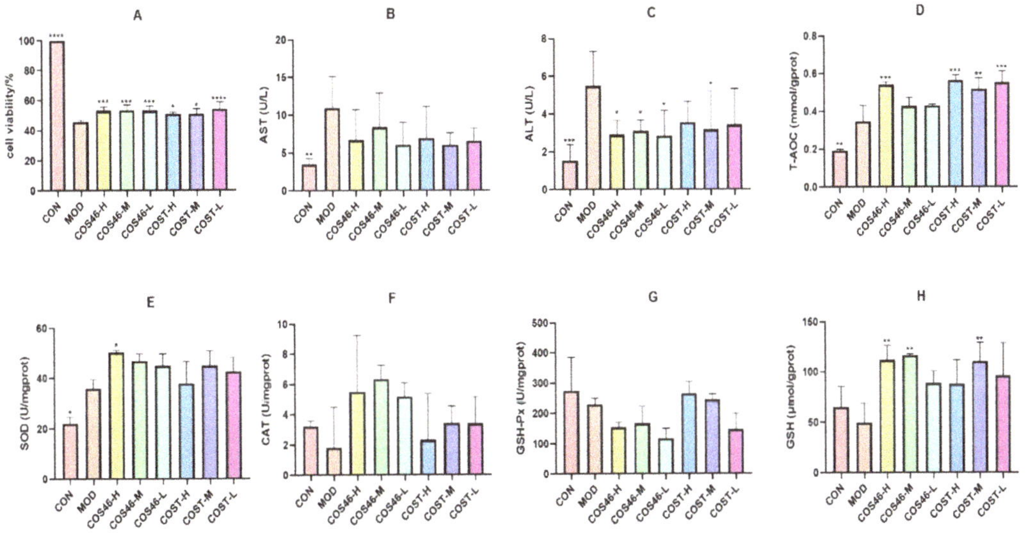

Figure 5. Effect of COS46 and COST on viability of L02 cells treated with CCl_4 (**A**); effect of COS46 and COST on AST (**B**) and ALT (**C**) levels in cell medium (n = 6, mean ± SD). Effects of COS46 and COST on T-AOC (**D**), SOD (**E**), CAT (**F**), GSH-Px (**G**), and GSH (**H**) in L02 cells (n = 3, mean ± SD); * $p < 0.05$, ** $p < 0.01$, *** $p < 0.001$, **** $p < 0.0001$.

The results of T-AOC, SOD, CAT, GSH-Px, and GSH levels in cells are shown in Figure 5D,H. Compared with the CON group, the T-AOC and SOD levels in the MOD group increased, and the values in each administration group were further increased than those in the MOD group. For T-AOC, COS46 showed significant differences only at high doses, while COST showed significant differences at all three doses. In terms of SOD measurement, only the COS46-H group showed a significant difference.

Compared with the CON group, the activity of CAT in the MOD group decreased, and the activity of CAT in different doses of COST and COS46 groups increased. Although there was no significant difference between the two groups, the increase in COS46 was higher than that of COST.

Compared with the CON group, the activity of GSH-Px in the MOD group decreased. After COS46 pretreatment, GSH-Px had not increased, while it had increased in COST-H,M groups, but there was no significant difference.

Compared with the CON group, the activity of GSH in the MOD group decreased, but after COS46 and COST pretreatment, the activity of GSH increased. There were significant differences between the COS46-H group and the COS46-M group, as well as between the COST-M groups.

In short, COS46 showed more obvious improvement in SOD, CAT, and GSH, but COST has a greater influence on T-AOC and the GSH-Px index. Both COS46 and COST can reduce the oxidative damage of CCl_4 to hepatocytes. Combined with the experimental results, COS46 has a more favorable improvement effect.

2.4. Effects of COS46 and COST on Mice with Liver Injury

2.4.1. Effects of COS46 and COST on Liver Appearance, Liver Index, and Serum AST and ALT

As shown in Figure 6A, the livers of mice in the CON group were ruddy in appearance and exquisite in texture, and no granular feeling was observed. The livers of the MOD group showed obvious yellow granules, a yellow surface, and decreased elasticity. Yellow particles were also observed on the liver surfaces of the positive drug group, but no yellowing or redness was restored compared with the MOD group. CCl_4-caused liver lesions were improved in all groups given chitosan oligosaccharides. As the dosage of COS46 increased, the effect became more obvious, and the particle sensation was lighter than that of COST.

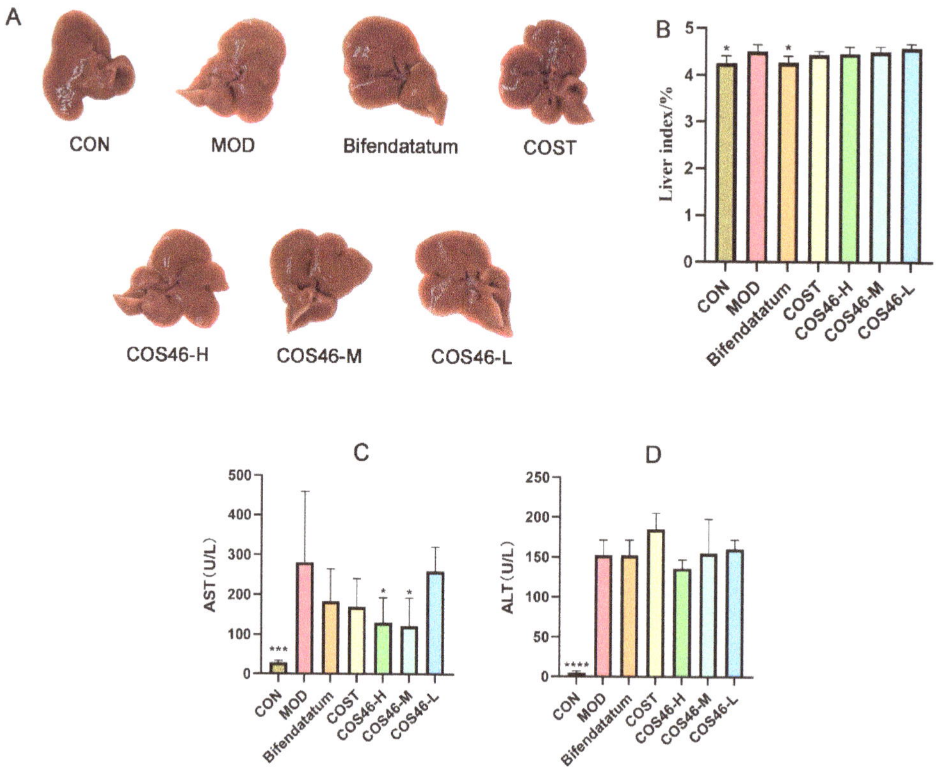

Figure 6. Appearance of mouse liver (**A**); liver index (**B**); AST (**C**) and ALT (**D**) in serum (n = 6, mean ± SD); * $p < 0.05$, *** $p < 0.001$, **** $p < 0.0001$.

As shown in Figure 6B, the liver index of the CON group was significantly lower than that of the MOD group, indicating that liver swelling and other conditions occurred after the administration of CCl_4. The liver index of COS46 dose groups decreased in a dose-dependent manner, but there was no significant difference compared with the MOD group.

As shown in Figure 6C,D, compared with the CON group, AST and ALT indexes in the MOD group significantly increased, indicating that CCl_4 caused liver cell damage. After pretreatment for 7 days, the serum AST value of each group decreased, and the COS46-H group and the COS46-M group showed significant differences. The ALT index did not decrease significantly, but the COS46 group showed dose dependency.

2.4.2. Mouse Liver Biochemical Indexes and HE Staining

Figure 7A–F shows the detection results of antioxidant enzyme activity and oxidative damage indexes in the livers of mice in each group. Compared with the CON group, the total antioxidant capacity of the MOD group significantly decreased, while T-AOC did not significantly increase in all administration groups, but the values of COS46-H and COS46-L groups increased slightly. There was no significant difference in SOD levels among all groups.

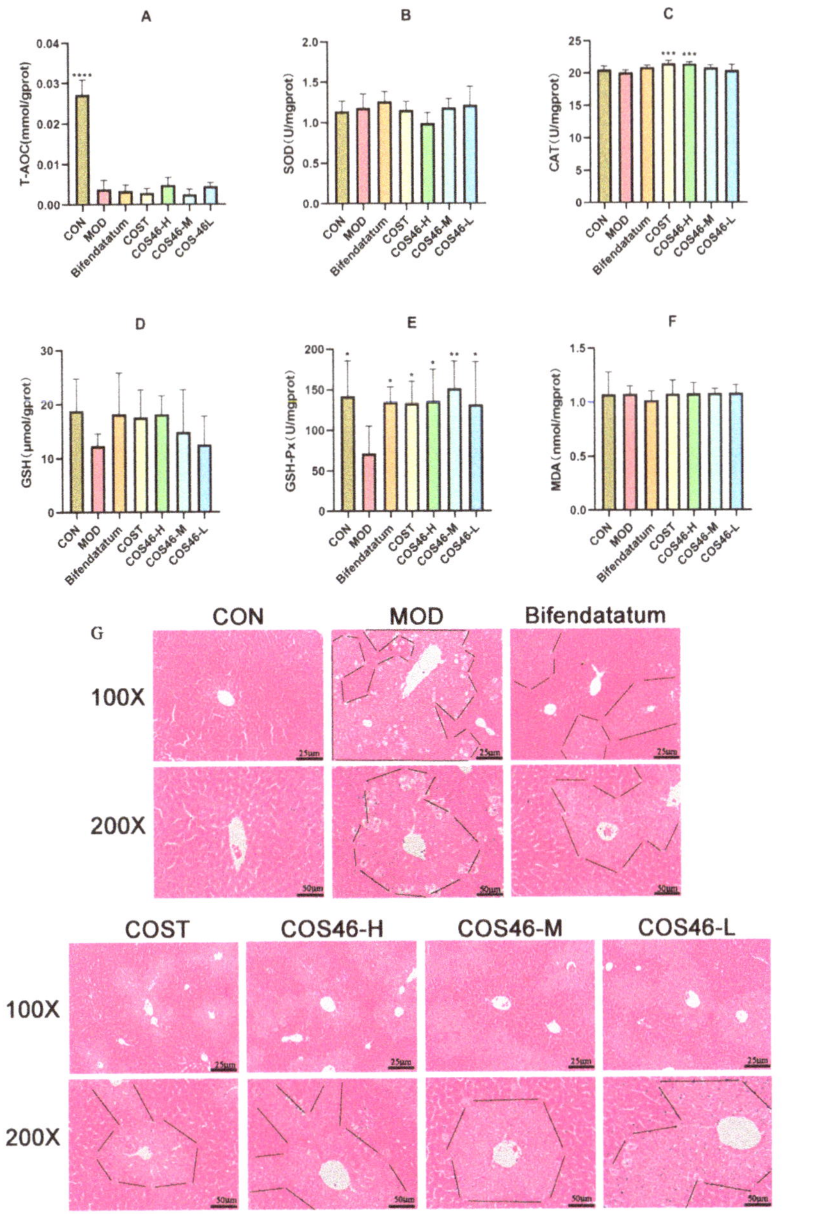

Figure 7. The antioxidant–oxidative index of liver (n = 6, mean ± SD), T-AOC (**A**), SOD (**B**), CAT (**C**), GSH (**D**), GSH-Px (**E**), and MDA (**F**); liver HE staining (**G**); * $p < 0.05$, ** $p < 0.01$, *** $p < 0.001$, **** $p < 0.0001$. Black indicates the outline of the damaged part.

Compared with the CON group, CAT enzyme activity in the MOD group decreased slightly, but there was no significant difference. In each administration group, the enzyme activity of the COST group and the COS46-H group increased compared with that of the MOD group and showed a significant difference. Meanwhile, COS46 showed a dose-dependent increase in CAT enzyme activity.

The content of GSH in the MOD group was much lower than that in the CON group, which showed no significant difference. In each administration group, the GSH content in the liver increased, and COS46 increased the GSH content in a dose-dependent manner, but there was no significant difference. The activity of the GSH-Px enzyme in the MOD group significantly decreased, while the activity of GSH-Px significantly increased in all administration groups, and COS46-M showed the best effect. In short, COS46 is better than COST in improving GSH and GSH-Px.

The levels of MDA in the CON group and the MOD group were close to each other, but there was no change trend in MDA content in each administration group.

The HE staining results of mouse livers in each group are shown in Figure 7G. The liver structure of the CON group was normal, and cells were arranged uniformly and in order, with the central vein as the center in a single radial arrangement. In the MOD group, there were obvious abnormalities, such as vacuolar degeneration, necrosis, nuclear ruptures, and cytoplasmic staining of hepatocytes. After 7 days of bifendatatum pretreatment, liver pathological sections showed some improvement in liver abnormalities and no obvious vacuole-like degeneration, and the degree of cytoplasmic redness was reduced. The COST group and the COS46-H group also had the effect of alleviating liver injury and the degree of the two groups was like that of the positive-drug group under visual observation, and both were slightly weaker than the positive-drug group. Similarly, the effects of COS46-M and COS46-L groups on liver injury were similar, with no significant difference, but compared with the MOD group, they could still improve the damage caused by CCl_4.

3. Materials and Methods

3.1. Materials

COST was purchased from AK Biotech Co., Ltd. (Chengdu, China) (detailed parameters can be found in the certificate of analysis in supplementary data).

The COS4 standard, COS5 standard, COS6 standard, and chitosan oligosaccharides mixed standard (DP1-7) were purchased from Huizhou Changlong Biotechnology Co., Ltd. (Guangdong, China).

The biochemical index kits used in this paper were purchased from Nanjing Jiancheng Bioengineering Institute.

The chemicals used are commercially available products.

3.2. COST Component Analysis and Purification of COS46

3.2.1. COST Component Analysis and DP4-6 Component Content Determination

(1) Qualitative analysis of COST components by TLC thin layer

COST and chitosan oligosaccharides mixed standards (dp1-7) were used to prepare the solution with the appropriate concentration. Plate: Merck high-efficiency silicone plate (1.05553). Development height: 9.5 cm. Developer system: isopropanol: ammonia: pure water = 15:7.5:1. After unfolding, the plate was dried, soaked in anisaldehyde sulfate colorant for 2 min, dried, and heated in an oven at 110 °C for 10 min to develop color.

(2) The content of DP4-6 components in COST was analyzed by LC-MS

The COST sample and chitosan oligosaccharides standard were weighed and a series of solutions were prepared as follows:

The COS4 standard was weighed as 2.80 mg, dissolved in (acetonitrile: water = 3:7) to make a 1.40 mg/mL solution, and diluted to 0.7, 0.35, 0.175, and 0.0875 mg/mL, respectively.

The COS5 standard was weighed as 2.10 mg, dissolved in (acetonitrile: water = 3:7) to make a 1.40 mg/mL solution, and diluted to 0.35, 0.175, 0.0875, and 0.04375 mg/mL, respectively.

The COS6 standard was weighed as 1.80 mg, dissolved in (acetonitrile: water = 3:7) to make a 1.40 mg/mL solution, and diluted to 0.45, 0.225, 0.1125, and 0.05625 mg/mL, respectively.

The COST standard was weighed as 3.4 mg and dissolved in (acetonitrile: water = 3:7) to make a 3.44 mg/mL solution.

Ten microliters of each solution were extracted for analysis and the standard curve was established.

Table 5 lists the experimental parameters of LC-MS.

Table 5. Experimental conditions of liquid chromatography–mass spectrometry (LC-MS).

	Item	Condition
LC	Column	GL Sciences Inertsil NH_2 3.0 mm × 150 mm, 5 μm
	T	35 °C
	μ	0.6 mL/min
	Mobile phase	CH_3CN:0.3%$NH_3 \cdot H_2O$ = 65:35
MS	Spray Voltage	3200 V
	Capillary Temperature	270.00 °C
	Sheath Gas	40.00 Arb
	Aux Gas	8.00 Arb
	Max Spray Current	100.00 μA
	Probe Heater Temp.	250.00 °C
	Ion Source	ESI + ms

3.2.2. Isolation and Identification of COS46

(1) GPC

The Sephadex G-15 was used as column packing, which was fully swelled in a 90 °C water bath and cleaned and had the gas removed, and the chromatographic column was installed by the wet method.

The G15 column was used at a flow rate of 3 mL/h to separate. Then it was collected and lyophilized. The collection frequency was once every 20 min until the reception was completed. This part of the operation is based on TLC operation described in Section 3.2.1 (1). The effluents containing chitosan oligosaccharides with DP = 4–6 were selected for consolidation and freeze-dried, and the yield was calculated according to the mass.

(2) LC-MS

The experimental conditions of LC-MS were the same as in Section 3.2.1 (2).

First, 1.82 mg COS46 was weighed and dissolved (acetonitrile: pure water = 3:7) into a 1.82 mg/mL solution, and 10 μL was injected for analysis.

(3) FT-IR

The COS46 and COST samples were mixed with dry potassium bromide, and then placed in an agate mortar. After grinding and mixing, the tablets were pressed on an oil press. The baseline was calibrated with dry potassium bromide. The scanning range was 400–4000 cm^{-1}, the resolution was 4 cm^{-1}, and the number of scans was 16. The infrared spectra of COS46 and COST were obtained.

3.2.3. Determination of Deacetylation Degree

The deacetylation degree of COS46 was determined by first-derivative ultraviolet spectrophotometry established in the laboratory [43].

(1) Determination of maximum absorption wavelength

GlcNAc (N-acetylglucosamine), GlcN (aminoglucose), and COS46 solutions were prepared in a 0.3 mol/L HCl solution. The scanning wavelength range was 200–400 nm, the slit width was 2 nm, the scanning speed was 20 nm/min, the time constant was 4S, and the recording speed was 10 cm/min. The optical path of the quartz cuvette is 1 cm. Then, under spectral scanning in the range of 200–215 nm, the absorbance of the solution was recorded, and the absorbance was converted into first-order differential calculation.

(2) First-derivative ultraviolet spectrophotometry was established

GlcNAc standard solutions with different concentrations were prepared in a HCl solution. The absorbance of each standard solution was recorded at the main wavelength of 204 nm and the baseline wavelength of 202 nm, and $\frac{\Delta A}{\Delta \lambda}$ (ΔA = A204 nm–A202 nm, Δλ = 2 nm) was calculated. The concentration of the standard solution was the abscissa, and $\frac{\Delta A}{\Delta \lambda}$ was the ordinate to establish the standard curve.

(3) Determination of deacetylation degree of COS46

$$DD(\%) = \frac{C_{46} - C_{GlcNAc}}{C_{46} - \frac{42}{203}C_{GlcNAc}} \times 100$$

C46 is the concentration of COS46 (μg/mL); C_{GlcNAc} is the concentration of the N-acetylglucosamine unit in the sample, which is obtained according to the standard curve (μg/mL); 203 is the molecular weight of the *GlcNAc* fragment in COS46, calculated according to $C_8H_{13}NO_5$; 42 is the relative molecular weight difference between *GlcNAc* and GlcN fragments.

3.2.4. Determination of Number Average Molecular Weight of COST and COS46

A 1 mg/mL glucosamine standard solution was prepared as solutions of different concentrations and added to different glass tubes. Each glass tube was filled with distilled water to 1 mL, 1 mL of the DNS reagent was added, it was placed in a boiling water bath for 5 min, 8 mL of distilled water was added, and 200 μL was removed and placed in a 96-well plate. *OD1* (optical density) was measured at 520 nm by a microplate reader.

Then, 1 mg/mL COST and COS46 solutions were prepared. The remaining operations were the same as above. The *OD1* value was determined at 520 nm with an enzyme-labeled instrument.

Next, we took 1 mL of 1 mg/mL COST and COS46, added 3 mL of 6 mol/L HCl, bathed it in boiling water for 2 h, neutralized the mixture with 6 mol/L NaOH, and diluted the solution to 10 mL with water. We then took 1 mL of diluted hydrolysate, added 1 mL of the DNS reagent, bathed the mixture in boiling water for 5 min, and added 8 mL of distilled water after cooling. Finally, we removed 200 μL and measured the *OD2* value with an enzyme-labeled instrument at 520 nm.

If both *OD1* and *OD2* are in the range of the scalar curve, the average degree of polymerization of the sample is:

$$\text{n} = 10 \times \frac{OD1}{OD2}$$

The average relative molecular mass of the sample is:

$$M_W = n \times 179 - (n-1) \times 18$$

3.3. *Determination of Oxidation Resistance of COS46 and COST*

3.3.1. Determination of Hydroxyl Radical Scavenging Ability

In each test tube, 1 mL of COS46 and COST (in concentrations of 0.2, 0.4, 0.6, 0.8, 1.0, 1.2, and 1.4 mg/mL) was added, followed by the sequential addition of 1 mL of 6 mmol/L $FeSO_4$, 2 mL of 6 mmol/L H_2O_2, and mixing. The mixture was then left in the dark at room temperature for 10 min. Subsequently, 1 mL of a 6 mmol/L salicylic acid-ethanol solution was added, and after mixing, the reaction was held in the dark at 37 °C for 60 min in a water bath. After cooling to room temperature, the absorbance was measured at 517 nm, yielding the absorbance value *A1*. For the blank group, an equivalent amount of water served as the substitute for the sample, yielding the absorbance value *A0*. Each group underwent three parallel measurements. The hydroxyl radical scavenging rate was then calculated using the formula below:

$$\text{Hydroxyl radical scavenging activity}(\%) = \frac{A_0 - A_1}{A_0} \times 100\%$$

3.3.2. Determination of DPPH Free Radical Scavenging Ability

First, we added 2 mL of the 0.1 mmol/L DPPH (1,1-Diphenyl-2-picrylhydrazyl radical)-EtOH solution and 2 mL of different concentrations of COS46 and COST (0.2, 0.4, 0.6, 0.8, 1.0, 1.2, 1.4, 1.6, 1.8, and 2.0 mg/mL) to each test tube, mixed well, and reacted the solution in a dark water bath at 37 °C for 60 min. After cooling to room temperature, we measured the absorbance at 517 and recorded it as *A1*. The control group used the same amount of absolute ethanol instead of the DPPH-EtOH solution, which was recorded as *A2*, and the blank group used the same amount of distilled water instead of the sample to be tested. We then calculated the DPPH radical scavenging rate according to the following formula:

$$\text{DPPH free radical scavenging activity}(\%) = \left(1 - \frac{A_1 - A_2}{A_0}\right) \times 100\%$$

3.4. Cell Experiment

3.4.1. L02 Cell Culture

L02 cells were cultivated in a constant-temperature and -humidity environment at 37 °C and 5% CO_2 in a complete medium containing RPMI-1640, fetal bovine serum (9:1, v/v), streptomycin (100 μg/mL), and penicillin (100 U/mL), with a 1% double-antibody ratio. The culture medium was replaced every two days, and the experiment was initiated when the cell fusion rate reached 80%.

3.4.2. Establishment of Liver Injury Model Induced by CCl4 (Determination of DMSO and CCl4 Concentrations)

First, 0.02%, 0.04%, 0.06%, 0.08%, 0.10%, 0.12%, 0.14%, 0.16%, and 0.18% DMSO were used to treat cells.

Next, 10, 20, 30, 40, 50, 60, 70, 80, and 90 mmol/L CCl_4 were used to treat cells. CCK-8 was used to determine the cell survival rate, and IC50 was calculated to determine the modeling concentration.

3.4.3. The Concentrations of COS46 and COST

First, 2, 4, 6, 8, 10, 12, 14, and 16 mg/mL COS46 and COST were given to cells, respectively, and then they were cultured for 24 h. The CCK-8 method was used to determine the cell inhibition rate and the dosage of COS46 and COST.

3.4.4. Effects of COS46 and COST on Survival Rate of L02 Cells Treated with CCl_4

L02 cells at the logarithmic growth stage were inoculated into 96-well culture plates with 2.0×10^4 cells per well. The CON group, MOD group, COS46-H/M/L group, and COST-H/M/L group were set up and cultured for 24 h. The administration group was added with different concentrations of drugs and continued to culture for 24 h. In addition to CON, each group was added with the modeling concentration of CCl_4 and continued to be treated for 24 h. CCK-8 measured the cell survival rate.

3.4.5. Culture Medium and Antioxidant and Oxidation Indices Detection

As described in Section 3.4.4, the supernatant of the medium was collected and the contents of ALT and AST were measured.

L02 cells at the logarithmic growth stage were inoculated into 6-well culture plates with 4×10^5 cells per well. The cells were then lysed, and the lysate was collected to determine the levels of T-AOC (Total antioxidant capacity), SOD (Superoxide dismutase), CAT (Catalase), GSH-Px (Glutathione peroxidase), and GSH (glutathione.).

3.5. Animal Experimental

3.5.1. Animal

Seventy SPF male C57BL/6J mice, 18–22 g, were purchased from Guangdong Sijia Jingda Biological Co., LTD., Production License No.: SCXK (Guangdong, China) 2020-0052; Reared in Animal Center of Guangdong Pharmaceutical University, License No. SYXK (Guangdong, China) 2022-0125. The feeding conditions of the animal experiment were 24.0 ± 2.0 °C, relative humidity of 54–65%, air change times > 15 times /h, and alternating light and dark for 12 h. The experimental animals are kept in the pathogen-free laboratory of the Experimental Animal Center of Guangdong Pharmaceutical University. This experiment was approved by the Experimental Animal Ethics Committee of Guangdong Pharmaceutical University and strictly followed the requirements of the Guidelines for Ethical Review of Experimental Animal Welfare (GB/T35892-2018) to fully protect the welfare of experimental animals.

3.5.2. Administration and Establishment of Acute Liver Injury Model

Seventy male C57BL/6J mice were randomly divided into 7 groups after 1 week of adaptive feeding. They were the blank control group (CON), model control group (MOD), positive drug control group (Bifendatatum), COST administration group (COST), COS46 high-dose group (COS46-H), COS46 medium-dose group (COS46-M), and COS46 low-dose group (COS46-L). There were 10 mice in each group, and they all had a free diet.

The COST dose was 250 mg/kg, and the COS46 high, medium and low doses were 500 mg/kg, 250 mg/kg, and 125 mg/kg, respectively. The blank control group and the model control group were intragastric with distilled water instead, and intragastric administration continued for 7 days. After the end of the 7th day, the liver injury model was induced by a one-time gavage of a 5 mL/kg 1% CCl_4 oil solution in each dose group and the model control group and the corresponding volume of olive oil was given to the blank control group, and then the tested animals were fasted for 16 h. The mice were then anesthetized by an intraperitoneal injection of a 60 mg/kg BW sodium pentobarbital solution, and blood was taken from the orbit. The animals were sacrificed by cervical dislocation after blood collection. The appropriate amount of liver was taken for embedding, fixed with 4%PFA for 24 h, and the remaining liver was stored at −80 °C for use.

(1) Liver index

Mouse body weight and liver weight were weighed, and the liver index was calculated.

$$\text{Liver index } (\%) = \frac{\text{liver weight (g)}}{\text{body weight (g)}}$$

(2) Biochemical index

Serum: ALT and AST

Liver: T-AOC, SOD, MDA, CAT, GSH, GSH-Px

The above indexes were measured by the Nanjing Jiancheng Kit.

(3) Liver HE staining

The liver was embedded in paraffin wax and cut into 4 μm pathological sections for HE staining (Leagene Biotechnology, Beijing, China, DH0006).

3.6. Statistical Analysis

The data obtained in this experiment were analyzed and mapped by GraphPad Prism 9.0, and the result of the data was expressed as the mean ± SD. The data significance analysis was conducted using One-Way ANOVA for multi-group data and T-tests for the data significance analysis between the two groups. $p < 0.05$ indicated statistical significance.

4. Discussions and Conclusions

Traditionally, chitosan oligosaccharides are a mixture prepared from chitin or chitosan. When using such chitosan oligosaccharides to explore their biological activity, the repeatability of the experiment is poor, and the results may be divided, so the study of separating chitosan oligosaccharides will definitely become a trend in the future. In addition, because chitosan oligosaccharides with different polymerization degrees have different biological activities [44], a detailed study on chitosan oligosaccharides with different polymerization degrees is conducive to the safer and more reasonable use of chitosan oligosaccharides and will further expand the application scope of chitosan oligosaccharides.

In this study, COS46 contains a tetramer, a pentamer, and a hexamer, and it is regarded as a whole as the research object, and the effect of each component is not analyzed in detail. However, according to the results provided by TLC, G-15 was selected as chromatographic column packing, according to its retention effect on chitosan oligosaccharides with different molecular weights. Reevaluating the height, diameter, and flow rate of the packing, it is possible to further separate the three components (COS4, COS5, and COS6) of COS46.

During the antioxidant activity test, it was observed that chito-oligosaccharides, both before and after separation, demonstrated a superior scavenging ability of free radicals. Hydroxyl free radicals, which are considered harmful to the human body, can form within the body. COS46 was found to be more effective in scavenging hydroxyl free radicals. When foreign substances stimulate the human body, hydrogen peroxide is produced. This then reacts with Fe^{2+} to generate reactive oxygen through the Fenton reaction. It is speculated that COS46 can chelate with Fe2+ more effectively to prevent their reaction [45,46]. Therefore, COS46 may have more potential as an antioxidant.

After using CCl_4 to establish liver injury modeling, this experiment only explored the effects of COST and COS46 on L02 hepatocyte injury, antioxidant and oxidative effects, and effects on the Keap-1/Nrf2/HO-1 pathway in L02 hepatocytes. The change in related indexes after drug administration was not discussed further. Based on the experiment, it was found that SOD and CAT enzyme activities were higher in the cell administration groups compared to the CON group. This, combined with the existing research, suggests that COS46 and COST treatment may activate the signaling pathway, increase the expression of related antioxidant proteins, and improve the activity of SOD, CAT, and other antioxidant enzymes to some extent. When cells are exposed to toxic substances, they can quickly develop protective mechanisms to reduce damage and prevent cell death. The MOD group showed higher T-AOC and SOD indexes compared to the CON group. This may be due to the insufficient concentration of CCl_4 used in the liver toxicity model to cause cell death in all cells, and the surviving cells were able to respond positively and significantly improve their enzyme activity, allowing them to survive in the presence of CCl_4.

Many studies have verified the protective effect of chitosan oligosaccharides on the liver [47–49]. However, their administration time is usually more than 4 weeks, and studies on the short-term effect of chitosan oligosaccharides are very rare. Therefore, in animal experiments, we chose to verify the protective effect of administration for one week on acute liver injury.

After seven days of pretreatment and gavage with CCl_4, the MOD group of mice exhibited liver damage, including an increased liver index, apparent granulation, and elevated serum AST and ALT levels. Each administration group was able to alleviate the injury's deterioration. Additionally, it was observed that administration could reverse the effects of CCl_4, leading to a reduction in liver GSH and Gsh-Px content and resistance to oxidative damage. However, the detection of related oxidation indexes yielded slightly different trends between the results of the cell experiment and the animal experiment. It is speculated that there may be differences between human hepatocytes and mouse cells in vitro, resulting in varying sensitivities to drugs and a slight deviation in the experimental results of the two cell types. It is also possible that the administration time in animal experiments is insufficient, which shows that chitosan oligosaccharides can only be used as a dietary supplement for a long time, and its short-term liver protection effect is not significant enough.

In summary, this paper demonstrates that the process of obtaining COS46 through GPC separation and purification has good process stability and high product purity. This method overcomes the current difficulties in preparing low-molecular-weight chitosan oligosaccharides to some extent. Simultaneously, COS46 and COST can act as exogenous antioxidants and improve the body's antioxidant levels. This can effectively prevent damage to human hepatocytes and mouse livers caused by CCl_4 to a certain extent. Low-molecular-weight COS46 is superior to COST in some aspects of antioxidation.

Supplementary Materials: The following supporting information can be downloaded at: https://www.mdpi.com/article/10.3390/md22030128/s1, Figure S1: 0.0875 mg/mL COS4 (A), 0.04375 mg/mL COS5 (B), 0.05625 mg/mL COS6 (C) chromatograms of standard products; Figure S2: 0.175 mg/mL COS4 (A), 0.0875 mg/mL COS5 (B), 0.1125 mg/mL COS6 (C) chromatograms of standard products; Figure S3: 0.350 mg/mL COS4 (A), 0.175 mg/mL COS5 (B), 0.225 mg/mL COS6 (C) chromatograms of standard products; Figure S4: 0.700 mg/mL COS4 (A), 0.350 mg/mL COS5 (B), 0.450 mg/mL COS6 (C) chromatograms of standard products; Figure S5: 1.400 mg/mL COS4 (A), 0.700 mg/mL COS5 (B), 0.900 mg/mL COS6 (C) chromatograms of standard products; Figure S6: Liquid phase diagram of 1.820 mg/mL COS46; Figure S7: Liquid phase diagram of 3.440 mg/mL COST; Figure S8: Mass spectrum of COS46 when RT = 3.60 min; Figure S9: Mass spectrum of COS46 when RT = 4.78 min; Figure S10: Mass spectrum of COS46 when RT = 6.44 min; Figure S11: Mass spectrum of COS46 when RT = 8.83 min; Figure S12: Mass spectrum of COS46 when RT = 12.65 min; Figure S13: Mass spectrum of COS46 when RT=18.10min; Figure S14: Mass spectrum of COST when RT = 3.55 min; Figure S15: Mass spectrum of COST when RT = 4.83min; Figure S16: Mass spectrum of COST when RT = 6.56 min; Figure S17: Mass spectrum of COST when RT = 9.01 min; Figure S18: Mass spectrum of COST when RT = 12.98 min; Figure S19: Mass spectrum of COST when RT = 18.67 min; Figure S20: Standard curve of photometric method; Table S1: Standard curves of retention time (RT) and concentrations of COS4, COS5 and COS6; Table S2: Liquid phase data of COS46 and COST target components.

Author Contributions: Methodology, software, data curation, writing—original draft preparation, K.W. and D.Y.; software, visualization, investigation, Y.B. and H.C.; supervision, J.G. and Z.S.; validation, Y.B. and H.C.; writing—reviewing and editing, K.W.; funding acquisition, Z.S. All authors have read and agreed to the published version of the manuscript.

Funding: This work was financially supported by the Science and Technology Program of Guangzhou, China (NO. 202103000089), the Science Foundation for Distinguished Young Scholars of Guangdong (2020B1515020026).

Institutional Review Board Statement: The animal study protocol was approved by the Ethics Committee of the Experimental Animal Center of Guangdong Pharmaceutical University for studies involving animals. The animal ethical review number is (No. SYXK (Guangdong) 2022-0125).

Data Availability Statement: The data from the present study are available in the article.

Conflicts of Interest: The authors declare no conflict of interest.

Abbreviations

COS	chitosan oligosaccharides
COST	chitosan oligosaccharides (Mw $\leq$ 1000)
COS46	chitosan oligosaccharides with degree of 4–6 polymerization
COS4	chitotetraose
COS5	chitopentaose
COS6	chitohexaose
DD	degree of deacetylation
DP	degree of polymerization
GlcNAc	acetylglucosamine
GlcN	aminoglucose

References

1. Pellis, A.; Guebitz, G.M.; Nyanhongo, G.S. Chitosan: Sources, Processing and Modification Techniques. *Gels* **2022**, *8*, 393. [CrossRef] [PubMed]
2. Zhao, X.-P.; Liu, J.; Sui, Z.-J.; Xu, M.-J.; Zhu, Z.-Y. Preparation and antibacterial effect of chitooligosaccharides monomers with different polymerization degrees from crab shell chitosan by enzymatic hydrolysis. *Biotechnol. Appl. Biochem.* **2023**, *70*, 164–174. [CrossRef] [PubMed]
3. Xia, W.; Wei, X.Y.; Xie, Y.Y.; Zhou, T. A novel chitosan oligosaccharide derivative: Synthesis, antioxidant and antibacterial properties. *Carbohydr. Polym.* **2022**, *291*, 119608. [CrossRef] [PubMed]
4. Zou, P.; Yuan, S.; Yang, X.; Guo, Y.; Li, L.; Xu, C.; Zhai, X.; Wang, J. Structural characterization and antitumor effects of chitosan oligosaccharides against orthotopic liver tumor via NF-κB signaling pathway. *J. Funct. Foods* **2019**, *57*, 157–165. [CrossRef]

5. Zhai, X.; Li, C.; Ren, D.; Wang, J.; Ma, C.; Abd El-Aty, A.M. The impact of chitooligosaccharides and their derivatives on the in vitro and in vivo antitumor activity: A comprehensive review. *Carbohydr. Polym.* **2021**, *266*, 118132. [CrossRef]
6. Wang, L.; Liu, X.; Tan, W.; Li, Q.; Guo, Z.; Zhang, J. Preparation and antioxidant activity of novel chitosan oligosaccharide quinolinyl urea derivatives. *Carbohydr. Res.* **2022**, *521*, 108678. [CrossRef]
7. Zhang, Y.; Wang, Y.; Liu, Y.; Gong, T.; Hou, M. The Anti-inflammatory Effect of Chitosan Oligosaccharide on Heart Failure in Mice. *Biomed. Res. Int.* **2022**, *2022*, 8746530. [CrossRef]
8. Antonov, Y.A.; Zhuravleva, I.L.; Bezrodnykh, E.A.; Berezin, B.B.; Kulikov, S.N.; Tikhonov, V.E. Complexation of oligochitosan with sodium caseinate in alkalescent and weakly acidic media. *Carbohydr. Polym.* **2023**, *302*, 120391. [CrossRef]
9. Jing, W.; Chen, C.; Wang, G.; Han, M.; Chen, S.; Jiang, X.; Shi, C.; Sun, P.; Yang, Z.; Shi, B.; et al. Metabolic Modulation of Intracellular Ammonia via Intravesical Instillation of Nanoporter-Encased Hydrogel Eradicates Bladder Carcinoma. *Adv. Sci.* **2023**, *10*, e2206893. [CrossRef]
10. Na, K.; Wei, J.; Zhang, L.; Fang, Y.; Li, X.; Lu, S.; Guo, X. Effects of chitosan oligosaccharides (COS) and FMT from COS-dosed mice on intestinal barrier function and cell apoptosis. *Carbohydr. Polym.* **2022**, *297*, 120043. [CrossRef] [PubMed]
11. Ji, X.; Zhu, L.; Chang, K.; Zhang, R.; Chen, Y.; Yin, H.; Jin, J.; Zhao, L. Chitooligosaccahrides: Digestion characterization and effect of the degree of polymerization on gut microorganisms to manage the metabolome functional diversity in vitro. *Carbohydr. Polym.* **2022**, *275*, 118716. [CrossRef]
12. Tzeng, H.P.; Liu, S.H.; Chiang, M.T. Antidiabetic Properties of Chitosan and Its Derivatives. *Mar. Drugs* **2022**, *20*, 784. [CrossRef]
13. He, N.; Wang, S.; Lv, Z.; Zhao, W.; Li, S. Low molecular weight chitosan oligosaccharides (LMW-COSs) prevent obesity-related metabolic abnormalities in association with the modification of gut microbiota in high-fat diet (HFD)-fed mice. *Food Funct.* **2020**, *11*, 9947–9959. [CrossRef]
14. Karadeniz, F. In Vitro Anti-HIV-1 Activity of Chitosan Oligomers N-Conjugated with Asparagine and Glutamine. *BioTech* **2023**, *12*, 18. [CrossRef] [PubMed]
15. Ouyang, Q.Q.; Zhao, S.; Li, S.D.; Song, C. Application of Chitosan, Chitooligosaccharide, and Their Derivatives in the Treatment of Alzheimer's Disease. *Mar. Drugs* **2017**, *15*, 322. [CrossRef] [PubMed]
16. Muanprasat, C.; Wongkrasant, P.; Satitsri, S.; Moonwiriyakit, A.; Pongkorpsakol, P.; Mattaveewong, T.; Pichyangkura, R.; Chatsudthipong, V. Activation of AMPK by chitosan oligosaccharide in intestinal epithelial cells: Mechanism of action and potential applications in intestinal disorders. *Biochem. Pharmacol.* **2015**, *96*, 225–236. [CrossRef]
17. Tao, W.; Wang, G.; Pei, X.; Sun, W.; Wang, M. Chitosan Oligosaccharide Attenuates Lipopolysaccharide-Induced Intestinal Barrier Dysfunction through Suppressing the Inflammatory Response and Oxidative Stress in Mice. *Antioxidants* **2022**, *11*, 1384. [CrossRef]
18. Sun, Y.; Ji, X.; Cui, J.; Mi, Y.; Zhang, J.; Guo, Z. Synthesis, Characterization, and the Antioxidant Activity of Phenolic Acid Chitooligosaccharide Derivatives. *Mar. Drugs* **2022**, *20*, 489. [CrossRef] [PubMed]
19. Mourya, V.K.; Inamdar, N.N.; Choudhari, Y.M. Chitooligosaccharides: Synthesis, characterization and applications. *Polym. Sci. Ser. A* **2011**, *53*, 583–612. [CrossRef]
20. Aghbashlo, M.; Amiri, H.; Moosavi Basri, S.M.; Rastegari, H.; Lam, S.S.; Pan, J.; Gupta, V.K.; Tabatabaei, M. Tuning chitosan's chemical structure for enhanced biological functions. *Trends Biotechnol.* **2022**, *41*, 785–797. [CrossRef]
21. Li, K.; Liu, S.; Xing, R.; Qin, Y.; Li, P. Preparation, characterization and antioxidant activity of two partially N-acetylated chitotrioses. *Carbohydr. Polym.* **2013**, *92*, 1730–1736. [CrossRef]
22. Lodhi, G.; Kim, Y.-S.; Hwang, J.-W.; Kim, S.-K.; Jeon, Y.-J.; Je, J.-Y.; Ahn, C.-B.; Moon, S.-H.; Jeon, B.-T.; Park, P.-J. Chitooligosaccharide and its derivatives: Preparation and biological a pplications. *BioMed. Res. Int.* **2014**, *2014*, 654913. [CrossRef]
23. Bonin, M.; Sreekumar, S.; Cord-Landwehr, S.; Moerschbacher, B.M. Preparation of Defined Chitosan Oligosaccharides Using Chitin Deacetyl ases. *Int. J. Mol. Sci.* **2020**, *21*, 7835. [CrossRef]
24. Xiong, C.; Wu, H.; Wei, P.; Pan, M.; Tuo, Y.; Kusakabe, I.; Du, Y. Potent angiogenic inhibition effects of deacetylated chitohexaose sepa rated from chitooligosaccharides and its mechanism of action in vitro. *Carbohydr. Res.* **2009**, *344*, 1975–1983. [CrossRef] [PubMed]
25. Sørbotten, A.; Horn, S.J.; Eijsink, V.G.H.; Vårum, K.M. Degradation of chitosans with chitinase B from Serratia marcescens. Pr oduction of chito-oligosaccharides and insight into enzyme processivit y. *FEBS J.* **2005**, *272*, 538–549. [CrossRef] [PubMed]
26. Wu, Y.; Lin, Q.L.; Chen, Z.X.; Wu, W.; Xiao, H.X. Preparation of chitosan oligomers COS and their effect on the retrogra dation of intermediate amylose rice starch. *J. Food Sci. Technol.* **2012**, *49*, 695–703. [CrossRef] [PubMed]
27. Dong, H.; Wang, Y.; Zhao, L.; Zhou, J.; Xia, Q.; Jiang, L.; Fan, L. Purification of DP 6 to 8 chitooligosaccharides by nanofiltration from the prepared chitooligosaccharides syrup. *Bioresour. Bioprocess.* **2014**, *1*, 20. [CrossRef]
28. Dong, X.; Shen, A.; Gou, Z.; Chen, D.; Liang, X. Hydrophilic interaction/weak cation-exchange mixed-mode chromatography for chitooligosaccharides separation. *Carbohydr. Res.* **2012**, *361*, 195–199. [CrossRef]
29. Le Dévédec, F.; Bazinet, L.; Furtos, A.; Venne, K.; Brunet, S.; Mateescu, M.A. Separation of chitosan oligomers by immobilized metal affinity chromat ography. *J. Chromatogr. A* **2008**, *1194*, 165–171. [CrossRef] [PubMed]
30. Wang, B.; Wang, L.; Qu, Y.; Lu, J.; Xia, W. Chitosan oligosaccharides exert neuroprotective effects via mod ulating the PI3K/Akt/Bcl-2 pathway in a Parkinsonian model. *Food Funct.* **2022**, *13*, 5838–5853. [CrossRef] [PubMed]
31. Cao, L.; Wu, J.; Li, X.; Zheng, L.; Wu, M.; Liu, P.; Huang, Q. Validated HPAEC-PAD Method for the Determination of Fully Deacetylated Chitooligosaccharides. *Int. J. Mol. Sci.* **2016**, *17*, 1699. [CrossRef] [PubMed]

32. Jiang, M.; Guo, Z.; Wang, C.; Yang, Y.; Liang, X.; Ding, F. Neural activity analysis of pure chito-oligomer components separated from a mixture of chitooligosaccharides. *Neurosci. Lett.* **2014**, *581*, 32–36. [CrossRef] [PubMed]
33. Zhu, X.-Y.; Zhao, Y.; Zhang, H.-D.; Wang, W.-X.; Cong, H.-H.; Yin, H. Characterization of the Specific Mode of Action of a Chitin Deacetylas e and Separation of the Partially Acetylated Chitosan Oligosaccharides. *Mar. Drugs* **2019**, *17*, 74. [CrossRef]
34. Hattori, T.; Anraku, N.; Kato, R. Capillary electrophoresis of chitooligosaccharides in acidic solution: Simple determination using a quaternary-ammonium-modified column and indirect photometric detection with crystal violet. *J. Chromatogr. B Anal. Technol. Biomed. Life Sci.* **2009**, *878*, 477–480. [CrossRef] [PubMed]
35. Beaudoin, M.-E.; Gauthier, J.; Boucher, I.; Waldron, K.C. Capillary electrophoresis separation of a mixture of chitin and chitos an oligosaccharides derivatized using a modified fluorophore conjugati on procedure. *J. Sep. Sci.* **2005**, *28*, 1390–1398. [CrossRef]
36. Boll, M.; Lutz, W.D.; Becker, E.; Stampfl, A. Mechanism of Carbon Tetrachloride-Induced Hepatotoxicity. Hepatocellular Damage by Reactive Carbon Tetrachloride Metabolites. *Z. Für Naturforschung C* **2001**, *56*, 649–659. [CrossRef]
37. Sun, J.; Wen, X.; Liu, J.; Kan, J.; Qian, C.; Wu, C.; Jin, C. Protective effect of an arabinogalactan from black soybean against carbon tetrachloride-induced acute liver injury in mice. *Int. J. Biol. Macromol.* **2018**, *117*, 659–664. [CrossRef]
38. Li, S.; Zhou, J.; Xu, S.; Li, J.; Liu, J.; Lu, Y.; Shi, J.; Zhou, S.; Wu, Q. Induction of Nrf2 pathway by Dendrobium nobile Lindl. alkaloids protects against carbon tetrachloride induced acute liver injury. *Biomed. Pharmacother.* **2019**, *117*, 109073. [CrossRef]
39. Unsal, V.; Cicek, M.; Sabancilar, I. Toxicity of carbon tetrachloride, free radicals and role of antioxidants. *Rev. Environ. Health* **2021**, *36*, 279–295. [CrossRef]
40. He, S.; Zhao, W.; Chen, X.; Li, J.; Zhang, L.; Jin, H. Ameliorative Effects of Peptide Phe-Leu-Ala-Pro on Acute Liver and Kidney Injury Caused by CCl_4 via Attenuation of Oxidative Stress and Inflammation. *ACS Omega* **2022**, *7*, 44796–44803. [CrossRef]
41. Ismail, A.F.; Salem, A.A.; Eassawy, M.M. Modulation of gamma-irradiation and carbon tetrachloride induced oxidative stress in the brain of female rats by flaxseed oil. *J. Photochem. Photobiol. B* **2016**, *161*, 91–99. [CrossRef]
42. Mazani, M.; Ojarudi, M.; Banaei, S.; Salimnejad, R.; Latifi, M.; Azizi, H.; Rezagholizadeh, L. The protective effect of cinnamon and ginger hydro-alcoholic extract on carbon tetrachloride-induced testicular damage in rats. *Andrologia* **2020**, *52*, e13651. [CrossRef]
43. Jiang, Y.; Fu, C.; Wu, S.; Liu, G.; Guo, J.; Su, Z. Determination of the Deacetylation Degree of Chitooligosaccharides. *Mar. Drugs* **2017**, *15*, 332. [CrossRef] [PubMed]
44. Hao, W.; Li, K.; Li, P. Review: Advances in preparation of chitooligosaccharides with heteroge neous sequences and their bioactivity. *Carbohydr. Polym.* **2021**, *252*, 117206. [CrossRef]
45. Hao, W.; Li, K.; Ma, Y.; Li, R.; Xing, R.; Yu, H.; Li, P. Preparation and Antioxidant Activity of Chitosan Dimers with Different Sequences. *Mar. Drugs* **2021**, *19*, 366. [CrossRef]
46. Liu, Y.; Zhong, Z.; Bao, L.; Wen, F.; Yang, H. The preparation and antioxidant activities of four 2-aminoacyl-chitooligosaccharides. *Carbohydr. Res.* **2022**, *521*, 108667. [CrossRef]
47. Xiang, J.; Wang, J.; Xie, H.; Liu, Y.; Bai, Y.; Che, Q.; Cao, H.; Huang, G.; Guo, J.; Su, Z. Protective effect and mechanism of chitooligosaccharides on acetaminop hen-induced liver injury. *Food Funct.* **2021**, *12*, 9979–9993. [CrossRef] [PubMed]
48. Liu, P.; Li, H.; Xu, H.; Gong, J.; Jiang, M.; Qian, J.; Xu, Z.; Shi, J. Chitooligosaccharides Attenuated Hepatic Encephalopathy in Mice throug h Stabilizing Gut-Liver-Brain Disturbance. *Mol. Nutr. Food Res.* **2023**, *67*, e2200158. [CrossRef] [PubMed]
49. Liu, P.; Li, H.; Gong, J.; Geng, Y.; Jiang, M.; Xu, H.; Xu, Z.; Shi, J. Chitooligosaccharides alleviate hepatic fibrosis by regulating the pol arization of M1 and M2 macrophages. *Food Funct.* **2022**, *13*, 753–768. [CrossRef]

marine drugs

MDPI

Article

Phenolic Acid Functional Quaternized Chitooligosaccharide Derivatives: Preparation, Characterization, Antioxidant, Antibacterial, and Antifungal Activity

Yan Sun [1,2,3,†], Jingmin Cui [1,2,†], Liguang Tian [4], Yingqi Mi [1,2,3] and Zhanyong Guo [1,2,*]

1 Key Laboratory of Coastal Biology and Bioresource Utilization, Yantai Institute of Coastal Zone Research, Chinese Academy of Sciences, Yantai 264003, China; sunyan20@mails.ucas.ac.cn (Y.S.); jmcui@yic.ac.cn (J.C.); yqmi@yic.ac.cn (Y.M.)
2 University of Chinese Academy of Sciences, Beijing 100049, China
3 Shandong Saline-Alkali Land Modern Agriculture Company, Dongying 257300, China
4 Yantai Agricultural Technology Extension Center, Yantai 265499, China; tianliguang@yt.shandong.cn
* Correspondence: zhanyongguo@hotmail.com; Tel.: +86-535-2109171; Fax: +86-535-2109000
† These authors contributed to the work equally.

Abstract: As a promising biological material, chitooligosaccharide (COS) has attracted increasing attention because of its unique biological activities. In this study, fourteen novel phenolic acid functional COS derivatives were successfully prepared using two facile methods. The structures of derivatives were characterized by FT-IR and ^{1}H NMR spectra. The in vitro antioxidant activity experiment results demonstrated that the derivatives presented stronger 1,1-Diphenyl-2-picryl-hydrazyl (DPPH), superoxide, hydroxyl radical scavenging activity and reducing power, especially the *N,N,N*-trimethylated chitooligosaccharide gallic acid salt (GLTMC), gallic acid esterified *N,N,N*-trimethylated chitooligosaccharide (GL-TMC) and caffeic acid *N,N,N*-trimethylated chitooligosaccharide (CFTMC) derivatives. Furthermore, the antifungal assay was carried out and the results indicated that the salicylic acid esterified *N,N,N*-trimethylated chitooligosaccharide (SY-TMC) had much better inhibitory activity against *Botrytis cinerea* and *Fusarium graminearum*. Additionally, the results of the bacteriostasis experiment showed that the caffeic acid esterified *N,N,N*-trimethylated chitooligosaccharide (CF-TMC) had the potential ability to inhibit *Escherichia coli* and *Staphylococcus aureus* bacteria. Altogether, this study may provide a neoteric method to produce COS derivatives with significantly increased biological activities, which have potential use in food, medicine, and health care products and other related industries.

Keywords: chitooligosaccharide; chitooligosaccharide quaternary ammonium salt derivatives; phenolic acid; antioxidant activity; antibacterial activity; antifungal activity

1. Introduction

Chitosan is the only natural cationic basic polysaccharide on earth and the second most abundant natural polysaccharide after cellulose [1]. Chitooligosaccharide (COS) is the degradation product of chitosan and usually produced by acid hydrolysis, physical hydrolysis or enzymatic degradation methods [2]. It is mainly a linear oligomer of glucosamine linked by a β-1,4 glycosidic bond, with a small amount of acetylglucosamine. The degree of polymerization ranges from 2 to 20 [3]. Due to the low degree of polymerization, COS has more excellent biological activities compared with chitosan. Previous research confirmed that COS not only has good water solubility [4], biocompatibility and biodegradability, but also has anti-inflammatory, anticancerogenic, antidiabetic, antimicrobial, anti-HIV-1, antioxidant, antiangiogenic, neuroprotective, and immunostimulatory effects [5–9]. Owning to these good activities, COS is becoming the focus of studies to develop products that can be used for humans, such as in the medicine, food, cosmetic, health care and agriculture industries [10]. Among the various studies, chemical modification is one of the most important research directions as the amino and hydroxyl groups of COS provide possible reaction sites to synthesize novel compounds [11]. And favorable substitution can further improve the biological activity and utilization value of COS.

Phenolic acids are natural products that are derived from plant secondary metabolism [12]. They are widely found in numerous plants and are easy to obtain [13]. Studies have shown that phenolic acids possess excellent antioxidant activity. The hydroxyl group of phenolic acids is the main functional group that exerts antioxidant activity because it can deliver hydrogen atoms to free radicals, interfere with the chain propagation reactions and chelate metal ions. In addition, the structural diversity of phenolic acids can affect the antioxidant capacity. Phenolic acids also have other biological activities, such as anti-inflammatory, immunoregulatory, anti-allergic, anti-atherogenic, anti-microbial, anti-thrombotic, cardioprotective, anti-cancer and antidiabetic properties [14–18]. Based on these multiple biological activities, phenolic acids are becoming a kind of potential biomaterial that can be used in the pharmaceutical, food and biomedicine industries, among others [19]. However, the low water solubility of phenolic acids limits their application.

So, based on the above characteristics of COS and phenolic acids, we want to synthesize new compounds with their combined activities. There has been much research on the synthesis of phenolic acid chitosan or COS derivatives. Tae-kil et al. selected eight phenolic acids (protocatechuic, 4-hydroxybenzoic, vanillic, syringic, *p*-coumaric, caffeic, ferulic and sinapinic acid) to synthesize eight acylated COS derivatives by the DCC/HOBt grafting system and demonstrated that caffeic acid-c-COS had the highest antioxidant activity [20]. Thanh et al. adopted a DCC grafting system to synthesize gallic acid acylated chitooligosaccharide and verified that it has a more enhanced inhibitory effect against allergic reactions in RBL-2H3 mast cells than COS [21]. Chen et al. prepared gallic acid grafted carboxymethyl chitosan through an ascorbic acid/hydrogen peroxide initiated graft copolymerization reaction and proved it can notably modulate intestinal microcirculation [22]. Sun et al. prepared 14 phenolic acid COS derivatives using seven phenolic acids and demonstrated that the derivatives had stronger antioxidant activity than COS [23]. Dai et al. used gallic acid to synthesize gallate-COS via carbodiimide and proved it had obvious antioxidant activity [24]. Mi et al. selected eight organic acids (cumaric acid, ascorbic acid, ferulic acid, p-coumaric acid, caffeic acid, gallic acid, salicylic acid, hydroxybenzoic acid) and synthesized eight hydroxypropyltrimethyl ammonium chitosan derivatives by the ion exchange method. The antioxidant activity test demonstrated these derivatives had dramatic enhancements in free radical scavenging activity [25]. Xing et al. prepared several organic acid COS salts by microwave method and tested their effect on NO secretion of macrophages. The results demonstrated that different salt chitooligosaccharides had different effects on promoting NO secretion [26]. To sum up, phenolic acid grafted chitosan or COS used to prepare new compounds can further enhance their biological activities, and the new compounds have great application potential. Moreover, quaternized chitosan is a widely used chitosan derivative. Because of the increased positive charge density, the quaternized chitosan has much better biocompatibility, solubility, antioxidant and antibacterial activities than chitosan [27]. Therefore, we speculate that the biological activity of derivatives prepared by quaternizing COS and grafting phenolic acids could also be improved.

In this study, we designed two synthetic routes and chose seven phenolic acids to prepare quaternized chitooligosaccharide phenolic acid derivatives. First, the *N,N,N*-trimethylated chitooligosaccharide phenolic acid salt derivatives were synthesized by ion exchange methods: the amino group of COS was first quaternized and then different phenolic acids were introduced through ion exchange. Seven derivatives were synthesized successfully: *N,N,N*-trimethylated chitooligosaccharide gallic acid salt (GLTMC), *N,N,N*-trimethylated chitooligosaccharide ferulic acid salt (FUTMC), *N,N,N*-trimethylated chitooligosaccharide *p*-coumaric acid salt (CMTMC), *N,N,N*-trimethylated chitooligosaccharide caffeic acid salt (CFTMC), *N,N,N*-trimethylated chitooligosaccharide protocatechuic acid salt (PCTMC), N,N,N-trimethylated chitooligosaccharide sinapic acid salt (SPTMC), N,N,N-trimethylated chitooligosaccharide salicylic acid salt (SYTMC). Second, the phenolic acid esterified N,N,N-trimethylated chitooligosaccharides were synthesized by the CDI catalytic system: the carboxyl group of phenolic acids was catalyzed by CDI and then reacted with the hydroxyl group of N,N,N-trimethylated chitooligosaccharide to form an ester bond. Seven derivatives were finally synthesized: gallic acid esterified N,N,N-trimethylated chitooligosaccharide (GL-TMC), ferulic acid N,N,N-trimethylated esterified chitooligosaccharide (FU-TMC), p-coumaric acid esterified N,N,N-trimethylated chitooligosaccharide (CM-TMC), caffeic acid esterified N,N,N-trimethylated chitooligosaccharide (CF-TMC), protocatechuic acid esterified N,N,N-trimethylated chitooligosaccharide (PC-TMC), sinapic acid esterified N,N,N-trimethylated chitooligosaccharide (SP-TMC) and salicylic acid esterified N,N,N-trimethylated chitooligosaccharide (SY-TMC). The structures and thermal stability of these derivatives were characterized by FT-IR, ^{1}H NMR spectra and thermal analysis methods. The degree of substitution in derivatives was also calculated and analyzed by ^{1}H NMR spectra. Moreover, the antioxidant activity of these derivatives was evaluated by the DPPH radical scavenging activity, superoxide radical scavenging activity, the hydroxyl radical scavenging activity and the reducing power experiments in vitro. Antifungal and antibacterial assay experiments were also implemented to detect the antimicrobial activities of derivatives. In addition, the cytotoxicity of derivatives was tested by L929 cells using the MTT method in vitro. This study may provide a novel method to synthesize COS phenolic acid derivatives, and the new green substances prepared in this paper can serve as antimicrobial or antioxidant agents for use in many fields.

2. Results and Discussion

2.1. Chemical Synthesis and Characterization

The synthesis procedure for phenolic acid TMC salt derivatives is shown in Route 1 of Scheme 1. In the first step of the reaction, in the presence of NaI and NaOH, TMCI was formed by the COS and CH_3I reaction. In the second step of the reaction, phenolic acids reacted with NaOH and formed a carboxylate anion. After TMCI was added, the carboxylic acid anion of phenolic acid replaced I^- and combined with the dissociated TMC cation by electrostatic attraction through ion exchange to form the desired products [28].

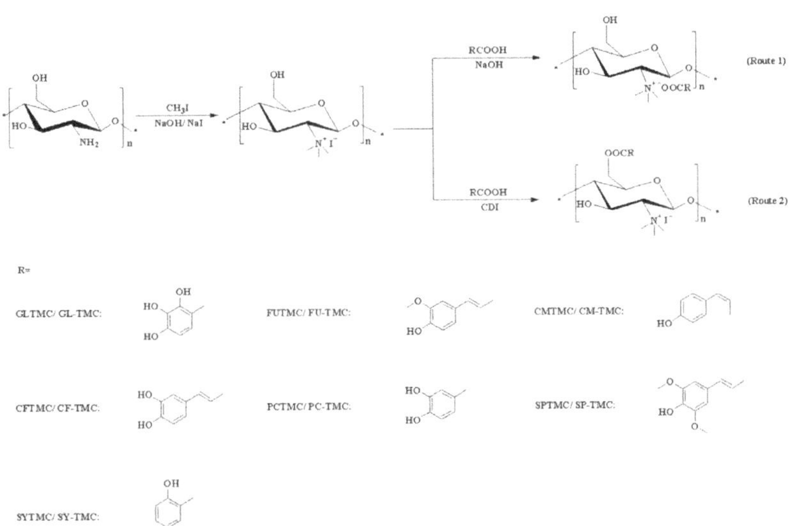

Scheme 1. Synthesis routes for phenolic acid COS derivatives.

The synthesis procedure for the phenolic acid esterified TMC derivatives is shown in Route 2 of Scheme 1. CDI is an imidazole derivative with high reactive activity due to the structure of double heterocycles. Herein, CDI reacted with the carboxyl groups in phenolic acids to form carbonyl imidazole. After TMCI was added, carbonyl imidazole reacted with the hydroxyl group of TMCI to form an ester bond; thus, the desired products were obtained [29,30].

2.2. FT-IR Spectra

The FT-IR spectra of the TMC phenolic acid salt derivatives are shown in Figure 1a. As exhibited in the figure, for COS, the wavenumber of 3395 cm^{-1} can be attributed to the O-H and N-H bending, and the 2924 cm^{-1} wavenumber is the stretching vibration of O-H. The band at 1582 cm^{-1} represents the amino group and the characteristic peak at 1415 cm^{-1} is the deformation vibration of $-CH_2$ and $-CH_3$. The absorption peak at 1084 cm^{-1} is the C-O-C stretching vibration [31]. For TMCI, a new spike appearing at 1472 cm^{-1} can be attributed to the $-N^+(CH_3)_3$, and the absorbance at 1644 cm^{-1} can be assigned to the band of $N-CH_3$ [32]. This demonstrates the successful synthesis of the TMCI derivatives. Moreover, for the TMC phenolic acid salt derivatives, the spikes at 1470 cm^{-1} and 1640 cm^{-1} are obvious, which prove the presence of $N^+(CH_3)_3$. In addition, the new spikes at 1549, 1516, 1554, 1554, 1553, 1554, 1504, 1596 and 1589 cm^{-1} of GLTMC, FUTMC, CMTMC, CFTMC, PCTMC, SPTMC and SYTMC can be attributed to the characteristic absorption of the C-C stretching vibration of the benzene ring [33]. Meanwhile, the peaks appearing at GLTMC (740, 798 cm^{-1}), FUTMC (702, 820, 852 cm^{-1}), CMTMC (721, 800 cm^{-1}), CFTMC (706, 820 cm^{-1}), PCTMC (790, 897 cm^{-1}), SPTMC (723, 828 cm^{-1}) and SYTMC (766, 813 cm^{-1}) are the C-H out-of-plane bending vibrations of the benzene ring [34]. Comprehensively, the FT-IR spectra characterization primarily demonstrates that the derivatives were successfully synthesized.

The FT-IR spectra of phenolic acid esterified TMC derivatives are shown in Figure 1b. It can be seen that the derivatives retain the characteristic absorption of COS and TMCI. The reaction displays an obvious characteristic spike around 1720 cm^{-1}, which can be attributed to the C=O stretching vibration of the ester bond. And the absorption at 1280–1050 cm^{-1} can be assigned to the C-O-C stretching vibration of the ester bond, which proves the successful synthesis of esterified derivatives. Furthermore, the spikes at 1555 (GL-TMC), 1508 (FU-TMC), 1506 (CM-TMC), 1554 (CF-TMC), 1555 (PC-TMC), 1504 (SP-TMC) and 1613 cm^{-1} (SY-TMC) represent the skeleton vibration absorption peak of the benzene ring, and the spikes at 700–850 cm^{-1} represent the C-H out-of-plane bending vibration absorption peak of the benzene ring. Therefore, the characteristic absorptions demonstrate the successful introduction of phenolic acids.

2.3. ^1H NMR Spectra

In order to further determine the structures of the products, ^1H NMR spectra were implemented in this study and the results are shown in Figure 2. The chemical shift at δ 4.79 ppm of the products can be attributed to the D_2O solvents. The chemical shift at δ 2.50 ppm of the products can be attributed to the DMSO solvents. In addition, the characterization of unmodified COS is as follows: the δ 4.48 ppm can be attributed to [H1], δ 2.69 ppm is [H2] and δ 3.21–4.08 ppm is [H2]-[H6]. As for TMCI, there is an obvious signal at δ 3.22 ppm, which can be assigned to the $N^+(CH_3)_3$ bond [35]. After the reaction with the phenolic acids, the ^1H NMR spectra of the derivatives obviously changes. The new shifts appearing at δ 6.0–9.0 ppm prove the successful synthesis of COS phenolic acid derivatives.

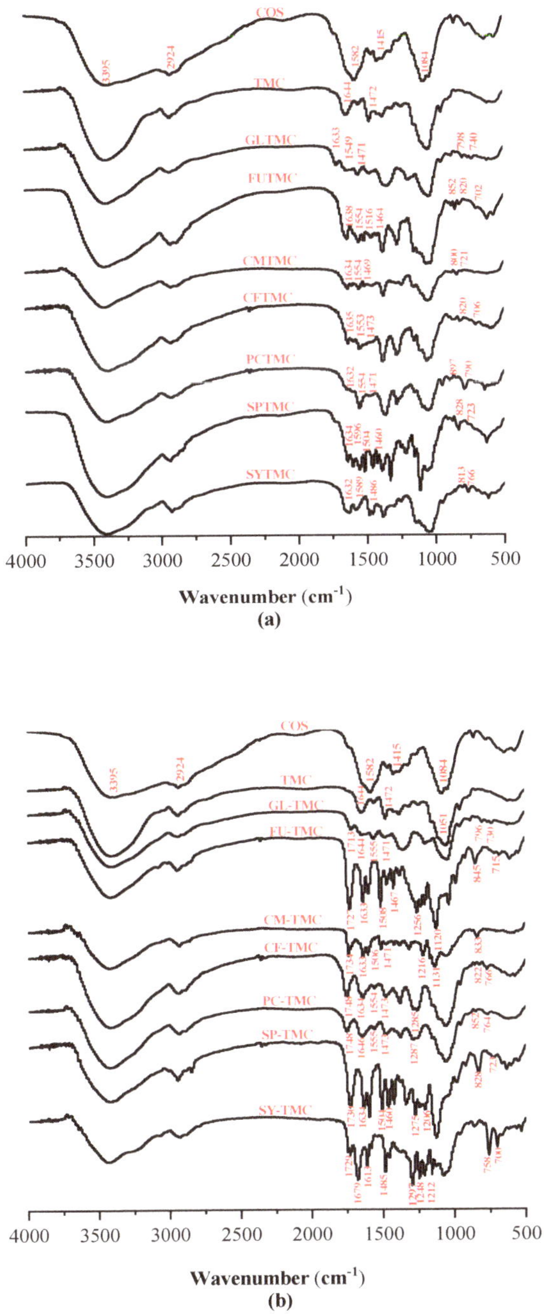

Figure 1. (**a**) FT-IR spectra of COS, TMCI and TMC phenolic acid salt derivatives; (**b**) FT-IR spectra of COS, TMCI and phenolic acid esterified TMC derivatives.

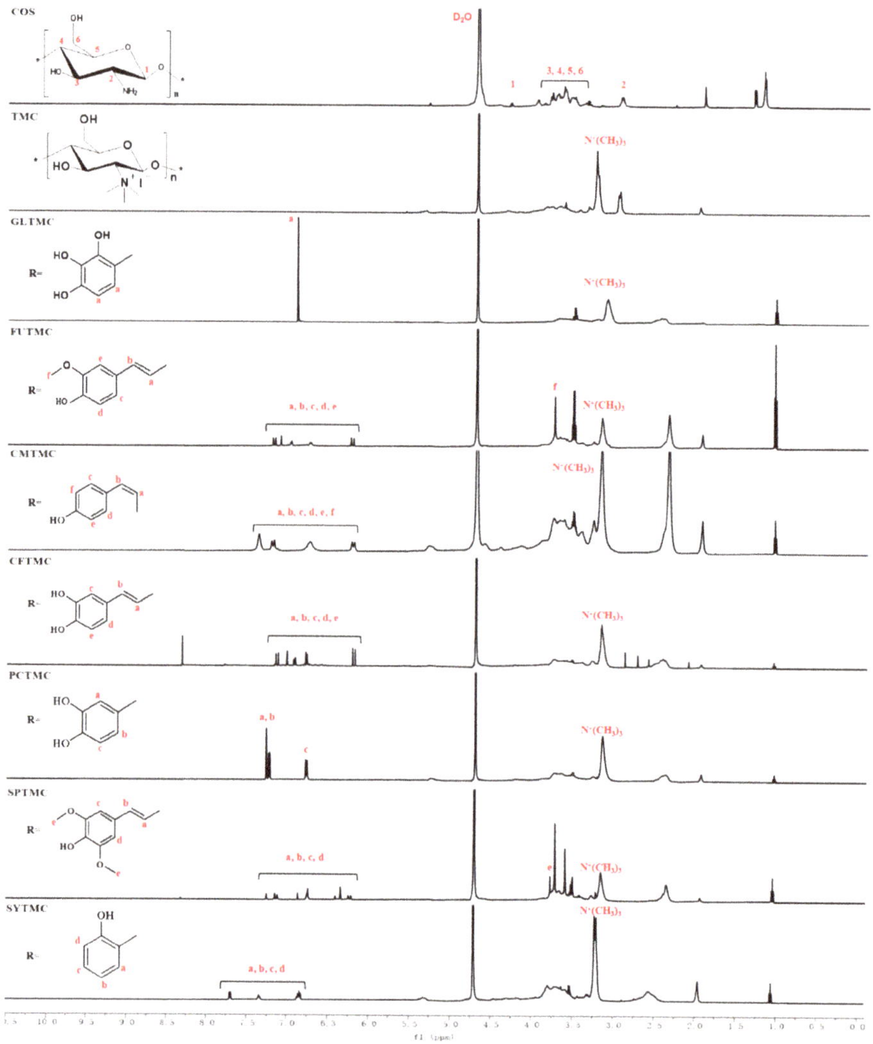

(a)

Figure 2. *Cont.*

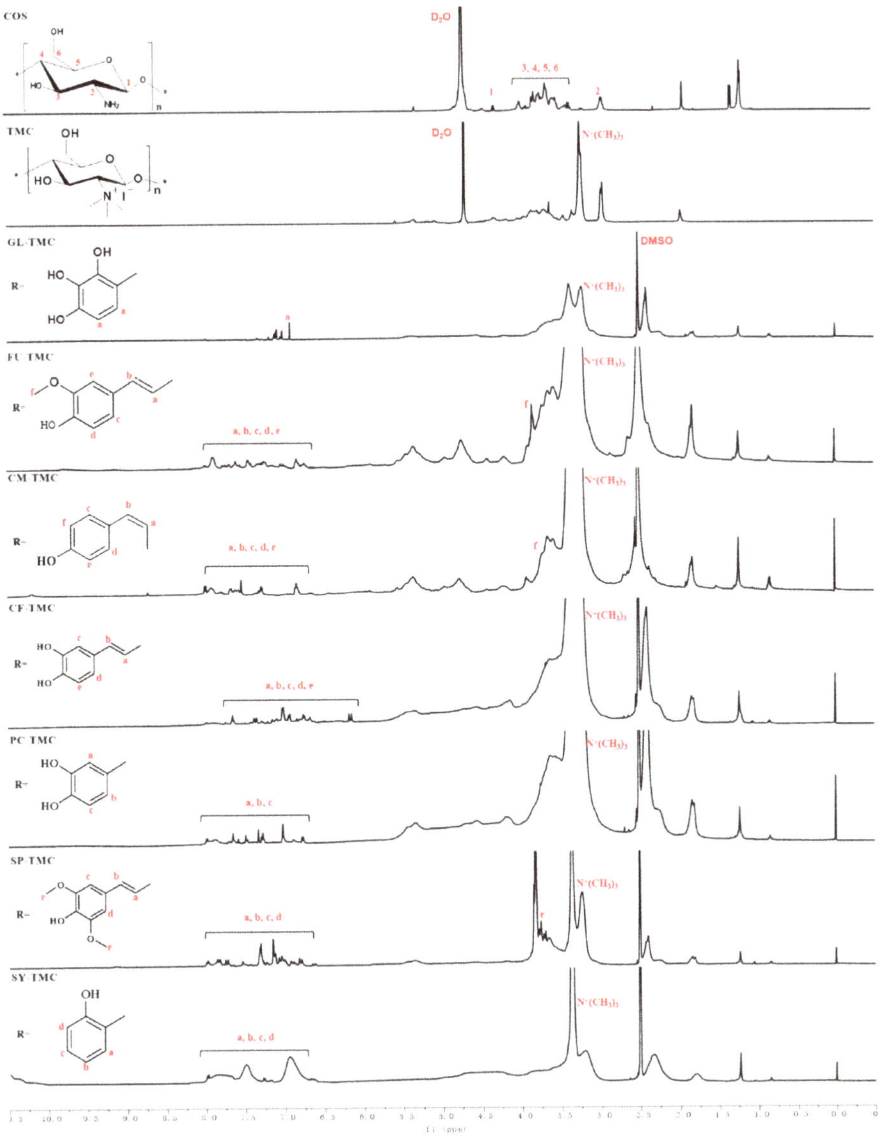

(b)

Figure 2. (**a**) ^{1}H NMR spectra of COS, TMCI and TMC phenolic acid salt derivatives; (**b**) ^{1}H NMR spectra of COS, TMCI and phenolic acid esterified TMC derivatives.

2.4. Yields and the DS of Chitosan Derivatives

After obtaining the desired products, we calculated their yields, and the result is shown in Table 1. In addition, according to the ^{1}H NMR spectra, the [H1] of COS was chosen as the integral standard peak and the DS of each derivative was determined by the ratio of the peak area. The result is also shown in Table 1.

Table 1. The yields and DS of COS derivatives.

Compounds	Yields (%)	DS (%)	Compounds	Yields (%)	DS (%)
COS	-	-	TMC	78.57	94.88
GLTMC	86.47	80.00	GL-TMC	58.96	32.00
FUTMC	24.20	66.00	FU-TMC	39.42	43.00
CMTMC	20.40	51.00	CM-TMC	32.40	41.00
CFTMC	68.97	54.00	CF-TMC	59.88	15.00
PCTMC	44.49	36.00	PC-TMC	53.16	12.00
SPTMC	65.42	46.00	SP-TMC	64.34	47.00
SYTMC	73.60	41.00	SY-TMC	73.40	43.00

2.5. Thermal Gravimetric Analysis (TGA) and Derivative Thermogravimetry (DTG)

With the temperature ranging from 25 to 700 °C, COS had three mass loss steps (Figure 3). The first stage in the range of 30–100 °C with about 8% weight loss was mainly caused by the evaporation of water. And the second stage from 100–200 °C with 17.80% weight loss was caused by the decomposition of the amine and hydroxyl group [36]. The third weight loss in the range of 200–700 °C was mainly caused by the decomposition of the saccharide rings [37]. The final weight loss of COS was about 70.90% of the initial weight. As for the derivatives, it can be seen from the figure that the decomposition temperature of the derivatives was higher compared with COS. In the first stage of 25–200 °C, they obviously showed more stability and less weight loss. In the stage of 200–700 °C, the decomposition rate of derivatives was lower than COS, which demonstrated the better thermal stability. The reason may be that after being introduced into the COS, the phenolic acid formed a hydrogen bond with COS, which strengthened the thermal stability of derivatives. The quaternary ammonium group and the amide group also enhanced the thermal stability of the derivatives. The final mass loss of each derivative was as follows: TMC: 75.80%, GLTMC: 71.74%, FUTMC: 50.1167%, CM-TMC: 63.14%, CFTMC: 66.16%, PCTMC: 69.32%, SPTMC: 50.90%, SYTMC: 71.08%, GL-TMC: 67.02%, FU-TMC: 73.30%, CM-TMC: 69.09%, CF-TMC: 74.81%, PC-TMC: 73.14%, SP-TMC: 70.94%, SY-TMC: 62.66%.

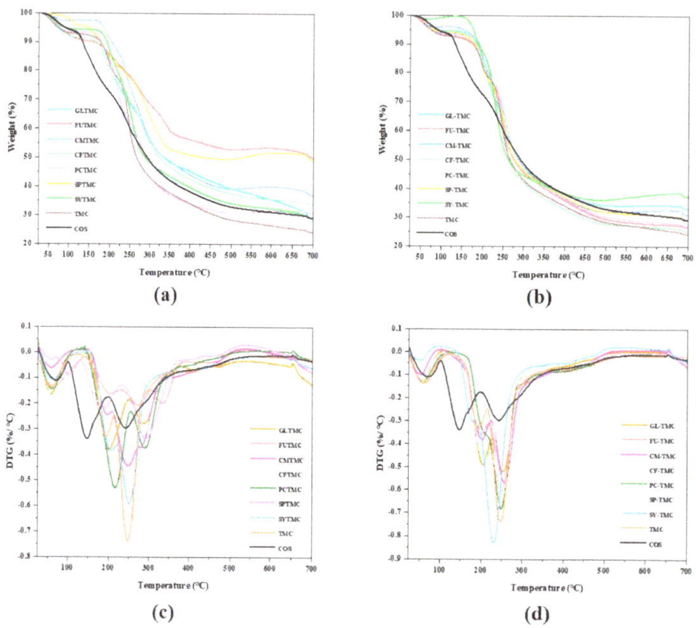

Figure 3. TGA (**a**,**b**) and DTG (**c**,**d**) curves of COS and its derivatives.

2.6. Antioxidant Activity

2.6.1. Scavenging Ability of the DPPH Radical

1,1-Diphenyl-2-picryl-hydrazyl (DPPH), as a nitrogen-centered synthetic radical with a lone pair of electrons, is one of the most common free radicals used to detect the antioxidant activity of substances. As a stable radical, DPPH radical ethanol solution is purple and has a maximum absorption wavelength at 517 nm. When added to the solution, antioxidants can deliver electrons or hydrogen to DPPH to generate the DPPH-H compound, which is more stable, and the color of the solution turns yellow. The degree of discoloration of the solution is quantitatively related to the number of electrons it accepts. The smaller the absorbance is, the stronger the antioxidant capacity of the substance is [38].

The DPPH radical scavenging activity of all products is concentration-related (Figure 4a,b). For TMC phenolic acid salt derivatives, it can be observed that the GLTMC, CMTMC, CFTMC, PCTMC, and SYTMC derivatives had obviously enhanced scavenging activity over COS within the range of tested concentrations. And GLTMC had the highest scavenging activity among these derivatives, achieving nearly 100%. So GLTMC is a potentially suitable antioxidant for DPPH radicals. As for FUTMC and SPTMC, at the concentration of 0.10 mg/mL, their scavenging effect reached more than 70%. But with the augmenting of concentration, the scavenging effect declined, and the possible reason may be as follows: with the increase in concentration, the color of the solution gradually deepened, which made its absorbance increase and the scavenging effect decline. As for phenolic acid esterified TMC derivatives, the DPPH radical scavenging activity of all products increased in a dose-dependent manner (Figure 4b). After being modified with phenolic acids, the DPPH radical scavenging activity of COS derivatives was significantly increased. Comprehensively, at the concentration of 1.60 mg/mL, the order of the DPPH radical scavenging activity of derivatives can be ranked as follows: VC > SY-TMC > SP-TMC > GL-TMC > PC-TMC > FU-TMC > CF-TMC > CM-TMC.

2.6.2. Scavenging Ability of Superoxide Radical

A superoxide anion is formed in almost all aerobic cells and is harmful to cellular components as a precursor to more active oxides. Superoxide anion radicals are more dangerous because they have a longer life span and can travel over greater distances than other oxygen radicals [39]. Therefore, it is necessary to develop antioxidants that can effectively scavenge this free radical. In this study, the NADH-PMS-NBT system was used in the superoxide radical scavenging assay experiment. In the presence of NADH and oxygen the in air, PMS can react with them and form a superoxide anion radical. Then the superoxide anion radical reacts with NBT to generate blue substances, which cause the solution to reach the maximum absorption wavenumber at 520 nm. But if there are antioxidants in the system, the superoxide anion radicals will react with the antioxidants first and the solvent will not turn blue, thus the absorbance at 520 nm will decrease. The smaller the absorbance of the solvent, the stronger the superoxide radical scavenging activity [39].

The superoxide radical scavenging ability of the samples is shown in Figure 4c,d. Several conclusions can be drawn from the figure. First, with the increase in concentration, the scavenging activity of all samples was enhanced. Moreover, for the phenolic acid TMC salt derivatives, all derivatives had a higher scavenging effect at the concentrations of 0.10, 0.20, 0.40 and 0.80 mg/mL. GLTMC and CFTMC in particular had outstanding scavenging effects, which reached 100%. The superoxide radical scavenging activity result of phenolic acid esterified TMC derivatives was similar to the salt derivatives. Compared to COS, the antioxidant activity of these derivatives was obviously enhanced. Moreover, among these derivatives, GL-TMC, CF-TMC and PC-TMC had significantly higher antioxidant activity than COS at all tested concentrations and could be developed into excellent antioxidants. The order of the scavenging activity at the concentration of 0.80 mg/mL can be sorted as follows: GLTMC = CFTMC = GL-TMC > CF-TMC > PCTMC > SPTMC > FUTMC > PC-TMC > CM-TMC > SY-TMC > SP-TMC > FU-TMC > CMTMC > SYTMC > TMC > COS. The analysis demonstrates the effective modification of the phenolic acids.

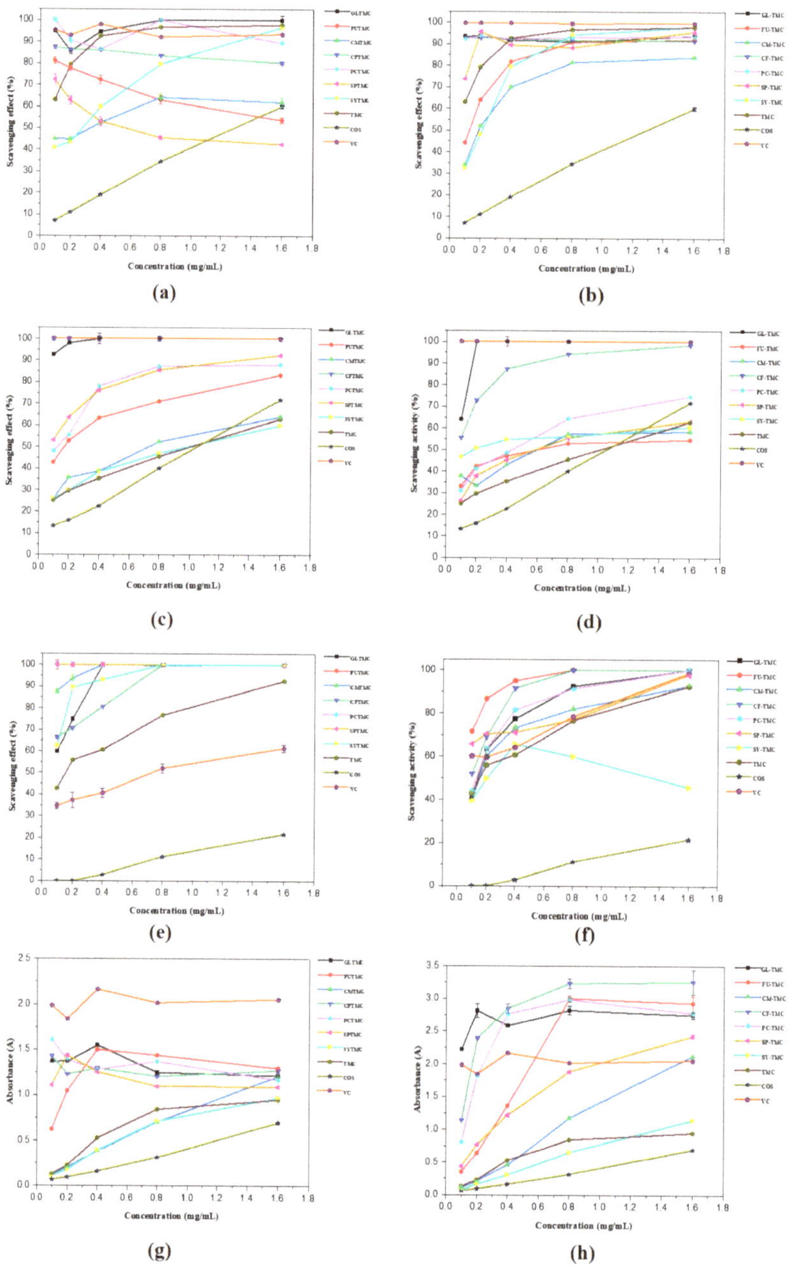

Figure 4. Antioxidant activity of COS and its derivatives: DPPH radical scavenging activity (**a**,**b**); Superoxide radical scavenging activity (**c**,**d**); Hydroxyl radical scavenging activity (**e**,**f**); and reducing power (**g**,**h**).

2.6.3. Scavenging Ability of Hydroxyl Radical

The hydroxyl radical is the most reactive free radical, which can react with non-selective compounds such as proteins, nucleic acids, unsaturated fatty acids and almost all biofilms. It can cause damage to macromolecules and organisms [40]. The hydroxyl radical scavenging activity of the samples is shown in Figure 4e,f. As we can see from the figure, augmentation of scavenging activity was identified for all the samples at increasing concentrations. COS had the lowest scavenging effect, which was only 21.66% at the concentration of 1.60 mg/mL. And the scavenging activity of all the derivatives was obviously enhanced at the tested concentrations compared to COS. For example, the scavenging activity of GLTMC reached 100% at the concentration of 0.40 mg/mL.

2.6.4. Reducing Power

Reducing power is usually used to measure the electron-donating capacity of substances, which is also a manifestation of their antioxidant activity [41]. Here the tested system consisted of ferric ions (Fe^{3+}). When substances with reducing power are added to the test system, they deliver electrons to Fe^{3+} and are reduced to Fe^{2+}. As a result, the color of the solution turns green and has the largest absorbance at a wavelength of 700 nm. The greener the color of the solution, the greater the absorbance value, which represents the stronger reducing power of the substance [42]. So, as the result shows (Figure 4g,h), most of the derivatives had higher absorbance than the COS at tested concentrations, which means a stronger reducing power than COS. On the whole, the reducing power of the derivatives at 1.60 mg/mL concentration can be ranked as follows: CF-TMC > FU-TMC > PC-TMC > GL-TMC > SP-TMC > CM-TMC > VC > FUTMC > CFTMC > GLTMC > CMTMC > PCTMC > SY-TMC > SPTMC > SYTMC > TMC > COS.

In this part, the antioxidant activity of products was measured in vitro by testing the scavenging activity of the DPPH radical, superoxide radical, hydroxyl radical and reducing power. As the above result shows, the antioxidant activity of both the TMC phenolic acid salt derivatives and the phenolic acid esterified TMC derivatives was increased to some extent compared with COS. This demonstrated the significant modification of phenolic acids into COS. Additionally, it can be seen that among these derivatives, GLTMC and GL-TMC had the strongest antioxidant activity, which proves that the number of hydroxyl groups is the key factor that determines the antioxidant activity of phenolic acids. The antioxidant activity of phenolic acids increases with the increase in the hydroxyl group number [43]. Moreover, the structural differences of phenolic acids also affect their antioxidant activity. The existence of the C=C bond in phenolic acids can make the formed free radicals more stable through the resonance effect. And the presence of the methoxy group is also beneficial to improving the antioxidant activity of phenolic acids [44]. So the CFTMC and CF-TMC have much higher antioxidant activity. Furthermore, quaternized COS with positive charges can react with free radicals and is also beneficial to improving the antioxidant activity. Therefore, through this part of experiment, we screened several derivatives with excellent antioxidant activity, which we hope to further develop into new antioxidants.

2.7. Antibacterial Activity

With the progress in science and technology and the exploration of human diseases, many antimicrobial drugs have been developed and utilized for humans, which have greatly improved the cure rate of diseases and are beneficial to human health. But because of enhanced bacterial resistance, existing antibacterial agents are facing a serious challenge. Development of novel antibacterial drugs is still a big challenge for humans. *E. coli* is a common Gram-negative bacterium that is widely found in the intestinal tract and other sites in human and animals. Studies have shown that *E. coli* can cause many diseases such as diarrhea, dysentery, urinary tract infection, meningitis and sepsis [45]. *S. aureus* is a typical Gram-positive bacterium that can cause infections and many diseases, seriously threatening human physical health [46]. In this part, we selected these two kinds of pathogenic bacteria and detected the antibacterial activity of COS and its derivatives. The result is shown in Tables 2 and 3.

Table 2. The antibacterial activity of COS, TMCI and TMC phenolic acid salt derivatives.

Bacterial Species		Samples								
		COS	TMC	GLTMC	FUTMC	CMTMC	CFTMC	PCTMC	SPTMC	SYTMC
S. aureus	MIC (mg/mL)	>8	>8	0.5	0.25	0.5	0.25	0.25	0.25	2
	MBC (mg/mL)	>8	>8	1	0.5	1	0.5	0.5	0.5	4
E. coli	MIC (mg/mL)	>8	>8	0.25	1	0.0625	0.5	1	0.5	>8
	MBC (mg/mL)	>8	>8	0.5	2	0.125	1	2	1	>8

Table 3. The antibacterial activity of COS, TMCI and phenolic acid esterified TMC derivatives.

Bacterial Species		Samples								
		COS	TMC	GL-TMC	FU-TMC	CM-TMC	CF-TMC	PC-TMC	SP-TMC	SY-TMC
S. aureus	MIC (mg/mL)	>8	>8	4	0.25	0.125	0.25	2	1	0.25
	MBC (mg/mL)	>8	>8	8	0.5	0.25	0.5	4	2	0.5
E. coli	MIC (mg/mL)	>8	>8	0.125	0.125	0.03125	0.5	1	0.5	0.25
	MBC (mg/mL)	>8	>8	0.25	0.25	0.0625	1	2	1	0.5

Several conclusions can be drawn from the data in Tables 2 and 3. First, COS and TMCI showed no antibacterial activity. However, after being characterized by phenolic acids, the antibacterial activity of the derivatives was greatly enhanced. Second, the antibacterial activity of different COS phenolic acid derivatives was different. Moreover, the inhibition activity of derivatives against *S. aureus* and *E. coli* was different. For example, the MIC value of GL-TMC against *S. aureus* was 4 mg/mL while against *E. coli* it was 0.125 mg/mL. The possible reason is the difference in the cell wall between *S. aureus* and *E. coli*. Furthermore, among these derivatives, CMTMC and CM-TMC exhibited the best antibacterial activity.

2.8. Antifungal Analysis

B. cinerea is a widely host-borne plant fungus that causes a variety of Botrytis cinerea diseases. It can infest the rhizomes, leaves, flowers and fruits of plants and can be pathogenic at all times of growth. It can cause more than 500 plant pathologies, affecting many economically valuable crops and causing great losses in the postharvest stage of plants. It is one of the most extensively studied necroptotic pathogens [47]. Figure 5a shows the TMC phenolic acid salt derivative inhibitory index of *B. cinerea*. As we can see, carbendazim had the optimal inhibitory effect, where the inhibitory index reached 100%. COS had the lowest inhibitory effect, which was only 25.65% at the concentration of 1.0 mg/mL. As for the derivatives, all the inhibitory indexes were obviously enhanced compared with COS. At the concentration of 0.1 mg/mL, the inhibitory index of the samples can be ranked as follows: SPTMC 48.95%, FUTMC 45.23%, GLTMC 44.17%, SYTMC 41.19%, CFTMC 40.99%, TMC 40.46%, CMTMC 39.73%, PCTMC 37.96%, COS 0.0%. The antifungal activity of the phenolic acid esterified TMC derivatives is shown in Figure 5b. Similar to the TMC phenolic acid salt derivatives, all the derivatives had more enhanced inhibitory activity than COS. Specifically among these derivatives, CM-TMC and SY-TMC presented outstanding inhibitory activity, achieving 85.49% and 89.71%, respectively.

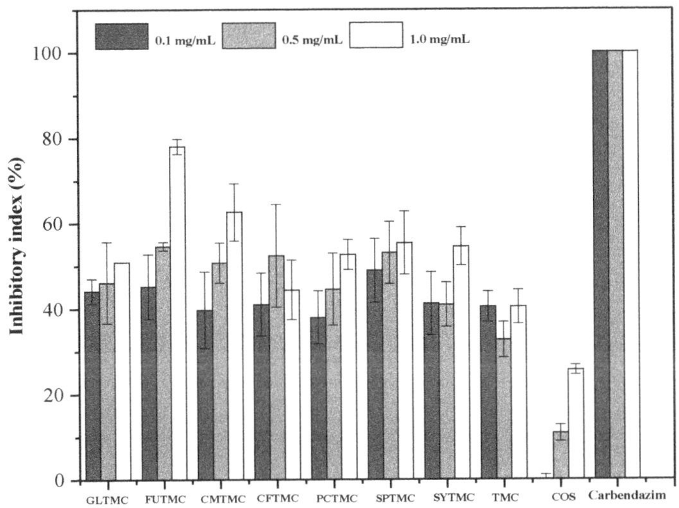

(a)

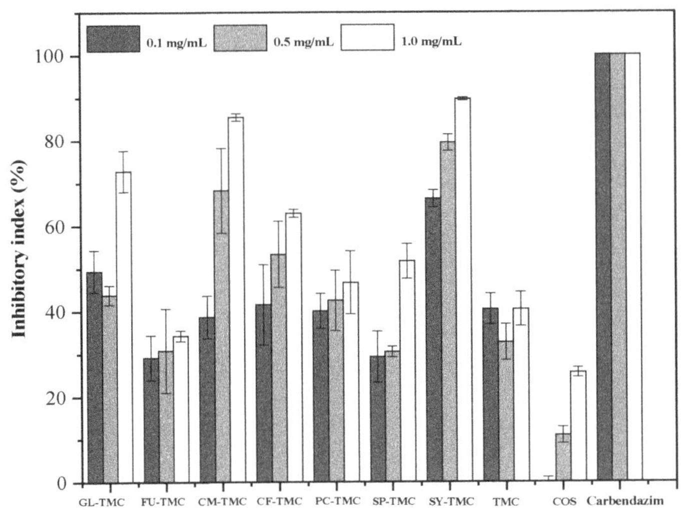

(b)

Figure 5. *Cont.*

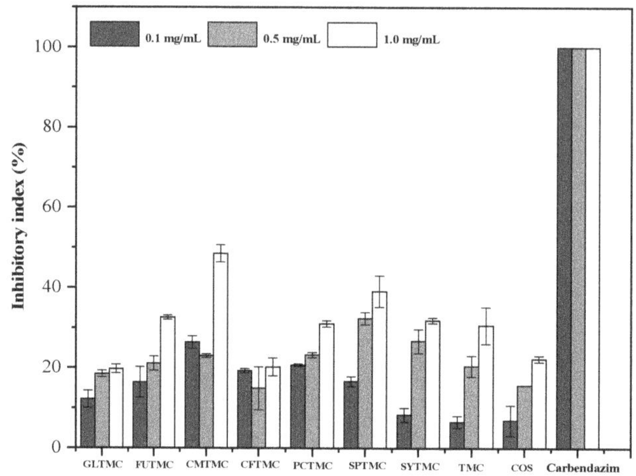

(c)

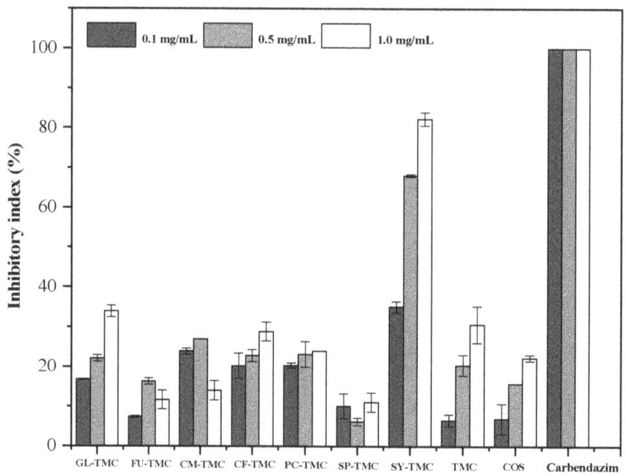

(d)

Figure 5. The antifungal activity of carbendazim, COS and its derivatives against *Botrytis cinerea* (**a**,**b**) and *Gibberella saubinetii* (**c**,**d**).

F. graminearum, once ranked as one of the top 10 phytopathogenic fungi worldwide, can cause diseases of wheat. Not only does it cause severe wheat yield loss, but its production of mycotoxins can also contaminate cereals, leading to poor grain quality. It causes immune function to decline after human and animal ingestion, causing acute or chronic poisoning, which can seriously harm zoonotic health [48]. The antifungal activity of COS phenolic acid derivatives against *F. graminearum* is shown in Figure 5c,d. As shown in the figure, carbendazim had the best inhibitory activity. The inhibitory index of COS at the concentrations of 0.1, 0.5 and 1.0 mg/mL was 6.86%, 15.24% and 22.86%, respectively. Compared to COS, the derivatives were increased in different degrees. For instance, at the concentration of 1.0 mg/mL, the inhibitory index of each derivative was as follows: GLTMC: 19.67%, FUTMC: 32.62%, CMTMC: 48.54%, CFTMC: 20.19%, PCTMC: 30.04%, SPTMC: 39.13%, SYTMC: 31.85%, GL-TMC: 33.90%, FU-TMC: 11.61%, CM-TMC: 14.01%, CF-TMC: 28.87%, PC-TMC: 23.60%, SP-TMC: 11.03%, SY-TMC: 82.25%, TMC: 30.62%. It can be seen from the data that the antifungal activities of different COS phenolic acid derivatives were different, and the activity of SY-TMC was the most significantly improved. The antifungal activity experiment demonstrated that the introduction of phenolic acids can effectively improve the antifungal activity of COS.

2.9. Cytotoxicity Analysis

The cytotoxicity of COS phenolic acid derivatives is shown in Figure 6a. As we can see from the figure, when the concentration of the derivatives varied from 62.5, 125, 250, 500 to 1000 μg/mL, the FUTMC, PCTMC and SYTMC derivatives showed a high cell viability, which proves that these derivatives have low cytotoxicity and good biocompatibility. As for other derivatives, CMTMC, CFTMC and SPTMC had high cell viability at concentrations of 62.5, 125, 250 and 500 μg/mL. GLTMC showed an obvious tendency of increasing cytotoxicity with the increase in concentration. The cytotoxicity of phenolic acid esterified TMC derivatives is shown in Figure 6b. It can be observed that FU-TMC, CMTMC, CFTMC and SP-TMC had high cell viability at all tested concentrations. And GL-TMC, SY-TMC had high cell viability at the concentrations of 62.5, 125, 250 and 500 μg/mL, which proves their low cytotoxicity and good biocompatibility. Furthermore, the cell viability of GL-TMC, PC-TMC and SY-TMC showed a descending trend with the increase in concentration. Totally speaking, through this part of the experiment, we screened several derivatives with high antioxidant, antibacterial and antifungal activity and good biocompatibility.

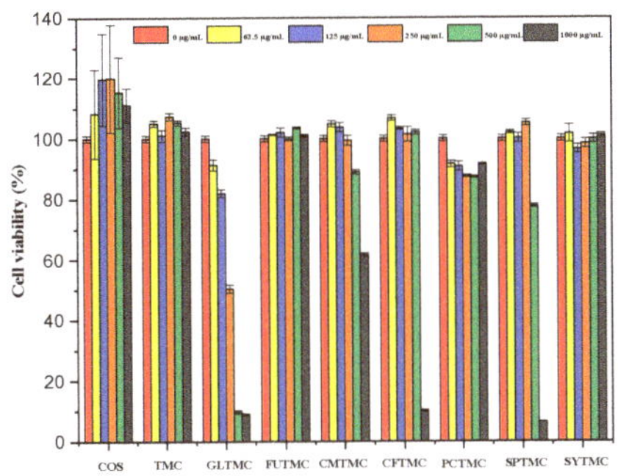

(a)

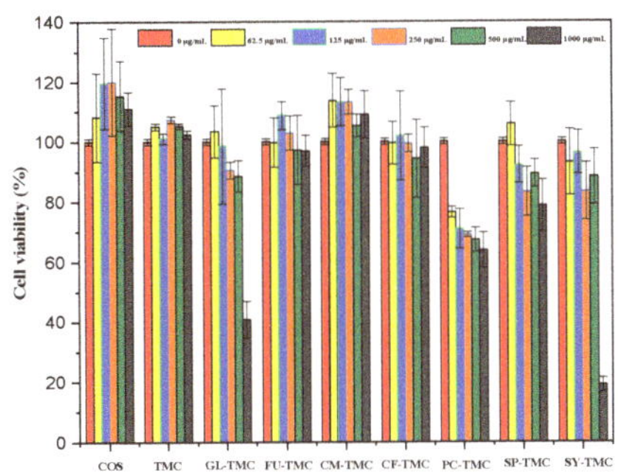

(b)

Figure 6. (**a**) the cytotoxicity of COS, TMCI and TMC phenolic acid salt derivatives; (**b**) the cytotoxicity of COS, TMCI and phenolic acid esterified TMC derivatives.

3. Materials and Methods

3.1. Materials

Chitooligosaccharide (molecular weight of 2000 Da, deacetylation degree of 95%) was purchased from Shandong Weikang Biomedical Technology Co., Ltd. (Linyi, China). L929 cells (GDC0034) were obtained from the China Center for Type Culture Collection. Gallic acid monohydrate, ferulic acid, *p*-coumaric acid, caffeic acid, protocatechuic acid, sinapic acid, salicylic acid, *N,N'*-Carbonyldiimidazole (CDI), sodium hydroxide 1-Methyl-2-pyrrolidinone (NMP) and other used chemical reagents were provided by Sigma-Aldrich Chemical Corp. (Shanghai, China). All chemical reagents were of analytical grade and no further purification was required.

3.2. Chemical Synthesis of COS Derivatives

3.2.1. Synthesis of *N,N,N*-Trimethylated Chitooligosaccharide

According to Zhang [49], the synthesis procedure of *N,N,N*-trimethylated chitooligosaccharide (TMCI) was slightly modified and implemented as follows: 20 mmol COS was distributed into 60 mL NMP with 30 mL 15% NaOH solution and 9 g NaI and stirred for 30 min in a round-bottom flask. Then 30 mL CH_3I was added into the flask drop by drop under freezing conditions. After that, the reaction system was refluxed at 60 °C for 2 h. The TMCI product was precipitated and washed by anhydrous ethanol and anhydrous diethyl ether and obtained after freeze-drying at −50 °C for 24 h.

3.2.2. Synthesis of COS Derivatives—TMC Phenolic Acid Salts

First, 30 mmol NaOH and 30 mmol phenolic acids (gallic acid monohydrate, ferulic acid, *p*-coumaric acid, caffeic acid, protocatechuic acid, sinapic acid, and salicylic acid) were dissolved in 10 mL deionized water and stirred for 2 h. Second, 10 mmol TMCI with 10 mL deionized water was added into the system drop by drop and stirred for 4 h at room temperature. After the reaction, anhydrous ethanol and anhydrous diethyl ether were used to precipitate and wash the products. Finally, the products were obtained after freeze-drying at −50 °C for 24 h [23,25].

3.2.3. Synthesis of COS Derivatives—Phenolic Acid Esterified TMC

First, 20 mmol of different phenolic acids (gallic acid monohydrate, ferulic acid, *p*-coumaric acid, caffeic acid, protocatechuic acid, sinapic acid, and salicylic acid) and CDI was dissolved in DMSO solution and stirred for 12 h at 60 °C. After that, 10 mmol TMCI was added to the reaction system and stirred for 12 h. Finally, the desired products were precipitated and washed with anhydrous ethanol and anhydrous diethyl ether. After freeze-drying at −50 °C for 24 h these derivatives were obtained.

3.3. Analytical Methods

3.3.1. Fourier Transform Infrared (FT-IR) Spectroscopy

Fourier transform infrared (FT-IR) spectroscopy (Nicolet IS 50 Fourier Transform Infrared Spectrometer, provided by Thermo Fisher Scientific, Thermo, Waltham, WA, USA) was used at 4000–400 cm^{-1} to analyze the FT-IR structure of the desired products. The products were mixed with KBr disks at a 1:100 ratio and scanned 32 times at 25 °C.

3.3.2. Nuclear Magnetic Resonance (NMR) Spectroscopy

The ^1H NMR spectra of the desired products were tested and recorded by a Bruke AVIII 500 Spectrometer (500 MHz, Bruker Switzerland, Fällanden, Switzerland) at 25 °C. The products were dissolved in D_2O or DMSO and the chemical shift values are given in δ (ppm).

3.3.3. Yields and Degree of Substitution (DS) Analysis

In order to estimate the reaction efficiency, the production yield and DS were calculated in this part. Yield represents the ratio (%) of the actual weight (g) of the products to the theoretical weight (g).

The calculation formulation of DS is as follows:

$$DS = H_s/H_1(\%)$$

where Hs represents the integral area of the hydrogen atom in the benzene ring of the derivatives, H_1 represents the integral area of the 1-position carbon-bonded hydrogen proton on the main chain of the chitosan.

3.3.4. Thermal Stability

The thermal stability characterization of COS and its derivatives was characterized by a Mettler 5 MP thermogravimetric analyzer (Mettler Toledo, Greifensee, Switzerland). After being vacuum-dried for 24 h, the samples were heated in a range of 30–700 °C at 10 °C /min under a nitrogen atmosphere.

3.4. Antioxidant Activity Assay

3.4.1. DPPH Radical Scavenging Activity Assay

According to Li [50], the 1,1-Diphenyl-2-picryl-hydrazyl (DPPH) scavenging activity of products was slightly modified and tested as follows. After lyophilizing to constant weight, the samples were prepared into a solution at an initial concentration of 10 mg/mL. A total of 17.75 mg 1,1-Diphenyl-2-picryl-hydrazyl was dissolved in anhydrous ethanol and placed in a 500 mL volumetric flask. The product solutions were finally diluted into different concentration solutions (0.1, 0.2, 0.4, 0.8, 1.6 mg/mL) with deionized water and 2.0 mL DPPH solution. Then they were incubated together for 20 min and the absorbance of the solution was measured and recorded at 517 nm by a microplate reader. As for the control group, DPPH was replaced by 2.0 mL of the anhydrous group, and the sample was replaced by 1.0 mL deionized water as the blank group. Three replicates for every sample concentration were measured and the scavenging effect was calculated by the following equation:

$$Scavenging\ activity(\%) = \left[\frac{A_{sample-}A_{control}}{A_{blank}}\right] \times 100$$

where A_{sample}, $A_{control}$, A_{blank} respectively represent the absorbance of the sample, the control and the blank at 517 nm.

3.4.2. Superoxide Radical Scavenging Activity Assay

The superoxide radical scavenging activity of the products was tested as follows. After lyophilizing to constant weight, the samples to be tested were prepared into a solution at an initial concentration of 10 mg/mL. Tris-HCl buffer (16 mM, pH = 8.0), nicotinamide adenine dinucleotide reduced (NADH, 338 μM NADH dissolved in Tris-HCl buffer), nitro blue tetrazolium (NBT, 72 μM NBT dissolved in Tris-HCl buffer), phenazine methosulfate (PMS, 30 μM PMS dissolved in Tris-HCl buffer) were prepared for the experiment. Measured sample solutions (0.03, 0.06, 0.12, 0.24, 0.48 mL) were added with 0.5 mL NADH, 0.5 mL NBT, and 0.5 mL PMS, and then the volume was fixed to 2 mL with deionized water. After that, the solution was incubated for 5 min under dark condition at room temperature and the absorbance was measured at 560 nm with a microplate reader. In the control group, the NADH was replaced with an equal amount of Tris-HCl buffer, and in the blank group, the sample solution was replaced with an equal amount of deionized water. The scavenging effect was calculated by the following equation:

$$Scavenging\ activity(\%) = \left[1 - \frac{A_{sample} - A_{control}}{A_{control}}\right] \times 100$$

where A_{sample}, $A_{control}$ and A_{blank} respectively represent the absorbance of the sample, the control and the blank at 560 nm [51].

3.4.3. Hydroxyl Radical Scavenging Activity Assay

The hydroxyl radical scavenging activity of products was tested as follows. After lyophilizing to constant weight, the samples were prepared into a solution at an initial concentration of 10 mg/mL. In addition, potassium phosphate buffer (pH = 7.4) H_2O_2 (60 μM dissolved in potassium phosphate buffer), EDTA-Fe^{2+} (220 μM, potassium phosphate buffer), and safranine O (0.23 μM, potassium phosphate buffer) were prepared for the experiment. Sample solutions (0.045, 0.09, 0.18, 0.36, 0.72 mL) with 0.50 mL EDTA-Fe^{2+}, 1.00 mL potassium phosphate buffer, 1.00 mL safranine O, 1.00 mL H_2O_2 solution were incubated without light at 37 °C for 30 min. Meanwhile, the control group contained 1.00 mL deionized water, 0.50 mL EDTA-Fe^{2+}, 2.00 mL potassium phosphate buffer, 1.00 mL safranine O, and the blank group constituted 1.00 mL deionized water, 0.50 mL EDTA-Fe^{2+}, 1.00 mL potassium phosphate buffer, 1.00 mL safranine O and 1.00 mL H_2O_2. Finally, the absorbance of all solutions was measured at 520 nm with a microplate reader, and the scavenging effect was calculated by the following equation:

$$Scavenging\ activity(\%) = \left[\frac{A_{sample} - A_{blank}}{A_{control} - A_{blank}}\right] \times 100$$

where A_{sample}, $A_{control}$ and A_{blank} respectively represent the absorbance of the sample, the control and the blank at 520 nm [52].

3.4.4. Reducing Power Assay

The reducing power of products was tested as follows. After lyophilizing to constant weight, the samples were prepared into a solution at an initial concentration of 10 mg/mL. In addition, sodium phosphate buffer (pH = 6.6), potassium ferricyanide solution (1%, dissolved in sodium phosphate buffer), trichloroacetic acid (10%, dissolved in sodium phosphate buffer), and ferric chloride solution (0.1%, dissolved in sodium phosphate buffer) were prepared for the experiment. First, sample solutions (0.03, 0.06, 0.12, 0.24, 0.48 mL) and 0.50 mL potassium ferricyanide reacted at 50 °C for 20 min, and then 0.50 mL trichloroacetic acid was added to terminate the reaction. Second, the solution was centrifuged and 0.75 mL supernatant was added into 0.60 mL deionized water with 0.15 mL ferric chloride solution. The reaction system was incubated at room temperature for 10 min. The absorbance of the reaction system was measured at 700 nm and the data represent the reducing power of products [53].

3.5. *Evaluation of Antibacterial Activity*

3.5.1. Preparation of Solutions

The samples were dissolved in sterile water at an initial concentration of 16 mg/mL. Then 100 μL of stock solutions was added to the 96-well plate and continuously diluted 11 times.

3.5.2. Minimum Inhibitory Concentration and Minimum Biocidal Concentration

The antibacterial activity of the samples was tested by the method of minimum inhibitory concentration (MIC) and minimum biocidal concentration (MBC). *Escherichia coli* (*E. coli*) and *Staphylococcus aureus* (*S. aureus*) bacteria were cultivated to test the antibacterial activity, as they were the epitome of Gram-positive bacteria and Gram-negative bacteria, respectively. The experimental process was implemented as follows. First, the bacteria species were incubated in Luria Bertani (LB) medium at 37 °C. And the well-grown bacteria were diluted to 105–106 cell/mL with LB medium. Then the bacterial suspension was added to the 96-well plates containing samples of different concentrations prepared in advance. The volume added to each well was 100 μL, and it was incubated at 37 °C for 24 h. In the meantime, the MIC and MBC were observed by the naked eye. The MIC is the lowest concentration of samples that can completely inhibit visible bacteria growth after 12 h and the MBC is the lowest concentration of the experimental samples that can completely inhibit visible bacteria growth after 24 h. All experimental operations were completed under sterile conditions [54].

3.6. *Antifungal Assay*

Herein, two kinds of plant fungus were cultured to test the antifungal activity of COS and its derivatives. The first one was *Botrytis cinerea* (*B. cinerea*) and the other was *Fusarium graminearum* (*F. graminearum*).

3.6.1. Preparation of Solutions

First, the samples were dissolved in sterile water at an initial concentration of 10 mg/mL. Then the potato dextrose agar (PDA) medium was prepared and sterilized at 120 °C for 25 min to prepare the sample medium. After that, sample solutions of different volumes (2.0, 1.0 and 0.2 mL) were mixed with different volumes of medium (18.0, 19.0 and 19.8 mL) and then poured into disposable Petri dishes. The sample culture medium at concentrations of 0.10, 0.50 and 1.00 mg/mL was formed after solidification. Here in the positive control group, the sample was replaced by carbendazim and the blank control group was deionized water.

3.6.2. Implementation of the Antifungal Experiment

The mycelial growth rate method was used in this experiment. The well-cultivated strains were coated on a PDA medium and cultured in a fungus incubator (28 °C, RH = 60%) for 48 h. After that, a fungi mycelia disk with a diameter of 5 mm was made and carefully inoculated on the sample medium. They were then placed into the fungus incubator for culture. When the mycelia of the blank group grew to the inner edge of the Petri dish, the diameter of the mycelia in all the sample culture media could be measured using the cross crossing method. Three values were measured in parallel for each sample concentration. All experiments were performed under aseptic conditions. Finally, the inhibition rate was measured as follows:

$$Inhibitory\ index(\%) = \left[1 - \frac{D_s - 5}{D_b - 5}\right] \times 100$$

where D_s is the diameter of the growth zone in the sample plates and D_b is the diameter of the growth zone in the blank control plates [37,55].

3.7. *Cytotoxicity Assay*

L929 cells were chosen in this part to test the cytotoxicity of these derivatives by MTT method in vitro. First, L929 cells were incubated in DEMO medium with 10% fetal calf serum and 1% mixture of penicillin and streptomycin at 37 °C with a 5% CO_2 atmosphere. After passing through twice, the well-incubated cells at a concentration of 5×10^4/mL were added to 96-well plates and incubated for 24 h. Then the derivative solutions at different concentrations were added to the wells and incubated until the cells in control wells multiplied to 90%; the MTT with 0.5 mg/mL was added to the wells and the absorbance was detected at 490 nm with a microplate reader after 4 h with DMSO solution. The cell viability was calculated by the following equation:

$$Cell\ viability = \left[\frac{A_{sample} - A_{blank}}{A_{control} - A_{blank}}\right] \times 100$$

where A_{sample}, $A_{control}$ and A_{blank} respectively represent the absorbance of the sample, the negative control and the blank group at 490 nm.

3.8. *Statistical Analysis*

All experiments above were conducted at least three times and data are presented as the mean ± standard deviation (SD). Significant differences between the groups were assessed using one-way ANOVA followed by Duncan's test using PASW Statistics 21.0 (SPSS Inc., Chicago, IL, USA). Differences were considered statistically significant at $p < 0.05$.

4. Conclusions

In this study, several COS phenolic acid derivatives were successfully synthesized by chemical modification. COS, as a leading compound, was the first to be quaternized and then modified with phenolic acids by two different methods. The structures and DS of these derivatives were characterized by FT-IR and ^{1}H NMR spectra. The thermal stability of the derivatives was studied and the results proved they were more stable than COS. Additionally, the antioxidant activity of the derivatives was explored in vitro according to four typical methods. The results demonstrated that the antioxidant activity of the derivatives was greatly improved. Subsequently, the antifungal and antibacterial activities of derivatives were also detected, and among them, derivatives of CMTMC, CM-TMC and SY-TMC had excellent antibacterial and antifungal activities. The cytotoxicity of these derivatives was explored by L929 cells in vitro and the results showed that the cytotoxicity of the derivatives exhibited concentration dependence. Therefore, this study confirms that introducing phenolic acid into COS can significantly improve its biological activity. The derivatives with outstanding activities can be further developed and utilized as antioxidants or bacteriostats in the fields of food, medicine, cosmetics, agriculture and environmental protection. However, the structure–activity relationship and antioxidant mechanism of the products still need further study.

Author Contributions: Formal analysis, Y.S. and J.C.; Funding acquisition, L.T. and Z.G.; Methodology, Y.S. and Z.G.; Project administration, Z.G.; Resources, L.T. and Z.G.; Software, Y.S., J.C. and Y.M.; Validation, Y.S., J.C. and Y.M.; Visualization, Y.S.; Writing—original draft, Y.S.; Writing—review and editing, Z.G. All authors have read and agreed to the published version of the manuscript.

Funding: This work was supported by the Shandong Key R&D Plan, Major Scientific and Technological Innovation Project (2022CXGC020413), the Science & Technology Specific Projects in the Agricultural High-Tech Industrial Demonstration Area of the Yellow River Delta (2022SZX01), the Key Deployment Project of the Center for Ocean Mega-Science of CAS (COMS2020J04) and the 2022 Central Guiding Local Science and Technology Development Fund Project in Shandong Province (YDZX2022007).

Institutional Review Board Statement: Not applicable.

Informed Consent Statement: Not applicable.

Data Availability Statement: All data contained in the manuscript are available from the authors.

Acknowledgments: The authors thank the support of the Yellow River Delta Saline-Alkali Agroecosystem Observation and Research Station.

Conflicts of Interest: The authors declare that there are no competing financial interests or conflicting relationships.

References

1. Joseph, S.M.; Krishnamoorthy, S.; Paranthaman, R.; Moses, J.A.; Anandharamakrishnan, C. A review on source-specific chemistry, functionality, and applications of chitin and chitosan. *Carbohydr. Polym. Technol. Appl.* **2021**, *2*, 100036. [CrossRef]
2. Naveed, M.; Phil, L.; Sohail, M.; Hasnat, M.; Baig, M.M.F.A.; Ihsan, A.U.; Shumzaid, M.; Kakar, M.U.; Khan, T.M.; Akabar, M.; et al. Chitosan oligosaccharide (COS): An overview. *Int. J. Biol. Macromol.* **2019**, *129*, 827–843. [CrossRef] [PubMed]
3. Liaqat, F.; Eltem, R. Chitooligosaccharides and their biological activities: A comprehensive review. *Carbohydr. Polym.* **2018**, *184*, 243–259. [CrossRef] [PubMed]
4. Zhai, X.; Li, C.; Ren, D.; Wang, J.; Ma, C.; Abd El-Aty, A.M. The impact of chitooligosaccharides and their derivatives on the in vitro and in vivo antitumor activity: A comprehensive review. *Carbohydr. Polym.* **2021**, *266*, 118132. [CrossRef] [PubMed]
5. Liu, X.; Xia, W.; Jiang, Q.; Yu, P.; Yue, L. Chitosan oligosaccharide-N-chlorokojic acid mannich base polymer as a potential antibacterial material. *Carbohydr. Polym.* **2018**, *182*, 225–234. [CrossRef] [PubMed]
6. Park, H.-H.; Ko, S.-C.; Oh, G.-W.; Jang, Y.-M.; Kim, Y.-M.; Park, W.S.; Choi, I.-W.; Jung, W.-K. Characterization and biological activity of PVA hydrogel containing chitooligosaccharides conjugated with gallic acid. *Carbohydr. Polym.* **2018**, *198*, 197–205. [CrossRef]

7. Oh, G.-W.; Kim, S.-C.; Kim, T.-H.; Jung, W.-K. Characterization of an oxidized alginate-gelatin hydrogel incorporating a COS-salicylic acid conjugate for wound healing. *Carbohydr. Polym.* **2021**, *252*, 117145. [CrossRef] [PubMed]
8. Karagozlu, M.Z.; Karadeniz, F.; Kim, S.-K. Anti-HIV activities of novel synthetic peptide conjugated chitosan oligomers. *Int. J. Biol. Macromol.* **2014**, *66*, 260–266. [CrossRef]
9. Zhang, J.; Dai, X.; Jiang, Z. Research Progress in Functions of Bioactive Chitooligosaccharides in Prevention and Treatment of Alzheimers Disease. *Food Sci.* **2013**, *34*, 316–320.
10. Liang, S.; Sun, Y.; Dai, X. A Review of the Preparation, Analysis and Biological Functions of Chitooligosaccharide. *Int. J. Mol. Sci.* **2018**, *19*, 2197. [CrossRef]
11. Je, J.-Y.; Kim, S.-K. Chitooligosaccharides as potential nutraceuticals: Production and bioactivities. *Adv. Food Nutr. Res.* **2012**, *65*, 321–336.
12. Boz, H. Ferulic Acid in Cereals—A Review. *Czech J. Food Sci.* **2015**, *33*, 1–7. [CrossRef]
13. Anantharaju, P.G.; Gowda, P.C.; Vimalambike, M.G.; Madhunapantula, S.V. An overview on the role of dietary phenolics for the treatment of cancers. *Nutr. J.* **2016**, *15*, 99. [CrossRef]
14. Bialecka-Florjanczyk, E.; Fabiszewska, A.; Zieniuk, B. Phenolic Acids Derivatives—Biotechnological Methods of Synthesis and Bioactivity. *Curr. Pharm. Biotechnol.* **2018**, *19*, 1098–1113. [CrossRef]
15. Insaward, A.; Duangmal, K.; Mahawanich, T. Mechanical, Optical, and Barrier Properties of Soy Protein Film As Affected by Phenolic Acid Addition. *J. Agric. Food Chem.* **2015**, *63*, 9421–9426. [CrossRef]
16. Perez-Jimenez, J.; Lluis Torres, J. Analysis of Nonextractable Phenolic Compounds in Foods: The Current State of the Art. *J. Agric. Food Chem.* **2011**, *59*, 12713–12724. [CrossRef] [PubMed]
17. Heleno, S.A.; Martins, A.; Queiroz, M.J.R.P.; Ferreira, I.C.F.R. Bioactivity of phenolic acids: Metabolites versus parent compounds: A review. *Food Chem.* **2015**, *173*, 501–513. [CrossRef] [PubMed]
18. Oluwole, O.; Fernando, W.M.A.D.B.; Lumanlan, J.; Ademuyiwa, O.; Jayasena, V. Role of phenolic acid, tannins, stilbenes, lignans and flavonoids in human health—A review. *Int. J. Food Sci. Technol.* **2022**, *57*, 6326–6335. [CrossRef]
19. Valanciene, E.; Jonuskiene, I.; Syrpas, M.; Augustiniene, E.; Matulis, P.; Simonavicius, A.; Malys, N. Advances and Prospects of Phenolic Acids Production, Biorefinery and Analysis. *Biomolecules* **2020**, *10*, 874. [CrossRef]
20. Eom, T.-K.; Senevirathne, M.; Kim, S.-K. Synthesis of phenolic acid conjugated chitooligosaccharides and evaluation of their antioxidant activity. *Environ. Toxicol. Pharmacol.* **2012**, *34*, 519–527. [CrossRef]
21. Vo, T.-S.; Ngo, D.-H.; Kim, S.-K. Gallic acid-grafted chitooligosaccharides suppress antigen-induced allergic reactions in RBL-2H3 mast cells. *Eur. J. Pharm. Sci.* **2012**, *47*, 527–533. [CrossRef]
22. Chen, D.; Bai, R.; Yong, H.; Zong, S.; Jin, C.; Liu, J. Improving the digestive stability and prebiotic effect of carboxymethyl chitosan by grafting with gallic acid: In vitro gastrointestinal digestion and colonic fermentation evaluation. *Int. J. Biol. Macromol.* **2022**, *214*, 685–696. [CrossRef] [PubMed]
23. Sun, Y.; Ji, X.; Cui, J.; Mi, Y.; Zhang, J.; Guo, Z. Synthesis, Characterization, and the Antioxidant Activity of Phenolic Acid Chitooligosaccharide Derivatives. *Mar. Drugs* **2022**, *20*, 489. [CrossRef]
24. Dai-Nghiep, N.; Kim, M.-M.; Kim, S.-K. Protective effects of aminoethyl-chitooligosaccharides against oxidative stress in mouse macrophage RAW 264.7 cells. *Int. J. Biol. Macromol.* **2012**, *50*, 624–631.
25. Mi, Y.; Tan, W.; Zhang, J.; Guo, Z. Modification of Hydroxypropyltrimethyl Ammonium Chitosan with Organic Acid: Synthesis, Characterization, and Antioxidant Activity. *Polymers* **2020**, *12*, 2460. [CrossRef]
26. Xing, R.; Xu, C.; Gao, K.; Yang, H.; Liu, Y.; Fan, Z.; Liu, S.; Qin, Y.; Yu, H.; Li, P. Characterization of Different Salt Forms of Chitooligosaccharides and Their Effects on Nitric Oxide Secretion by Macrophages. *Molecules* **2021**, *26*, 2563. [CrossRef]
27. Belalia, R.; Grelier, S.; Benaissa, M.; Coma, V. New bioactive biomaterials based on quaternized chitosan. *J. Agric. Food Chem.* **2008**, *56*, 1582–1588. [CrossRef]
28. Zhang, J.; Tan, W.; Luan, F.; Yin, X.; Dong, F.; Li, Q.; Guo, Z. Synthesis of Quaternary Ammonium Salts of Chitosan Bearing Halogenated Acetate for Antifungal and Antibacterial Activities. *Polymers* **2018**, *10*, 530. [CrossRef]
29. Chikkulapalli, A.; Aavula, S.K.; Mona, R.N.P.; Karthikeyan, C.; Kumar, V.C.H.; Sulur, M.G.; Sumathi, S. Convenient N-acetylation of amines in N,N-dimethylacetamide with N,N-carbonyldiimidazole. *Tetrahedron Lett.* **2015**, *56*, 3799–3803. [CrossRef]
30. Engstrom, K.M. Practical Considerations for the Formation of Acyl Imidazolides from Carboxylic Acids and N,N′-Carbonyldiimidazole: The Role of Acid Catalysis. *Org. Process Res. Dev.* **2018**, *22*, 1294–1297. [CrossRef]
31. Liu, Y.; Liu, Z.; Gao, Y.; Gao, W.; Hou, Z.; Zhu, Y. Facile Method for Surface-Grafted Chitooligosaccharide on Medical Segmented Poly(ester-urethane) Film to Improve Surface Biocompatibility. *Membranes* **2021**, *11*, 37. [CrossRef]
32. Zhang, J.; Tan, W.; Wang, G.; Yin, X.; Li, Q.; Dong, F.; Guo, Z. Synthesis, characterization, and the antioxidant activity of N,N,N-trimethyl chitosan salts. *Int. J. Biol. Macromol.* **2018**, *118*, 9–14. [CrossRef]
33. Witt, J.D.; Gabriel, M.K.; Julian, R.L. Gc-Ft-Ir Analysis Of Commercial Divinyl Benzene. *J. Chromatogr. Sci.* **1979**, *17*, 445–448. [CrossRef]
34. Sedman, J.; Butler, I.S. Polarized Ft-Ir Spectra of Benzene and Anthracene Oriented in A Nematic Liquid-Crystal Solvent. *Appl. Spectrosc.* **1985**, *39*, 621–625. [CrossRef]
35. Li, Q.; Wei, L.; Zhang, J.; Gu, G.; Guo, Z. Significantly enhanced antioxidant activity of chitosan through chemical modification with coumarins. *Polym. Chem.* **2019**, *10*, 1480–1488. [CrossRef]

36. Wang, Y.; Cen, C.; Chen, J.; Fu, L. MgO/carboxymethyl chitosan nanocomposite improves thermal stability, waterproof and antibacterial performance for food packaging. *Carbohydr. Polym.* **2020**, *236*, 116078. [CrossRef]
37. Cui, J.; Sun, Y.; Wang, L.; Miao, Q.; Tan, W.; Guo, Z. Preparation of l-Arginine Schiff Bases Modified Chitosan Derivatives and Their Antimicrobial and Antioxidant Properties. *Mar. Drugs* **2022**, *20*, 688. [CrossRef]
38. Ak, T.; Gulcin, I. Antioxidant and radical scavenging properties of curcumin. *Chem. -Biol. Interact.* **2008**, *174*, 27–37. [CrossRef] [PubMed]
39. Sekine, T.; Masumizu, T.; Maitani, Y.; Nagai, T. Evaluation of superoxide anion radical scavenging activity of shikonin by electron spin resonance. *Int. J. Pharm.* **1998**, *174*, 133–139. [CrossRef]
40. Gill, S.S.; Tuteja, N. Reactive oxygen species and antioxidant machinery in abiotic stress tolerance in crop plants. *Plant Physiol. Biochem.* **2010**, *48*, 909–930. [CrossRef] [PubMed]
41. Sun, T.; Yao, Q.; Zhou, D.; Mao, F. Antioxidant activity of N-carboxymethyl chitosan oligosaccharides. *Bioorganic Med. Chem. Lett.* **2008**, *18*, 5774–5776. [CrossRef]
42. Singhal, M.; Paul, A.; Singh, H.P. Synthesis and reducing power assay of methyl semicarbazone derivatives. *J. Saudi Chem. Soc.* **2014**, *18*, 121–127. [CrossRef]
43. Yang, X.; Lan, W.; Sun, X. Antibacterial and antioxidant properties of phenolic acid grafted chitosan and its application in food preservation: A review. *Food Chem.* **2023**, *428*, 136788. [CrossRef] [PubMed]
44. Mu, K.; Yao, Y.; Wang, D.; Kitts, D.D. Prooxidant capacity of phenolic acids defines antioxidant potential. *Biochim. Et Biophys. Acta-Gen. Subj.* **2023**, *1867*, 130371. [CrossRef]
45. Mansouri, M.; Mobarez, A.M.; Nojoomi, F. Comprehensive study of common Enterogenic E. coli in Iran during 2010–2020: A systematic review. *Gene Rep.* **2021**, *25*, 101316. [CrossRef]
46. Zhang, J.; Wang, J.; Jin, J.; Li, X.; Zhang, H.; Shi, X.; Zhao, C. Prevalence, antibiotic resistance, and enterotoxin genes of Staphylococcus aureus isolated from milk and dairy products worldwide: A systematic review and meta-analysis. *Food Res. Int.* **2022**, *162*, 111969. [CrossRef] [PubMed]
47. Munhuweyi, K.; Lennox, C.L.; Meitz-Hopkins, J.C.; Caleb, O.J.; Opara, U.L. Major diseases of pomegranate (*Punica granatum* L.), their causes and management-A review. *Sci. Hortic.* **2016**, *211*, 126–139. [CrossRef]
48. Dean, R.; Van Kan, J.A.L.; Pretorius, Z.A.; Hammond-Kosack, K.E.; Di Pietro, A.; Spanu, P.D.; Rudd, J.J.; Dickman, M.; Kahmann, R.; Ellis, J.; et al. The Top 10 fungal pathogens in molecular plant pathology. *Mol. Plant Pathol.* **2012**, *13*, 414–430. [CrossRef]
49. Zhang, J.; Tan, W.; Wei, L.; Chen, Y.; Mi, Y.; Sun, X.; Li, Q.; Dong, F.; Guo, Z. Synthesis of urea-functionalized chitosan derivatives for potential antifungal and antioxidant applications. *Carbohydr. Polym.* **2019**, *215*, 108–118. [CrossRef] [PubMed]
50. Li, Q.; Zhang, C.; Tan, W.; Gu, G.; Guo, Z. Novel Amino-Pyridine Functionalized Chitosan Quaternary Ammonium Derivatives: Design, Synthesis, and Antioxidant Activity. *Molecules* **2017**, *22*, 156. [CrossRef] [PubMed]
51. Wei, L.; Li, Q.; Tan, W.; Dong, F.; Luan, F.; Guo, Z. Synthesis, Characterization, and the Antioxidant Activity of Double Quaternized Chitosan Derivatives. *Molecules* **2017**, *22*, 501. [CrossRef] [PubMed]
52. Sun, X.; Zhang, J.; Mi, Y.; Miao, Q.; Tan, W.; Li, Q.; Guo, Z. Synthesis, characterization, and antioxidant activity of carboxymethyl chitosan derivatives containing sulfonium salt. *J. Oceanol. Limnol.* **2022**, *40*, 284–295. [CrossRef]
53. Tan, W.; Li, Q.; Li, W.; Dong, F.; Guo, Z. Synthesis and antioxidant property of novel 1,2,3-triazole-linked starch derivatives via 'click chemistry'. *Int. J. Biol. Macromol.* **2016**, *82*, 404–410. [CrossRef] [PubMed]
54. Sahariah, P.; Masson, M.; Meyer, R.L. Quaternary Ammoniumyl Chitosan Derivatives for Eradication of Staphylococcus aureus Biofilms. *Biomacromolecules* **2018**, *19*, 3649–3658. [CrossRef]
55. Mi, Y.; Li, Q.; Miao, Q.; Tan, W.; Zhang, J.; Guo, Z. Enhanced antifungal and antioxidant activities of new chitosan derivatives modified with Schiff base bearing benzenoid/heterocyclic moieties. *Int. J. Biol. Macromol.* **2022**, *208*, 586–595. [CrossRef]

MDPI AG
Grosspeteranlage 5
4052 Basel
Switzerland
Tel.: +41 61 683 77 34

Marine Drugs Editorial Office
E-mail: marinedrugs@mdpi.com
www.mdpi.com/journal/marinedrugs

www.ingramcontent.com/pod-product-compliance
Lightning Source LLC
Chambersburg PA
CBHW062050220926
51850CB00010B/2722
* 9 7 8 3 7 2 5 8 5 3 7 3 1 *